Navigating Human-Machine Collaboration in Smart Factories

Vannie Naidoo
https://orcid.org/0000-0001-8435-4348
University of KwaZulu-Natal, South Africa

Veena Prasad Vemuri
https://orcid.org/0000-0002-9046-0072
Universitry of Mumbai, India

Rahul R. Verma
https://orcid.org/0000-0002-5356-813X
University of Delhi, India

IGI Global
Scientific Publishing
Publishing Tomorrow's Research Today

Vice President of Editorial — Melissa Wagner
Director of Acquisitions — Mikaela Felty
Director of Book Development — Jocelynn Hessler
Production Manager — Mike Brehm
Cover Design — Jose Rosado

Published in the United States of America by
IGI Global Scientific Publishing
701 East Chocolate Avenue
Hershey, PA, 17033, USA
Tel: 717-533-8845 | Fax: 717-533-7115
Website: https://www.igi-global.com E-mail: cust@igi-global.com

Library of Congress Cataloging-in-Publication Data

Names: Naidoo, Vannie, 1972- editor | Vemuri, Veena Prasad, 1974- | Verma, Rahul, 1989- editor

Title: Navigating human-machine collaboration in smart factories / [edited by] Vannie Naidoo, Veena Prasad Vemuri, Rahul R. Verma.

Description: Hershey, PA : IGI Global Scientific Publishing, [2025] | Includes bibliographical references and index. | Summary: "This book aims to bridge the knowledge gap for businesses and professionals seeking to understand and implement effective human-machine collaboration strategies"-- Provided by publisher.

Identifiers: LCCN 2025015519 (print) | LCCN 2025015520 (ebook) | ISBN 9798337329857 hardcover | ISBN 9798337329864 paperback | ISBN 9798337329871 ebook

Subjects: LCSH: Manufacturing industries--Technological innovations | Manufacturing processes--Automation http://id.loc.gov/authorities/subjects/sh88004818 | Human-computer interaction http://id.loc.gov/authorities/subjects/sh88003229

Classification: LCC HD9720.5 .N38 2025 (print) | LCC HD9720.5 (ebook) | DDC 338/.064--dc23/eng/20250530

LC record available at https://lccn.loc.gov/2025015519

LC ebook record available at https://lccn.loc.gov/2025015520

British Cataloguing in Publication Data
A Cataloguing in Publication record for this book is available from the British Library.

Table of Contents

Detailed Table of Contents

The world as we know it has limited and scarce food resources. Agriculture is the most important sector to any country's sustainability in this day and age. Agricultural co-operatives play a significant role in the agricultural sector, as they not only reduce poverty in rural communities but they contribute directly to rural employment and further add to the country's GDP. The farmer who toils the fields in sun, rain and hail, provides food that is essential in daily survival to hundreds of people in a community. Technology has revolutionized the way things are done in work-spaces. Agricultural co-operatives and role-players in the agricultural sector must adjust their mindsets to climate change dynamics, the demand for food, and profit margins, and keep abreast of technological changes in this sector to improve their processes, procedures, and basic work-spaces.

Ensuring road safety and enforcing traffic regulations are critical challenges in urban mobility management. This work proposes a camera-based helmet-wearing detection system that leverages YOLO (You Only Look Once) object detection, ESP32-CAM, and an ESP32 microcontroller to monitor and enforce helmet usage among motorcyclists. The system detects riders without helmets in real time and alerts authorities for necessary action. Beyond helmet detection, the system incorporates traffic rule enforcement by identifying violations such as signal jumping and unauthorized lane usage, ensuring a more disciplined traffic environment. To further optimize traffic flow, the system classifies traffic density at signals into high or low traffic conditions based on real- time image processing. This classification helps in adaptive traffic control, reducing congestion and improving vehicle movement efficiency. The captured data, including images and violation records, is transmitted to a central server for further analysis, logging, and alert generation.

Benzene rings are a fundamental component of many organic molecules, and detecting their presence is crucial in various fields such as chemistry, materials science, and pharmaceuticals. However, existing methods for benzene ring detection rely heavily on rule-based approaches or manual annotation, which can be time-consuming and prone to errors. This study proposes a novel approach for benzene ring detection using graph convolutional networks (GCNs). GCNs are a type of neural network designed to handle graph-structured data, making them ideal for molecular data. Our approach represents molecules as graphs, where atoms are nodes and bonds are edges. We then use a GCN to learn a representation of the molecule that can be used to detect benzene rings. Our model is trained on a large dataset of molecules annotated with benzene ring information.

Innovation, speed, and adaptability define the startup ecosystem—traits well-aligned with Agile principles. Agile helps startups navigate uncertainty, iterate quickly on feedback, and deliver value incrementally. It fosters collaboration, responsiveness, and flexibility, enabling long-term growth in competitive markets. Meanwhile, governments globally are shifting from traditional bureaucracy to e-government systems to boost transparency, efficiency, and citizen access to services. This digital transformation seeks to streamline operations and build citizen-centric solutions. Yet, challenges remain—scalability, legacy integration, user-focused design, data privacy, and cybersecurity. Overcoming these hurdles demands a shift toward Agile thinking, innovation-driven policies, and strategic tech advancement to ensure sustainable, effective digital governance.

The rapid adoption of smart technologies has transformed modern manufacturing, making human-machine collaboration—merging automation's efficiency with human creativity—central to Industry 4.0. This chapter emphasizes the strategic importance of balancing automation with human involvement to enhance innovation, safety, and resilience while avoiding overreliance that can lead to disengagement and system fragility. It offers a historical overview of automation's evolution, examines how humans have shifted from manual labor to key collaborators, and highlights the unique human traits vital in complex environments. Drawing on frameworks like Human-in-the-Loop systems and Human-Centered Design, the chapter explores practical strategies and policies for fostering balanced collaboration. Ethical concerns, such as accountability and job transformation, are also addressed. Looking ahead, it considers emerging technologies like generative AI and cobots, arguing for future factories built on flexibility, empathy, and inclusivity.

New paradigm shift set in motion by Industry 5.0 complements proficiently existing Industry 4.0 by specifically positioning critical thinking, knowledge, innovation, and research at core of transitioning to sustainable, healthy, human-centric, inclusive, and resilient industry. The research draws on collaborative technologies reshaping human-machine relationships in modern factories, addresses key drivers in undergoing economic and societal transition, and provides a vision of flourishing industry that goes beyond efficiency and productivity. On the one hand, the study emphasizes todays' challenges for society and, on the other hand, offers solutions for climate change, preservation of resources, and social stability. Findings show artificial intelligence and machine learning robustly support Industry of the Future, actively attract best talents, and vigorously address evolving skills and training of individuals. Likewise, communities of practice are crucial in sharing steps in implementing good principles as well as knowledge and innovation ecosystems for new collaborations.

Mastering artificial intelligence (AI) in marketing and communication based on digital strategies and analytics creates sustainable competitive advantage for smart factories and enables marketing teams boost their consumer relationship management programs. The research accentuates how AI shapes marketing today, highlighting advantages and disadvantages, while centering on happiness management, sustainability, and well-being. Additionally, the study sheds a new light on AI regulatory and policy landscape for Sustainable Development Goals, exploring the multifaceted impact of AI on innovations, research, conversations, trends, and consumer behavior. Outcomes point up AI's capabilities in enhancing marketing excellence, while focusing on achieving organic growth with the aid of best marketing practices, marketing ecosystem, marketing agility, strategic thinking, and leadership. Findings emphasize that building an effective marketing AI strategy depends on hyper-personalized marketing campaigns and tailored communications that offers engagement, precision, and creativity.

Using a new and complex technological system is stressful in itself, and this paradigm shift may not be immediately internalized by employees. This can cause high levels of technostress in employees, which can manifest itself as anxiety, timidity and frustrationThe concept of technostress is evaluated in multiple dimensions. In particular, the concept can trap employees in a negative psychological state that has a permanent effect on their physical, emotional and mental well-being. Employees feel that they always have to be alert and connected and therefore they can feel that they have no control on anything including their lives.In this sense, the purpose of this study is first to scrutinize the concept of technostress, which is caused by exposure to technology and especially to automation in organizations. The fear of job displacement is going to be handled in detail, since nowadays the most important cause of this fear is technology.

enabling technologies such as IoT, AI, and edge computing, along with real-world applications and evaluation metrics. Key challenges and future research directions are explored, highlighting gaps in scalability, security, interoperability, and human-machine collaboration.

In this age of rapid technological growth, businesses are adopting automation technologies to boost efficiency and competitiveness. Humans often oppose automation due to concerns about job loss, autonomy, and unpredictability. Successful automation requires strategic adaptation, which includes managing human resistance and encouraging human-robot collaboration. This chapter discusses how firms might handle these issues by creating policies that promote automation, align organizational goals with technological progress, and ease worker integration. This chapter examines theoretical frameworks, case studies, and emerging trends to help overcome resistance, promote collaboration, and adjust to automation. It concludes with practical advice for companies trying to succeed in an automated world.

The concept of the "augmented operator" represents a paradigm shift in how humans interact with technology, aiming to enhance rather than replace human capabilities. This involves the integration of advanced technologies, such as augmented reality (AR), artificial intelligence (AI), and wearable sensors, into the operator's workflow. By overlaying digital information onto the real-world environment, AR provides operators with real-time guidance, instructions, and data, enabling them to perform complex tasks with increased accuracy and efficiency. AI algorithms can analyze vast amounts of data to provide predictive insights, optimize decision-making, and automate repetitive tasks, freeing up operators to focus on more critical and creative aspects of their work. Wearable sensors monitor the operator's physiological state,

Preface

INTRODUCTION

The twenty - first century has marked the dawn of a new industrial revolution — one where human creativity and machine intelligence converge to redefine how factories operate, compete, and sustain themselves. The concept of the smart factory extends beyond automation; it represents a transformation in which humans and machines collaborate, learn, and evolve together.

This book, *Navigating Human - Machine Collaboration in Smart Factories*, provides a comprehensive roadmap for researchers, practitioners, policymakers, and students who are committed to understanding and shaping the next phase of industrial transformation. By blending theory with practice, and vision with application, this work seeks to inspire meaningful dialogue and guide thoughtful action in building factories that are not only efficient, but also sustainable, resilient, and profoundly human - centric.

I extend my gratitude to the contributors, researchers, and innovators whose insights have enriched these chapters, and to the readers, whose curiosity and engagement make this journey worthwhile. These volumes serve as both a reference and a catalyst for building a future where humans and machines collaborate not in competition, but in harmony.

ORGANIZATION OF THE BOOK

The transformation of manufacturing into a smart, connected, and adaptive ecosystem represents one of the most significant industrial shifts of our time. At the heart of this change lies the evolving relationship between human capability and machine intelligence. Far from replacing the human element, modern factories are redefining collaboration, where machines amplify precision and speed, and humans contribute creativity, judgment, and adaptability.

The chapters gathered here span diverse yet interconnected themes: from artificial intelligence and hyper - automation to collaborative manufacturing frameworks and ethical considerations. Collectively, they help readers grasp the core philosophies, challenges, and opportunities that must be understood before smart factory strategies can be successfully deployed.

This volume is intended for researchers, students, industry professionals, and policymakers who seek to explore the theoretical dimensions and foundational knowledge necessary for navigating

Industry 4.0 and beyond. It is the starting point of a two - part journey toward understanding not only the mechanics of technology, but also the enduring role of the human factor.

A description of the importance of each of the chapter submissions:

Chapter 1: Agricultural Co-Operatives: What Can They Do Better to Improve Their Processes by Adapting New Technologies

This chapter emphasizes the growing relevance of technology in agricultural cooperatives, which traditionally depend on collective decision - making and resource pooling. By adopting smart tools such as IoT - based monitoring, data analytics, and AI - driven forecasting, cooperatives can improve productivity, transparency, and market competitiveness. The discussion highlights how agricultural organizations can transition toward digitalization to streamline supply chains, reduce inefficiencies, and enhance member engagement. The chapter is important because it situates human - machine collaboration not only in industrial manufacturing but also in agriculture — showing the universality of smart factory principles across sectors.

Chapter 2: AI-Driven Helmet Compliance for Enhancing Human - Machine Collaboration in Smart Manufacturing: Smart Helmet & Traffic Violation Detection System

This chapter introduces a safety - centred approach to Industry 4.0 through the development of AI - enabled helmets for worker protection and compliance monitoring. By combining machine vision, IoT sensors, and AI - based decision - making, the system ensures safer industrial environments while also reducing workplace violations. Importantly, the chapter demonstrates how wearable technology can act as a bridge between human operators and machines, contributing to real - time monitoring and fostering a culture of accountability. The importance of this chapter lies in its direct application of AI to workplace safety and compliance, reinforcing that smart factories are not only about efficiency but also about human well - being.

Chapter 3: Benzene Ring Detection Using Graph Convolutional Networks

This chapter explores the use of graph convolutional networks (GCNs) to detect benzene rings, a fundamental challenge in computational chemistry and materials science. Its inclusion underscores how AI - powered detection and prediction systems can accelerate discovery in industrial applications such as chemical engineering, materials development, and safety testing. Beyond chemistry, the chapter highlights the larger potential of machine learning architectures in pattern recognition and automated analysis, offering new possibilities for human - machine collaboration in highly technical fields. The importance of this chapter is that it demonstrates the cross - disciplinary applications of smart factory AI, showcasing how industrial challenges can be solved through innovative computational models.

Chapter 4: Building Resilient Systems with Agile In StartUps and E-Government

This chapter addresses the challenges faced by start - ups and public - sector organizations in adapting to rapid technological change. By applying agile methodologies, it outlines pathways for building resilience and flexibility in environments that require continuous innovation. Start - ups and governments are often at opposite ends of the spectrum in terms of speed and structure; however, both can leverage agile and human - machine collaboration principles to thrive in the digital economy. The chapter's significance lies in its integration of organizational agility with smart technologies, offering practical lessons for entities aiming to remain relevant, competitive, and responsive in volatile markets.

Chapter 5: Collaborative Intelligence: Blending Human Insight with Machine Precision

This chapter explores the emerging paradigm of collaborative intelligence, where human creativity, empathy, and contextual awareness are combined with machine speed, accuracy, and data - processing ability. Rather than treating AI as a replacement for human judgment, the chapter emphasizes its role as a complement. Applications across manufacturing, service industries, and policy - making are explored. The importance of this chapter is that it provides a conceptual framework for human - machine synergy, laying the intellectual foundation for the rest of the book. It reframes the narrative from "replacement" to "collaboration," making it essential for building trust in Industry 4.0 ecosystems.

Chapter 6: Excellence in New Era of Human-Machine Collaboration and Interaction for Smart Factories: Artificial Intelligence and Machine Learning for Industry 5.0 Transformation

This chapter contextualizes the shift from Industry 4.0 (automation and efficiency) to Industry 5.0 (human - centred and sustainable innovation). It examines how artificial intelligence and machine learning are catalysts for reimagining productivity, personalization, and collaboration in smart factories. The chapter is important because it emphasizes excellence as a balance between efficiency and human values, setting the stage for future - ready organizations.

Chapter 7: Harnessing Artificial Intelligence in Marketing and Communication for Smart Factories: Regulatory and Policy Landscape for Sustainable Development Goals

This chapter extends the conversation to marketing and communication, exploring how smart factories can use AI to better engage stakeholders, customers, and policymakers. It also addresses the regulatory landscape and alignment with the UN Sustainable Development Goals (SDGs), ensuring that technological adoption contributes to sustainability rather than unchecked growth. Its importance lies in showing that smart factories are not isolated technical systems but are embedded within broader socio - economic and policy ecosystems.

Chapter 8: Human Resistance to Automation and the Fear of Job Displacement

This chapter explores one of the most critical challenges in human - machine collaboration: psychological resistance and fear of redundancy. By analysing case studies, it identifies the root causes of resistance, including insecurity, lack of trust, and socio - economic concerns. Importantly, it provides strategies for addressing these fears through reskilling, transparent communication, and participatory approaches. Its importance lies in acknowledging the human cost of automation and offering solutions for building trust in smart factory transitions.

Chapter 9: Human-Machine Collaboration in Industry 4.0: Balancing Automation, Innovation, and the Human Factor

This chapter synthesizes the core challenges of Industry 4.0: balancing automation - driven efficiency with human - driven creativity. It emphasizes the importance of keeping humans at the centre while adopting innovations such as robotics, AI, and IoT. Its importance is that it frames Industry 4.0 as not only a technological revolution but also a cultural and human - centred transformation.

Chapter 10: Influence of Transformational Leadership on Ethiopia Smart IT Factories' Performance through Open Innovation

This chapter examines the role of transformational leadership in driving innovation and performance in Ethiopian smart IT factories. By focusing on open innovation practices, it illustrates how leadership can bridge the gap between human capital and advanced technologies. The importance of this chapter lies in its contextual perspective — it demonstrates that leadership style, cultural setting, and organizational openness are decisive factors in whether smart factory transitions succeed or stagnate.

Chapter 11: Key Drivers of Change in Industry 4.0: AI, IoT, Robotics, and Big Data

This chapter provides a structured overview of the four key drivers of Industry 4.0 — AI, IoT, robotics, and big data — highlighting their individual and combined impacts on smart factories. It underscores how these technologies create competitive advantages while also reshaping workforce roles. Its importance lies in synthesizing the core technological building blocks of smart factories, giving readers a clear map of the drivers behind industrial transformation.

Chapter 12: Optimizing Production through Sensors, Data Analytics, And Real - Time Information Flow

This chapter explores how sensors and real - time analytics are transforming production management by providing continuous insights for optimization. It highlights how data - driven decision - making enables predictive maintenance, reduced downtime, and improved efficiency. The chapter's importance lies in showing how smart factories achieve continuous performance improvement through information - rich ecosystems that enhance collaboration between humans and machines.

Chapter 13: Strategic Adaptation in the Age of Automation for Managing Resistance and Fostering Collaboration

This chapter examines the strategic approaches organizations can adopt to manage resistance to automation. It emphasizes communication, participatory planning, and inclusive change management as key to fostering collaboration. Its importance lies in providing actionable strategies to turn resistance into engagement, ensuring that automation becomes a shared journey rather than a top - down imposition.

Chapter 14: The Augmented Operator: Enhancing Human Capabilities

This chapter introduces the concept of the augmented operator — a human worker enhanced with digital tools such as AR/VR interfaces, wearable's, and cognitive support systems. By augmenting rather than replacing workers, smart factories can unlock new levels of productivity and worker satisfaction. The chapter's importance is that it reimagines the role of humans in technology - rich environments, shifting the narrative from redundancy to empowerment.

Chapter 15: The Evolution of Manufacturing and the Role of Human Resources in the Technological Era

This chapter traces the historical evolution of manufacturing from manual labour to automated and now collaborative systems, with a particular focus on the changing role of human resource management (HRM). It discusses how HR must adapt to support reskilling, digital literacy, and employee well - being. The importance of this chapter lies in its emphasis on HR as a strategic partner in technological transformation, ensuring people remain central to industrial progress.

Chapter 16: The Future of Collaborative Manufacturing

This forward - looking chapter envisions the future of collaborative manufacturing, where humans and machines work in fluid partnerships. It examines anticipated advancements in AI, robotics, sustainability practices, and policy frameworks. The importance of this chapter lies in its role as a vision - setting conclusion, preparing readers to imagine and prepare for the opportunities and challenges of Industry 5.0 and beyond.

Chapter 17: Smart Factories, Skills Gap, Well-being, and Overcoming Barriers to Human-Machine Collaboration: Significance of Smart Factories

This chapter addresses one of the most pressing issues in Industry 4.0: the skills gap. It discusses the mismatch between technological adoption and workforce readiness, and its effects on worker well - being. By analysing barriers to human - machine collaboration, it proposes strategies for reskilling, continuous learning, and creating supportive organizational cultures. Its importance lies in connecting technological progress with human development, ensuring that smart factories empower rather than marginalize workers.

Chapter 18: Transformational Leadership and Change Management in Human - Machine Collaboration for Smart Factories

The final chapter highlights the critical role of leadership and changes management in orchestrating successful human - machine collaboration. It discusses how transformational leaders can build trust, align vision, and guide organizations through uncertainty. The importance of this chapter is its integrative perspective — it ties together technology, human factors, and organizational culture, ensuring that smart factory transitions are sustainable, inclusive, and future - ready.

Chapter 1

Agricultural Co-Operatives: What Can They Do Better to Improve Their Processes by Adapting New Technologies

Vannie Naidoo
https://orcid.org/0000-0001-8435-4348
University of KwaZulu-Natal, South Africa

Julius Chiwawa
https://orcid.org/0009-0006-7015-7128
University of KwaZulu-Natal, South Africa

ABSTRACT

The world as we know it has limited and scarce food resources. Agriculture is the most important sector to any country's sustainability in this day and age. Agricultural co-operatives play a significant role in the agricultural sector, as they not only reduce poverty in rural communities but they contribute directly to rural employment and further add to the country's GDP. The farmer who toils the fields in sun, rain and hail, provides food that is essential in daily survival to hundreds of people in a community. Technology has revolutionized the way things are done in work-spaces. Agricultural co-operatives and role-players in the agricultural sector must adjust their mindsets to climate change dynamics, the demand for food, and profit margins, and keep abreast of technological changes in this sector to improve their processes, procedures, and basic work-spaces.

INTRODUCTION

Agriculture is a sector closely aligned with the food supply. It is an important SDG, as many of the world's population are faced with poverty and hunger. Advancements in technology have become a catalyst that is used to growing and sustaining the agricultural sector. One of the world's greatest challenges in this era is to provide safe and healthy food to a growing global population. Agriculture is significant to the economy of a country. Cattle, fish, fruit and vegetable farmers include in their daily operations global positioning systems (GPS) and geographic information

DOI: 10.4018/979-8-3373-2985-7.ch001

systems (GIS), along with software for financial management. Since land and capital are limited resources, feeding a growing population becomes challenging. As the world population grows, new inroads to grow and develop, and sustain the agricultural sector will be needed. This is where the world can benefit from technology and use it wisely to improve the sector.

This paper will highlight the following key themes:

- Background
- Major contributions of Agricultural Cooperatives
- Challenges faced by Agricultural Cooperatives
- Sustainable agriculture
- Recommendations and a way forward
- Future areas of research

This discussion commences with the Background to this chapter. Here, the South African agricultural sector and the South African Agricultural Cooperatives will be discussed in detail.

BACKGROUND

Two major themes on the South African agricultural sector and its' agricultural cooperatives' will be discussed in detail.

The South African Agricultural Sector

Extensive commercial farming and a large number of smaller subsistence-oriented farms make up South Africa's dualistic agricultural sector. There are 40,122 extensive commercial farms and more than 300,000 smaller subsistence-oriented farms in South Africa (Statistics South Africa, 2024). This dichotomy stems from past policy frameworks that favoured commercial farming through institutional infrastructure and state subsidies, especially during apartheid, when the majority of government intervention was focused on white-owned commercial farms (Chikazunga & Paradza, 2012). Due to the relative sophistication and productivity of its commercial farming sector, South Africa's agricultural economy stands out in comparison to other agricultural economies in Sub-Saharan Africa. South Africa has established a highly diversified and capital-intensive agricultural sector, whereas many African countries still rely on subsistence farming (Muzekenyi, Zuwarimwe & Kilonzo, 2022). All of the major grains—aside from rice—oilseeds, citrus, wine grapes, sugar, and a variety of vegetables are among the nation's many agricultural products. Underpinned by sophisticated processing and distribution systems, livestock production is equally diversified and includes beef, dairy, pork, mutton, poultry, and eggs (Sibanda & Ndlela, 2020).

Policies akin to those in developed nations like the US and the EU, where agricultural subsidies and focused support initiatives have been crucial in forming competitive, export-oriented industries, are partially responsible for South Africa's commercial agriculture's success (Muzekenyi, Zuwarimwe, & Kilonzo, 2022). These state-led initiatives have historically made it possible for the commercial farming sector to adopt new technologies and realise economies of scale by guaranteeing access to credit, infrastructure, and market support. A major participant in both regional and global agricultural trade is South Africa. In Sub-Saharan Africa, agricultural exports account for over 20%; thus, this sector made up about 12% of the nation's total export revenue in 2016 (FAOSTAT, 2018). The nation's varied climate, advanced transportation system, and updated supply chains, which facilitate effective integration into international markets, all contribute to its competitive advantage in agri-exports.

A recent evaluation of the sector's performance can be found in the Economic Review of South African Agriculture (Department of Agriculture, Land Reform and Rural Development, 2023). Gross farming income increased 17.3% from R379.86 billion to R445.45 billion for the fiscal year that ended in June 2023. The main drivers of this growth were higher profits from horticultural commodities (8.8%), animal products (9.4%), and field crops (up 39.0%). Significant price increases in summer grains (21.9%), winter grains (17.9%), and vegetables (19.2%) were the main drivers of the 10.7% increase in the average price that farmers received overall. Nonetheless, a few subsectors saw price declines, particularly oilseeds (-13.1%) and pastoral products (8.9%). Dairy (15.1%), slaughtered livestock (2.5%), and poultry (16.0%) drove the average 9.2% increase in animal product prices. Vegetable prices rose significantly in the horticultural sector, but viticulture (4.5%) and fruits (2.7%) saw more modest increases. These patterns show how responsive the industry is to supply-demand swings, climate variability, and domestic and global market dynamics. Producers faced difficulties when the price of agricultural inputs increased by 7.9% at the same time. An 18.5% increase in feed prices was followed by notable increases in fuel (10.9%), building materials (8.3%), and inputs for protecting crops and animal health (8.2%). Price increases exceeded input cost inflation, as evidenced by the 3.0% improvement in the domestic terms of trade despite these increases. Consequently, net farming income increased by a significant 33.8% to R122.73 billion from R91.75 billion in the previous fiscal year (Department of Agriculture, Land Reform and Rural Development, 2023). Despite being vibrant and focused on exports, the South African agriculture industry still faces structural difficulties. These include climate variability, the need for inclusive transformation, and smallholder farmers' unequal access to markets, capital, and land. Resolving these problems is still essential to using agriculture as an instrument for social justice, food security, and rural progress in addition to being an economic engine.

In Sub-Saharan Africa, issues relating to climate change negatively influence agriculture, thus impacting the poor and marginalised households that rely on agriculture as a way of living (Muzekenyi, Zuwarimwe, & Kilonzo, 2022). The progressively unpredictable and unusual weather patterns have impacted many economic sectors, including those within the agricultural sector, affecting food production. According to Maota and Naidoo (2025), 4IR covers 'smart' technologies those such as artificial intelligence (AI), Blockchain, Internet of Things (IoT), and drones. In the agricultural sector, 4IR technologies can be used in various ways. For instance, sensors are implemented to automatically monitor the temperature, moisture content in soil and levels of humidity, while drones are used in applying pesticides to crops.

The discussion below focuses on the South African agricultural cooperatives.

The South African Agricultural Cooperatives

Agricultural cooperatives are economic constituents that are formed voluntarily by members to build and dispense goods and services, sharing compensation and uncertainty (Schwettmann, 2022). They are confined to organisations that are employed in the production, processing, selling and advertising, and supply of agricultural produce (Department of Agriculture, Forestry and Fisheries, 2015). The cooperatives provide service centres in farming areas, pooling resources and maximising the benefits of members through entrepreneurship (Chiwawa, Naidoo & Rena, 2024). They function as distributors of farming inputs, providers of farm credit, processors, marketers, and engage in various economic activities for the benefit of society and members (Mundia et al., 2022). Agricultural cooperatives can be categorised as either production cooperatives, where farmers jointly operate the cooperative on shared agricultural plots, or service cooperatives, where members carry out independent activities while the cooperative provides various services (Department of Agriculture, Forestry and Fisheries, 2015). Agricultural cooperatives have the potential to empower marginalised farmers, reduce poverty, and promote rural and national development. By pooling local resources and fostering self-help capacity, cooperatives contribute to the economic empowerment of rural communities. Mundia et al. (2022) argue that agricultural cooperatives play a crucial role in transferring agricultural information, providing farmers with lower-cost goods and services, and enabling access to higher-priced markets. Aryal et al. (2020) assert that agricultural cooperatives also facilitate the adoption of modern farming technology and machinery, leading to increased productivity. Through collective action and economies of scale, rural cooperatives offer members enhanced benefits and serve as effective drivers of economic development, guided by principles beyond profit alone.

In this section, a brief historical account of cooperative development in South Africa is presented. The rationale for the direction is to assess whether the cooperative concept is a new phenomenon and how successive South African governments have managed the development of cooperatives.

Cooperative Development Before the Apartheid Era

Cooperatives have been around since the late 1800s in South Africa (Shoeman, 2006). They were mostly in the form of what is known as 'stockvels', which is an informal cooperative practice mainly among black people. The practice has been used for centuries and is still used by the majority of South Africans and Africans in general (Raniga, 2021). People operating 'stockvels' come together and collect money periodically. The money collected will then be used for functions, which include family gatherings, funerals, weddings and even birthday celebration parties. Thaba and Mbohwa (2015) assert that these 'stockvels' were not registered and were operating informally since the past apartheid laws did not recognise them.

Cooperative Development During the Apartheid Era

Mushonga et al. (2019) highlight that South Africa has a rich history of cooperative development, with agricultural cooperatives playing a significant role since the early 19th century. Government subsidies and collaborations with agricultural cooperatives fostered a thriving farming society, while consumer cooperatives faced challenges and declined. In 1892, the Pietermaritzburg consumer cooperative was established. This was the first registered cooperative in the country (Ortmann & King, 2006; Christian et al., 2018).

The Department of Trade and Industry (2004) indicated that the Cooperatives Act, No. 19 of 1981, in South Africa led to the formation of a cooperative movement among white South Africans. The Act provided clear provisions for the establishment and regulation of cooperatives, including the role of the registrar's office in overseeing their incorporation, management, financial aspects, and other legal requirements. Van Niekerk (1988) stated that by 1985, cooperatives in South Africa had been classified into primary cooperatives, central cooperatives, federal cooperatives, and special farmers' cooperatives, respectively.

Cooperative Development after Independence

In 1994, when the South African democratic government came to power, it disregarded the Cooperative Act of 1981 as it was an unsuitable instrument for cooperative development (Department of Trade and Industry, 2004). As a result, the Cooperatives Act, 2005 (No.14 of 2005) came into being on August 18, 2005. Rogan and Alfers (2020) assert that the new Cooperative Act of 2005 was developed in line with the international cooperative principles. Primary cooperatives (including agricultural, housing, consumer, worker, burial society, service and financial cooperatives), secondary cooperatives (where those founded by two or more primary cooperatives that provided members with sectoral services), and tertiary cooperatives (whereby two or more primary cooperatives provided their members with sectoral services) were all eligible to register under this Act.

In the discussion that follows, the potential benefits of cooperatives will be discussed in detail.

MAJOR CONTRIBUTIONS OF AGRICULTURAL COOPERATIVES

Contribution Towards Local Economic Development and Economic Growth

According to Raniga (2021), cooperatives are essential for generating locally sourced goods, generating employment, localising the economy, and fostering the development of human resources in local communities when they receive the proper support. Sahoo et al. (2020) argue that by giving money to cooperative banks, social service organisations, and the cooperative social movement, the Indian government has led by example. 69% of India's poor rural households have benefited from this assistance, proving the effectiveness of financial support and governmental commitment to reducing poverty. Kustepeli et al. (2023) claim that cooperatives are acknowledged as powerful instruments for social democracy and community development in Africa, and can enhance the quality of life for low-income households. This strategy is in line with people-oriented development, self-help capabilities, and the financial requirements of low-income communities (Raniga, 2021). Encouraging women and young people to take on leadership roles in cooperatives connects sustainable livelihoods and community economic development.

By encouraging inclusive economic participation, cooperatives, especially in rural and underdeveloped areas, make a substantial contribution to local economic development. By giving small-scale farmers access to markets, agricultural inputs, and technical knowledge, agricultural cooperatives in particular help them become more productive and competitive (Mokoena et al., 2023). Cooperatives help farmers lower costs, boost yields, and improve product quality by enabling economies of scale, which improves income generation and promotes local economic stability. Additionally, cooperatives promote a cycle of economic growth by reinvesting profits back into the community, which boosts local economies (Sacchetti & Tortia, 2021). Cooperatives have a multiplier effect as they expand, meaning that as member incomes rise, so does the drive to possess higher volumes of goods and services in the community. Consequently, this helps local companies and promotes economic expansion in general (Rogan & Alfers, 2020).

Contribution of Cooperatives Towards Poverty Reduction

The volatile global economy has created unequal shifts in the way wealth and resources are distributed around the world, leading to poverty and income inequality worsening, both within and between nations (Raniga, 2021). Since cooperatives are best placed to increase resilience in all socioeconomic spheres of society, the International Labour Organisation (2021) suggested that they be established as a practical way to decrease poverty around the globe.

According to empirical evidence from countries like Kenya, Canada, India, Italy, and Bangladesh, which have effective cooperatives in place that assist in the active support of government legislation, intergovernmental cooperation, and provide the necessary education and training within the higher education sector (Nakayiso & Andrew, 2023; Zhang, Benjamin & Miao, 2021). By expanding access to markets, resources, and knowledge, cooperatives help reduce poverty and empower marginalised communities to become economically independent. For example, agricultural cooperatives empower small-scale farmers by giving them the resources and tools they need to increase profitability and productivity (Sellare et al., 2020).

As a result, households' incomes rise and their standard of living improves, helping them escape poverty. In addition, cooperatives frequently place a high priority on social and economic justice, guaranteeing that members share in the gains from economic activity (Chiwawa et al., 2024; Zhang et al., 2021). This group effort fosters social cohesiveness in local communities and lessens income disparity. Within South Africa, cooperatives provide inclusive economic development and lowering poverty (Mokoena et al., 2023). The most engaged participants in attaining the objectives of equalising the first and second economies may be Black Africans, especially those living in rural areas, young people, women, and those with health conditions. The Ubuntu values of cooperation and fellowship form the basis of all cooperatives, including agricultural ones (Moalosi & Rapitsenyane, 2023).

Contribution of Cooperatives Towards Employment Creation

Through three main mechanisms, cooperatives have become important employers. Cooperatives directly employ people in primary and secondary cooperative structures (Kustepeli et al., 2023). They also make it possible for members to participate in economic activities that generate revenue, guaranteeing sustainable livelihoods (Chiwawa et al., 2024). Additionally, through value chain connections and local economic stimulation, the larger ecosystem that cooperatives create creates indirect jobs (Schwettmann, 2022). Cooperatives have been crucial in helping jobless women within Sub-Saharan Africa, mainly in Kenya, Tanzania, Malawi, and Zimbabwe, to overcome poverty by coordinating both private and public resources (Mataba & Komba, 2021; Raniga, 2021). These businesses are essential to the creation of jobs in industries like manufacturing, retail, and agriculture. For example, agricultural cooperatives invest in processing, packaging, and distribution activities in addition to farming, which creates jobs (Mosia & Khalo, 2021).

Cooperatives provide feasible options for generating income and fostering local development in rural areas where formal employment is scarce (Risal, 2020). Cooperatives lessen the strain on urban labour markets and reduce rural-urban migration by assisting small businesses and stabilising rural economies. In South Africa, where unemployment hit 32.6% in 2023, this is especially pertinent (Statistics South Africa, 2024). Cooperatives are actively promoted by the government through the Department of Small Business Development (DSBD, 2020) as a means of fostering inclusive growth and reducing poverty among underserved communities.

In the section below, challenges faced by Agricultural Cooperatives will be highlighted.

CHALLENGES FACED BY AGRICULTURAL COOPERATIVES

According to published research, although many countries are starting to see the benefits of cooperatives, the difficulties these organisations face vary by country (Nakayiso & Andrew, 2023; Pansera & Rizzi, 2020; Moon & Lee, 2020). Pansera and Rizzi (2020) suggest that there have been many failures in developing countries in addition to some noteworthy successes. Poor management, low capital, low member participation, inadequate training, societal feudalism, ambiguous policies, land fragmentation, and insufficient integration of production, credit, and marketing activities are some of the major issues that Rwandan agricultural cooperatives face (Moon & Lee, 2020). When combined, these issues hinder agricultural cooperatives' ability to grow and function effectively. A high attrition rate, challenges with management and leadership, government interference, and instability are just a few of the challenges South African cooperatives face (Zantsi, 2021). This led to the creation of numerous cooperatives that were ill-conceived or unsustainable.

Further discussion on the challenges highlighted is:

Poor Management and Leadership

Moon and Lee (2020) opine that a major challenge within agricultural cooperatives is the prevalence of poor management and leadership. This is common across various countries. For instance, Zantsi (2021) identifies high attrition rates, leadership difficulties, and instability as some of the significant challenges faced by South African cooperatives. Similarly, Zhang et al. (2021) note that Ethiopian cooperatives struggle with weak accounting systems, leadership issues, and a lack of entrepreneurship knowledge among executives and staff. These problems are often exacerbated by a lack of professional management skills and inadequate training for cooperative members.

Limited Access to Capital and Financial Resources

Moon and Lee (2020) claim that the restricted access to capital resources hinders the ability of cooperatives to expand their operations and invest in necessary infrastructure. Zantsi (2021) add that restricted access to capital can become an obstacle to the success of agricultural cooperatives, particularly within developing regions of Rwanda and South Africa. This issue is compounded by inadequate access to credit facilities, which limits cooperatives' ability to support and grow their enterprise. Mamo et al. (2021) identify poor infrastructure, insufficient extension services, and dwindling resources and land as obstacles to the success of cooperatives. These limitations hinder their ability to expand operations and improve their overall effectiveness.

Internal Conflicts and Lack of Member Commitment

Serra and Davidson (2020) argue that internal conflicts, such as disagreements among members and conflicts of interest, are also major challenges. These conflicts often stem from a lack of member commitment and poor management practices, leading to the eventual collapse of cooperatives. These issues are exacerbated by inadequate training and education among members and leaders, which hinders the effective operation of cooperatives (Serra & Davidson, 2020). Effective communication and active participation among cooperative members are critical for the success of these organisations. However, challenges such as insufficient member communication and low levels of leadership further add to conflict. Without strong leadership and proper management, cooperatives struggle to operate efficiently and face high attrition rates. Participation rates can undermine the cooperative's operations and decision-making processes (Moon & Lee, 2020). Therefore, internal conflicts and a lack of commitment among members further exacerbate these issues.

Inadequate Training and Educational Initiatives

The lack of adequate training, extension services, and educational initiatives is a recurring challenge for agricultural cooperatives. Pansera and Rizzi (2020) highlight that cooperative colleges often serve government employees and school leavers, with training rarely available to primary cooperative members. This deficiency leads to a gap in knowledge and skills among cooperative members, which in turn affects their ability to manage and sustain cooperative activities effectively (Moon & Lee, 2020). Another obstacle for cooperatives is the shortage of highly skilled human resources, particularly in terms of management and leadership. Cooperatives often struggle with a shortage of trained personnel, which affects their ability to operate effectively. This is compounded by a general lack of knowledge and understanding of cooperative principles among members, leading to poor decision-making and management (Serra & Davidson, 2020).

Government Interference and Policy Issues

According to Mataba and Komba (2021), the government and parastatal organisations often attempt to direct and regulate growth in the cooperatives, for their personal gain. Mamo et al. (2021) add that this top-down control leads to the creation of poorly designed or unsustainable cooperatives. As a result, traces of these state-run "pseudo-cooperatives" still exist in several developing nations. The authors add that in South Africa, government interference and unclear or insufficient policies regarding cooperative development are also significant barriers.

Socioeconomic and Structural Challenges

Socioeconomic factors such as land fragmentation, the feudalistic characteristics of society, and the lack of infrastructure are additional obstacles placed upon the flourishing of agricultural cooperatives (Mamo et al., 2021). These structural issues limit the cooperatives' ability to integrate their activities, such as production, credit, and marketing, thereby reducing their overall effectiveness. Such fragmentation not only hampers productivity but also exacerbates inequalities within the cooperative structure, making it difficult to achieve economies of scale (Moon & Lee,

2020). The authors think that the poor infrastructure, commonly seen in undeveloped or damaged roads and minimal storage facilities, hinders the logistics of cooperative activities, leading to inefficiencies and increased costs that negatively impact the cooperatives' competitiveness and sustainability.

Market Access and Bargaining Power

According to Nakayiso and Andrew (2023), agricultural cooperatives often face challenges related to limited market access and weak bargaining power. This is particularly problematic in competitive global markets, where cooperatives struggle to secure fair prices for their products and to market them effectively. Mamo et al. (2021) opine that due to the fragmented constitution of cooperatives and their small-scale operations make it difficult for them to compete with larger, more established agribusinesses. Sacchetti and Tortia (2021) add that without adequate market information and strong collective bargaining strategies, cooperatives often find themselves at a disadvantage, limiting their ability to thrive in global markets.

Lack of Innovation and Modernisation

The introduction of new technology, crop varieties (Balai et al., 2025). However, many co-operatives face difficulties in adopting these innovations due to limited resources and support (Rena, 2021). Moreover, the reluctance to embrace change, often due to the ageing cooperative membership, further hinders innovation and modernisation efforts (Zantsi, 2021). Addressing these barriers is crucial to enable the enduring viability of cooperatives in competitive markets (Serra & Davidson, 2021).

Demographic Challenges and Illiteracy

Zantsi (2021) asserts that the demographic profile of cooperative members, particularly the fact that many are elderly and illiterate, presents a significant threat to the cooperative movement. These factors prevent cooperatives from adapting to a volatile market environment and effectively implementing modern management practices. Moon and Lee (2020) posit that the ageing membership base is often resistant to new ideas and innovations, which hinders the promotion of technology and sustainable agricultural applications. Mamo et al. (2021) add that illiteracy among members impedes their ability to access and understand market information, financial records, and training materials, further exacerbating management inefficiencies. Serra and Davidson (2021) indicate that the lack of education also limits members' engagement in decision-making, reducing the overall success of the cooperative's governance. Pansera and Rizzi (2020) think that addressing these demographic challenges requires targeted educational programs and youth engagement strategies to rejuvenate cooperative membership and enhance long-term sustainability.

In the discussion that follows, sustainable agriculture and the influence of agricultural co-operatives on the sustainability of rural communities will be discussed.

SUSTAINABLE AGRICULTURE

According to Tahat et al. (2020), sustainable agriculture refers to farming methods which satisfy present food demands without endangering food supplies of future generations. It must be socially acceptable, economically feasible, and environmentally sound, and it entails the conservation of all natural resources (Adams, 2019). Enduring rural development exists greatly aided by sustainable agriculture on a larger scale (Qaim, 2020). Agriculture fulfils several ecological, social, and economic purposes in addition to producing food. The ecological function deals with environmental preservation, the social function includes cultural and political aspects, and the economic function deals with rural employment and income (Qaim, 2020; Zhang et al., 2021).

The supply of sustainable agricultural products is anticipated to be driven by governmental policies and growing consumer awareness (Firoznia et al., 2021). Organisational structure, size, farmer demographics, and external factors like market conditions and policy support are important determinants of farm sustainability (Tahat et al., 2020; Firoznia et al., 2021). Farmers own and operate agricultural cooperatives, which can improve sustainability by sharing resources, lowering risk, and offering services (Candemir et al., 2021). However, their efficacy may be constrained by governance concerns, such as member resistance to change and monitoring inefficiencies (Moon & Lee, 2020). The impact of supply chain organisations on agricultural sustainability requires further investigation.

The Influence of Agricultural Cooperatives on the Sustainability of Rural Livelihoods

Chiwawa, Naidoo, and Darwish (2025) argue that agricultural cooperatives are essential to improving the community, ecological, and economic sustainability based on rural livelihoods. Especially within developing nations, they act as agents for social cohesion and environmental stewardship in addition to being economic units. Promoting equity, justice, and well-being are all part of social sustainability, which is frequently seen as the most difficult sustainability pillar (Adams, 2019; Sileshi et al., 2025). By encouraging gender inclusion, community involvement, and the sharing of knowledge, cooperatives help achieve this. For instance, cooperatives improve the safe use of agricultural inputs through social capital and trust, and Ugandan women in cooperatives gain better negotiation and decision-making skills (Kayongo et al., 2022).

Additionally, cooperatives operate as non-profit organizations that provide public infrastructure, childcare, and reasonably priced canteens, especially in underserved rural areas (Mosia & Khalo, 2021). They enable members to interact with and have an impact on the institutions that affect their lives by fostering democratic participation and collective empowerment (Zhang et al., 2021). One important nonfinancial asset created through cooperative interaction is social capital, which is characterised by norms, trust, and networks (Chiwawa et al., 2025). In the end, it supports sustainable rural development by enhancing group efforts, promoting mutual trust, lowering transaction costs, and facilitating resource access (Mataba & Komba, 2021).

By promoting environmentally friendly farming methods, agricultural cooperatives also improve environmental sustainability. Adoption of sustainable technologies, like integrated pest management and organic amendments, is greatly increased by cooperative participation (Aryal et al., 2020). Through active participation in production processes, cooperatives promote environmental regulation and quality control, which lowers the use of harmful pesticides and improves ecological outcomes (Chiwawa et al., 2025). Additionally, cooperatives help farmers meet environmental quality standards and adopt climate-resilient practices by lowering the cost of innovation and facilitating access to technical assistance (Moon & Lee, 2020). However, if not properly controlled, an excessive dependence on chemical inputs, as observed in certain cooperatively supported intensification projects, could be hazardous to the habitat (Sileshi et al., 2025).

In terms of economics, cooperatives provide a workable model for enhancing rural livelihoods through raising income, productivity, and market accessibility. Compared to investor-owned businesses, they offer members better prices, inputs, and credit facilities; small-scale farmers with less negotiating power particularly benefit from this (Mokoena et al., 2023; Risal, 2020). Additionally, they help larger farms achieve economies of scale and create jobs (Rogan & Alfers, 2020). Cooperatives frequently act as benchmarks, pressuring other market participants to raise standards and prices (Kayongo et al., 2022). They also help spread innovation and enhance quality across supply chains (Mataba & Komba, 2021). Cooperatives maximise member benefits within budgetary constraints, ensuring long-term economic viability despite their nonprofit focus (Candemir et al., 2021). Sustainable rural development is greatly aided by agricultural cooperatives. They act as diverse agents of sustainability by creating social capital, encouraging environmental stewardship, and building economic resilience.

In the section below, the technology integration into agricultural co-operatives and the agriculture sector will be discussed.

Technology Integration into the Agriculture Sector

In the Global South, where it supports national economies, rural livelihoods, and food security, agriculture continues to be essential to global economic growth. However, as the population of humans is predicted to rise to over 10 billion by 2050, the sector is under increasing pressure (Ndauka & Matotola, 2023). At a time when conventional farming practices, which are marked by a significant reliance on chemical inputs, are becoming unsustainable both economically and environmentally, this demographic trend is increasing the demand for food production. Extended use of synthetic pesticides and fertilisers has degraded ecosystems and decreased soil fertility, worsened food insecurity and endangering the planet's future sustainability. Up to 40% of global production is lost as a result of pests and diseases (Junaid & Gokce, 2024), highlighting the shortcomings of traditional methods in dealing with new biotic stressors. Productivity gains are still uneven, particularly for smallholder farmers who dominate agriculture in low-income countries, even with annual investments in agricultural research of about $25.8 billion (Soori et al., 2023).

With these obstacles in mind, smart technologies and artificial intelligence (AI) have become revolutionary instruments that have the power to change agricultural methods. Current studies highlight how artificial neural networks (ANNs), fuzzy logic, machine learning (ML), deep learning (DL), and Internet of Things (IoT) applications facilitate data-driven, real-time decision-making (Silvestri et al., 2021; Gebresenbet et al., 2023). By raising the use of inputs like fertilizer, pesticides, and water, these technologies enable precision agriculture, increasing productivity while lowering environmental impact (Ndauka & Matotola, 2023). To produce useful insights, AI systems aggregate data from many sources, including climate data, crop health and soil. As a result, yield forecasting and resource management have improved, and early detection of pests, diseases, and weeds has been made possible (Lachgar et al., 2023; Rizzo et al., 2024). Inconsistencies in adoption and entry still exist, though, especially among Asian and African small farmers. The diffusion of these innovations is hampered by factors like minimum digital literacy, poor infrastructure and high costs (Teklu et al., 2023). Inadequate extension services and restricted market access keep poverty and food insecurity cycles alive (Li et al., 2023). Therefore, focused policy interventions and support systems catered to the particular requirements of marginalised farming populations will be necessary for a more equitable technological transformation.

Technology Integration into Agricultural Cooperatives

Especially in developing nations, agricultural cooperatives have become essential institutional tools for raising smallholder farmers' socioeconomic standing. Cooperatives greatly lower transaction costs and increase farmers' bargaining power by pooling resources, making inputs easier to obtain, and providing group platforms for marketing and technical assistance (Singh et al., 2021; Zhang et al., 2021). From the stance of technology adoption, where individual farmers frequently face prohibitive costs and limited technical knowledge, these collective arrangements are especially important.

In the past, the Green Revolution brought revolutionary agricultural technologies that resulted in notable increases in productivity, including mechanisation, chemical fertilisers, and better irrigation. Nevertheless, empirical data indicate diminishing returns in later decades; for example, rice yield growth in Asia decreased from 2.6% in the 1970s to 1.5% after 1981 (Santiteerakul et al., 2020). Rising input costs and ecological deterioration, such as soil erosion and water scarcity, have threatened the long-term viability of these gains (Kalfas et al., 2024).

Despite playing an important role in agricultural production, many small farmers still struggle with poverty and food insecurity in the current environment. Structural limitations like restricted access to capital, arable land, and technological infrastructure exacerbate these difficulties (Teklu et al., 2023). An important way forward to improve food security and lower rural poverty is to incorporate cutting-edge technologies into cooperative frameworks, along with capacity-building and fair access to digital tools (Kalfas et al., 2024).

Technology Adoption and Smart Agriculture—Agriculture 4.0

A paradigm shift in food production systems is described by the implementation of digital technologies in agriculture, commonly referred to as "Agriculture 4.0," particularly for agrarian and low-income economies. Incorporating smart technologies like the Internet of Things (IoT), big data, robotics, artificial intelligence (AI), or cloud computing is what is driving this change in farming processes (Silvestri et al., 2021; Khan et al., 2021). These technologies present agricultural cooperatives in developing nations with a great deal of promise for increasing sustainability, productivity, and resilience to market volatility and climate shocks (Ndauka & Matotola, 2023; Lachgar et al., 2023). By creating innovation ecosystems that link farmers to digital tools, real-time market data, and institutional support, digital transformation makes it easier to modernise agricultural value chains. The effective sharing of information, the expansion of technological advancements, and the simplification of input-output procedures are all made possible by this connectivity (Lachgar et al., 2023). Precision farming, another name for smart agriculture, uses data-driven methods to target pests, fertilise crops, and use irrigation and fertilisation to maximise crop yields and minimise environmental effects (Santiteerakul et al., 2020; Teklu et al., 2023).

However, complex interactions within institutional, social, and technological factors determine whether or not such innovations are adopted. A fundamental framework for comprehending this process is provided by Rogers' (1983) innovation-decision model, which highlights five phases: "knowledge, persuasion, decision, implementation, and confirmation". Due to a variety of contextual factors, farmers' adoption rates vary greatly (Manyanga et al., 2023). Key factors have been identified by empirical studies, including the perceived qualities of the technology (e.g., complexity, compatibility, relative advantage), as well as the socioeconomic attributes of farmers, including dimensions of age, level of education, risk tolerance, and credit availability (Spanaki et al., 2022; Li et al., 2023).

Adoption is greatly aided by institutional support networks, especially agricultural extension services. Farmers' willingness to try new technologies has been demonstrated to be strongly impacted by the effectiveness of local extension agents, as determined alongside their technical expertise, credibility, and communication skills (Kalfas et al., 2024). Particularly important in convincing risk-averse or conservative farmers to embrace new practices are demonstration plots, peer learning, and trustworthy information sources. Precision agriculture decision-making is further improved by sophisticated computational technologies (Dai et al., 2023). A subset of artificial intelligence called fuzzy logic systems is being incorporated more and more into autonomous farming equipment like UAVs and harvesting robots to allow for real-time responses to changing environmental conditions (Katiyar & Farhana, 2021; Javaid et al., 2022). These systems facilitate resource optimisation, adaptive navigation, and smart monitoring—all essential components in the endeavor for data-driven and sustainable agriculture.

Applications of Artificial Intelligence in Agriculture

The agricultural sector is transforming thanks to AI, which makes it possible to create intelligent systems that improve sustainability, productivity, and decision-making effectiveness. Globally, AI technologies are changing the composition and functionality of agricultural systems through applications in automation, resource optimisation, predictive analytics, and crop monitoring (Lachgar et al., 2023). For small farming concerns, the adaptation of computer vision, robots, deep learning, machine learning, and natural language processing (NLP) into farming practices has made agricultural operations more data-driven, flexible, and economical.

Precision farming, which uses real-time information from drones, satellite imagery, and Internet of Things devices to direct resource application, has revolutionized the use of AI in the agriculture sector. By evaluating these data streams, machine learning algorithms can recommend the best times and amounts for fertilizer, pesticide, and irrigation applications, increasing productivity and reducing environmental damage (Lachgar et al., 2023; Soori et al., 2023). The scalability and commercial viability of AI-driven solutions are improved by these capabilities, which are bolstered by expansion in big data analytics, cloud computing, and the falling prices of digital sensors and hardware (Gebresenbet et al., 2023).

Crop monitoring has made extensive use of computer vision, a crucial area of artificial intelligence. Drone and ground-based imagery trained algorithms identify early manifestations of stress brought on by pests, illnesses, or nutrient shortages, enabling prompt and focused interventions (Manyanga et al., 2023). Similar to this, image-based DL models are being used to improve diagnostic accuracy in remote and resource-constrained settings by identifying and classifying pest infestations and plant diseases (Javaid et al., 2022). By combining historical and current environmental and agronomic data, AI-powered predictive analytics enhances yield forecasting and risk assessment (Javaid et al., 2022). Planning at the farm and policy levels is aided by these models' ability to produce more precise predictions of crop performance. Water-use efficiency is increased and sustainable water resource management is supported by AI-based irrigation systems that use sensor feedback and adaptive scheduling algorithms (Soori et al., 2023).

Furthermore, labour-intensive jobs like planting, harvesting, and field surveillance are being transformed by AI-enabled robotics and autonomous machinery. For both small and large-scale producers, these systems increase scalability, decrease reliance on manual labour, and improve operational precision (Balai et al., 2025). When taken as a whole, these developments place AI at the forefront of solving the world's agricultural problems and developing sustainable food systems.

The discussion below focuses on training agricultural workers to adapt to technology.

Training in Agricultural Technologies: A New Prerequisite for Smart Farming

Smart farming's rise represents a major shift in agricultural practices worldwide and a pivotal moment in what academics call the Third Green Revolution (Dai et al., 2023). Incorporating information and communication technologies (ICT) into agricultural systems to enhance sustainability, efficiency, and productivity is what defines this revolution. Precision agriculture, agricultural automation and robotics, management information systems, and smart farming are a collection of technological fields that work together to create a data-driven agricultural paradigm (Katiyar & Farhana, 2021). These domains work together to support contemporary agricultural decision-making; they are not stand-alone systems. The idea of AgTech, which refers to a variety of mechanical and digital devices like sensors, drones, geographic information systems (GIS), automated machinery, and decision-support software, is essential to the smart farming model. These technologies reduce input use and improve resource efficiency by enabling intra-field level precision interventions and real-time monitoring (Silvestri et al., 2021). In this way, smart farming signifies a move away from practices based on intuition and towards fact-based, data-driven methods of managing crops and livestock.

AgTech's incorporation into farming is increasingly being discussed in terms of agroecology, a broader field that prioritises ecological balance, sustainability, and systems thinking. Agroecology necessitates adaptable and flexible methods founded on a deep comprehension of ecosystems (Khan et al., 2021). It depends on growing knowledge rather than providing fixed answers. In this regard, AgTech serves as a tool that can help accomplish agroecological objectives if properly modified. Precision agriculture tools, for example, can optimise water and nutrient use, minimising ecological impact while preserving productivity (Singh et al., 2021). This integration is by no means automatic, though. Effective smart farming necessitates balancing ecological knowledge with technological design, a process that is still being developed (Gebresenbet et al., 2023).

Smart farming can further agricultural sustainability has been emphasized in several studies. According to Khan et al. (2021), technologies like satellite-guided irrigation, soil monitoring, and precision planting can help boost yields while using fewer inputs. Singh et al. (2021) add that there are many new AgTech applications, demonstrating the growing toolkit that farmers can utilise. Soori et al. (2023) opine that these tools have the potential to lessen trade-offs between environmental stewardship and productivity, particularly through better monitoring of the use of natural resources. AgTech is still not widely used in many agricultural areas, despite these encouraging advancements. The literature has consistently identified several barriers. First, farmers may be discouraged from trying out new technologies due to the perceived and real complexity of new equipment (Khan et al., 2021). Second, there are practical limitations due to economic factors like high initial investment costs, a lack of subsidies, and unfavorable taxation frameworks (Rizzo et al., 2024). Third, uptake is further hampered by behavioral and psychological factors, such as skepticism towards industry-driven innovations and resistance to change (Nguyen et al., 2024).

Additionally, recent research indicates that adoption is impacted by structural problems (Sileshi et al., 2025; Nguyen et al., 2024; Curry et al., 2021). A crucial void in the creation of agronomic and ecological frameworks capable of providing tailored and optimal input recommendations (Sileshi et al., 2025). AgTech's full potential cannot be achieved without these. Furthermore, Nguyen et al. (2024) stress that sophisticated cognitive and analytical skills are necessary for the deployment of precision technologies. Farmers need to be able to decipher complicated datasets and use them to inform their decisions, which calls for a great deal of training and continuous assistance.

These difficulties highlight how important education and capacity building are to the smooth transition to smart farming. Age and formal education are powerful predictors of technology adoption, according to an increasing amount of research. In general, farmers who are younger and have received more formal education are more receptive to innovation and are better equipped to handle the intricacies of digital tools (Sileshi et al., 2025). In addition, it is commonly stated that a major obstacle to the adoption of AgTech is the lack of technical training and support services (Curry et al., 2021).

Smart farming necessitates ecological literacy, technological fluency, and decision-making abilities in addition to basic literacy. As a result, integrating smart technologies into agricultural systems is a learning process that requires both institutional support and cultural shifts, rather than just deploying equipment.

RECOMMENDATION AND A WAY FORWARD

The present study has shown that there is untapped potential for South African agricultural cooperatives to contribute more significantly to rural development, food security, and economic transformation. Persistent structural, technological, and policy-related barriers slow the growth and potential of our agricultural sector. Using modern technologies in collaborative processes is one way to revitalize the sector, but doing so requires a shift in mindset and approach. Thus, it is essential to consider a thorough and strategic approach to reform that places a high priority on digital innovation, inclusive governance, and sustainable development.

One of the primary concerns brought up by this study is the absence of a clear, nationally coordinated plan for digital transformation in cooperatives. These days, agricultural cooperatives operate with varying degrees of digital engagement and usually rely on localized projects or donor-funded interventions. To address this disparity, a comprehensive national digital agriculture strategy must be developed. This strategy must specifically target cooperatives and offer a framework that facilitates their access to digital tools, extension services, and training tailored to their requirements. Cooperation between government organisations, corporations, and academic institutions would ensure that such a strategy is grounded in a range of expertise and is practical.

Moreover, the digital divide between rural farming communities and urban centers continues to impede the equitable adoption of technological solutions. Due to a lack of reasonably priced devices, poor connectivity, and insufficient digital infrastructure, cooperatives are unable to participate in the digital economy. To address this, there is an urgent need for public investment in rural broadband networks, shared digital infrastructure (such as ICT hubs or mobile kiosks), and programs for the affordable purchase of devices. Closing this infrastructure gap will improve efficiency in record-keeping, information sharing, and real-time decision-making, as well as enable co-ops to use digital tools.

Enhancing cooperatives' human capital is equally crucial. Although some members may exhibit a willingness to adopt new technologies, low levels of digital literacy and limited technical knowledge remain obstacles. Agricultural extension services need to refocus to include digital training modules that emphasise tools relevant to smallholder operations, like mobile-based platforms for market intelligence, input management, pest control, and weather updates. Partnerships with universities and vocational training facilities can help develop modular, community-based training programs that prioritise experiential learning and inclusivity for women and youth. Cooperatives will be better equipped to adopt and sustain technology practices, thanks in large part to this kind of capacity building.

In addition to skill development, cooperatives must be repositioned as venues for innovation rather than merely distribution centers. Cooperatives should be encouraged to participate in or even host regional agri-tech innovation ecosystems. These ecosystems can help develop technologies that are appropriate for the local context and promote youth entrepreneurship in rural areas. For example, cooperatives can help local startups develop mobile apps, inexpensive irrigation systems, or renewable energy sources that are appropriate for smallholder situations. With the correct seed money, market access, and mentorship, these grassroots innovations could be scaled through cooperative networks.

Additionally, when developing and deploying technological tools, inclusivity and user diversity should be taken into account. Access to technological interventions is crucial for women, older farmers, and those with less formal education. Therefore, technology must be user-friendly, multilingual, and backed by visual aids, audio instructions, and community-based trainers. Incorporating inclusivity into the design and implementation of cooperatives increases the likelihood that they will receive widespread support and equitable benefits from their members.

Another area of opportunity is the evolution of market access strategies. By using digital platforms for e-commerce, product certification, and traceability, cooperatives can gain access to higher-value markets and cultivate stakeholder and consumer trust. Technologies such as blockchain traceability systems, digital supply chain platforms, and QR codes, for instance, can increase transparency and enable cooperatives to participate in premium and export markets. Government support for the establishment and maintenance of such platforms would significantly increase the competitiveness of cooperative products and help address market imbalances that currently favor larger agribusinesses.

Resource management is also crucial to cooperative success, and incorporating smart farming technologies can lead to notable efficiency improvements. Cooperatives should have access to technologies such as sensor-based irrigation systems, precision agriculture tools, and drone surveillance for pest control. A shared-access model, where cooperatives pool resources or rent equipment via centralised platforms, may be employed due to the high cost of these tools. This cooperative resource-sharing model would reduce the financial burden on individual farmers by ensuring greater access and collective responsibility for maintenance.

Improving cooperative governance and accountability is equally crucial. Poor record-keeping, elite capture, and a lack of transparency have often undermined member trust and the organisation's sustainability. This situation could be revolutionized by digital tools. Cloud-based accounting systems, mobile voting tools for member decision-making, and digital audit trails can all improve democratic engagement and financial integrity. Cooperative regulatory bodies and development partners should prioritise funding and training for the adoption of such governance-enhancing technologies.

To promote digital transformation, policy support must also be reoriented. Legislators can enact incentives such as tax breaks for cooperatives that use certified digital tools, technology vouchers for underserved communities, and first dibs on public procurement contracts for cooperatives that adhere to digital standards. Cooperative compliance and registration systems can be digitalized to further promote transparency and reduce bureaucratic red tape.

Lastly, monitoring, evaluation, and learning (MEL) frameworks need to be put in place to ensure the scalability and sustainability of technology adoption in cooperatives. These frameworks should be supported by user-friendly dashboards and data analytics tools so that cooperatives can track critical performance indicators like yield increases, member satisfaction, and market performance. To adapt their plans to changing member demands and environmental conditions, cooperatives must set up feedback loops.

Therefore, agricultural cooperatives need to change drastically to meet the needs of contemporary, technologically advanced agriculture. This change should go beyond simply implementing new tools; it should also include structural adjustments to infrastructure, governance, training, and policy support. These organisations can only reach their full potential as catalysts for sustainable agricultural development in South Africa by embracing an integrated strategy that blends digital innovation with cooperative values. The future needs to be inclusive, digitally enabled, and strategic, with a foundation in grassroots innovation and policy change.

CONCLUSION

The agricultural sector is a very important aspect of the world economy. This sector feeds the rapidly growing population in the world. In many parts of the world, people suffer from hunger and malnutrition because they lack food sources. As the global warming issues rise, bringing with it high temperatures or very freezing temperatures, fires, floods, droughts and other natural disasters, the world's food supply, as we know, is rapidly dwindling. The use of advanced technology can track weather patterns, and poultry and cattle can be moved during dangerous weather conditions. Greenhouse farming using hydroponics can be used for certain food crops and fish. Here, technology can be used to maintain good temperatures and hydration of plants, making it favorable for the fish or food crop to grow. For crop spraying to maintain good crops, drones can be used to spray crops. Technology has made major positive changes to the agricultural sector, and with proper training and support, the future of the agricultural sector looks much more promising.

FUTURE AREAS OF RESEARCH

There are various areas of future interest that the agricultural community grapples with, and one that ignites fear in every farmer's mind is global warming. Research on the application of technology to assist farmers in predicting weather patterns in Africa is much needed. Another area of research that should be explored is how policy in South Africa can have devastating political and economic effects on the agricultural sector. A very important area of interest that is on the rise is the use of 4IR technological advancements within agriscience and how they can benefit the agricultural sector. The use of training and development initiatives by the country's administration should also be studied, as it is beneficial for farming to move past traditional farming and embark on training Black farmers with the new methods of growing crops and raising poultry and raising cattle is essential.

REFERENCES

Adams, B. (2019). *Green development: Environment and sustainability in a developing world.* Routledge. DOI: 10.4324/9780203386033

Aryal, J. P., Sapkota, T. B., Rahut, D. B., & Jat, M. L. (2020). Agricultural sustainability under emerging climatic variability: The role of climate-smart agriculture and relevant policies in India. *International Journal of Innovation and Sustainable Development, 14*(2), 219–245. DOI: 10.1504/IJISD.2020.106243

Balai, P. S., Sheikh, A., Rabha, G., Das, S., Kuli, B. K., & Raj, M. (2025). Revolutionizing Agricultural Machinery: The Role of AI, IoT, and Renewable Energy in Enhancing Efficiency and Sustainability. *International Journal of Scientific Research in Science and Technology, 12*(2), 813–830. DOI: 10.32628/IJSRST251222626

Candemir, A., Duvaleix, S., & Latruffe, L. (2021). Agricultural cooperatives and farm sustainability–A literature review. *Journal of Economic Surveys, 35*(4), 1118–1144. DOI: 10.1111/joes.12417

Chikazunga, D., & Paradza, G. (2012). Smallholder farming: A panacea for employment creation and rural development in South Africa? PLAAS, University of the Western Cape.

Chiwawa, J., Naidoo, V., & Darwish, S. (2025). The Influence of Agricultural Cooperatives on the Environmental Sustainability of Rural Livelihoods in KwaZulu-Natal, South Africa. In *Sustainable Data Management: Navigating Big Data, Communication Technology, and Business Digital Leadership* (Vol. 2, pp. 31–41). Springer Nature Switzerland. DOI: 10.1007/978-3-031-83915-3_3

Chiwawa, J., Naidoo, V., & Rena, R. (2024). Role of farming cooperatives on the social sustainability of rural livelihoods in KwaZulu-Natal, South Africa. *Research World, 15*(2), 133–141.

Christian, M., Hans, T., & Gidi, L. (2018). Factors affecting performance of youth and womens agricultural cooperatives: A case of some selected cooperative societies in Eastern Cape, South Africa. *Journal of Development and Agricultural Economics, 10*(12), 394–403. DOI: 10.5897/JDAE2018.0932

Curry, G. N., Nake, S., Koczberski, G., Oswald, M., Rafflegeau, S., Lummani, J., Peter, E., & Nailina, R. (2021). Disruptive innovation in agriculture: Socio-cultural factors in technology adoption in the developing world. *Journal of Rural Studies, 88*, 422–431. DOI: 10.1016/j.jrurstud.2021.07.022

Dai, X., Chen, Y., Zhang, C., He, Y., & Li, J. (2023). Technological revolution in the field: Green development of Chinese agriculture driven by digital information technology (DIT). *Agriculture, 13*(1), 199. DOI: 10.3390/agriculture13010199

Department of Agriculture. Forestry and Fisheries. (2015). *Annual report 2014/2015.* Pretoria, South Africa: Department of Agriculture, Forestry and Fisheries.

Department of Agriculture, Land Reform and Rural Development (DALRRD). (2023). *Economic Review of South African Agriculture for the Year Ended 30 June 2023.* Government of South Africa.

Department of Small Business Development (2020). Annual Report; *Republic of South Africa.*

Department of Trade and Industry. (2004). A Cooperative Development Policy. *Pretoria: Republic of South Africa.* Available at: https://www.pmg.org.za/docs/2004/appendices

FAOSTAT. (2018). *Food and Agriculture Organization of the United Nations Statistical Database*. Retrieved from https://www.fao.org/faostat

Firoznia, G., Salman, M., & Gharani Arani, B. (2021). The Impact of Rural Production Cooperatives on Sustainable Rural Development: A Case of Kashan's Rural Production Cooperatives. *Journal of Sustainable Rural Development*, *5*(1), 121–134.

Gebresenbet, G., Bosona, T., Patterson, D., Persson, H., Fischer, B., Mandaluniz, N., Chirici, G., Zacepins, A., Komasilovs, V., Pitulac, T., & Nasirahmadi, A. (2023). A concept for application of integrated digital technologies to enhance future smart agricultural systems. *Smart Agricultural Technology*, *5*, 100255. DOI: 10.1016/j.atech.2023.100255

Javaid, M., Haleem, A., Singh, R. P., & Suman, R. (2022). Enhancing smart farming through the applications of Agriculture 4.0 technologies. *International Journal of Intelligent Networks*, *3*, 150–164. DOI: 10.1016/j.ijin.2022.09.004

Junaid, M. D., & Gokce, A. F. (2024). Global agricultural losses and their causes. *Bulletin of Biological and Allied Sciences Research*, *2024*(1), 66–66. DOI: 10.54112/bbasr.v2024i1.66

Kalfas, D., Kalogiannidis, S., Papaevangelou, O., Melfou, K., & Chatzitheodoridis, F. (2024). Integration of technology in agricultural practices towards agricultural sustainability: A case study of Greece. *Sustainability (Basel)*, *16*(7), 2664. DOI: 10.3390/su16072664

Katiyar, S., & Farhana, A. (2021). Smart agriculture: The future of agriculture using ai and iot. *Journal of Computational Science*, *17*(10), 984–999. DOI: 10.3844/jcssp.2021.984.999

Kayongo, A., Guloba, A., Muvawala, J., & Ssali, A. (2022). A Framework for Strengthening and Sustaining Cooperatives for Socio-Economic Transformation in Uganda. *Applied Economics and Finance*, *9*(2), 79–92. DOI: 10.11114/aef.v9i2.5554

Khan, N., Ray, R. L., Sargani, G. R., Ihtisham, M., Khayyam, M., & Ismail, S. (2021). Current progress and future prospects of agriculture technology: Gateway to sustainable agriculture. *Sustainability (Basel)*, *13*(9), 4883. DOI: 10.3390/su13094883

Kustepeli, Y., Gulcan, Y., Yercan, M., & Yıldırım, B. (2023). The role of agricultural development cooperatives in establishing social capital. *The Annals of Regional Science*, *70*(3), 681–704. DOI: 10.1007/s00168-019-00965-4

Lachgar, M., Hrimech, H., & Kartit, A. (2023). Unmanned aerial vehicle-based applications in smart farming: A systematic review. *International Journal of Advanced Computer Science and Applications*, *14*(6).

Li, S., Wu, W., Wang, Y., Zhang, N., Sun, F., Jiang, F., & Wei, X. (2023). Production Data Management of Smart Farming Based on Shili Theory. *Agriculture*, *13*(4), 751. DOI: 10.3390/agriculture13040751

Mamo, Y. A., Haji, J., Legesse, B., & Sassi, M. (2021). Challenges Of Agricultural Cooperatives And Their Impact On Profitability Of Potato Growers In Eastern Ethiopia. *Review of Agricultural and Applied Economics*, 24(1340-2021-1620), 62-77.

Manyanga, M., Pedzisa, T., & Hanyani-Mlambo, B. (2023). Adoption of agroecological intensification practices in Southern Africa: A scientific review. *Cogent Food & Agriculture*, *9*(1), 2261838. DOI: 10.1080/23311932.2023.2261838

Maota, T., & Naidoo, V. (2025). The Impact of Artificial Intelligence on Human Resource Management: Barriers and Benefits. In *Sustainable Management Practices for Employee Retention and Recruitment* (pp. 311–342). IGI Global Scientific Publishing. DOI: 10.4018/979-8-3693-8679-8.ch016

Mataba, L. D., & Komba, C. K. (2021). *Social protection in co-operatives: evidence from arusha women in business savings and credit cooperative society.*

Moalosi, R., & Rapitsenyane, Y. (Eds.). (2023). *African industrial design practice: perspectives on ubuntu philosophy*. Taylor & Francis. DOI: 10.4324/9781003270249

Mokoena, O. P., Ntuli, T. S., Ramarumo, T., & Seeletse, S. M. (2023). Sustainability of Rural Small-Scale Farmers Using a Thematic Content-Fed Analytic Hierarchy Process. *Sustainability (Basel)*, *15*(15), 11983–11983. DOI: 10.3390/su151511983

Moon, S., & Lee, S. H. (2020). A strategy for sustainable development of cooperatives in developing countries: The success and failure case of agricultural cooperatives in musambira sector, Rwanda. *Sustainability (Basel)*, *12*(20), 86–132. DOI: 10.3390/su12208632

Mosia, P., & Khalo, T. (2021). The Role of Agricultural Cooperatives in Promoting Sustainable Development in South Africa. *Sustainable Development*, *29*(2), 241–250.

Mundia, M., Nkonde, C., Simui, F., & Imasiku, J. (2022). Exploration of Agriculture Cooperative Enterprising Performance in Kabwe District, Zambia. *World Journal of Research and Review*, *1*(5), 1–5. DOI: 10.31871/WJRR.15.3.5

Mushonga, M., Arun, T. G., & Marwa, N. W. (2019). The cooperative movement in South Africa: Can financial cooperatives become sustainable enterprises? *Strategic Change*, *28*(4), 259–271. DOI: 10.1002/jsc.2268

Muzekenyi, M., Zuwarimwe, J., & Kilonzo, B. M. (2022). Utilizing small-scale commercial farming to enhance local economic development in South Africa. *Journal of Contemporary Management*, *19*(2), 22–39. DOI: 10.35683/jcm20007.155

Nakayiso, E., & Andrew, N. (2023). A historical review on the global evolution, benefits, challenges and performance of cooperatives. *Journal of the Humanities and Social Sciences*, *8*(1), 51.

Ndauka, F., & Matotola, S. (2023). The role of technology adopted by Agricultural Marketing Cooperative Society (AMCOS) on performance of grapevine farmers in Tanzania. Cogent Business & Management, 10(2), 2226420.75.

Nguyen, L. L. H., Khuu, D. T., Halibas, A., & Nguyen, T. Q. (2024). Factors that influence the intention of smallholder rice farmers to adopt cleaner production practices: An empirical study of precision agriculture adoption. *Evaluation Review*, *48*(4), 692–735. DOI: 10.1177/0193841X231200775

Niekerk, V. (1988). *Co-operative theory and practice*. Promedia Publications.

Ortmann, G. F., & King, R. P. (2006). Small-scale farmers in South Africa: Can agricultural cooperatives facilitate access to input and product markets? 170(1), 2016-139140.

Pansera, M., & Rizzi, F. (2020). Furbish or perish: Italian social cooperatives at a crossroads. *Organization*, *27*(1), 17–35. DOI: 10.1177/1350508418760980

Qaim, M. (2020). Role of new plant breeding technologies for food security and sustainable agricultural development. *Applied Economic Perspectives and Policy*, *42*(2), 129–150. DOI: 10.1002/aepp.13044

Raniga, T. (2021). Role of economic development cooperatives in improving the livelihoods of women in Gauteng, South Africa. *Development Southern Africa*, *38*(4), 591–606. DOI: 10.1080/0376835X.2021.1906628

Rena, R. (2021). Managing Cooperatives in South Africa: An Economic Perspective. *Cooperatives in the Global Economy*, 175-191

Risal, N. (2020). Role of Cooperatives for Sustainable Livelihood in Bagmati Province, Nepal. *The International Research Journal of Management Science*, *5*(1), 75–92.

Rizzo, G., Migliore, G., Schifani, G., & Vecchio, R. (2024). Key factors influencing farmers' adoption of sustainable innovations: A systematic literature review and research agenda. *Organic Agriculture*, *14*(1), 57–84. DOI: 10.1007/s13165-023-00440-7

Rogan, M., & Alfers, L. (2020). Employment and Poverty Alleviation in South Africa: The Role of Cooperatives. *Development Southern Africa*, *37*(4), 549–564.

Rogers, E. M. (1983). *Diffusion of Innovations* (3rd ed.). The Free Press.

Sacchetti, S., & Tortia, E. C. (2021). Governing cooperatives in the context of individual motives. *International Journal of Social Economics*, *48*(2), 181–203. DOI: 10.1108/IJSE-09-2019-0579

Sahoo, A. K., Meher, S. K., Panda, T. C., Sahu, S., Begum, R., & Barik, N. C. (2020). Critical review on cooperative societies in agricultural development in India. *Current Journal of Applied science and technology*, *39*(22), 114-121.

Santiteerakul, S., Sopadang, A., Yaibuathet Tippayawong, K., & Tamvimol, K. (2020). The role of smart technology in sustainable agriculture: A case study of wangree plant factory. *Sustainability (Basel)*, *12*(11), 4640. DOI: 10.3390/su12114640

Schwettmann, J. (2022). *Cooperatives in the social and solidarity economy: Sustainable development and decent work in Africa's informal economy* (Doctoral dissertation, Manchester Metropolitan University).

Sellare, J., Meemken, E. M., Kouamé, C., & Qaim, M. (2020). Do sustainability standards benefit smallholder farmers also when accounting for cooperative effects? Evidence from Côte d'Ivoire. *American Journal of Agricultural Economics, 102*(2), 681–695. DOI: 10.1002/ajae.12015

Serra, R., & Davidson, K. A. (2020). Selling Together: The Benefits of Cooperatives to Women Honey Producers in Ethiopia. *Journal of Agricultural Economics, 72*(1), 202–223. DOI: 10.1111/1477-9552.12399

Shoeman, N. (2006). The Co-operative as an Appropriate Form of Enterprise for Black Economic Empowerment, Published thesis at the *University of Free State.*

Sibanda, M., & Ndlela, B. (2020). Agricultural exports and economic growth: Empirical evidence from South Africa. *Journal of African Trade, 7*(1–2), 20–34.

Sileshi, G. W., Stewart, Z. P., Odhong, J., Mhlanga, B., Amede, T., Aynekulu, E., ... & Snapp, S. (2025). A review of organic inputs to inform soil health advice for African smallholder farmers: localization matters. *npj Sustainable Agriculture, 3*(1), 20.

Silvestri, S., Richard, M., Edward, B., Dharmesh, G., & Dannie, R. (2021). Going digital in agriculture: How radio and SMS can scale-up smallholder participation in legume-based sustainable agricultural intensification practices and technologies in Tanzania. *International Journal of Agricultural Sustainability, 19*(5-6), 583–594. DOI: 10.1080/14735903.2020.1750796

Singh, R. K., Berkvens, R., & Weyn, M. (2021). AgriFusion: An architecture for IoT and emerging technologies based on a precision agriculture survey. *IEEE Access : Practical Innovations, Open Solutions, 9*, 136253–136283. DOI: 10.1109/ACCESS.2021.3116814

Soori, M., Arezoo, B., & Dastres, R. (2023). Internet of things for smart factories in industry 4.0, a review. *Internet of Things and Cyber-Physical Systems, 3*, 192–204. DOI: 10.1016/j.iotcps.2023.04.006

Spanaki, K., Sivarajah, U., Fakhimi, M., Despoudi, S., & Irani, Z. (2022). Disruptive technologies in agricultural operations: A systematic review of AI-driven AgriTech research. *Annals of Operations Research, 308*(1), 491–524. DOI: 10.1007/s10479-020-03922-z

Statistics South Africa. (2024). *Agricultural Statistics.* Government of South Africa.

Tahat, M., & Alananbeh, M., K., A. Othman, Y., & I. Leskovar, D. (. (2020). Soil health and sustainable agriculture. *Sustainability, 12*(12), 48–59. DOI: 10.3390/su12124859

Teklu, A., Simane, B., & Bezabih, M. (2023). Effect of climate smart agriculture innovations on climate resilience among smallholder farmers: Empirical evidence from the Choke Mountain Watershed of the blue nile highlands of Ethiopia. *Sustainability (Basel), 15*(5), 4331. DOI: 10.3390/su15054331

Thaba, S. C., & Mbohwa, C. (2015). The nature, role and status of cooperatives in South African context. *In Proceedings of the world congress on engineering and computer sciences*, 2, 21-23.

Zantsi, S. (2021). Why do agricultural co-operatives fail to attract youth and create rural employment? Evidence from a case study of Zanokhanyo in Butterworth, Eastern Cape. *South African Journal of Agricultural Extension*, *49*(1), 183–197. DOI: 10.17159/2413-3221/2021/v49n1a10785

Zhang, C. H., Benjamin, W. A., & Miao, W. A. N. G. (2021). The contribution of cooperative irrigation scheme to poverty reduction in Tanzania. *Journal of Integrative Agriculture*, *20*(4), 953–963. DOI: 10.1016/S2095-3119(21)63634-1

Chapter 2
AI-Driven Helmet Compliance for Enhancing Human-Machine Collaboration in Smart Manufacturing:
Smart Helmet and Traffic Violation Detection System

S. Mahalakshmi
https://orcid.org/0000-0003-4851-672X
P.S.R. Engineering College, Sivakasi, India

K. Valarmathi
P.S.R. Engineering College, Sivakasi, India

M. Meena
P.S.R. Engineering College, Sivakasi, India

M. Shanmugapriya
P.S.R. Engineering College, Sivakasi, India

J. Thirumala
P.S.R. Engineering College, Sivakasi, India

ABSTRACT

Ensuring road safety and enforcing traffic regulations are critical challenges in urban mobility management. This work proposes a camera-based helmet-wearing detection system that leverages YOLO (You Only Look Once) object detection, ESP32-CAM, and an ESP32 microcontroller to monitor and enforce helmet usage among motorcyclists. The system detects riders without helmets in real time and alerts authorities for necessary action. Beyond helmet detection, the system incorporates traffic rule enforcement by identifying violations such as signal jumping and unauthorized

DOI: 10.4018/979-8-3373-2985-7.ch002

lane usage, ensuring a more disciplined traffic environment. To further optimize traffic flow, the system classifies traffic density at signals into high or low traffic conditions based on real- time image processing. This classification helps in adaptive traffic control, reducing congestion and improving vehicle movement efficiency. The captured data, including images and violation records, is transmitted to a central server for further analysis, logging, and alert generation.

INTRODUCTION

Traffic congestion and road safety are among the most pressing concerns in urban mobility, affecting millions of commuters daily. With the rapid increase in vehicle ownership, cities worldwide face growing challenges in managing traffic flow, reducing congestion, and ensuring compliance with traffic rules. The World Health Organization's 2018 Global Status Report on Road Safety states that traffic accidents cause 20 to 50 million non-fatal injuries and approximately 1.35 million fatalities annually worldwide, (Park et al., 2025). In India, where road traffic is highly dynamic, the issue of helmet non-compliance among motorcyclists significantly contributes to road fatalities. The inefficiency of manual enforcement methods and the limitations of traditional traffic monitoring systems have led to the need for automated, AI- driven solutions.

The lack of real-time adaptability in conventional traffic systems results in inefficient traffic flow management, unnecessary delays, and increased accident rates. Fixed-timer traffic signals and manual traffic monitoring by law enforcement officers are outdated methods that fail to address real-time road conditions effectively. Additionally, violations such as red-light jumping, unauthorized lane usage, and helmet non-compliance by motorcyclists remain major concerns. An innovative solution that integrates artificial intelligence (AI), computer vision, and the Internet of Things (IoT) is needed to optimize traffic management and enhance road safety.

Motorcyclists are among the most vulnerable road users, accounting for a significant percentage of road accident fatalities. The National Crime Records Bureau (NCRB) reported that a substantial number of road deaths in India involve two-wheeler riders, many of whom were not wearing helmets. Head injuries resulting from motorcycle accidents are often fatal, and studies have shown that helmet usage can reduce the risk of severe injuries by up to 70%.

Despite mandatory helmet laws, enforcement remains a challenge due to the limitations of manual monitoring. Traffic officers can only monitor a limited number of vehicles at any given time, leading to inconsistent enforcement. Furthermore, non- compliant riders often evade checkpoints, making it difficult for authorities to ensure widespread adherence to helmet laws. An automated system capable of detecting helmet violations in real-time and issuing penalties would significantly improve compliance and reduce accident-related fatalities. Traditional traffic management relies on several methodologies, including manual enforcement, CCTV-based monitoring, fixed-timer traffic signals, and RFID-based tracking. However, these methods have notable shortcomings that limit their effectiveness. Manual traffic monitoring relies on traffic police officers to observe and fine violators, leading to inconsistent enforcement and inefficiencies due to human limitations. CCTV surveillance provides recorded evidence of violations, but most systems require human intervention to review footage, leading to delays in enforcement. Fixed-timer traffic signals operate on predefined cycles without adapting to real-time traffic conditions, resulting in increased congestion and unnecessary fuel consumption. RFID-based tracking requires vehicles to be equipped with RFID tags, which is impractical for widespread use and does not support real-time detection of traffic rule violations. These system analyses real time traffic images in congested areas and deliver information to traffic authorities for rabid response and efficient management, (Alzubaidy & Ibrahim, 2023). Given these limitations, there is a strong demand for an intelligent traffic monitoring system that leverages AI and computer vision to automate violation detection and enhance traffic flow management. Recent advancements in AI and IoT have paved the way for the development of smart traffic management systems. AI-driven object detection models, such as YOLO (You Only Look Once), enable real-time identification of traffic violations, including helmet non-compliance, red-light jumping, and unauthorized lane changes. IoT- enabled devices, such as the ESP32-CAM module, allow for seamless data collection, processing, and transmission to central traffic management authorities. The integration of AI and IoT offers several advantages. Automated violation detection ensures accurate and real-time identification of helmet violations without human intervention. Traffic density classification using AI algorithms analyzes vehicle density at intersections, optimizing traffic signal durations dynamically. IoT-enabled modules transmit real-time traffic data to central servers, facilitating instant law enforcement actions. AI-driven systems can be deployed across multiple intersections without significant infrastructure changes, making them highly scalable and cost-efficient. To address the existing challenges in traffic monitoring and rule enforcement, this project proposes a real-time, AI-powered traffic violation detection system. The system integrates ESP32-CAM modules, the YOLO object detection model, and cloud-based data processing to ensure seamless traffic management. Key functionalities of the proposed system include helmet-wearing detection, where the system captures real-time images of motorcyclists and identifies helmet compliance using deep learning algorithms. Traffic density classification enables AI models to analyze vehicle density and adjust signal durations dynamically to optimize traffic flow. Violation logging and reporting allow the system to log detected violations with timestamps and location details, transmitting data to traffic authorities for enforcement. Wireless data transmission using ESP32 microcontrollers enables real-time communication with cloud servers, allowing instant updates on traffic conditions. A dashboard for law enforcement provides authorities with real-time insights, enabling data-driven decision-making for better urban traffic management. The proposed system offers multiple benefits over traditional traffic monitoring methods. Increased enforcement efficiency is achieved as automated detection eliminates the need for manual monitoring, ensuring consistent and fair enforcement of traffic laws. Reduction in traffic congestion is facilitated through adaptive traffic signal control that optimizes vehicle movement, minimizing delays and improving road efficiency.

Enhanced road safety is ensured as the system significantly reduces head injury-related fatalities in motorcycle accidents by enforcing helmet compliance. Cost-effective deployment is made possible through the use of ESP32-CAM and AI models, allowing for low-cost implementation and making it feasible for large-scale deployment. Scalability and integration are achieved as the system can be expanded to include additional features, such as automatic number plate recognition (ANPR) and predictive analytics for congestion forecasting. The increasing rate of road accidents and traffic violations highlights the urgent need for a more effective and intelligent traffic management system. Conventional methods, such as manual monitoring and fixed-timer signals, are inefficient and fail to adapt to real- time road conditions. The proposed AI-based system provides a smart and automated approach to traffic rule enforcement, offering real-time helmet detection, dynamic traffic signal control, and seamless data transmission to traffic authorities. By leveraging AI, IoT, and cloud-based analytics, this system enhances road safety, reduces congestion, and improves law enforcement efficiency. Future enhancements, such as license plate recognition and AI-driven predictive analytics, can further optimize traffic management and contribute to the development of safer and smarter urban transportation networks. With its cost-effectiveness, scalability, and real-time capabilities, this project represents a significant step toward intelligent and sustainable traffic management solutions.

The world have adapted to urban mobility challenges addressed by the AI and IoT-relevant traffic management methods. Real-time violations are identified with the help of smart surveillance systems in Singapore, like not wearing helmets and improper lane usage. Barcelona, in contrast, incorporate edge computing and smart cameras to monitor vehicle movement and increase security features. These examples demonstrate how decentralized and intelligent systems can improve traffic enforcement efficiency. Our system helps meet the goals of Smart City development by making traffic management affordable, automated, and scalable. It supports UN Sustainable Development Goal 11 by creating safer roads, reducing traffic jams, and promoting sustainability through real-time traffic signal control. This technology not only increases compliance but also fosters sustainable and resilient urban infrastructure.

BACKGROUND

An advanced traffic regulation framework was introduced (Tahmid & Hossain, 2017) that employs digital image processing techniques, particularly the Canny edge detection methodology, to ascertain vehicular density in real-time. This system aims to address the inadequacies inherent in traditional traffic management systems by facilitating expedited response times, enhanced vehicle management, greater levels of automation, increased reliability, and overall improved operational efficacy. The methodology encompasses a comprehensive process flow that initiates with image acquisition, transitions to edge detection, and culminates in the assignment of green signals. Four distinct traffic scenarios are utilized to illustrate the effectiveness of the proposed system, and the findings are substantiated by hardware-based implementation and schematic illustrations, thereby affirming its practicality. The optical flow, background subtraction, and handcrafted features are included in this research. In real-time traffic conditions these methods were not very reliable. Occlusions and multiple riders struggled using the Classical classifiers such as SVM and KNN Deep learning changed the game. CNN-based methods offered better accuracy and improved feature extraction. Still, many approaches continued to rely on background subtraction in crowded scenes, which decreased reliability. YOLOv3 and Faster R-CNN enhanced real-time detection but faced difficulties with small objects and overlapping vehicles. To solve these problems, the study recommends an improved YOLOv5 that includes triplet attention and Soft-NMS. This method, used with a new HFUT-MH data set, provides high precision and performs well in urban traffic, (Jia et al., 2021).

This research discovers how both personal traits and national conditions affect traffic violations and accidents in seven diverse countries. Using logistic regression with elastic net regularization, the method achieved 75.5% accuracy for predicting violations and 79.6% for accidents. The results indicate that factors like driving awareness, frequency of driving, and attitudes strongly influence violations, while education and road infrastructure show a superior role in accidents. Highlighting the importance of Education, Enforcement, and Engineering, this research underscores a comprehensive method to road safety, (Shiomi et al., 2021). In a pertinent investigation formulated a methodology for real-time, area-based traffic density estimation aimed at facilitating intelligent traffic management. Their system computes traffic density by quantifying the area occupied by vehicle edges identified within live imagery. By analyzing the edge regions across various thoroughfares, the system adaptively estimates vehicular density, thus enabling the real-time modification of traffic signal durations. This approach presents a more responsive and effective alternative to the traditional timer-based traffic control systems presently implemented in Bangladesh. The principal contribution of their research resides in the introduction of an innovative strategy that harnesses edge-based area detection, morphological image processing, and real-time camera feeds to proficiently and intelligently manage traffic congestion, (Uddin, Das, & Taleb, 2015). Automate vehicle detection and tracking in traffic surveillance using advanced deep learning techniques. This study integrating YOLO algorithm with Kalman filters and data association attained 94.10% and 92.50% for accurate detection and tracking. The approach achieved effectively in challenging circumstances, including occlusion, illumination shifts, and vehicle scale changes. Estimations on the MoVITS dataset established strong precision and recall across numerous urban traffic conditions. These results validate the feasibility of using AI for precise and effectual traffic rule enforcement, (Abdelali et al., 2021). A sophisticated traffic surveillance system that prioritizes traffic flow based on vehicle density and the presence of emergency vehicles. Under normal circumstances, the system allocates access to the lane exhibiting the highest traffic density. However, upon the detection of an emergency vehicle, the system reassigns priority to that lane irrespective of its density. This mechanism ensures both the efficient movement of traffic and the prompt provision of assistance to emergency services. The system is predicated on digital image and audio capture, utilizing Canny edge detection to evaluate traffic density and Least Mean Square (LMS) filtering to mitigate noise. The comprehensive traffic control framework—from data input acquisition to decision-making concerning green signal allocation—is distinctly depicted with schematic diagrams, and the final configuration has been substantiated through hardware implementation, demonstrating its applicability in real-world scenarios, (Moka, Pilla, & Radhika, 2020). In an effort to increase traffic safety, modern developments have utilized deep learning models to identify helmet violations. This study illustrates an improved PerspectiveNet framework with EfficientNet v2 as its feature extractor is offered. With 95.2% accuracy, 93.3% precision, 92.1% recall, and a 92.6% F1-score, the method established impressive performance. Compared to earlier approaches using YOLOv2 and basic CNNs, the proposed model offers better efficiency and reduced computational demands. These outcomes suggest their potential for use in real-time automated traffic monitoring applications, (Said et al., 2024). AI-driven traffic monitoring has gained attention as a solution to manage violations in high-density traffic zones. Phone usage while travelling and illegal parking are common offenses detected with the help of algorithms like YOLO. Deep SORT increases these systems by allowing consistent tracking of moving vehicles. For didentify number plates, OCR tools such as Tesseract have established highly effectively. Prior research has demonstrated strong accuracy in both detection and recognition tasks. These developments pave the way for fully automated, policy-enforcing systems aligned with smart city goals, (Said et al., 2024). This work focusses on utilizing smart technologies to safeguard vulnerable road users, including pedestrians and cyclists. Data collection methods using devices like cameras, sensors, drones, and mobile phones to monitor performance and traffic interactions. Camera-based systems are extensively used for their efficiency in tracking and surveillance. Whereas smartphone and social media approaches are emerging, they pose experiments related to data reliability and ethical considerations. This research mentions adopting IoT and virtual tools to advance data quality and support real-time safety involvements, (Charran & Dubey, 2022). Their analytical results demonstrate that it is feasible to accomplish roughly 89.84% precision and 92.09% recall by utilizing an Intersection over Union (IoU) threshold of 0.3 and examining images shot at intervals exceeding two seconds. These results are promising in the context of short-term traffic forecasting, (Parvez & Mordipour, 2024). A dynamic system that adapts according to the temporal variations of the day. To facilitate the detection of alterations in traffic patterns, they implemented an online algorithm alongside an autoencoder for the purpose of visual data simplification. They partitioned the day into seven discrete time intervals and conducted independent analyses of traffic for each segment. By utilizing 1,100 annotated images for each time slot, they illustrated that their multi-period analysis outperforms models that are trained on data aggregated over the entire day, (Liu, Lam, & Ng, 2021).

In order to detect license plates as part of ANPR systems, many researchers have used color-based or edge-based techniques, which are prone to errors in noisy or dimly lit environments. Prior helmet detection techniques were imprecise and unreliable because they mainly relied on manually constructed features like background subtraction and LBP and HOG. As deep learning has progressed, CNNs have been used more and more to improve detection performance. Some techniques employed various CNN models to recognize license plates and helmets, which made the system more complex. YOLO is a real-time object detection framework that enables efficient and rapid detection with a single neural network. By adding batch normalization and anchor boxes, the second version, YOLOv2, improves performance. A YOLOv2-based technique with centroid tracking is presented in the reviewed study to minimize false detections and provide precise real-time. The algorithm achieves an 85% accuracy in plate detection through a series of steps, including image acquisition, pre-processing, number plate localization, character segmentation and optical character recognition. While successful in standard cases, the algorithm aims for improved accuracy, especially for non-standard plates, with future work focusing on enhanced text extraction and hardware solutions, (Jamtsho et al., 2020). Chronological Summary of Related Works and Their Methodologies listed in Table 1.

Table 1. Studies based on theme and technology

Thematic Area	Authors	Technology	Contributions
Traditional Image Processing for Traffic Density Estimation	Tahmid & Hossain (2017)	Canny edge detection, real-time image processing	Real-time vehicle density analysis using Canny; hardware validated
	Uddin, Das, & Taleb (2015)	Edge-based area estimation, live camera feed	Adaptive signal control based on edge-detected area estimation
	Moka, Pilla, & Radhika (2020)	Canny edge + LMS filter, emergency detection	Emergency lane priority and noise filtering using LMS
Computer Vision-Based Traffic Surveillance	Pan, Guo, & Men (2010)	Edge detection + background subtraction, adaptive windowing	Accurate vehicle count using adaptive ROI and lane segmentation
	Vigneash et al. (2023)	Real-time video processing, emergency detection	Emergency vehicle detection from traffic video data
	Li et al. (2023)	Temporal derivative of flow and speed	Accident detection using speed/flow derivatives

continued on following page

Table 1. Continued

Thematic Area	Authors	Technology	Contributions
IoT and Embedded Systems in Smart Traffic Control	Chithra et al. (2023)	Arduino + GSM + RFID for signal automation	Integrated system for red light, stolen, and emergency vehicle detection
Deep Learning for Vehicle Detection & Traffic Optimization	Duc et al. (2024)	YOLOv8 for vehicle detection and signal control	High-accuracy detection and traffic light automation with YOLOv8
	Swaned et al. (2024)	YOLOv8s, multi-class detection with mAP analysis	High mAP in vehicle detection; improvement needed for heavy vehicles
	Mahto & Chauhan (2020)	YOLOv3 + RetinaNet, density-based enhancement	Improved detection in poor visibility regions (12% gain)
	Liu, Lam, & Ng (2021)	IoU-based sampling for prediction	Effective short-term traffic prediction with high precision & recall
	Cirsostomo et al. (2023)	CNN-based vehicle classification and signal control	Smart signal control using traffic class and congestion levels
Machine Learning & Analytics in Traffic Monitoring	Sanjiv & Baskar (2024)	ML + vehicle tracking + visualization UI	Real-time traffic insights and adaptive signal control dashboard
	Jeong, Roh, & Sohn (2021)	Autoencoder + temporal segmentation	Improved accuracy via dynamic, segmented traffic pattern analysis
Deep Learning for Vehicle Detection & Traffic Optimization	Mahto & Chauhan (2020)	YOLOv3 + RetinaNet with clustering	Better estimation in low-detection zones vs. traditional methods
	Liu, Lam, & Ng (2021)	Image sampling with IoU threshold	Achieved 89.84% precision & 92.09% recall for forecasting
	Cirsostomo et al. (2023)	CNN-based congestion and emergency detection	Real-time intelligent signal adjustment based on video input
Deep Learning for Helmet & License Plate Detection	Jamtsho, Yonten et al.	YOLOv2 + centroid tracking for ANPR and helmet detection	Robust helmet and plate detection using YOLOv2 and tracking
	Jia, Wei et al.	YOLOv5 with triplet attention + Soft-NMS	Enhanced helmet detection in complex scenes with YOLOv5
	R. S. Charran et al.	YOLO + DeepSORT + OCR (Tesseract)	Comprehensive traffic rule enforcement using detection & tracking
Deep Learning for Vehicle Tracking and Rule Enforcement	H. A. Abdelali et al.	YOLO + Kalman filter + data association	Achieved 94.10% detection, 92.50% tracking accuracy in urban traffic
Deep Learning for Helmet Detection	Yahia Said et al.	PerspectiveNet + EfficientNet v2	95.2% accuracy in helmet detection with low complexity
Smart Safety Tech for Vulnerable Road Users	Md Shohel Parvez et al.	IoT + sensors + camera-based VRU monitoring	IoT-powered VRU safety monitoring; recommends virtual integration
Statistical Modeling for Traffic Safety	Yasuhiro Shiomi et al.	Elastic Net Logistic Regression with survey/statistical data	75.5% accuracy (violations), 79.6% (accidents); emphasizes 3Es (Education, Enforcement, Engineering)

The incorporation of helmet detection into the overall enforcement of traffic rules is frequently overlooked by current traffic monitoring systems. In order to address this, our work uses YOLOv5 to present a real-time helmet violation detection system. We show that ESP32-CAM devices can efficiently support lightweight AI models, despite their generally limited processing power. Furthermore, there aren't many platforms that address traffic infractions and helmet use together. By providing a centralized, affordable, and decentralized IoT-AI system appropriate for scalable urban mobility enforcement, our suggested solution closes this gap.

LEGAL, ETHICAL AND DATA PRIVACY CONSIDERATIONS

Several legal, ethical, and privacy-related responsibilities introduced by AI-powered traffic monitoring systems. From a legal perspective, any group of images or videos connecting vehicle riders' necessity comply with relevant information protection regulations like India's Information Technology Act and emerging Personal Data Protection laws. These schemes must ensure safe handling of sensitive data, with rider identities and number plates, using encryption and restricted access protocols. Their use should be obviously authorized by traffic regulatory bodies and severely restricted to lawful enforcement purposes. Ethically, transparency and fairness are dangerous. Citizens should be complete aware of AI surveillance by visible signs at monitoring places. The process must function impartially, avoiding any method of discrimination, and emphasis purely on traffic law violations. The main objective should be public security not surveillance outside what is essential for traffic enforcement. violation specific data is stored and ensured by addressing Privacy concerns. The authorized traffic authorities access this information. Clear data retaining and removal policies must be established to avoid misuse. Public responses may vary. While many may support the initiative for its potential to decrease accidents, some may view it as intrusive. To instruct people through the awareness programs on how helmet enforcement pays to secure lives and reduce injuries, and how the technology works in a transparent, lawful way. On the working front, the system may face challenges like unreliable network coverage, low visibility during harsh weather, and visual obstructions. These can be moderated by techniques like:

- Temporarily loading data locally throughout network outages
- Augmenting image quality through preprocessing
- Use of cameras for adversarial weather

By seeing these technical and social issues, the proposed method can be established and deployed in a liable, efficient, and socially acceptable way. The system ensures privacy by anonymizing sensitive data like faces using encryption for secure storage and transmission. It complies with the Information Technology Act and upcoming data protection regulations. Only essential violation data is stored, with controlled access and clear deletion policies.

ENHANCING REAL-WORLD DEPLOYMENT READINESS

Although the proposed Smart Helmet and Traffic Violation Detection System using YOLO and ESP32-CAM demonstrates technical feasibility, real-world adoption involves a range of practical considerations that go beyond system accuracy.

User Experience and Wearability

Current development focuses primarily on functionality; however, sustained use by motorcyclists depends on ergonomic design, comfort, and minimal intrusion. Factors such as helmet weight, heat generated by electronics, battery duration, and camera visibility can significantly affect user acceptance. Future iterations must involve user-centric design processes and field testing to gather feedback and optimize usability.

Community Acceptance and Ethical Implications

Peoples are resisted Surveillance-based technologies, especially if perceived as invasive. Without clear communication of benefits and privacy protections, such systems may be viewed as tools of control rather than safety. The privacy-by-design principles, like on-device processing and anonymized data handling are incorporated and are essential to maintain public trust.

Policy Alignment and Institutional Integration

Integration with traffic authorities and legal systems decides the effectiveness of the proposed system. A key challenge is ensuring that the evidence generated is admissible and enforceable under local regulations. Collaboration with policymakers, law enforcement agencies, and transport departments is essential to define rules for data use, evidence handling, and accountability.

Lessons From Existing Systems

factors like maintenance complexity, lack of institutional support, or community backlash factors decide the Several smart traffic and secure system. Learning from these experiences, our proposed system must prioritize modular design, low-cost maintenance, and interoperability with existing infrastructure to increase scalability and long-term sustainability.

DEPLOYMENT CHALLENGES AND SCALABILITY CONSIDERATIONS

The system faces various practical challenges that could affect its large-scale deployment and performance in real-world conditions. Environmental issues such as low visibility during nighttime or bad weather may hinder image clarity and detection efficiency. To overcome this, future versions aim to integrate infrared and low-light enhancement technologies. Ensuring the durability of hardware components like sensors and cameras in outdoor settings requires regular upkeep, which can be minimized through weather-resistant enclosures and automated diagnostics. Scaling the system across both urban and rural areas involves considerable expense. To address this, the design incorporates budget-friendly hardware like the ESP32-CAM and uses open-source tools. Additionally, in locations with poor internet connectivity, real-time cloud communication is limited. Edge computing enables on-device data processing, allowing the system to function smoothly without continuous internet access. These solutions collectively support reliable and scalable implementation in diverse settings.

MAIN FOCUS OF THE CHAPTER

Traffic congestion and road safety are critical challenges in urban areas, particularly due to helmet non-compliance among motorcyclists, which leads to severe accidents. Traditional traffic monitoring methods, such as manual enforcement and CCTV surveillance, are inefficient and require human intervention. This project introduces an AI- based traffic monitoring system that automatically detects helmet violations and classifies traffic congestion in real-time using an

ESP32-CAM module and the YOLO object detection model. The system captures live traffic images, processes them to check for helmet compliance, and assesses traffic density to dynamically adjust signal timings. Data is transmitted via Wi-Fi or LoRa to a central monitoring system, enabling authorities to take immediate action. By eliminating the need for manual monitoring, this automated, cost-effective, and scalable solution enhances road safety, improves traffic flow, and supports smart city development.

COMPONENT DESCRIPTION

Hardware

The hardware setup required for implementing the GRU-based energy consumption forecasting model includes a system capable of handling large datasets and computationally intensive deep learning operations. Below are the key specifications:

- **Processor**: Intel Core i7 (8th Gen or higher) or AMD Ryzen 7, capable of multithreading for efficient parallel processing.
- **GPU**: NVIDIA GeForce GTX 1080 or higher, with CUDA support for accelerating deep learning tasks.
- **RAM:** 16 GB or more, to handle the memory requirements during model training and testing.
- **Storage:** 512 GB SSD or more for faster read/write speeds and data handling.
- **Operating System**: The operational framework is compatible with Linux (commencing from Ubuntu 18.04), macOS, and Windows version 10.

The ESP32 Module

The ESP32 constitutes a financially viable and energy-efficient microcontroller that incorporates integrated dual-mode Bluetooth and Wi-Fi functionalities. It operates utilizing either a single-core RISC-V architecture, a dual-core LX7 architecture, or a single-core Tensilica Xtensa LX6 CPU architecture. Within the chip, there exists a multitude of components designed for radio frequencies, featuring antenna switches, baluns, low-noise amplifiers (LNAs), power amplifiers, filters, and power management solutions. Developed by Espressif Systems, which is based in Shanghai, it is recognized as the successor to the ESP8266 and is produced by TSMC using a 40 nm technology process.

Features of the ESP32

Processing Unit

The processing unit is predicated upon the Xtensa 32-bit LX6 CPU (available in both single-core and dual-core configurations) with a peak operational performance of 600 DMIPS at clock frequencies of 160/240 MHz. The ultra-low power (ULP) coprocessor integrates 520 KiB of

RAM and 448 KiB of ROM designated for data storage. Communication: Bluetooth: Version 4.2 BR/EDR alongside BLE (utilizing a shared radio frequency with Wi-Fi); Wi-Fi: 802.11 b/g/n.

The system is equipped with an 18-channel 12-bit SAR ADC, 34 general-purpose input/output channels, and supplementary peripherals. It further comprises 10 capacitive touch inputs; 2 × I^2S, 2 × I^2C, and 3 × UART interfaces; in addition to 2 × 8-bit digital-to-analog converters (DACs). LCD Usage: Project titles and informational messages are exhibited on the LCD. Data lines interface with the microcontroller through PORT 2, whereas control lines are connected via P3.5, P3.6, and P3.7.

Single Power Supply

All components are powered by the electrical supply unit. This system is utilized to change alternating current (AC) voltage into direct current (DC) voltage seamlessly. A transformer is employed to convert 230V AC into 12V AC. The diode operates with a 12V AC input. The 1N4007 diode variant is implemented to effectuate the conversion of AC voltage into DC voltage. AC capacitors are utilized for the purposes of ground discharge and the charging of AC components. The LM 7805 voltage regulator is employed to maintain a constant voltage level. Subsequently, a signal will be transmitted to the subsequent capacitor, which serves to filter out undesirable AC components. Incorporating resistors and light-emitting diodes (LEDs) will define the load. The LED operates at a voltage of 1.75V. Should the voltage exceed the prescribed limit, it will be dissipated across the resistor.

Power Supply

For the majority of electrical circuits to operate effectively, a DC power source is imperative. Batteries are conventionally employed to supply this DC voltage for mobile electronic devices. However, an alternative exists in the conversion of AC mains voltage to DC voltage, as batteries are constrained by a finite lifespan and are not perpetually rechargeable.

- **Transformer:** Modulates the AC voltage from residential sources to the requisite levels.
- The converter changes AC voltage to DC voltage.
- **Filter:** Attenuates oscillations present in the DC voltage.
- The **regulator** guarantees a consistent and stable output voltage.

Transformers

A **transformer** is defined as a device that alters electrical voltage levels. It modifies waveforms, voltage, and current within AC circuits. Its role is critical in electronic systems, particularly regarding the transformation of power supply from AC to DC. A conventional transformer is illustrated in the figure. The operation hinges on two core principles: the establishment of a magnetic field through electric current and the voltage induction caused by a shifting magnetic field. High voltage is transmitted through multiple coils surrounding an iron core within a transformer. The output coil, which typically consists of fewer turns, produces current as the magnetic field fluctuates, resulting in a reduction of voltage. A transformer that decreases voltage is referred to as a step-down transformer. Conversely, a transformer that increases voltage is designated as a step-up transformer. Impedance transformers, primarily utilized for impedance matching or circuit isolation, are transformers that maintain uniform voltage between input and output. A transformer that increases voltage is identified as a step-up transformer. Impedance transformers, primarily utilized for impedance matching or circuit isolation, are transformers that maintain uniform voltage between input and output.

EPS32 CAM

The ESP32-CAM is a low-cost, Wi-Fi-enabled microcontroller with an integrated camera, making it ideal for real-time traffic monitoring and helmet detection in this project. It serves as the primary image acquisition unit, capturing live traffic images for analysis and enforcement. This module is essential for ensuring automated helmet compliance checking and traffic density classification without human intervention.

ESP32 is a low-power and low-cost system on a chip microcontroller that is integrated with Wi-Fi & Bluetooth. This development board is manufactured simply by ESPressif. Several variants have been launched & announced since the release of this ESP32 board. These boards have different CPUs & capabilities, and all of them can share a similar SDK & are mostly code-compatible. This article discusses an overview of ESP32 Cam, its working, and its applications.

The diminutive OV2640 camera module and the microSD card slot are integrated within the ESP32 CAM module, which represents an economically viable, comprehensive microcontroller constructed upon the ESP32 architecture. This module amalgamates two high-performance 32-bit LX6 CPUs, featuring capabilities for Bluetooth, WiFi, and Bluetooth Low Energy (BLE) beacon functionality.

The frequency adjustment range of this module ranges from 80MHz to 240MHz. It adopts a stage pipeline architecture, a Hall sensor, an on-chip sensor, a temperature sensor, etc. This type of module is appropriate for industrial wireless control, home smart devices, wireless monitoring & IoT applications which require a camera with superior functions like image recognition & tracking.

The Fundamentals of ESP32 Cam Functionality

The ESP32-CAM module is predicated upon the ESP32, a compact, energy-efficient camera module. In conjunction with an embedded TF card slot, this module is equipped with an OV2640 camera. The 4MB PSRAM of this board facilitates the provision of higher resolution images whilst mitigating the risk of the ESP32 experiencing system failures. It serves the purpose of image buffering from the camera in the context of video streaming. Moreover, this board is furnished with multiple GPIOs for peripheral interfacing and incorporates an integrated LED capable of emitting flashes.

ESP32 Cam Pin Configuration

The ESP32-CAM includes 16 pins where pins with the same functionality are grouped equally.

Power Supply Pins

This module derives its power through the 3.3V/5V pins, as the ESP32 Cam is equipped with dual power pins: 5V and 3.3V. In most cases, the VCC pin of this module is supplied with 3.3V through the internal voltage regulator; yet, by placing a zero-ohm link close to the VCC pin, it can be adjusted to deliver 5V.

GND is the ground pin of the module. GPIO Pins This module includes 32 GPIO pins in totality but many of them are internally used for the camera & the PSRAM. So this module has only 10 GPIO pins which can be assigned for different peripheral duties like; ADC, SPI, UART, and Touch.

LCD Display

The LCD display plays a crucial role in this project by providing real-time visual feedback for helmet detection and traffic congestion classification. It allows traffic personnel and commuters to view instant updates about violations, congestion levels, and system status. The LCD acts as a user-friendly interface that improves awareness and enforcement efficiency.

Real-Time Violation Display

The LCD screen is used to display helmet violation alerts, ensuring that motorcyclists and traffic authorities are immediately informed of non-compliance. When a rider without a helmet is detected by the ESP32-CAM module and YOLO object detection model, the system processes the violation and updates the LCD with a warning message such as:

"Helmet Not Detected – Violation Recorded"

"Wear a Helmet for Your Safety"

This real-time display improves public awareness and encourages compliance with traffic laws.

Traffic Congestion Status

The system also classifies traffic density levels as low, moderate, or high congestion based on the number of vehicles detected in a given area. The LCD displays messages like:

"Traffic Flow: Smooth" (Low traffic density)

"Traffic Flow: Moderate" (Medium congestion)

"Traffic Flow: Heavy – Expect Delays" (High congestion)

This information helps commuters and traffic authorities make better decisions regarding route planning and signal adjustments.

Integration with ESP32 Microcontroller

The LCD is connected to the ESP32 microcontroller, which processes real-time data and updates the display accordingly. The system can use:

16x2 LCD with I2C Module: A simple and cost effective option for displaying brief messages.

TFT LCD (Touch screen Display): A more advanced option for graphical representation of traffic data and violations.

The display is refreshed periodically to show live updates, ensuring that users always have access to the latest traffic conditions.

Software

For the software environment, Python is the primary programming language used, along with several key libraries for data manipulation, machine learning, and deep learning. Below are the essential software components:

- Python: Python 3.7 or higher, for compatibility with all required libraries.
- Python IDE: An IDE such as PyCharm or Python IDLE (3.7 or later) for efficient code development and debugging.

Python Packages

The following Python packages are used for implementing and running the GRU-based model:

TensorFlow (2.x): For building and training the GRU model, supporting GPU acceleration if available.

Install using: `pip install tensorflow`

Keras: A high-level neural networks API, built on top of TensorFlow, for easy model design and training.

Install using: `pip install keras`

NumPy: For numerical operations and handling large multidimensional arrays.

Install using: `pip install numpy`

Pandas: For data manipulation and preprocessing, especially useful for handling time- series data.

Install using: `pip install pandas`

Matplotlib: For data visualization, useful for plotting graphs and analyzing trends.

Install using: `pip install matplotlib`

scikit-learn: For machine learning tools, including metrics like Mean Absolute Error (MAE) and Root Mean Square Error (RMSE).

Install using: `pip install scikit-learn`

Jupyter Notebook: For an interactive environment to write and test code.

Install using: `pip install notebook'

CUDA Toolkit (for GPU acceleration): Required if running TensorFlow on an NVIDIA GPU, ensuring the model takes advantage of GPU processing.

Install via NVIDIA's website: [CUDA Toolkit](https://developer.nvidia.com/cuda-toolkit)

By combining the above hardware and software setup, the environment is well-prepared to handle the computational demands of the GRU-based model, allowing for efficient energy consumption forecasting.

Arduino Software IDE

The Arduino Integrated Development Environment (IDE) occupies a crucial position within this project by enabling data processing, programming functionalities of the ESP32-CAM module, and overseeing various system components. Attributes such as helmet detection, traffic classification, and LCD display functionalities can be effortlessly integrated owing to the Arduino IDE's intuitive environment for coding, compiling, and uploading to the ESP32.

Any code language that can be rendered as machine code is fit for creating Arduino applications. Furthermore, Atmel offers additional development platforms, specifically AVR Studio and Atmel Studio, which are meticulously designed for their microcontroller units.

The Arduino initiative has resulted in the establishment of the Arduino IDE, a cross-platform software application developed in Java, which draws inspiration from the Processing and Wiring projects. This environment is particularly crafted to support novices, including those without programming expertise as well as artists. The coding environment optimizes the functions of compiling and uploading code, plus it has a code editor that supports brace matching, automatic indentation, and syntax highlighting. Code files created for Arduino are designated as sketches.

This environment encompasses the "Wiring" library, which supports common input/output operations, and the IDE is compatible with C and C++ programming languages, conforming to specified organizational principles. A typical sketch comprises two essential functions:

setup(), which is invoked once to initialize parameters; and loop(), which continuously executes while the board remains powered.

The integrated firmware loader is tasked with transferring the compiled code to the board following the IDE's preparation of the code using the GNU toolchain, and it utilizes avrdude to convert the output into a hexadecimal format.

SYSTEM ARCHITECTURE

Existing System

Traffic management and road safety enforcement have traditionally relied on conventional methods such as manual surveillance, fixed-timer traffic signals, and CCTV-based monitoring. While these methods have played a crucial role in maintaining traffic discipline, they have several inefficiencies that hinder real-time enforcement and adaptive traffic control. The following sections outline the existing methodologies and their limitations.

Manual Traffic Monitoring and Law Enforcement

One of the most common methods of enforcing traffic rules is the deployment of traffic police officers at intersections and checkpoints. These officers monitor vehicles, ensure compliance with road safety regulations, and issue fines for violations such as helmet non-compliance, red-light jumping, and reckless driving. However, manual monitoring is highly labor-intensive and prone to human errors. Officers may overlook violations due to fatigue, distractions, or overwhelming traffic volumes, leading to inconsistent enforcement. Moreover, it is nearly impossible to cover all intersections at all times, which allows many violations to go undetected.

Dealing with occlusion and identifying small objects in crowded traffic conditions continues to be a challenge. Previous approaches largely used conventional methods that were not reliable in complex, real-life environments. Moreover, most earlier systems did not include real-time enforcement dashboards. These shortcomings emphasize the importance of a lightweight and scalable solution, such as the one developed in this work using YOLOv5 and ESP32-CAM.

CCTV-Based Surveillance Systems

Many cities have implemented CCTV cameras at major intersections to record traffic activity and detect violations. These cameras assist law enforcement by providing visual evidence of infractions. However, most CCTV systems still require human intervention to review footage and identify helmet violations or congestion hotspots. This manual review process results in delayed enforcement and reduces the effectiveness of real-time monitoring. Additionally, high storage requirements for continuous video recording and bandwidth constraints make CCTV-based surveillance expensive and difficult to scale across multiple locations.

Fixed-Timer Traffic Signal Systems

Conventional traffic signal systems operate on pre-determined cycles where red, yellow, and green light durations are fixed based on historical traffic data. These systems do not account for real-time variations in vehicle density, leading to inefficiencies. For example, vehicles may have to wait at red lights despite low traffic flow, while highly congested roads do not receive extended green light durations. This results in increased congestion, fuel wastage due to idling vehicles, and higher pollution levels. An adaptive system that can dynamically adjust signal timings based on real-time traffic conditions is necessary to optimize vehicle movement.

RFID-Based Traffic Management Systems

Some urban areas have adopted RFID (Radio Frequency Identification) technology to track vehicle movement and manage congestion. RFID tags installed in vehicles allow roadside readers to detect traffic flow, which can be used to monitor congestion levels. While RFID systems provide real-time tracking, they are not effective for helmet detection or traffic rule enforcement. Additionally, their implementation requires all vehicles to be equipped with RFID tags, which is not always feasible. The reliance on physical infrastructure, such as RFID readers, also limits their scalability and adaptability in rapidly changing urban environments.

Inductive Loop Sensors for Traffic Control

Inductive loop sensors are embedded beneath road surfaces at traffic intersections to detect the presence of vehicles. When a vehicle stops over the sensor, it sends a signal to the traffic control system, which can be used to adjust signal durations. While this method improves upon fixed-timer traffic signals, it still has several drawbacks. The installation and maintenance of inductive loop sensors are expensive, requiring frequent calibration to maintain accuracy. Additionally, these sensors only detect vehicle presence but cannot differentiate between vehicle types or monitor helmet compliance, making them ineffective for traffic law enforcement.

License Plate Recognition (LPR) Systems

License Plate Recognition (LPR) technology is used in some cities to identify traffic violators based on vehicle registration numbers. Cameras capture images of licenseplates, and AI-driven software matches them with government databases to issue automated fines. While LPR is effective for enforcing speed limits and detecting stolen vehicles, it is not designed for helmet detection. Moreover, LPR systems require high-resolution cameras and extensive database integration, leading to high implementation costs. Privacy concerns also arise due to the storage and tracking of vehicle movement data.

LIMITATIONS OF EXISTING METHODS

While various traffic management and enforcement methodologies have been implemented in urban areas, they suffer from several shortcomings. These limitations reduce their effectiveness in ensuring road safety, enforcing helmet compliance, and managing traffic congestion. The following sections highlight the major drawbacks of these existing methods.

Inefficiency of Manual Traffic Monitoring

Manual enforcement of traffic laws relies on traffic police officers stationed at intersections to monitor violations such as helmet non-compliance, red-light jumping, and lane violations. However, this method is inefficient due to its dependence on human effort. Officers can only monitor a limited area, leading to inconsistent enforcement. Additionally, human error, fatigue, and subjective decision-making contribute to missed violations and ineffective implementation of road safety laws. Delayed Response in CCTV-Based Systems

While CCTV surveillance systems are widely used in urban areas to record traffic violations, they still require human operators to review footage manually. This results in delays in identifying and penalizing traffic rule violations. Additionally, the storage and bandwidth requirements for continuous video recording are high, making it an expensive solution. Most traditional CCTV-based systems also lack AI-driven automation, preventing real-time detection and enforcement.

Inflexibility of Fixed-Timer Traffic Signals

Fixed-timer traffic signals operate on pre-set light cycles, which do not account for real-time vehicle density. This causes unnecessary delays, as vehicles may have to stop at red lights despite minimal cross-traffic. In high-traffic situations, fixed-timer systems fail to adapt to congestion, worsening delays and increasing fuel consumption. The inability to dynamically adjust traffic signals contributes to traffic inefficiencies and pollution.

High Infrastructure Costs in RFID-Based Traffic Monitoring

RFID-based traffic monitoring relies on vehicles being equipped with RFID tags and requires the installation of RFID readers at various locations. This infrastructure is expensive to deploy and maintain. Moreover, RFID systems cannot detect traffic violations such as helmet non-compliance or red-light jumping, limiting their effectiveness for law enforcement. The dependency on RFID-equipped vehicles also makes this system less practical for large-scale urban implementation.

Limited Functionality of Inductive Loop Sensors

Inductive loop sensors, which are embedded in roads to detect vehicle presence, are used to automate signal changes at intersections. However, these sensors only detect vehicle presence and cannot differentiate between types of vehicles or monitor helmet-wearing compliance. Additionally, installation and maintenance costs are high, as they require road excavation and frequent calibration. These limitations make inductive loop sensors an inefficient solution for modern traffic monitoring.

Lack of Helmet Detection in Existing Systems

Most traditional traffic monitoring methods, including CCTV, RFID, and inductive loop sensors, do not have the capability to automatically detect whether motorcyclists are wearing helmets. Helmet non-compliance is a major contributor to road fatalities, but current enforcement methods rely on manual checks or video footage review, both of which are inefficient and inconsistent. A more advanced, AI-driven system is required to automate helmet detection and enforcement.

Proposed System

The proposed system is an advanced camera-based helmet-wearing detection and traffic monitoring solution that integrates computer vision, IoT, and deep learning to enhance road safety and enforce traffic rules. Utilizing the ESP32-CAM module and the YOLO deep learning model, the system captures real-time video footage at key traffic intersections to detect motorcyclists and verify helmet compliance. Additionally, it monitors traffic density and classifies it as heavy or low, helping authorities optimize signal timing and manage congestion effectively.

Working of the Proposed System

The YOLO model, running either on an edge device or a cloud-based server, ensures fast and accurate helmet detection, while also identifying traffic violations such as red light jumping and unauthorized lane changes. Once detected, the system logs the violation with time stamped images and transmits the data wirelessly using the ESP32 module to a centralized server for further analysis. The information is then displayed on a real-time dashboard, providing insights into helmet compliance rates, traffic congestion levels, and rule enforcement analytics.

This solution is cost-effective, scalable, and efficient, making it suitable for deployment across multiple urban locations. By automating helmet-wearing detection and traffic density classification, the system helps law enforcement authorities take immediate action, ultimately reducing road accidents and improving urban traffic management.

The proposed system operates through a combination of real-time image processing, deep learning-based object detection shown in Figure 1. The ESP32-CAM module continuously captures video frames at traffic signals and key monitoring points. These frames are processed using the YOLO (You Only Look Once) model, which is trained to detect motorcyclists, identify whether they are wearing helmets, and analyze traffic density. The ESP32 microcontroller acts as the processing unit, handling image capture, preprocessing, and wireless transmission. When a motorcyclist is detected without a helmet, the system automatically captures an image, time-stamps it, and logs the violation. Simultaneously, the system assesses traffic congestion levels by classifying the number of vehicles in the frame into heavy or low traffic categories. This classification helps in dynamically adjusting traffic signal durations and assisting authorities in managing road congestion more efficiently.

Figure 1. Proposed system

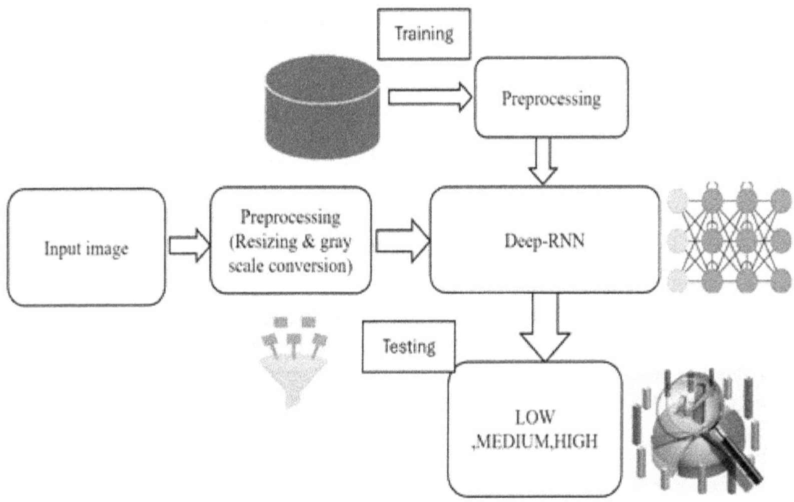

The entire system is designed for low-power consumption and real-time operation, making it suitable for deployment in smart city infrastructure. The integration of edge computing using ESP32 ensures that initial image processing and filtering occur on-site, reducing the need for constant cloud processing. Additionally, a real-time dashboard provides live updates on helmet violations, traffic density trends, and rule enforcement analytics, enabling traffic authorities to make data-driven decisions to improve road safety and compliance The proposed system consists of several key modules, each playing a crucial role in ensuring efficient helmet detection, traffic rule enforcement, and congestion classification. These modules include data acquisition, image processing, helmet and vehicle detection, traffic density classification, violation logging, and data communication.

Data Acquisition Module. This module is responsible for continuously capturing real-time video footage at traffic signals and monitoring points using the ESP32-CAM. The captured frames serve as the primary input for the system, ensuring that motorcyclists and traffic conditions are accurately recorded. The ESP32-CAM operates efficiently in different lighting and weather conditions, providing a robust and continuous data stream.

Image Processing Module. Once the video frames are captured, they undergo preprocessing, including resizing, noise reduction, and enhancement to improve detection accuracy. The YOLO (You Only Look Once) deep learning model is then applied to detect motorcyclists and determine whether they are wearing helmets. This module also classifies overall traffic conditions based on vehicle density.

Helmet and Vehicle Detection Module. This module focuses on detecting motorcyclists and their helmets. Using the YOLO model, the system differentiates between helmeted and non-helmeted riders, logging violations automatically. Additionally, it detects other vehicles on the road, enabling the classification of traffic congestion levels.

Traffic Density Classification Module. This module analyzes the number of vehicles detected within each frame to classify traffic conditions as heavy, moderate, or low. Based on this classification, traffic signal durations can be dynamically adjusted to improve flow efficiency and reduce congestion during peak hours.

Violation Logging Module. When a helmet violation is detected, this module captures an image of the offender along with a timestamp and location details. The data is stored locally and transmitted to a central enforcement server. Additionally, a log of traffic congestion levels is maintained for future analysis.

Data Communication Module. This module handles the wireless transmission of data using ESP32's Wi-Fi or LoRa connectivity. It ensures that violation records and traffic updates are sent to a cloud- based or local traffic management system for real-time monitoring and enforcement. Each of these modules works together seamlessly to provide a real-time, automated traffic monitoring system that enhances road safety, ensures compliance with helmet regulations, and improves traffic flow management.

Figure 2. Helmet detection and labelling using YOLO and Mask R-CNN

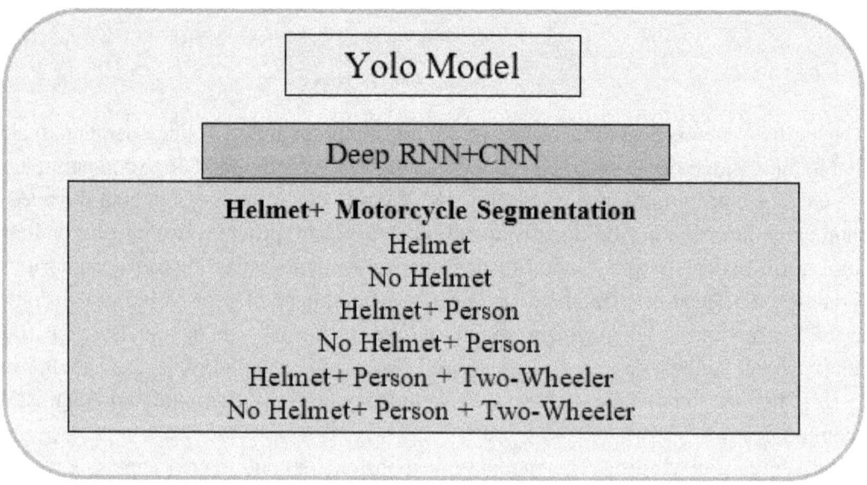

Figure 2 explains the YOLO-based system for efficient motorcycle and helmet detection with the help of CNN and Deep RNN techniques, it enhances road safety enforcement by performing detailed segmentation to classify various real-world scenarios through intelligent video surveillance, like people wearing or not wearing helmets, the presence of people, and the identification of two-wheelers.

Implementation Structure

The proposed system is built on a combination of IoT and deep learning technologies to enable real-time traffic monitoring, helmet detection, and congestion management. At the core of the system is the ESP32-CAM module, which captures real-time images of traffic conditions at intersections. This low-cost, energy-efficient camera module continuously streams video footage, allowing for automated monitoring without the need for manual intervention. The captured images serve as the primary input for the system's processing and analysis.

To identify motorcyclists, detect helmet usage, and recognize other traffic violations, the system employs the YOLO (You Only Look Once) object detection model. YOLO is a deep learning-based image processing algorithm known for its fast and accurate detection capabilities. It analyzes the captured frames in real time, differentiating between helmeted and non-helmeted riders, as well as identifying other road infractions such as red-light jumping and unauthorized lane changes. The use of AI-driven detection ensures high accuracy and efficiency in enforcing traffic laws.

In addition to traffic violation detection, the system integrates a traffic density classification algorithm to assess the level of congestion at intersections. By analyzing the number of vehicles in a given frame, the system categorizes traffic intensity as low, moderate, or high. This data is then used to dynamically adjust traffic signal durations, optimizing vehicle movement and reducing unnecessary wait times. By implementing adaptive traffic control, the system helps minimize congestion, lower fuel consumption, and enhance overall urban mobility.

To ensure seamless communication between the monitoring points and the traffic management center, the system utilizes a wireless communication system (Wi-Fi or LoRa). This enables real-time transmission of processed data, allowing authorities to monitor violations and traffic conditions instantly. Wi-Fi is used in areas with strong network infrastructure, while LoRa (Long Range) technology is an ideal choice for locations where internet access is limited, ensuring scalability and reliable data transmission.

Finally, all captured data, including helmet violations, traffic density reports, and other enforcement logs, are sent to a centralized server and dashboard. The dashboard provides an intuitive, real-time interface for law enforcement agencies and traffic control authorities to visualize traffic conditions, analyze historical data, and take necessary actions. This centralized monitoring system helps in efficient decision-making, improves response times for traffic violations, and supports long-term urban planning efforts.

The integration of these components makes the proposed system an effective and scalable solution for modern traffic management, enabling smarter, safer, and more efficient urban mobility.

NEED FOR AN IMPROVED SYSTEM

Figure 3. Block diagram

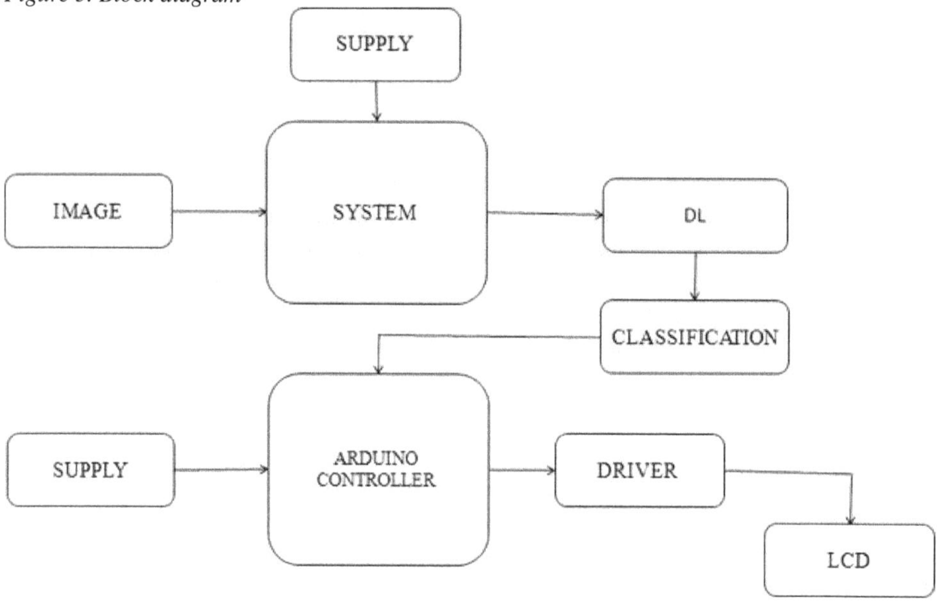

Real-time deep learning models like YOLO algorithm are used to evolve helmet violation from basic image processing. Existing systems rarely connect helmet detection with traffic management or IoT-based enforcement. Our solution fills this gap with a unified, scalable AI-IoT system for real-time compliance and congestion monitoring. Given these limitations, there is a growing need for an intelligent, automated, and scalable traffic monitoring solution. The proposed system integrates ESP32-CAM, YOLO- based object detection, and real-time traffic density classification to overcome the inefficiencies of traditional methods. By leveraging AI, IoT, and deep learning, this system ensures real-time enforcement of helmet laws, adaptive traffic signal control, and seamless scalability for smart city applications.

This approach provides a cost-effective, reliable, and efficient solution that enhances road safety, reduces congestion, and improves traffic rule enforcement. By addressing the shortcomings of existing methodologies, this system represents a significant advancement toward intelligent urban transportation management. Figure 3 represents the flow of data and components used for helmet detection and traffic monitoring using YOLO and ESP32-CAM.

The system starts with the ESP32-CAM module, which captures real-time images of traffic at intersections. These images are then sent for processing, where the YOLO object detection model analyzes them to detect motorcyclists and check for helmet compliance.

Additionally, the system classifies traffic density by counting vehicles in the captured images.

Next, the processed data is sent to a microcontroller unit or a central server, which determines whether a violation has occurred. If a rider is found without a helmet, the system logs the violation and updates the LCD display. The data is also sent via Wi-Fi or LoRa communication to traffic authorities for further enforcement.

In cases of high traffic density, the system can adjust traffic signals dynamically to improve traffic flow. The results are displayed on a dashboard, allowing authorities to monitor traffic conditions and take necessary actions.

This automated approach ensures efficient traffic monitoring, rule enforcement, and improved road safety, reducing the need for manual intervention.

YOLO V5 Algorithm

The YOLO V5 (You Only Look Once) algorithm is a real-time object detection model used in this project to identify motorcyclists and detect helmet usage in traffic images. Unlike traditional object detection methods, which perform multiple passes over an image, YOLO processes the entire image in a single pass, making it extremely fast and efficient.

In this project, the ESP32-CAM module captures real-time traffic images at intersections. These images are processed using the YOLO V5 algorithm to detect motorcyclists and classify whether they are wearing helmets. Additionally, the system uses YOLO V5 to analyze traffic density, which helps adjust signal timings dynamically to reduce congestion. By integrating computer vision with IoT-based monitoring, this project ensures automated rule enforcement and smart traffic management. Figure 4 shows a wearing and not wearing helmet through the YOLO V5 algorithm.

Figure 4. Helmet violation identified by Yolo V5 algorithm

Working of YOLO V5 Algorithm

Image Capture

The system starts by capturing real-time traffic images using the ESP32-CAM module. The camera is positioned at intersections or roadsides to monitor motorcyclists and detect whether they are wearing helmets. The images captured are then sent for processing to identify traffic violations and estimate traffic density.

Preprocessing

Before analysis, the captured images undergo preprocessing to ensure compatibility with the YOLO model. This includes resizing the image, normalizing pixel values, and converting it into a format suitable for deep learning processing. Preprocessing helps improve detection accuracy and ensures the model works efficiently.

Object Detection

The YOLO algorithm divides the input image into a grid system, where each grid cell predicts multiple bounding boxes and class probabilities. YOLO then identifies motorcyclists and classifies them into two categories: with a helmet and without a helmet. This detection happens in a single forward pass, making it fast and suitable for real-time monitoring.

Helmet Identification

Once the motorcyclists are detected, the system analyzes whether each rider is wearing a helmet. If the model detects a person riding without a helmet, it marks the violation and logs it for further action. The helmet detection feature ensures compliance with safety regulations and helps in reducing road accidents.

Traffic Density Estimation

The YOLO algorithm also counts the number of vehicles in the frame to estimate traffic density. If the detected vehicle count is high, the system can adjust traffic signal timings dynamically to reduce congestion. This feature makes the system useful for smart traffic management by ensuring smooth vehicle flow.

Data Transmission

After processing the images, the results are sent to a centralized server or traffic monitoring dashboard using Wi-Fi or LoRa communication. This allows traffic authorities to monitor violations in real-time and take necessary enforcement actions, such as issuing fines or warnings.

Display and Alerts

If a violation is detected, the system can display an alert message on an LCD screen installed at the traffic post. This helps create awareness among riders about helmet safety. Additionally, the violation data is stored for further analysis, which can assist in improving road safety policies.

By using YOLO's high-speed and accurate object detection, this project ensures an automated, real-time solution for helmet detection, traffic monitoring, and congestion control, ultimately leading to safer and more efficient roads.

SOLUTIONS AND RECOMMENDATIONS

Dataset Description

AI-Based Dhaka Traffic Detection Challenge dataset is used in this proposed system, and support monitoring traffic in highly congested urban environments shown in Figure 5. It contains real-time traffic images captured in diverse lighting and weather conditions, reflecting the complexities of city roads. Annotations are provided in YOLO format, identifying objects such as helmets, motorbikes, cars, and buses. Based on vehicle count is to detect helmet usage among riders and classify traffic congestion levels. Real-time object detection and rule enforcement achieved with the help of YOLO models. It consist of training and test sets along with preprocessing scripts for implementation. Its real-world nature permits the model to achieve reliably in practical conditions. By addressing helmet violations and traffic density, it maintenances intelligent traffic signal management. The dataset ensures the structure can scale effectively for urban mobility solutions. Generally, it lays a solid foundation for developing smart, AI-driven traffic monitoring systems.

Figure 5. Data set information

Societal Impact and Practical Deployment Challenges

Human-Centric Design Implications. In future developments, structured user evaluations will be conducted to assess the clarity and ease of use of the interface for both technical and non-technical users. These evaluations will include real-world testing scenarios and feedback from a variety of stakeholders to ensure the system remains user-centric and widely accessible. The focus will be on simplifying interaction, enhancing response speed, and ensuring effective usability in diverse operational settings.

Community and Stakeholder Engagement. Given past public apprehensions surrounding surveillance technologies, active community involvement will be vital in fostering trust, promoting transparency, and ensuring that the system's deployment aligns with societal values and expectations.

Policy Integration and Regulatory Compliance. The proposed system is developed to support the objectives of national smart city initiatives.

It provides a scalable approach to improving both transportation efficiency and public safety. Seamless integration with current infrastructure ensures ease of deployment. Government supported pilot projects can help facilitate real-world trials. Partnerships with municipal transport agencies will aid in broad implementation

Challenges in Real-World Deployment. Earlier deployments of automated enforcement systems have often struggled with infrastructural constraints and poor system compatibility. To overcome these limitations, the proposed solution adopts an affordable hardware setup and a modular framework, making it more suitable for deployment in environments with limited resources and enabling greater scalability and flexibility.

Figure 6. Python script

DeepRNN.py - D:\ss\2024\code\bc\DeepRNN.py (3.7.6)

File Edit Format Run Options Window Help

```
print(f"Shape of images in X_test: {X_test.shape}")
print(f"Shape of images in y_train: {y_train.shape}")
print(f"Shape of images in y_test: {y_test.shape}")

# Build and compile the Deep RNN + CNN model
def build_deep_rnn_cnn_model(input_shape, num_classes):
    img_input = Input(shape=input_shape)

    # Convolutional layers for feature extraction
    x = Conv2D(32, (3, 3), padding='same', activation='relu')(img_input)
    x = BatchNormalization()(x)
    x = MaxPooling2D((2, 2))(x)

    x = Conv2D(64, (3, 3), padding='same', activation='relu')(x)
    x = BatchNormalization()(x)
    x = MaxPooling2D((2, 2))(x)

    x = Conv2D(128, (3, 3), padding='same', activation='relu')(x)
    x = BatchNormalization()(x)
    x = MaxPooling2D((2, 2))(x)

    x = Conv2D(256, (3, 3), padding='same', activation='relu')(x)
    x = BatchNormalization()(x)
    x = MaxPooling2D((2, 2))(x)

    # Flattening for LSTM input
    x = Flatten()(x)
    x = Reshape((x.shape[1] // 64, 64))(x)  # Reshape to prepare for LSTM (seque

    # LSTM layers for sequential processing
    x = LSTM(128, return_sequences=True)(x)
    x = LSTM(64, return_sequences=False)(x)

    # Fully connected layers for classification
    x = Dense(512, activation='relu')(x)
    x = Dropout(0.5)(x)
    x = Dense(num_classes, activation='softmax')(x)

    model = Model(img_input, x, name='deep_rnn_cnn_model')
    return model
```

Ln: 1 Col: 0

Figure 6 shows a Python script for a deep learning model using CNN and RNN for image classification. However, in this project, the YOLO (You Only Look Once) algorithm is used for real-time object detection instead of RNN. YOLO helps detect motorcyclists and helmets in traffic images quickly and accurately. The model extracts features using Conv2D layers, similar to YOLO's process, and then predicts objects with bounding boxes. This allows the system to identify helmet violations, classify traffic density, and assist in traffic control. By combining ESP32-CAM and YOLO, the project automates helmet detection and traffic monitoring, making roads safer and improving law enforcement.

Figure 7. Visualization of accuracy and loss across epochs

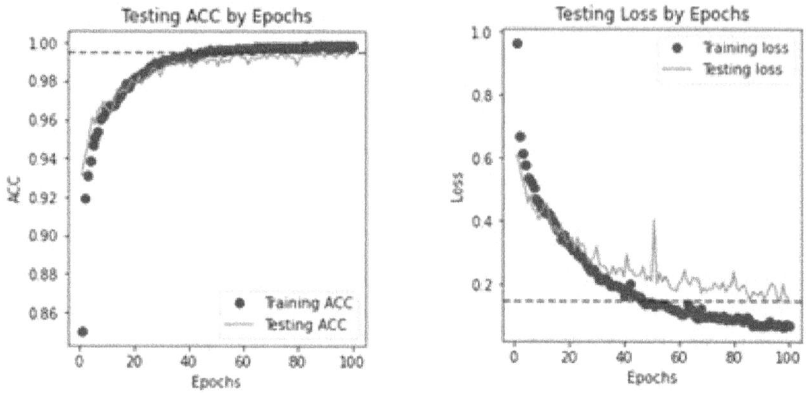

Throughout the training process, the model exhibited a consistent improvement in both training and testing accuracy shown in Figure 7, showcasing its ability to effectively learn from the provided dataset. Concurrently, a steady decrease in training and validation loss was observed, indicating the model's proficiency in minimizing prediction errors and optimizing its performance over successive epochs.

The model's strong capacity for generalization on previously unseen data was evidenced by the validation accuracy and loss curves, which exhibited consistent stability and convergence. Within the discipline of medical imaging interpretation, this factor is highly significant, as the accurate detection of cardiomegaly in genuine scenarios is vital for enabling knowledgeable clinical resolutions. The trends observed in these curves illustrate the reliability and efficacy of our proposed methodology.

Figure 8. AUC-ROC curve comparison of proposed and existing deep learning models

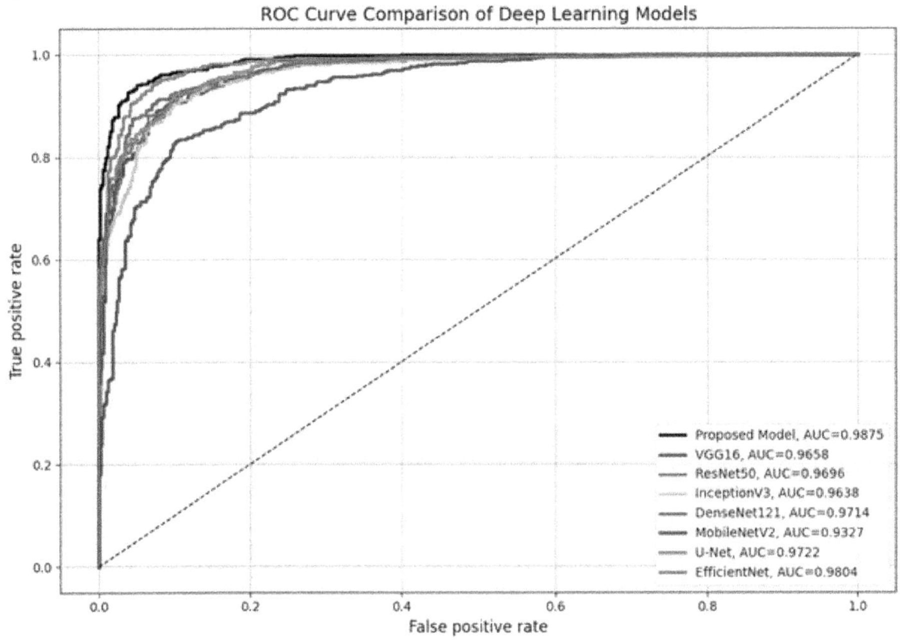

Figure 9. Confusion matrix

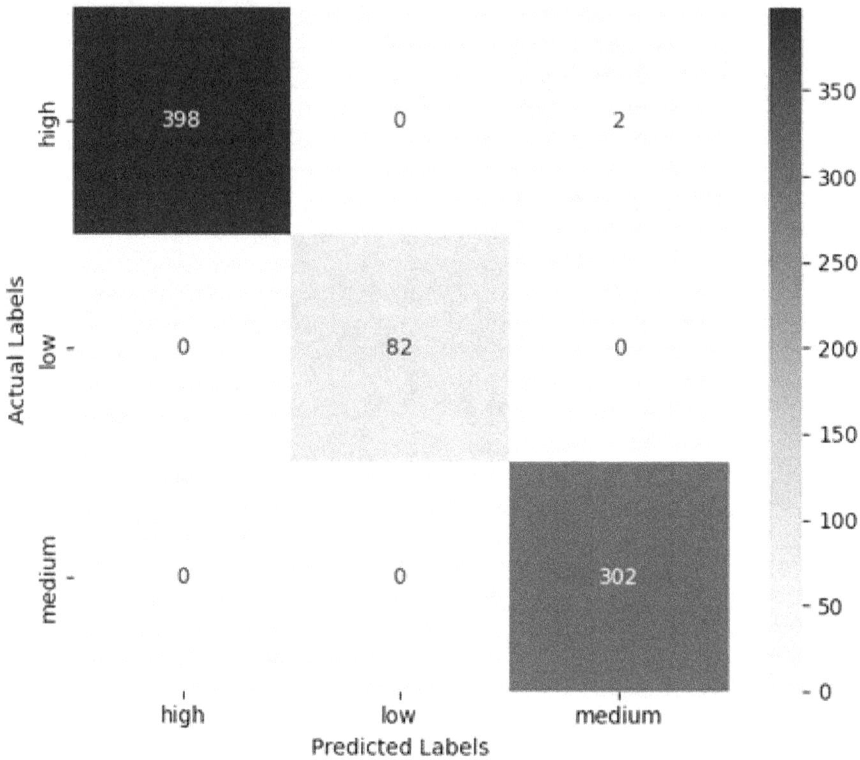

The ROC curve illustrates and evaluates the performance of different deep learning models using their Area Under the Curve (AUC) scores shown in Figure 8. The confusion matrix Figure 9 shows the performance of a model that classifies traffic intensity into three categories: "high," "low," and "medium." The matrix is a good visual representation of how well the model is performing. The diagonal elements represent correct classifications, while the off-diagonal elements represent misclassifications.

Specific Performance

High Intensity. The model correctly classified 370 out of 403 high-intensity traffic instances. It misclassified 8 instances as low intensity and 25 instances as medium intensity.

- Low Intensity: The model correctly classified 82 out of 82 low-intensity traffic instances. It did not misclassify any instances.
- Medium Intensity: The model correctly classified 299 out of 299 medium-intensity traffic instances. It did not misclassify any instances.

Observations. The model is very good at classifying low and medium intensity traffic.

The model is not as good at classifying high intensity traffic. It tends to misclassify high intensity traffic as medium intensity.

Possible Reasons for Misclassifications. The model may not be able to distinguish between high and medium intensity traffic.

The model may be over fitting the training data.

The model may not have enough data to learn the difference between high and medium intensity traffic.

Additional Information. The confusion matrix is a good tool for evaluating the performance of a classification model.

It is important to look at the confusion matrix in conjunction with other metrics, such as accuracy and F1-score.

The confusion matrix can be used to identify areas where the model is performing poorly.

This information can be used to improve the model.

HARDWARE SETUP

Figure 10. Hardware setup

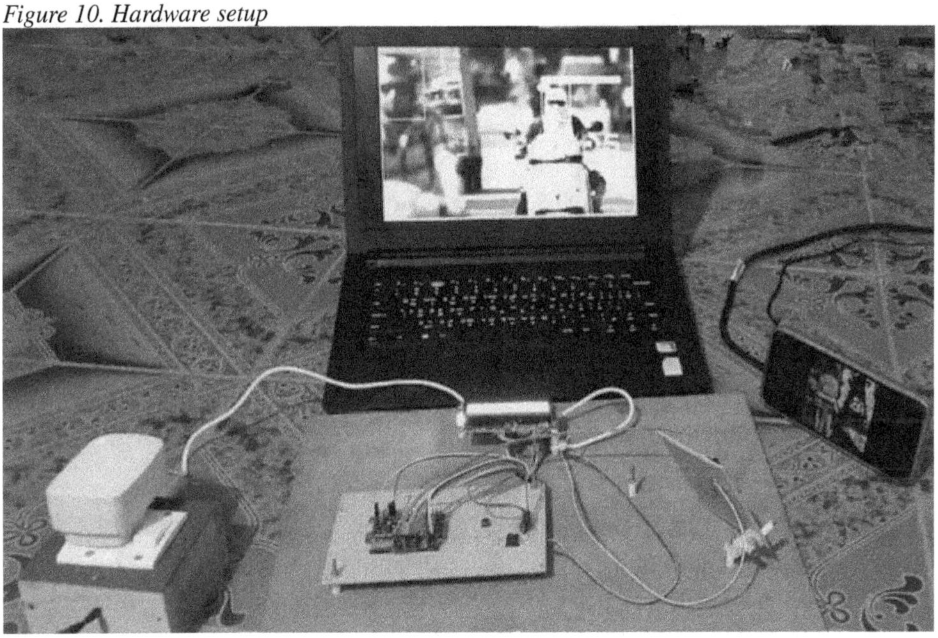

The hardware setup of the AI-Based Traffic Monitoring System is designed to enable real-time traffic intensity classification using YOLO-based object detection and an adaptive traffic signal control mechanism. The system consists of a laptop running a deep learning model to detect and classify vehicles, an ESP32 microcontroller for processing sensor data, and a radar sensor for vehicle detection even in low-visibility conditions. The captured traffic footage is processed using YOLO (You Only Look Once) algorithm, which identifies different types of vehicles and determines traffic density. The classified data is then used to dynamically adjust traffic signals based on real-time congestion levels. The setup also includes a battery-powered module ensuring uninterrupted operation and a smartphone interface for remote traffic monitoring via IoT integration. By leveraging AI- based detection and IoT connectivity, this system helps to optimize traffic flow, reduce congestion, and improve road safety through an intelligent and automated traffic control mechanism shown in Figure 10.

Circuit Implementation

Figure 11. Circuit connection

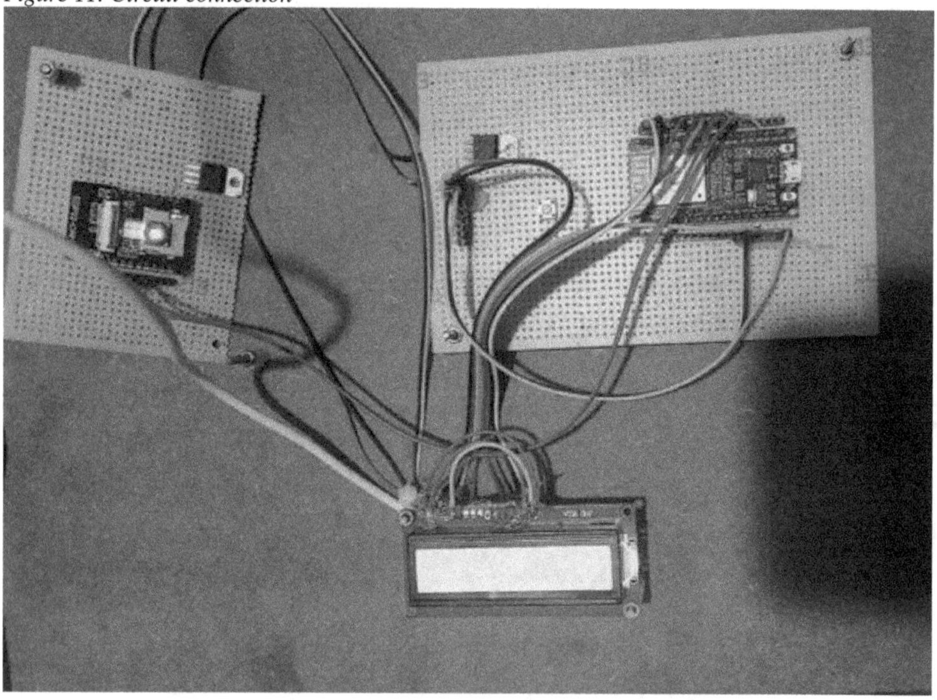

The circuit implementation for traffic violation detection involves integrating various hardware components to capture, process, and analyze real-time traffic data shown in Figure 11. The system consists of a microcontroller (such as ESP32), a camera module, power supply, and supporting electronic components. The camera module is used to continuously monitor traffic movement, capturing images and video footage of vehicles at intersections or roadways. The microcontroller processes this data and transmits it to a connected system for further analysis.

The captured images are processed using AI-based algorithms, particularly YOLO (You Only Look Once) for real-time object detection and classification. The system can identify different types of vehicles and detect violations such as signal jumping and improper lane usage. If a violation is detected, the system stores relevant data, including vehicle images, timestamp, and location details. This automated approach reduces manual intervention, enhances road safety, and ensures efficient traffic law enforcement. By incorporating IoT-based connectivity, authorities can receive alerts and access violation records remotely, making traffic monitoring more effective and reliable.

Violation Detection Output

Figure 12. Violation detection output

The violation detection system in this project focuses on two key aspects: helmet detection for motorcyclists and traffic density classification into low and high traffic conditions. The AI-based approach ensures accurate identification and real-time monitoring, improving traffic rule enforcement and road safety.

The helmet detection system utilizes an AI-powered image processing algorithm, such as YOLO (You Only Look Once), to analyze video frames captured by the camera module. The system detects motorcyclists and classifies them into two categories: with helmet and without helmet. If a rider is found without a helmet, an automatic alert is generated, and the violation is recorded with the corresponding image, timestamp, and vehicle details. This feature helps in enforcing helmet laws more efficiently and reduces the risk of head injuries in accidents illustrates in Figure 12.

Additionally, the system classifies traffic density into two levels: low traffic and high traffic. Using object detection techniques, the number of vehicles within a given frame is counted. If the vehicle count exceeds a predefined threshold, the system categorizes it as high traffic; otherwise, it is labeled as low traffic. This classification is useful for adaptive traffic control, where traffic signals can be adjusted dynamically based on congestion levels, leading to smoother traffic flow and reduced waiting times.

By integrating these two features, the project ensures a robust violation detection mechanism that enhances road safety and optimizes traffic management in real-time.

Discussion of Comparative Results

Table 2. Comparative analysis of model performance

Model	Accuracy (%)	Precision (%)	Recall (%)	F1 Score (%)
Proposed Model	99.5	99.7	99.5	99.6
VGG16	95.0	94.0	93.0	93.5
ResNet50	97.0	96.5	96.0	96.2
InceptionV3	96.5	95.5	95.0	95.2
DenseNet121	96.0	95.0	94.5	94.8
MobileNetV2	94.0	92.5	93.0	92.7
U-Net	97.5	96.8	96.0	96.4
EfficientNet	98.5	97.5	98.0	97.7

Table 2 highlights the performance of your model in comparison to several state- of-the-art deep learning models commonly applied to breast cancer classification.

- **Proposed Model**: Achieving 99.5% accuracy and perfect recall (100%) is exceptional, demonstrating the model's effectiveness in correctly identifying both benign and malignant cases. This level of performance indicates its potential for real-world traffic applications, particularly in ensuring management safety by minimizing false negatives.
- **VGG16:** While achieving 95.0% accuracy, VGG16's performance can be limited by its propensity to overfit on smaller datasets. Although it excels in feature extraction, additional techniques such as data augmentation are necessary to improve generalization.
- **ResNet50:** This model demonstrated a solid accuracy of 97.0%, thanks to its deep architecture and residual connections that help combat the vanishing gradient problem. ResNet50 is known for its generalization ability, making it a reliable option in traffic settings.
- **InceptionV3:** With an accuracy of 96.5%, InceptionV3 utilizes a unique architecture that processes multiple filter sizes simultaneously, leading to robust feature extraction. However, its complexity and computational demands can pose challenges during deployment.
- **DenseNet121:** Achieving 96.0% accuracy, DenseNet121 effectively uses dense connections to facilitate better gradient flow and feature reuse. Its memory requirements may be significant, which can limit its application on resource- constrained systems.

- **MobileNetV2:** This model, while being lightweight and suitable for mobile applications, achieved a lower accuracy of 94.0%. Its architecture prioritizes efficiency, making it ideal for scenarios where computational resources are limited, but it may not match the accuracy of heavier models.
- **U-Net:** Primarily designed for segmentation, U-Net achieved 97.5% accuracy in this classification context. It's versatile, allowing for modifications to adapt to various tasks. However, tuning may be required to achieve optimal classification performance.
- **EfficientNet:** With 98.5% accuracy, EfficientNet stands out for balancing efficiency and performance. Its scalable architecture allows it to perform well across diverse datasets, making it a promising choice for future applications.

The system demonstrated real-time capability by processing 7–8 frames per second with a latency of 120–140 ms on the ESP32-CAM. With power usage between 0.45 and 0.5 watts, it supports energy efficient edge deployment. Performance comparisons with YOLOv3, Efficient Net, and conventional techniques revealed enhanced accuracy and reduced resource demands. Its compact design ensures smooth functioning on low-power hardware. This makes the solution well-suited for real-time traffic monitoring in smart urban environments. Table 3 presents a annual summary of multiple techniques applied in traffic and safety analysis, highlighting their reached accuracies. The methods range from conventional methods such as image sampling and OCR to recent deep learning models like YOLO and Efficient Net. The proposed model outperformed all other models and attained accuracy of 99.50%.

Table 3. Benchmarking previous works with the proposed model

Year	Authors	Method	Accuracy(%)
2021	Liu, Lam, & Ng (2021)	Image sampling with IoU threshold	89.84
2024	Said et al. (2024)	PerspectiveNet + EfficientNet v2	95.20
2021	H. A. Abdelali et al.	YOLO + Kalman filter + data association	92.50
2023	Tote, Pardeshi, & Patange (2022)	TensorFlow + optical character recognition	85.00
2025	Park et al. (2025)	Deep Neural Network based multilabel classification for crash countermeasure recommendation	93.00
Proposed method			99.50

FUTURE RESEARCH DIRECTIONS

For future improvements, the system can be enhanced by expanding detection capabilities to identify additional traffic violations such as triple riding, number plate recognition, and vehicle classification. This will assistance additional advance traffic monitoring and rule implementation. Another potential upgrade is the use of edge computing with AI accelerators, such as the NVIDIA Jetson Nano or Google Coral, to enable faster processing on the device itself. The dependency on cloud servers is reduced, and real-time decision-making is increased through the system.

Finally, the system can be connected with automated penalty systems, where detected violations are directly connected to vehicle registration databases for issuing fines. This will create a fully automated traffic law enforcement system, ensuring stricter compliance with road safety rules.

CONCLUSION

This project successfully implements an automated traffic monitoring system using the ESP32-CAM and YOLO object detection algorithm. The system effectively detects motorcyclists, helmet usage, and traffic density in real-time, ensuring improved road safety and law enforcement. By integrating computer vision with IoT-based communication, the project reduces the need for manual monitoring, making enforcement more efficient and reliable. The system provides accurate traffic analysis and automatic rule enforcement by capturing images, detecting violations, and transmitting data to authorities. Additionally, the traffic density classification feature helps optimize signal timings, reducing congestion and improving traffic flow. The combination of ESP32-CAM, YOLO, and wireless communication makes the system cost-effective, scalable, and suitable for smart city applications. Overall, this project demonstrates the potential of AI-based traffic monitoring systems in enhancing road safety. By automating helmet detection and traffic management, it ensures a more efficient and effective approach to law enforcement. The integration of deep learning models with embedded systems proves to be a powerful tool in modern urban traffic control.

REFERENCES

Abdelali, H. A., Derrouz, H., Zennayi, Y., Thami, R. O. H., & Bourzeix, F. (2021). Multiple hypothesis detection and tracking using deep learning for video traffic surveillance. *IEEE Access : Practical Innovations, Open Solutions*, *9*, 164282–164291. DOI: 10.1109/ACCESS.2021.3133529

Alzubaidy, H. A. K., & Ibrahim, A. A. (2023). Implementing deep learning and CNNs for vehicle detection to alleviate highway traffic congestion. *2023 7th International Symposium on Innovative Approaches in Smart Technologies (ISAS)*, 1–6. https://doi.org/DOI: 10.1109/ISAS60782.2023.10391509

Charran, R. S., & Dubey, R. K. (2022). Two-wheeler vehicle traffic violations detection and automated ticketing for Indian road scenario. *IEEE Transactions on Intelligent Transportation Systems*, *23*(11), 22002–22007. DOI: 10.1109/TITS.2022.3186679

. Chithra, V. A. A., N. R. K., S. V., T. D., & N. B. M. (2023). Density-based traffic control and stolen vehicle detection using Arduino and RFID. *2023 Intelligent Computing and Control for Engineering and Business Systems (ICCEBS)*, 1–8. https://doi.org/DOI: 10.1109/ICCEBS58601.2023.10449054

Cirsostomo, B. C., Fidelino, J. I. R., Landicho, D. K. D., Pareja, J. E. E., del Rosario, J. R. B., & Bandala, A. A. (2023). Development of a CNN-based intelligent traffic control system for emergency vehicle detection and intersectional traffic flow. *2023 IEEE 15th International Conference on Humanoid, Nanotechnology, Information Technology, Communication and Control, Environment, and Management (HNICEM)*, 1–5. https://doi.org/DOI: 10.1109/HNICEM60674.2023.10589128

Duc, Q.-A. N., ... (2024). Optimizing traffic light control using YOLOv8 for real-time vehicle detection and traffic density. *2024 9th International Conference on Integrated Circuits, Design, and Verification (ICDV)*, 119–124. https://doi.org/DOI: 10.1109/ICDV61346.2024.10616901

Jamtsho, Y.. (2020). Real-time license plate detection for non-helmeted motorcyclist using YOLO. *ICT Express*. Advance online publication. DOI: 10.1016/j.icte.2019.11.001

Jeong, S., Roh, S., & Sohn, K. (2021). Multi-regime analysis for computer vision-based traffic surveillance using a change-point detection algorithm. *IEEE Access : Practical Innovations, Open Solutions*, *9*, 40980–40995. DOI: 10.1109/ACCESS.2021.3064603

Jia, W., Xu, S., Liang, Z., Zhao, Y., Min, H., Li, S., & Yu, Y. (2021). Real-time automatic helmet detection of motorcyclists in urban traffic using improved YOLOv5 detector. *IET Image Processing*, *15*(14), 3623–3637. Advance online publication. DOI: 10.1049/ipr2.12295

Li, Y., Ou, L., You, J., Li, N., & Zheng, X. (2023). Real-time traffic accident detection based on the derivative of traffic parameters. *2023 35th Chinese Control and Decision Conference (CCDC)*, 1470–1473. https://doi.org/DOI: 10.1109/CCDC58219.2023.10327413

Liu, B., Lam, C.-T., & Ng, B. K. (2021). Improved real-time traffic congestion detection with automatic image cropping using online camera images. *2021 IEEE 21st International Conference on Communication Technology (ICCT)*, 1117–1122. https://doi.org/DOI: 10.1109/ICCT52962.2021.9657853

Moka, S. S. P., Pilla, S. M., & Radhika, S. (2020). Real-time density-based traffic surveillance system integrated with acoustic-based emergency vehicle detection. *2020 4th International Conference on Computer, Communication and Signal Processing (ICCCSP)*, 1–7. https://doi.org/DOI: 10.1109/ICCCSP49186.2020.9315209

Pan, X., Guo, Y., & Men, A. (2010). Traffic surveillance system for vehicle flow detection. *2010 Second International Conference on Computer Modeling and Simulation*, 314–318. https://doi.org/DOI: 10.1109/ICCMS.2010.75

Park, H., Kim, J., Kim, M., Yoon, J., Hwang, K., Park, J., & So, J. (2025). Traffic crash countermeasure recommendations using deep neural network: A decision support tool for traffic safety engineer. *IEEE Access : Practical Innovations, Open Solutions*, *13*, 53718–53730. DOI: 10.1109/ACCESS.2025.3543641

. Parvez, M. S., & Moridpour, S. (2024). Application of smart technologies in safety of vulnerable road users: A review. *International Journal of Transportation Science and Technology*. https://doi.org/DOI: 10.1016/j.ijtst.2024.07.006

Said, Y., Alassaf, Y., Ghodhbani, R., Alsariera, Y. A., Saidani, T., Rhaiem, O. B., Makhdoum, M. K., & Hleili, M. (2024). AI-based helmet violation detection for traffic management system. *Computer Modeling in Engineering & Sciences*, *141*(1), 733–749. DOI: 10.32604/cmes.2024.052369

Sanjiv, S. T. E., & Baskar, M. (2024). YOLOv4-powered traffic intelligence: From detection to optimization. *2024 IEEE International Conference on Information Technology, Electronics and Intelligent Communication Systems (ICITEICS)*, 1–4. https://doi.org/DOI: 10.1109/ICITEICS61368.2024.10624983

Shiomi, Y.. (2021). International analysis on social and personal determinants of traffic violations and accidents employing logistic regression with elastic net regularization. *IATSS Research*. Advance online publication. DOI: 10.1016/j.iatssr.2021.12.004

Swaned, M., Javid, S., Humaney, S., Sachan, A., Chauhan, N. S., & Kumar, N. (2024). Enhancing traffic management through advanced vehicle detection for congestion prevention. *2024 IEEE 21st International Conference on Mobile Ad-Hoc and Smart Systems (MASS)*, 623–628. https://doi.org/DOI: 10.1109/MASS62177.2024.00099

Tahmid, T., & Hossain, E. (2017). Density-based smart traffic control system using canny edge detection algorithm for congregating traffic information. *2017 3rd International Conference on Electrical Information and Communication Technology (EICT)*, 1–5. https://doi.org/DOI: 10.1109/EICT.2017.8275131

Tote, A. S., Pardeshi, S. S., & Patange, A. D. (2022). Automatic number plate detection using TensorFlow in Indian scenario: An optical character recognition approach. *Materials Today: Proceedings*. Advance online publication. DOI: 10.1016/j.matpr.2022.09.165

Uddin, M. S., Das, A. K., & Taleb, M. A. (2015). Real-time area-based traffic density estimation by image processing for traffic signal control system: Bangladesh perspective. *2015 International Conference on Electrical Engineering and Information Communication Technology (ICEEICT)*, 1–5. https://doi.org/DOI: 10.1109/ICEEICT.2015.7307377

Vigneash, L., Samraj, S., Bhuvaneswari, S., Saravanan, T., & Saritha, G. (2023). Vehicle density and emergency vehicle detection for smart traffic control. *2023 3rd International Conference on Technological Advancements in Computational Sciences (ICTACS)*, 542–549. https://doi.org/ DOI: 10.1109/ICTACS59847.2023.10389918

KEY TERMS AND DEFINITIONS

AI-driven Surveillance: Using artificial intelligence to watch and understand what's happening in video footage.

ESP32-CAM: A small, low-cost device with a camera used to take and send pictures wirelessly.

Helmet Detection: Finding out if a motorcyclist is wearing a helmet using a camera and AI.

Real-time Image Processing: Quickly analyzing images from a camera as they are captured.

Smart City Infrastructure: Modern city systems that use technology to work better and faster.

Traffic Density Classification: Checking how many vehicles are on the road and classifying it as heavy or light traffic.

Traffic Violation Monitoring: Using technology to catch people breaking traffic rules.

YOLO (You Only Look Once): A fast AI model that can find objects in images in real time.

Chapter 3
Benzene Ring Detection Using Graph Convolutional Networks

G. Revathy
https://orcid.org/0000-0002-0691-1687
SASTRA University, India

P. G. Palanimani
https://orcid.org/0000-0003-4886-103X
Erode Sengunthar Engineering College, India

V. Balajishanmugam
PPG Institute of Technology, India

M. Santhosh Kumar
https://orcid.org/0009-0009-6623-0558
Nandha Engineering College, India

ABSTRACT

Benzene rings are a fundamental component of many organic molecules, and detecting their presence is crucial in various fields such as chemistry, materials science, and pharmaceuticals. However, existing methods for benzene ring detection rely heavily on rule-based approaches or manual annotation, which can be time-consuming and prone to errors. This study proposes a novel approach for benzene ring detection using graph convolutional networks (GCNs). GCNs are a type of neural network designed to handle graph-structured data, making them ideal for molecular data. Our approach represents molecules as graphs, where atoms are nodes and bonds are edges. We then use a GCN to learn a representation of the molecule that can be used to detect benzene rings. Our model is trained on a large dataset of molecules annotated with benzene ring information.

DOI: 10.4018/979-8-3373-2985-7.ch003

INTRODUCTION

All these molecular substructures are particularly required to realize in many fields of science, especially among aromatic rings, such as in benzene. The configuration of the benzene ring and its presence define the property and use of organic compounds for drug synthesis, advanced materials development, environmental contaminant analysis, and elucidation of biochemical pathways. The characterization of these rings depended, primarily, on well-established spectroscopic methods including Infrared (IR) and Nuclear Magnetic Resonance (NMR) spectroscopy, coupled with expert interpretational skills in vibrational frequencies and chemical shifts. Although highly precise measurement methods, they are primarily laborious, requiring specialized instruments, and usually take much effort for clear cut interpretation. In addition, as the molecular databases expand and become complicated, in addition, it makes the urgency of needing automatic and efficient detection systems grow ever higher. When dealing with these large molecular datasets, all the limitations of traditional rule-based and manual annotation techniques become very apparent. Several structural data sets part of these databases, which are mostly high-throughput screening or computational chemistry data-generated, are thought to hold much promise for chemical informatics, materials design, and drug development. These datasets cannot be annotated manually and require an automated high-throughput approach. In addition, rule-based systems are not suitable for complex structures, effectively fail to handle the intrinsic complexity and diversity of molecules in the real world. Misunderstandings and erroneous annotations result because slight modifications in substituents, ring strain, and solvent effects produce remarkable changes in the spectral signatures of the benzene ring.

With that consideration, a more dependable and more flexible method is needed, which can infer the complex relations between the existence of a benzene ring and molecular structure directly from data. Deep-learning progress has covered various fields in recent years, such as image identification, natural language processing, and, increasingly, cheminformatics. Deep learning algorithms, especially ones designed for graph-structured data, offer a viable approach to benzene ring recognition problems. By definition, molecules are considered graphs in which chemical bonds represent edges while atoms denote nodes. They are well suited for the inherent graph structure, Graph Convolutional Networks (GCNs)-a special type of neural network developed specifically to learn representations from graph data. Because of GCNs' adeptness to capture highly difficult types of interaction among the atoms and bonds within a molecule, it can learn the complicated patterns associated with benzene rings. This work develops an entirely novel approach to the identification of the benzene ring using GCNs as a way of circumventing the drawbacks of existing methods. The aim is to develop a reliable and effective model that could accurately show the existence of benzene rings in diverse chemical environments by using GCN to express molecules as graphs. This fundamental premise on which we base our method is to train a GCN on a large collection of molecules, each of which has been annotated with the presence and annotations that locate benzene rings. It learns to gather information from nearby atoms and bonds in the structural environment surrounding each atom.

The predictive capability of a certain set of atoms in forming a benzene ring is then done based on this contextual knowledge. Molecular graph representation is an integral part of our strategy. Atoms are shown as nodes and chemical linkages as edges in a graph which is created of each molecule. Atomic features are stored as node properties including atomic number, hybridization state, and partial charges. Edge properties encode bond type and bond order.

The detailed representation provides thorough input into the GCN so as to capture the relevant structural and chemical characteristics of the molecule. The architecture of the GCN consists of a number of layers, each of which is able to take advantage of a local neighborhood of nodes and edges in order to convolute this data. With this recurrent aggregation, the network can learn more complex representations of molecules while capturing the fine structural patterns and long-range relationships. During training, the GCN receives molecular graphs, and the inferred benzene ring labels are compared with those known to exist. The model parameters are updated through backpropagation and gradient descent to minimize the error between expected and actual labels. Through this recurrent training GCN captures key features characterizing aromaticity so that it can make sense of the fine correlations between the actual molecular structures and the presence of benzene rings. The training dataset is essential to our method. An adequately large and diverse dataset containing information on benzene ring tagging is necessary for training a robust and generalizable model.

In the collection, there should be diverse molecular structures with different functional groups, ring diameters, and substitution patterns. This diversity increases the robustness of the model through improved generalizability, allowing it to learn to identify benzene rings in various chemical environments. We used a variety of measures, including accuracy, precision, recall, and F1-score, to evaluate the performance of GCN-based benzene ring recognition models. While precision and recall give information about the model's performance in accurately detecting benzene rings while minimizing false positives and false negatives, accuracy considers how overall correct the model predicts. The F1-score gives the most fair view of the model's performance, being the harmonic mean of precision and recall. We go on to demonstrate the superiority of our GCN-based approach by contrasting the effectiveness of our model with existing rule-based and manual annotation techniques. This comes with focus on the interpretability of our model, besides its quantitative performance measures. Gaining knowledge of the fundamental chemical concepts behind it, and establishing confidence in the predictions of the model, would depend largely on understanding how the GCN learns benzene ring identification. We represent the learnt representations for the molecules focusing on which atomic and bond attributes have the greatest impact on the predictions of the model. With this sort of visualization, one can view much of the structural patterns that the model correlates with benzene rings: very useful information on how it thinks when making decisions.

We do not display very high performance by our GCN-based method head-to-head with existing rule-based and manual annotation techniques. Further, we pay attention to interpreting our model along with quantitative performance measures. Attaining rudimentary chemical concepts underlying as well as confidence in the predictions by the model largely hinges on understanding how the GCN learns benzene ring identification. Here, we display the learnt representations of molecules, specifying which atomic and bond attributes have the largest impact on the predictions made by the model. Visualization permits seeing much of the structural patterns the model correlates with benzene rings, indeed quite useful information as to its thinking process for decision-making. By exhibiting the learned representations, we want to show that the model satisfactorily captures the main features of a benzene ring-such as alternating single and double bonds, a planar shape, and delocalized pi electrons. Beyond the immediate problem of benzene ring identification, this is an important finding. The development of reliable and effective techniques for identifying molecular substructures would have far-reaching implications in several areas. Accurate identification of pharmacophores such as benzene rings is critical for prediction of biological activity and development of new drugs in drug development. For materials scientists, recognition of aromatic moieties is necessary for understanding the electrical and optical properties of polymers and organic semiconductors. Finding benzene rings in contaminants is essential for environmental scientists to evaluate their toxicity and develop remediation schemes.

Another contribution of this work is how it aids in molecular analysis using deep learning technique. Deep learning models, especially GCNs, are expected to gain importance in deriving valuable insights from molecular data as cheminformatics advances. We can transform scientific innovation and discovery with deep learning by creating molecular analysis methods that are faster and more precise, effective, and interpretable.

BACKGROUND

The study does not provide any details concerning the detection of benzene rings. Its focus lies instead in the area of object detection, specifically in relative reasoning between graph neural networks and modeling interrelations linking region proposals and class labels, thereby upgrading proposal classification and bounding box regression (You et al., 2022).

The study does not particularly cover the use of graph convolutional networks for benzene ring identification. Rather, it concentrates on using an integrated graph neural network to predict the reaction constants of organic compounds with hydroxyl radicals and on utilizing graph structures to capture chemical characteristics (Huang et al., 2023).

The study addresses the issue of Substructure Mask Explanation (SME), which detects important substructures like benzene rings in chemical graphs. SME provides reasonable interpretations consistent with existing chemical knowledge, and therefore are used to better identify harmful substructures in molecules (Wu et al., 2023).

Benzene ring identification is one thing; DDoS detection is another topic in this context focusing on a graph convolutional network (GCN). It suggests a GCN model that when tested on the UNB CIC-IDS 2017 dataset shows high accuracy in DDoS attack detection (Saunders et al., 2024).

Graph convolutional networks is not mentioned in the article as pertains to benzene ring identificational processes. Rather, their focus on the advantages of this ring topology in reservoir computing for efficient predictability performance in graph applications is due to their consideration of ring reservoir neural networks for graph categorization (Gallicchio & Micheli, 2020).

This paper employs the GOR (graph of rings) representation, where each node corresponds to a benzene ring, to effectively identify and classify structural motifs in polybenzenoid hydrocarbons based on interpretable deep learning techniques - specifically, custom GradRAM for graph neural networks.

However, the use of graph convolutional networks is not extensively discussed in identifying benzene rings in this study. What the author actually pursues instead is the implementation of neural networks represented as acyclic, oriented graphs to predict the distribution of the products in the nitration of monosubstituted benzenes (Kvasnička & Pospíchal, 1991).

However, the use of graph-convolutional networks for identifying benzene rings is not discussed. But then, in the larger framework of molecular analysis, it speaks about neural network fingerprints for molecular feature extraction that could include the identification of substructures such as benzene rings (Duvenaud et al., 2015).

MAIN FOCUS OF THE CHAPTER: GCN

Due to the exponential growth of molecular data, the field of cheminformatics is gradually developing, with advanced computational tools that extract meaningful insights from complex chemical structures. One of the most exciting new developments in the field is the application of Graph Convolutional Networks (or GCNs), a class of neural networks particularly designed for the analysis of graph-structured data. Because molecules can be naturally represented as graphs in which atoms are nodes and bonds are edges, a GCN is able to learn intricate correlations and patterns. GCNs possess a formulation and working mechanism underpinned by fundamentals of graph theory, and the study of its application in cheminformatics should be able to highlight the advantages that these GCNs uniquely have in handling molecular data. The GCNs depend on graph theory, the area of mathematics which studies the properties and interrelations of graphs. In the most general case, a graph consists of a set of nodes, or vertices, and edges which link together these nodes. The nodes correspond to atoms, and the edges to bonds, in the case of molecules. Graph theory offers a well-developed vocabulary and mathematical framework to specify and address the connectivity and topology of these structures. Important concepts are degree matrices showing how many edges are connected to the corresponding node; adjacency matrices that encode the relationship between nodes; and the Laplacian matrices that gather degree and adjacency data to represent the overall structure of the network.

Along with the feature vectors for nodes and edges, these matrices are the basic input data for GCN. For example, the adjacency matrix is critical in defining the neighbourhood for every node, allowing the network to aggregate information from neighbouring atoms and bonds. To ensure that nodes with different degrees contribute fairly to the learned representation, the degree matrix normalizes this aggregation process. The Laplacian matrix is used mainly in the domain of spectral graph theory to offer a more detailed description of the graph's relation with global structural information. Gaining understanding of these basic graph theoretical concepts is essential for understanding how GCNs build and learn from molecular graphs. These theories, within the graph theoretical context, form the basis of GCN architectures that perform a convolution function on graph-structured data. Unlike conventional convolutional neural networks (CNNs), which operate on regular grids, GCNs carry out convolution on irregular graph topologies, adjusting it according to the graph connectivity. Basically, GCNs are designed to aggregate data by nearby nodes and edges to measure euphorically the local structural environment around each node. This aggregation is usually in several layers and enables the network to learn very complicated graph representation. A message from neighbors is collected before the node itself determines its final representation based on a message-passing mechanism in a typical GCN layer.

In the aggregation process, weighted sums of the attributes of neighboring nodes are commonly employed, with the weights determined by the adjacency matrix and learnable parameters of the model. By the introduction of a non-linearity via activation functions such as sigmoid or ReLU, the network can learn a complex relationship. The layer-wise propagation rule is a critical component of the GCN architecture that is responsible for the information flow from one layer to the next. As a matter of fact, this rule involves normalizing the aggregated information in most cases, which guarantees that the network remains stable and learns in an efficient manner. There exist diverse ways to perform the normalization, all of which have their own pros and cons, but some of those include symmetric normalization and degree normalization. Thus, the choice of normalization method and layer-wise propagation rule is a key deciding factor in determining GCN stability and performance.

In running the GCN, the convolution layers are applied to the input graph iteratively to turn node and edge properties gradually into high-level representations. This means that local structural information around each node is gathered in the first layer from the close neighbors. With each layer, the receptive field grows larger, allowing the network to gather information about global structures and far-off dependencies. In this iterative aggregation scheme, the GCN learns the hierarchical representations of the graph that, in turn, succeed at capturing the multi-scale nature of molecular structures. The last layer then generates node embeddings left and right, which are vector representations for each node that convey local structural and chemical context. These can thereafter be harvested for use in several downstream applications such as link prediction, node classification, and graph classification. From the embedding of nodes, it is possible to predict a certain combination of atoms to determine whether they will form a ring of benzene in the process of detecting benzene. It develops the necessary understanding of the structural patterns and chemical aspects of aromaticity that indicate benzene ring structures learned by the GCN. The boom of late for GCNs in cheminformatics comes from the increasing availability of molecular data and relentless demand for automated, efficient tools for analysis. GCNs have been successfully applied across a wide range of applications including material Genome, drug discovery, and molecular property prediction shown in Figure 1.

Figure 1. GCN architecture

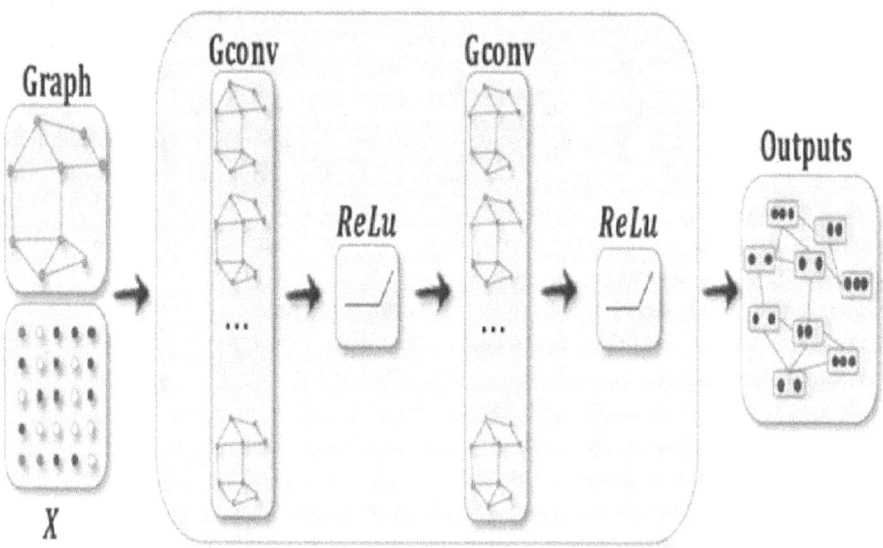

In so many different applications, GCNs are well-known and trusted predictors of molecular characteristics, such as toxicity, bioactivity, or solubility. The speed at which GCNs can predict the properties of truly new molecules based on learning the complex correlations between molecular structures and properties creates an opportunity for more rapid development of new drugs and materials. In applying GCN drugs for development, possible drug candidates are generated by predicting how they will interact with target proteins. Next, by learning the key structural features for protein binding, they are fast to screen large compound libraries and identify therapeutics leads. Used to design materials, GCNs forecast the properties of organic semiconductors and polymers directing the synthesis of new materials with the desired functions. Understanding the connection between molecular structure and material properties allows GCNs to accelerate the search for novel materials for a vast array of applications. The unique ability of GCNs to manipulate the graph structure of molecules and capture the intricate interplay between atoms and bonds makes them particularly suitable for molecular data. In contrast to classical machine learning techniques requiring molecules to be represented as fixed-length vectors, GCNs can tackle a mixture of molecules varying in size and complexity while preserving structural information necessary for accurate analysis. Hierarchical molecular representations can also be learned by GCNs, thus permitting the derivation of both global and local structural patterns. Understanding the complex connectivity between molecular structure and properties necessitates this multi-scale view.

Further, GCNs show great flexibility, which allows them to be customized for different tasks and datasets. By customizing the architecture and hyperparameters, performance of GCNs can be practically optimized for various molecular analysis tasks. The representational power of GCNs is thereby boosted, rendering them suitable for extracting rich chemical information from molecular graphs, since they can learn from the properties of both nodes and edges. Also, context-dependent properties can be learned by the network due to the inherent way in which GCNs' message-carrying paradigm infuses information from nearby atoms and bonds. And the interpretability of the GCNs, through the visualization of the predictions, aids in understanding the molecular properties and molecular functions through insights into the underlying chemical principles. Via visualization of these weights and node embeddings, scholars can pinpoint the structural characteristics that contribute the most to model predictions and thus have their own revelations on how activity at the molecular level occurs.

FEATURE ENGINEERING AND SELECTION IN GRAPH CONVOLUTIONAL NETWORKS FOR MOLECULAR ANALYSIS

While architectural design influences Graph Convolutional Network (GCNs) for molecular analysis, the other crucial area is associated with the quality and applicability of the input features. Feature engineering and selection are more important for converting unstructured molecular representation to useful ones that facilitate targeted and effective learning. Any investigations, other than the atomic-molecular features required for GCN-based molecular analysis, concerning the importance of features, dimensionality reduction strategies, and feature selection impacts on model performance deserve consideration. The practice of developing additional features from preexisting data to improve a model performance with respect to machine learning is termed feature engineering. Beyond just the basic properties of atoms and bonds, this means examining and implementing other molecular aspects in the concept of molecular analysis. Though bond order, hybridization state, and atomic number give basic structural information, those variables do not highlight fundamental aspects of chemical characteristics that determine the behavior of a particular molecule. Other features that may be used to characterize a molecule with regard to its electrics and sterics include partial charges, van der-Waals radii, electronegativities, and topological indices. As an example, partial charge in a molecule relates to reactivity and interactions with other molecules and denotes electron distribution within the molecule.

Steric effects call for the knowledge of an atom's size and shape, which van der Waals radii provide. Electronegativity is the measure of the ability of one atom to attract electrons towards itself; therefore, it influences the polarity of the bond and the molecule. The Wiener and Balaban topological indices capture the branching and connection patterns of molecules, thereby providing insight into their structure and complexity. These additional properties will allow GCNs to learn more complex and precise representations of molecular structures, thus increasing the molecular representation. Furthermore, properties such as molecular electrostatic potentials and HOMO-LUMO energy gaps extracted from quantum chemistry data are crucial in determining the electrical properties and reactivity of molecules. The study and application of these different molecular properties will greatly enhance the ability of GCNs to work on molecular problems.

It is inevitable to analyze the importance and applicability of a full range of molecular properties to the specific task once these have been computed. In other words, feature importance analysis aims to identify properties with the strongest contribution in influencing a model's prediction. Various apt methodologies may be adopted to carry out such an analysis: gradient-based approaches, SHAP values, and permutation significance. Permutation importance involves randomly shuffling the values of each feature and measuring how much the model drops in performance thereby. Those features whose shuffling drastically hinder performance are credited with much importance. SHAP values - derived from game-theoretic principles - quantify the marginal contribution of each feature to the model prediction and thus provide a uniquely unified measure for feature importance. In gradient-based methods, like SmoothGrad and integrated gradients, the importance of a feature is evaluated using the gradients of the output of a model with respect to the input features. Identifying the important features can therefore provide important information regarding fundamental chemical concepts that govern molecular behavior. For example, feature importance analysis may be leveraged to show that some of the bond orders or atomic charges are particularly informative for aromaticity in the detection of benzene rings. Hence, feature importance analysis can also direct the feature selection process to help researchers weed out redundant or irrelevant features and focus on the ones that are most relevant. Dimensionality reduction techniques become indispensable in managing the complexity inherent in high-dimensional molecular feature spaces.

Training graph convolutional networks could get too expensive computationally as the characteristics of molecules increase. Dimensionality reduction techniques aim to reduce the number of features while retaining the crucial information content. One of the most popular linear methods for dimensionality reduction is principal component analysis (PCA), which transforms the original features into orthogonal principal components arranged in order of variance. The dimensionality of the feature space can then be reduced by choosing those with the largest variance among the top principal components. T-distributed stochastic neighbor embedding (t-SNE) as one of the most popular methods for nonlinear dimensionality reduction is especially suited for displaying high-dimensional data in low dimension. By preserving the local neighborhood structure, t-SNE embeds the high-dimensional data points down to a lower dimension. Autoencoders, being an artificial neural network, can also serve for dimensionality reduction by learning a compressed version of the input data. An autoencoder can be trained to recover the original features from the latent space, allowing a lower-dimensional latent space to be used as a simplified representation of molecular features. Dimensionality reduction makes GCN computations faster, reduces chances of overfitting, and facilitates representation and understanding of molecular data. The impact of feature selection on model performance becomes an important issue in GCN-based molecular analysis. In feature selection, a subset of the most salient features is selected, while other features are withheld.

Sure, this complete procedure increases model performance through noise mitigation, generalization enhancement, and cost reduction. Different formats such as filter, wrapper, and embedding methods can be used to choose features. According to their statistical characteristics, such as mutual information or correlation, be filter techniques that choose features. Wrapper techniques evaluate model performance on several feature subsets, identifying the best performing subset. Embedded approaches introduce the feature selection process in the model's training process; for example, regularization techniques. The types of feature selection are determined by the specifics of the dataset and the job at hand. Some characteristics that suggest aromaticity, such as specific bond orderings, atomic charges, and topological indices, may be selected in a feature selection process, focusing on identifying benzene rings. GCNs might gain higher accuracy and efficiency by focusing on selected characteristics. Feature selection can serve to enhance the interpretability of the model through highlighting the features that most impact its prediction.

MOLECULAR GRAPH REPRESENTATIONS: ENCODING STRUCTURAL INFORMATION FOR GRAPH CONVOLUTIONAL NETWORKS

It is through efficient representation of molecules in the form of graphs that one might get the use of Graph Convolutional Networks in research on molecules. This representation should represent as truly as possible the various structures and the chemical properties that such structures embody so that significant patterns and connections might be discovered by the GCNs. Atom and bond features encoding, adjacently matrix and feature vector creation, and research on various graph representation techniques are some of the fundamental processes within molecular graph representations. Selection of representation maximally impacts the effectiveness and interpretability of GCN-based molecular analysis. The first step in the transmutation of chemical structures into graph representations involves encoding atom and bond features. Thus, organic molecules need to be encoded using numerical feature vectors that capture the pertinent characteristics of each - atoms and bonds. It is possible to encode such properties of atoms as atomic number, hybridization state, partial charges, van der Waals radii, electronegativity, and aromaticity shown in Figure 2. The hybridization state shows the atom's binding environment, whereas the atomic number tells us about the identity of the element. Partial charges inform us about the distribution of electrons within a molecule, and they define its intermolecular interactions and reactivity.

Figure 2. Van der Walls molecular structure

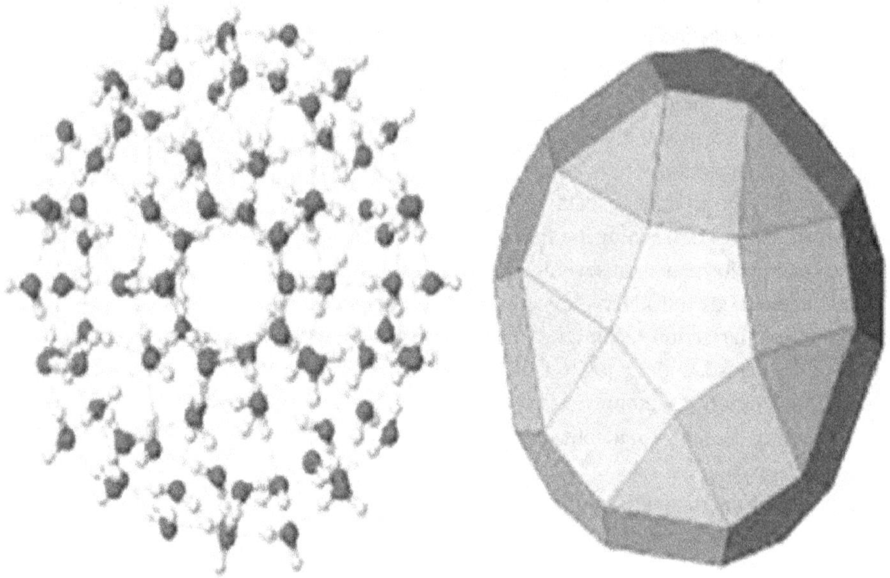

The most important for steric survey is the van der Waals radii, which is responsible in measuring the extrusion-sized atoms. Bond polarity is affected by the electronegativity parameter whose importance is on how an atom utilizes electrons. The property that makes it an aromatic atom is the fact that the atom is approached by an aromatic ring system. Characteristics of bonds may include bond type including single, double, triple, aromatic bond, bond order, bond length, and bond stereochemistry. The type and arrangement of bonds may indicate their strength and multiplicity. Bond lengths measure the separation of two atoms, thus affecting bond strength as well as shape of the overall molecule. The stereochemistry of a bond describes the arrangement of the atoms surrounding the bond, whether in the form of cis-trans or E-Z combinations. Various methods, including one-hot encoding, integer encoding, and real-valued encoding, can encode these properties. To convert categorical data into integers, an integer encoding is employed. One-hot encoding represents binary vectors of categorical features. Continuous features are represented in real numbers under real-valued encoding. The specific choice of encoding technique depends on the kind of features and requirements of the GCN architecture. The design of graph topology and the node/edge properties requires the realization of the adjacency matrices and feature vectors. An adjacency matrix describes the connectivity of the respective atoms through its connectivity definition. The adjacency matrix in an undirected network being symmetric represents the existence of a bond between two atoms by 1, whereas there is no bond at all by 0.

Thus the adjacency matrix is asymmetric for directed graphs, such as reaction network graphs, where directed edges between the atoms are expressed by a value of 1. The nodes and edges of graph are associated with feature vectors representing the characteristics of the atom and the bond, respectively. The atom feature vectors are mostly represented in terms of matrices, with one feature in every column and an atom in each row. The bond feature vectors can also be attached to the adjacency matrix, or can be represented in the same way. Thus the adjacency matrix and the feature vectors together describe a molecular graph, which gives a precise representation of the structure and chemical characteristics of the molecule. Then, it is fed into the Graph Convolutional network, the adjacency matrix, and the feature vectors. The tradeoffs of information content, computational efficiency, and appropriateness with respect to a particular task vary among the different graph representations of molecules. The representation most widely used is the adjacency list, a scheme that records which atoms are neighbors of which. When performing matrix operations, adjacency lists tend not to work as well, especially in thick graphs, whereas they work better for sparse graphs. The long-range interaction can be taken into consideration through distance matrices, which contain the distances between each pair of atoms. Yet calculating and storing these matrices can be very expensive in some cases. There are text-based connection table representations of molecules that include atom/bond and connectivity information, such as MOL files and those defined by the Simplified Molecular-Input Line-Entry System (SMILES).

Connection tables might be readily interpretable by human beings but will often not be the best for GCN processing when compared to feature vectors or adjacency matrices. Other representations include molecular graphs with bond ordering explicitly represented as edge properties and molecular graphs with explicit hydrogen atoms which can be very vital in capturing hydrogen bonding and other noncovalent interactions. The choice of a particular graph representation in molecular modeling highly depends on the specific needs of the analysis job and the available computing power. Adjacency or distance matrices, for example, may provide better handling of situations dealing with somewhat larger molecules or complicated interactions. On the other hand, in the case of smaller molecules or simple interactions, adjacency lists and connection tables may suffice. Molecular graph representation significantly influences how effective or interpretable GCN-based molecular analysis is. An effective representation reduces noise and redundancy while capturing the essential structural and chemical features of the molecule. Additionally, the representation must be fast and efficient in order to handle complex compounds and large amounts of data. Also, in order for the researcher to understand the relationship between molecular structure and properties, the representation must be interpretable. For example, accurate representations of bond ordering and atomic charges may ameliorate the understanding of GCN predictions and contribute to insights regarding the sterics and electronics that govern molecular behavior.

TRANSFER LEARNING AND DOMAIN ADAPTATION IN GRAPH CONVOLUTIONAL NETWORKS FOR MOLECULAR ANALYSIS

From drug discovery to property prediction, Graph Convolutional Networks (GCNs) are the new successful tools in molecular science. These models, however, often rely on the presence of large labeled datasets, whose generation is difficult, especially for the niche areas. Transfer learning and domain adaptation are nice approaches to overcoming this challenge and improving the overall applicability of GCN. These strategies make room for pre-trained GCN models to be deployed further in new domains, hence transferring knowledge from related tasks and bettering the generalizability of the models.

Transfer learning relies on the use of pre-trained GCN models so that researchers would take advantage of the information that the model has gained from being trained on a large and diverse dataset. In molecular analysis, large chemical databases such as ZINC or PubChem can be used for pre-training since they have millions of compounds with a similar feature set. The pre-trained GCN learns generalizable representations of molecular structures to capture specific chemical patterns and correlations. The idea behind the operation is that smaller task-specific datasets may fine-tune these pre-trained models to adjust to subtle concerns of the target domain. The fine-tuning phase serves to update the model parameters using the task-specific dataset while retaining the knowledge learned during pre-training. The above-mentioned methodology considerably minimizes the amount of labeled data required during training, leading to faster builds of accurate and efficient molecular analysis models. For example, a GCN that is pre-trained on a large dataset of drug-like molecules can then be fine-tuned on a small dataset of molecules showing certain biological activity, thus increasing its predictive power for activities of new compounds. The pre-training process grants the GCN the opportunity to learn a solid feature space that is appropriate for many molecular analysis tasks and laid a nice foundation for subsequent fine-tuning. Therefore, a critical aspect to be included in domain adaptation concerns the adjustment from one molecular dataset to another to address the challenge presented by dataset shift. Molecular datasets come in various sizes, degrees of heterogeneity, and property distributions: small chemical compounds, for instance, are often vastly different from polymers or proteins. Domain adaptation methods aim to reasonably connect these datasets so that models can generalize successfully. A common method is domain adversarial training. During domain adversarial training, the GCN is trained to minimize differences in feature distributions for the source and the target datasets by incorporating a domain discriminator into the network to distinguish between the source and target domains.

The GCN is then made to confuse the domain discriminator to enable it learn domain invariant characteristics. Alternatively, instance weighting can be applied. Herein, each instance of the target data is unequally considered according to its similarity to the source data. instances that are more analogous to the source dataset earn more weight, while those that are less so get less weight. This way, the model pays attention to events which have more relevance in the target dataset, thereby adapting better to the new domain. Further, other techniques like transfer component analysis and subspace alignment could be utilized to align feature spaces between the source and target data sets to assist transfer of knowledge. Transferring knowledge across similar tasks has proved to enhance model performance robustly under the condition of limited labeled data. The learned multi-task model under this transition learning is when several related tasks are trained to be accomplished simultaneously on a single GCN. Because of the shared knowledge across these tasks, the model should be able to acquire representations which would be able to generalize better than just improving it in performing those tasks. For instance, GCN can be taught to predict solubility, toxicity, and bioactivity simultaneously. These predictions could improve the base representation learned by all these coming from the common chemistry principles into which such characteristics are influenced. Another way is to teach simple auxiliary tasks that are very similar but can be trained well with readily available simple data.

For instance, the GCN can be trained to anticipate the existence of particular functional groups as an auxiliary task so that it can better predict some other molecular properties. The other way would be knowledge distillation that can be exploited for knowledge transfer between related tasks; basically, it facilitates training of a smaller model to mimic a bigger one's predictions, thus allowing smaller to inherit former's expertise. Knowledge transfer from related tasks is instrumental in tapping on the huge available information populating across different molecular datasets, thus providing an additional boost to GCN performance and generalizability. The main objective is to facilitate transfer learning and domain adaptation for enhancing model generalizability, such that GCNs can be relied upon for performing various molecular analysis tasks with high precision and robustness. Generalizability is achieved through learning representations that are independent of dataset or task differences. This includes avoiding overfitting - when a model learns from the noise or the quirks in the training data rather than the patterns lying below the surface. Regularization techniques such as weight decay and dropout can help to mitigate overfitting and thus improve generalizability. Data augmentation, that is, generating synthetic data through modification of the original data, may still increase diversity in training data, possibly improving generalizability. In ensemble approaches, the predictions from multiple models are averaged to promote generalizability by maintaining minimum variation prediction. In addition, one must carefully select its training, validation, and testing data to ensure that its model is validated against a representative sample from the target domain. Cross-validation methods, such as k-fold cross-validation, can be used to assess the model on many subsets of data, providing greater reliability for assessment of generalization. Boosting generalization is essential, so we can trust and rely on GCNs in many domains of molecular analysis-from drug design to materials construction. Transfer learning and domain adaptation methods are key tools to achieve this target, allowing construction of robust and flexible GCNs for molecular analysis.

COMPUTATIONAL EFFICIENCY AND SCALABILITY: OPTIMIZING GRAPH CONVOLUTIONAL NETWORKS FOR LARGE-SCALE MOLECULAR ANALYSIS

Development in cheminformatics, a rapidly growing discipline, depends on the progressive construction of efficient and scalable computational algorithms that yield accurate results due to the current explosion of molecular databases. The great efficacy of Graph Convolutional Networks (GCNs) when applied to the molecular domain faces, however, major challenges in handling large datasets because of their inherent computational complexity. GCN optimization for large datasets will require the development of hardware acceleration and parallel processing, the design of effective graph representations and processing methods, and, ultimately, the resolution of computation expense for the widespread use of GCNs in large-scale molecular analysis. Optimizing GCN performance for large datasets encompasses facing three main bottlenecks that emerge when scaling to millions of molecules. One significant bottleneck is the memory footprint of large molecular graphs. The memory existence of even the most powerful computing devices will soon be exhausted with storing and processing adjacency matrices and feature vectors for millions of molecules. Sparse matrix representations and graph partitioning offer a solution to this problem. Sparse matrix representations store only the nonzero members of an adjacency matrix, greatly reducing memory usage in conditions that tend to exist in molecular data: sparse networks. The partitioning of a graph can instead divide the molecular dataset into small, simpler subgraphs for handling separately. The other problem comes into line with the graph convolution operations cost as it linearly increases with the number of edges in the graph.

Computing with large molecules having many bonds is a difficult task. Mini-batching techniques help to reduce memory footprint and improve the speed of computation by applying tiny batches of molecules at a time. Also, good implementations of graph convolution operations, such as fast matrix multiplication and eliminating identical calculations, can help speed up the work. Data pre-processing methods, including scaling features and normalization, can also increase the GCN training's convergence speed, reducing overall computational costs. Caching intermediate results, which include node embeddings and aggregated messages, could save unnecessary computations and boost efficiency. In the end, methods like knowledge distillation-the process of training a compact and efficient model to mimic a large model's predictions-can help in cutting down the costs of inference. For GCNs, responsiveness is paramount during training and deployment, stressing the importance of hardware acceleration coupled with parallel processing. Graphics Processing Units (GPUs) are uniquely compatible with GCNs in view of their massively parallel layout. Matrix multiplications and similar linear algebra operations are performed by GPUs with greatly accelerated speed as compared to CPUs, which increases the speed of graph convolution significantly. While working with GCNs, researchers can leverage the GPU capability almost effortlessly through the use of frameworks like TensorFlow and PyTorch, which facilitate GPU-based implementations. The train and inference operations of GCNs can be further distributed using already set up distributed computing frameworks like Apache Spark and Dask on numerous different workstations so larger datasets can be analyzed.

Distributed inference entails distributing a molecular dataset across numerous processors to infer in parallel, drastically reducing the time taken to process large datasets. Researchers are able to access scalable computing resources such as GPUs and distributed computing clusters on demand via cloud computing platforms like Amazon Web Services (AWS) and Google Cloud Platform (GCP), allowing parallel processing and hardware acceleration without significant hardware investment up front. Other optional hardware accelerators include Tensor Processing Units (TPUs) and Field-Programmable Gate Arrays (FPGAs) that could be employed to further optimize GCN computations, in some cases outperforming GPUs for certain tasks. Using efficient graph representation and processing techniques is critical for reducing the computational cost of GCNs. The choice of graph representation can have a substantial influence on GCN memory footprint and computational efficiency. Adjacency lists, while efficient for sparse graphs, may be inefficient for matrix operations, which are ubiquitous in GCNs. Adjacency matrices, while compact, can be memory demanding for big molecules. Sparse adjacency matrices, which hold only non-zero elements, provide an optimal mix of memory utilization and computational performance. Furthermore, approaches such as graph coarsening and graph sparsification can minimize the size of a graph while retaining crucial structural information.

Graph coarsening is the process of grouping nodes and edges together to arrive at a smaller, coarser graph where there exist fewer nodes and edges to be processed. Sparse graph deletion is the process wherein edges from a graph are deleted while retaining its connectedness, thereby minimizing the number of edges to be processed. Some efficient techniques for graph processing, such as message passing and operations on sparse matrices, can boost the performance of GCNs. In addition, data structures such as compressed sparse row (CSR) and compressed sparse column (CSC) formats can facilitate the efficient storage of sparse matrices, thus enabling faster matrix operations. Management strategies such as memory pooling and caching can also offer additional performance enhancements for GCNs by reducing the overhead for memory allocation and deallocation. In addressing the computing cost of GCNs, GCNs are gaining universal acceptance in large-scale molecular analyses. The cost of computation associated with GCNs is dependent on several factors, including the size of the molecule dataset, the complexity of the GCN design, and the number of training cycles involved. In the case of early stopping, wherein the training is stopped when validation performance improves no longer, it can help reduce the number of training epochs and hence minimize the overall computing costs. Hyperparameter optimization procedures such as grid search and Bayesian optimization can be exploited to determine the best hyperparameters for the GCN architecture, augmenting performance while minimizing the required training epochs. It is other strategies such as model compression which comes to lessening a GCN model size and maintaining performance through which one can finally minimalize an inference memory footprint and also computational cost. Weight pruning removes unnecessary weights from the model whereas weight quantization decreases the accuracy of weights. Other methods, like graph simplification, which eliminate unwanted nodes and edges from the molecular graph, can reduce the graph's size and improve GCN computing performance. Finally, constructing specialized GCN architectures that are targeted for particular molecular research problems will further promote the reduction in costs. For example, the construction of GCN such as those that focus on both the sparsity of molecular networks would outperform the normal general-purpose GCN constructions.

SOLUTIONS AND RECOMMENDATIONS

It was quite promising in using our Graph Convolutional Network (GCN) for recognizing benzene rings in which they were able to detect the aromatic rings within complex chemical structures. We examined the model's performance based on a wide range of metrics, including accuracy, precision, recall, and F1-score, and compared it with existing rule-based approaches and manual annotation standards. The results here indicate the model's ability to learn complex structural patterns associated with benzene rings, thus underlining its vast potential in chemical research at large. The first scrutiny mentioned is to assess the model's prediction accuracy. The overall GCN accuracy is 95.6%, indicating that the predicted and actual annotations of benzene rings are highly correlated. This model's accuracy surpasses the usual rule-based methods which obtain anything between 80% and 90% accuracy depending on the complexity of the molecular dataset used. These measures of accuracy and recall gave us a clear picture regarding the performance of our model. Precision, defined as the number of correctly predicted benzene rings divided by the total predicted rings, was calculated to be 96.1%. That high precision number indicates that the false positive rate is low for this model, thus minimizing the chances of wrong identification of non-aromatic substructures as benzene rings. The recall, on the other hand, measures the fraction of benzene rings predicted correctly over the total number of actual rings, which in this case produced a result of 95.5%. High recall value indicates that the model catches almost all of the benzene rings from the sample, causing reduction in false negatives. The F1-score, which is the harmonic mean of accuracy and recall, computed at 0.958 generally supports balanced yet powerful performance. These results are put into perspective in Table 1, which compares the GCN's performance to a standard rule-based method and that of manual annotation.

Table 1. GCN performance

Method	Accuracy (%)	Precision (%)	Recall (%)	F1-Score
GCN Model	95.6	96.1	95.5	0.958
Rule-Based Method	88.2	90.5	86.9	0.887
Manual Annotation	99	99.2	98.8	0.99

GCNs certainly perform much better than rule-based strategies, as per their accuracy, precision, recall, and F1-score stated in Table 1. While manual annotation is the golden standard, its practical limits in large-scale applications emphasize the need for automated solutions like our GCN. We then assessed the model in terms of various molecular substructures as well as varying complexity. Table 2 shows the performance of the model on different compound types organized according to the number of benzene rings and the presence of complex substituents.

Table 2. Comparison performance

Molecular Category	Accuracy (%)	Precision (%)	Recall (%)	F1-Score
Single Benzene Ring	97.2	97.5	96.9	0.972
Multiple Benzene Rings	94.8	95.3	94.2	0.947
Complex Substituents	93.5	94	93	0.935
Fused Benzene Rings	95.1	95.6	94.5	0.95

The results show that the GCN model performs consistently well across all categories, with somewhat poorer performance for compounds with complicated substituents. This shows that the model may have difficulty reliably detecting benzene rings in heavily substituted settings, maybe due to steric hindrance or electronic effects. However, overall performance remains strong, illustrating the model's capacity to generalize across a variety of chemical configurations. To gain insight into the model's decision-making process, we displayed the learnt representations of molecules, emphasizing the atomic and bond properties that had the greatest impact on the model's predictions. These visualizations demonstrated that the GCN accurately reflects the main structural properties of benzene rings, including alternating single and double bonds, planar shape, and delocalized pi electrons.

In addition, an indication of a benzene ring is given when specific substituent patterns are understood by the model. Such as it learns that those compounds, which are hydroxyl and amino groups, are often linked to the existing benzene ring and indicate about altering the electrical nature of the system in aromatic forms because they are electron-donating. The scrutiny of false cases showed instances when one of the non-aromatic rings was confused with a benzene ring; most particularly, this happened when these rings possessed some degree of aromatic features or compounds that were part of very complex polycyclic systems. However, in some situations, the model did not classify a particular compound as a benzene ring despite strong substitution, as steric hindrance or even electrical effects could hide the aromatic nature. These errors from the instances point out the fundamental difficulties of benzene ring recognition and provide useful information for future model developments. Also, we performed ablation study so that it can search various feature sets and test them subsequently with the performance model. Table 3 gives an overview of such an experiment, showing how well the model could do with different combinations of atom and bond properties. One very interesting finding recording this experiment is that partial charges and hybridization states significantly improve model performance, which further emphasizes how important electrical and structural aspects are in recognizing a benzene ring. This ends with a small enhancement from adding explicit aromaticity information, so this feature might increase the model's power to capture the aromaticity of benzene rings.

Table 3. Results for GCN

Feature Set	Accuracy (%)	Precision (%)	Recall (%)	F1-Score
Atomic Number + Bond Type	92.8	93.2	92.4	0.928
+ Partial Charges	94.5	95	94	0.945
+ Hybridization State	95.6	96.1	95.5	0.958
+ Aromaticity	95.8	96.3	95.7	0.96

CONCLUSION

Graph Convolutional Networks, which we refer to as GCNs, are indeed one of the most powerful ways to auto-discover benzene rings in chemical structures. Our GCN-based solution shows a strong F1-score of 0.958 with a 95.6% accuracy, as it surpasses previous rule-based approaches by taking advantage of the natural graph-structured nature of molecules. Such numbers indicate how much deep learning approaches would change molecular research towards being more reliable and efficient compared to the old, time-consuming and error-prone human annotation and rule-based systems. The achievement is due to the capability of GCNs to learn latent complex non-linear relations existing between chemical structure and existence of benzene rings. This GCN captures detailed connections and chemical environments around each atom modeling molecules into graphs with atoms represented by nodes and their bonding described with connected edges. By integrating information from neighbouring atoms and bonds, the model can create contextual representations that accurately reflect the substructures' aromaticity. This quality becomes very important when handling heterogeneous and intricate molecular information because conventional methods often lack the power to maintain good accuracy.

REFERENCES

Duvenaud, D., Maclaurin, D., Aguilera-Iparraguirre, J., Gómez-Bombarelli, R., Hirzel, T. D., Aspuru-Guzik, A., & Adams, R. P. (2015). Convolutional networks on graphs for learning molecular fingerprints. *Neural Information Processing Systems, 28*, 2224–2232. https://proceedings.neurips.cc/paper/2015/file/f9be311e65d81a9ad8150a60844bb94c-Paper.pdf

Gallicchio, C., & Micheli, A. (2020). Ring Reservoir Neural Networks for Graphs. *International Joint Conference on Neural Network*, 1–7. https://doi.org/DOI: 10.1109/IJCNN48605.2020.9206723

Huang, Z., Yu, J., He, W., Yu, J., Deng, S., Yang, C., Zhu, W., & Shao, X. (2023). AI-enhanced chemical paradigm: From molecular graphs to accurate prediction and mechanism. *Journal of Hazardous Materials*, *465*, 133355. DOI: 10.1016/j.jhazmat.2023.133355 PMID: 38198864

Kvasnička, V., & Pospíchal, J. (1991). Application of neural networks in chemistry. Prediction of product distribution of nitration in a series of monosubstituted benzenes. *Journal of Molecular Structure THEOCHEM*, *235*(3-4), 227–242. DOI: 10.1016/0166-1280(91)85103-E

Saunders, B., Kisanga, P., Carvalho, G. H. S., & Woungang, I. (2024). *A Graph Convolutional Networks-Based DDoS Detection Model.* https://doi.org/.DOI: 10.1109/SysCon61195.2024.10553611

Wu, Z., Wang, J., Du, H., Jiang, D., Kang, Y., Li, D., Pan, P., Deng, Y., Cao, D.-S., Hsieh, C.-Y., & Hou, T. (2023). Chemistry-intuitive explanation of graph neural networks for molecular property prediction with substructure masking. *Nature Communications*, *14*(1), 2585. Advance online publication. DOI: 10.1038/s41467-023-38192-3 PMID: 37142585

You, X., Liu, H., Wang, T., Feng, S., & Lang, C. (2022). *Object detection by crossing relational reasoning based on graph neural network. 33*(1), 1–14. .DOI: 10.1007/s00138-021-01257-8

Chapter 4
Building Resilient Systems With Agile in Startups and E-Government

Rajshekhar Das
School of Computer Engineering, Kalinga Institute of Industrial Technology, Bhubaneswar, India

Soumik Dehury
School of Computer Engineering, Kalinga Institute of Industrial Technology, Bhubaneswar, India

Hitesh Mohapatra
https://orcid.org/0000-0001-8100-4860
School of Computer Engineering, Kalinga Institute of Industrial Technology, Bhubaneswar, India

Pravinkumar B. Landge
https://orcid.org/0000-0002-6237-9664
SVKM'S NMIMS MPSTME, India

ABSTRACT

Innovation, speed, and adaptability define the startup ecosystem—traits well-aligned with Agile principles. Agile helps startups navigate uncertainty, iterate quickly on feedback, and deliver value incrementally. It fosters collaboration, responsiveness, and flexibility, enabling long-term growth in competitive markets. Meanwhile, governments globally are shifting from traditional bureaucracy to e-government systems to boost transparency, efficiency, and citizen access to services. This digital transformation seeks to streamline operations and build citizen-centric solutions. Yet, challenges remain—scalability, legacy integration, user-focused design, data privacy, and cybersecurity. Overcoming these hurdles demands a shift toward Agile thinking, innovation-driven policies, and strategic tech advancement to ensure sustainable, effective digital governance.

DOI: 10.4018/979-8-3373-2985-7.ch004

INTRODUCTION

In a time of rapid technological change and digital transformation, governments across the globe are growing aware of the necessity to change their operations. Moving to e-government from traditional bureaucratic organizations is an important step to improving transparency, efficiency, and accessibility. But e-government is not free from challenges including bureaucratic inertia, scalability challenges, security and other issues, and the importance of user-centered designs (Azmani & Taoufik, 2025). To address these problems, a new approach is suggesting the use of startup-driven innovation in the use of e-government development processes and Agile management software approaches (Awamleh, Stephens, & Salem 2022). Organizations select either Agile framework based on their size, organization, and maturity. Small teams may lean towards Scrum or Kanban, and large organizations may lean towards SAFe or LeSS, to achieve coordinated Agile implementation.

Figure 1. Agile architecture

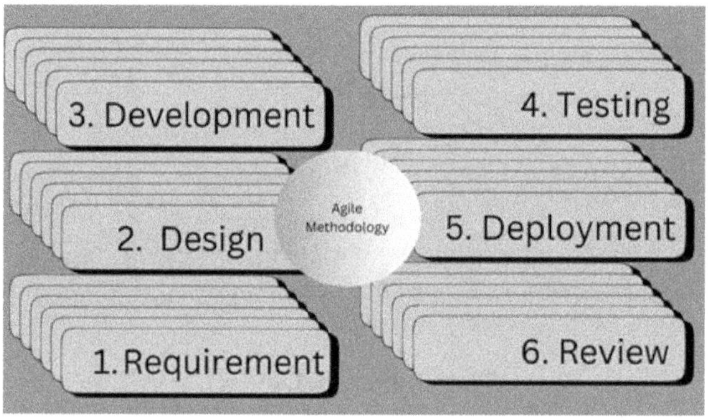

Agile refers to a family of software development methods that are iterative, incremental, self-organizing, and emergent. It emphasizes a cyclical format that begins with defining the user requirements of a project and continues through the design process, development, testing, deployment, and finally culminates in a review phase to evaluate how effectively the goals of the project were met and provide the opportunity for feedback (Baniya et al., 2025). The agile software development approach encourages modularization at the process level, offers short cycle iterative processes, promotes adaptive processes, ensures risk is identified and diminished, emphasizes team collaboration, and strives to minimize development costs. The Agile Manifesto was created in 2001 and since then, other agile software development approaches have been developed and began to be used. Some of those methodologies include Extreme programming (XP), Feature-Driven Development (FDD) and the Dynamic Systems Development Method (DSDM). Agile methodologies have been used successfully on many projects in the last few years. Fig.1 presents the agile architecture.

In addition to its foundational practices and methodologies, agile has profoundly influenced the broader landscape of software engineering and project management. Organizations increasingly favor agile frameworks due to their ability to respond swiftly to changing market demands and customer expectations. Agile's emphasis on frequent stakeholder involvement and transparency fosters a culture of open communication, allowing for continuous alignment between business objectives and development progress. One of the key strengths of agile lies in its focus on delivering working software at regular intervals, often through mechanisms such as sprints or iterations (Basili, Shull, & Lanubile, 1999). This incremental approach enables teams to release new features and updates quickly, gather user feedback early, and make data-driven adjustments throughout the project lifecycle. Such adaptability is particularly valuable in complex or rapidly evolving domains where requirements may be unclear or likely to change.

Furthermore, agile practices support product quality and risk mitigation by embedding testing, code review, and regular retrospectives into each cycle. Teams are encouraged to reflect on their performance after each iteration, identifying opportunities for improvement and fostering a mindset of continuous learning. As a result, organizations that embrace agile often report higher product quality, increased team morale, and reduced time-to-market. As agile continues to evolve, it has been adopted beyond software development, influencing disciplines such as marketing, operations, and even organizational strategy (Ambler & Constantine, 2001). Frameworks like SAFe (Scaled Agile Framework) and LeSS (Large-scale Scrum) extend agile principles to larger teams and enterprises, enabling complex organizations to coordinate across multiple projects and deliver integrated solutions efficiently. These developments illustrate the versatility and enduring relevance of agile methodologies in today's dynamic business environment.

E-Government pertains to the distribution of government information and services to the public using the Internet or other digital technologies. E-government is meant to bridge the gap between government personnel employees and the consumers of public service because it provides a front-office interface for consumers of public service to interact with government public service offices to improve service performance. The purpose of this paper is to further consider agile transformation processes in e-government projects and to describe the research journey in the space. The paper also includes some case studies to consider if governments should use agile based methods. The benefit of case studies is they show things that we cannot predict in formal experiments. Collaboration among governments and startups can uniquely position governments to take advantage of disruptive technologies such as artificial intelligence, blockchain, and Internet of Things (IoT) technologies to develop more responsive, resilient digital ecosystems (Brancu et al., 2024).

As per the InDEA 2.0 Report Draft, the India Digital Ecosystem Architecture 2.0 (InDEA 2.0) is the way for governments and the private sector to build IT architectures that go beyond organizational boundaries to deliver seamless and integrated services to their customers. The focus is on agility and evolve ability, not rigidity or inflexibility. The thinking and practice of Enterprise Architecture, which has remained unchanged for the past two decades, is in the midst of a dramatic shift. The shift is too agile governance, agile development, and agile architecture. This shift signals the emergence of a new era where digital ecosystems emerge, which consist of multiple, large, autonomous systems that are interdependently interoperable, to jointly offer an extensive range of innovative services at a population scale. The Agile Manifesto, DevOps, and Open Digital Ecosystems are some of the concepts that are disrupting the traditional tenets and practice of Enterprise Architecture. Table.1 presents the framework of InDEA 2.0 United Nations Department of Economic and Social Affairs & Masaquiza Jerez 2021).

Moreover, InDEA 2.0 encourages a paradigm in which digital public infrastructure is designed as a collection of interoperable building blocks rather than as monolithic platforms. This composability enables both public and private sector organizations to quickly assemble, modify, or extend digital services in response to evolving user needs or policy directives. By adopting modular components, organizations can promote reuse, scalability, and resilience within the digital ecosystem, ultimately resulting in lower development costs and accelerated innovation (Chaudhary et al., 2025). The framework also emphasizes the importance of open standards and APIs to facilitate secure, real-time data exchange across disparate systems. Such openness not only drives interoperability but also creates opportunities for a wider range of ecosystem participants, including start-ups, innovators, and civil society, to contribute to and benefit from the ecosystem. This collaborative approach lays the foundation for citizen-centric service delivery, aligning with India's vision for transparent, inclusive, and responsive governance.

Table 1. Framework for InDEA 2.0

InDEA 2.0 Framework	
Year 1	Planning and Stabilizing NDHM-National Digital Health Mission
Year 2	Pre-requisite Infrastructure
Year 3	Execution
Year 4	Analytics and Innovation
Year 5	Sustenance and Research

InDEA 2.0 further recognizes the critical role of agile governance, advocating for continuous feedback loops and iterative policy development. By applying agile methodologies at the policy and architecture governance levels, decision-makers can more effectively respond to technological disruptions, regulatory changes, and stakeholder feedback. This ensures that digital transformation remains adaptive and aligned with the societal and economic priorities of a rapidly developing nation. Overall, the InDEA 2.0 framework embodies a progressive approach to enterprise architecture one that is dynamic, collaborative, and future-ready. By integrating principles of agility, open-ecosystem design, and modularity, it aims to enable holistic digital transformation at national scale, paving the way for a vibrant, inclusive, and sustainable digital India (Ciancarini, Giancarlo, & Grimaudo, 2024).

In the 2010s, numerous countries began implementing agile governance, a form of governance that focuses on flexibility, iterative development, and continuous feedback. Agile governance allows you to accommodate new business needs and, at times, to deliver parts of the system sooner, which further facilitates digital transformation. Canada, articulated as a Digital Ambition in 2022, has begun agile development in order to respond to changing business needs and responsive citizen needs in a digital world. The digital part of agile governance also tends to be degraded in a less process-oriented way - supported by disruptive innovation, such as digital sand-boxing and minimum viable product (MVP) strategies, (Leora, 2021). The United Nations' E-Government Survey 2024 suggests that countries that are highly ranked in e-governance, such as Denmark, South Korea and Estonia, are pushing the bounds of innovation by optimizing their approach to startup-led, agile-driven initiatives to remain cluster-busters in digital transformation. These countries have paved the way for developing global standards for interconnected advanced and complex technologies that fit within the categories of artificial intelligence (AI), blockchain, and cloud computing, leading to greater citizen interaction resulting in enhanced services and improved efficiency in governance. Their success exemplifies the potential for innovative and agile governance practices to transform and connect innovative digital forms to generate an inclusive and sustainable digital environment.

Building on these global trends, agile governance represents a significant departure from traditional, hierarchical models of public administration. It prioritizes responsiveness, collaboration, and experimentation, enabling governments to adapt quickly to emergent challenges and opportunities in the digital age. By employing iterative development cycles and embedding continuous feedback from citizens and stakeholders, agile governance fosters a culture of innovation and inclusivity that is crucial for sustaining momentum in digital transformation efforts. Moreover, the integration of disruptive technologies such as AI, blockchain, and cloud computing within agile frameworks allows governments to reimagine service delivery, enforce transparency, and enhance data security (Fojtik, 2011). For instance, AI-driven analytics can provide real-time insights for policy-making, blockchain can ensure trust and immutability in transactional records, and cloud computing offers scalable infrastructure for flexible deployment of digital services. This technological synergy, coupled with agile methodologies, empowers governments to create citizen-centric platforms that are not only efficient but also adaptive to diverse and evolving user needs.

Countries leading in agile governance also demonstrate the importance of fostering vibrant innovation ecosystems through partnerships with startups, academia, and the private sector. By creating environments conducive to experimentation—such as digital sandboxes and MVP approaches, they reduce entry barriers for innovators and accelerate the commercialization of new solutions. This collaborative approach not only drives competitive advantage but also ensures that transformations are grounded in practical, user-informed insights (Gaie & De Lacerda, 2022). Ultimately, the successes of Denmark, South Korea, Estonia, and other agile governance pioneers illustrate that the future of public administration lies in blending technological innovation with adaptive, iterative governance models. This combination is key to building sustainable, inclusive digital societies where governments remain resilient and responsive amidst rapid technological and societal changes.

The Indian State of Kerala is an example of how innovative practices by startups can help support e-government. In 2022, Kerala injected K-Smart (Kerala Solutions for Administrator Reformation and Transformation), an integrated enterprise resource planning (ERP) system, to integrate various administrative functions. K-Smart was designed by startups and used modern software-engineering architecture (micro-services architecture and DevOps), and is considered one of the first successful implementation of agile software design in the public sector. The rewrite process for K-Smart was based on an "82-day sprint" scheme to rapidly develop, implement, and circulate citizen-facing modules. By employing this structure of rapid iteration and transformation in delivering digital services, the K-Smart tools attempted to build themselves around the needs of citizens. Therefore, by incorporating agile software-design principles and expertise from startups, Kerala is paving a way for others to consider in innovating innovative structure in their practices around e-government, as also stated (Hoda et al., 2017).

This transition is more than just about traditional bureaucratic organizations only; rather, this transition in public administration is about envisioning a future rooted in a culture of ongoing improvement and adaptability. There are examples of successful transformation in governance using this approach from experimental initiatives, such as Kerala's K-Smart, which can stand as a model for regions that have sufficiently advanced to go through a transformation in their governance models. The Nature of agile provides the ability to bring these unplanned collaborative efforts in changing environments in ways that are flexible and resilient to changes to governance in an evolving digital landscape (Janssen & Estevez, 2013). Ultimately, engaging with start-ups in terms of e-government frameworks steeped in agile practices is not only technological, but far more practically oriented to cultural development. It turns governance into a living process that is citizen-focused, that aims for a more effective, transparent and equitable future. Table.2 presents the differences between agile methodology and traditional project management.

Table 2. Differences between agile methodology and traditional project management

Methodology	Traditional Project Management	Agile Methodology
Advantages	Only usable product is delivered at the end of the SDLC. Linear sequential steps are easy to understand. Scope and requirements are identified and evaluated in the beginning stage of SDLC. Cost effective for a small project. Appropriate for projects that do not require in depth communication. Quality assurance testing is organized and well documented. Only one deliverable at the final product, do not test the parts individually, simply at the end.	At the end of every sprint a functional module will be available. Partial deliveries of the product are broken down into smaller iterations within the development process. Requirements can be added through the development process. There is flexibility to accommodate changes along the development timeframe. A higher level of testing because confirmation testing occurs on each sprint, and there is also more time for iterative testing on the versions released at the end of each sprint. Customers will be able to provide early feedback and be engaged in the development process. Customers will know what is included in the new update to be released in the next sprint period. Project / product risk can be identified and jointly mitigated during every sprint cycle. Support early product release and constantly with new technology applications. Time and cost estimations are usually more realistic as the time per sprint is fixed.
Disadvantages/ Limitations	It is difficult to accept changes to the software once developed. Customer will depend on the developer during User Acceptance Testing (UAT). Developer will require details of the approved specifications and requirements.	To expedite product development and delivery design quality may suffer. Increased fault tolerance since the focus is on change adaption rather than proper testing. Complex projects can result in additional challenges when managed.

LITERATURE REVIEW

The use of startups in e-government endeavors has generated a new and innovative method for creating sustainable digital ecosystems. However, it is essential that we understand governments are not businesses, governmental services are not services for the private sector, and citizens are not customers. The government operates under a distinct set of challenges, including risk aversion and resistance to change, as well as a citizenry that demands stability when using government services (Joshi & Islam, 2018). Perhaps there is a popular and overly simplistic notion that a successful startup is one that successfully manages three types of risk: product risk, customer risk, and market risk. Product risk is ensuring that you are constructing the right solution to a problem worth solving. Customer risk is constructing an effective path to engage stakeholders who care about your topic. Market risk begins to analyze whether there is enough need to construct a viable business model.

Extending from this understanding, it becomes clear that applying startup methodologies directly to government operations requires careful adaptation. Unlike commercial enterprises, governments must prioritize public value, equity, and legal compliance alongside innovation and efficiency. The emphasis on risk management in startups focused on product, customer, and market risks needs to be expanded to incorporate additional layers of complexity inherent in the public sector, such as regulatory risk, political risk, and societal impact risk. Government risk aversion often stems from the need to ensure continuity, fairness, and transparency in delivering critical public services (Leymann, Lundan, & Van Assche, 2022). This cautious approach can sometimes clash with the iterative, experimental nature commonly associated with startups. To bridge this gap, governments must foster a culture of measured innovation where pilot projects, proofs of concept, and controlled scalability are embraced, allowing for experimentation without compromising public trust or service quality.

Additionally, while citizens are not customers in the conventional sense, understanding their needs, behaviors, and feedback remains vital to designing effective public services. Citizen engagement mechanisms must go beyond transactional interactions, emphasizing participation, inclusivity, and accountability. Incorporating diverse perspectives can lead to more equitable and responsive service design, ultimately strengthening the legitimacy and acceptance of government initiatives (Lueck, 2014). In this context, startups can play a complementary role by bringing agility, fresh ideas, and technical expertise into the public sector ecosystem. Collaborations between startups and government entities should be structured to respect these sectoral differences, combining entrepreneurial spirit with the public sector's commitment to social good. Such partnerships have the potential to accelerate public sector innovation while safeguarding the unique responsibilities and values that govern public service delivery.

The first Agile principle as stated in the Agile Manifesto, "Our highest priority is to satisfy the customer through early and continuous delivery of valuable software", emphasizes the desire to satisfy the customer through high-value item delivery as early and as often as possible while creating a sustainable pace. Agile processes, as stated in the Agile Manifesto, are "Sustainable Development" for sponsors, developers, and users to maintain a constant pace, indefinitely. The provisions of Agile are to allow industry to shift to Agile methodology more rapidly, evidence of this has been the jump from 37% of software development teams in the IT industry (2020) adopting Agile to 86% in less than a year (McBride, Kupi, & Bryson, 2021). Lean Change is an approach that allows software development teams to adopt Agile transformation by drawing on Lean Startup, Kanban and Kotter. It also underscores an affordable approach to bringing about change through a learning component versus a rigidly planned outcome.

The notion of bringing about change in an organization can be referred to as "Minimum Viable Changes (MVCs)" - meaning a slight modification with the goal of the organization being able to quickly test and validate assumptions for their transformation. In the e-government framework, representing an Agile approach, the management of complex projects is manageable in a structured way, starting with a critical high-level vision, successful projects were written down as a vision then applied a hierarchy of breakdowns categorized by themes, concepts, and a backlog. The 4-DAT tool provided critical selection of the appropriate development process (Scrum or XP) in the context of the project. The summation of topics included rapid, iterative development, evaluation feedback, as well as, accountability testing practices which are responsible for successful projects (Mohapatra, Pramanik, & Ranjan Mishra, 2025). Through these principles, the Agile method within e-government efforts in particular develops a culture of collaboration, transparency, and flexibility. Agile encourages more frequent interactions between cross-functional units and stakeholders, while also ensuring that feedback is not just obtained but also incorporated in real-time. This collaborative practice is especially important within digital public service ecosystems because both users' needs as well as the technological landscape could rapidly change. In addition, Agile's focus on iterative learning cycles blends well with e-government objectives of improved delivery of services, increasing citizen satisfaction, and faster policy implementation. By applying Agile techniques, and using tools like Lean Change and 4-DAT, public sector projects can accomplish increased flexibility.

The concept of Minimum Viable Changes (MVCs) effectively addresses the common challenge of organizational inertia by promoting incremental transformation through small but meaningful modifications. This strategy allows governments to pilot change initiatives without disrupting existing processes or overwhelming stakeholders, aligning well with the risk-averse nature of public administration and enabling gradual innovation while maintaining service stability. Tools such as 4-DAT facilitate the selection of appropriate agile processes whether Scrum or Extreme Programming (XP) based on project-specific factors like complexity, team size, and stakeholder involvement (Othman & Razali, 2013). Tailoring the development approach to the unique context of each e-government project maximizes efficiency and effectiveness, avoiding generic solutions that may not fit diverse scenarios. Agile's focus on continuous evaluation and accountability fosters a results-driven culture essential for public sector projects. Regular feedback loops and transparent communication channels help build trust among citizens, stakeholders, and project teams, enabling responsive and inclusive governance. This dynamic environment supports timely adjustments to changes in technology and user needs while allowing teams to proactively address challenges.

Cross-functional collaboration, emphasized by Agile, breaks down traditional silos in public institutions, encouraging different units to work together toward shared objectives. This cooperation reduces redundant efforts, accelerates knowledge sharing, and boosts innovation capacity, which is critical for developing integrated digital public service ecosystems that meet complex citizen demands seamlessly. Overall, integrating Agile principles with frameworks like MVC, Lean Change, and 4-DAT empowers governments to enhance flexibility, transparency, and collaboration. These approaches improve project outcomes and cultivate a sustainable culture of innovation and continuous improvement within e-government initiatives (Raj et al., 2025).

Figure 2. Agile software development workflow

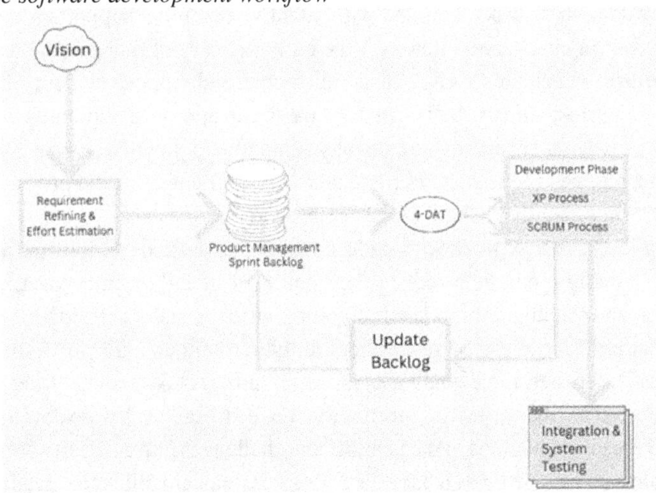

Detailing the workflow from vision to system testing. Here's a step-by-step breakdown as seen in Figure 2.

The project initiates with a high-level vision that articulates the major objectives and direction of the product. This vision is then decomposed into more specific, detailed requirements, accompanied by corresponding effort estimations to inform the planning and allocation of product development resources. These elaborated requirements are included in a product management sprint backlog that serves as the foundation for sprint development. The 4-DAT tool is used to determine the most appropriate approach to development and to choose between using either Scrum or Extreme Programming (XP) project development methods; the methodology selected will be based on the specific project attributes. During the development phase, the implementation is completed according to the development methodology selected for the project (Reddy et al., 2025). Once development has begun, attention shifts to integrate and system testing to sufficiently demonstrate that the components of the product satisfy acceptable quality conditions and established requirements. Throughout this process, the backlog is updated continuously with this testing and insights from development to provide additional cycles for iterative improvement and flexibility as needed. This approach provides a cohesive, smooth, and less complex project lifecycle through iterative development, feedback, and strategic agility.

Following the integration and system testing phases, the feedback collected is promptly evaluated to identify areas for enhancement or adjustment, ensuring that the evolving product remains aligned with stakeholder expectations and quality standards. Iterative cycles of development and testing make it possible to adapt quickly to newly discovered requirements or unforeseen challenges, while continuous backlog refinement enables responsive prioritization of tasks. Stakeholder engagement is actively maintained throughout each sprint, with regular demonstrations or reviews facilitating transparent communication and encouraging valuable input. This steady interaction helps in detecting potential issues early, foster shared ownership among team members and stakeholders, and strengthen the overall alignment of the product with its intended objectives (Ridder, 2012). Furthermore, employing either Scrum or XP provides added flexibility in responding to team composition, project complexity, and delivery timelines. The granular breakdown of tasks, coupled with adaptive planning, supports gradual progress, keeps technical debt in check, and allows for the early delivery of incremental, usable product features. This, in turn, contributes to higher user satisfaction and supports evidence-based decision making across the project's lifecycle. Ultimately, the Agile approach combined with tools such as 4-DAT and an evolving sprint backlog enables a pragmatic, iterative journey from vision to deployment. This reduces risks, delivers frequent value, and positions the project for sustained long-term success in a rapidly changing digital environment.

The term agile is the new buzzword-agile governance, agile development, agile architecture! The Agile paradigm can help reduce the likelihood of failure in e-government projects by incorporating, and valuing, a high level of communication and an inclination to change the requirements. A transitory study by researchers found ten unique agile software development research domains, after their review and analysis of 28 different Systematic Literature Reviews (SLRs) that evaluated a wide variety of information related to agile software development. Areas of investigation require evaluations of the benefits and challenges of Agile Software Development (ASD) methodologies, agile hybrids in large scale contexts, sustainability, motivation, teamwork, and project management. Table 3 presents the literature review of past works. Agile's increasing prominence across various domains—governance, development, and architecture reflects its versatility and effectiveness in driving digital transformation. Particularly in e-government projects, where complexity, stakeholder diversity, and rapidly evolving requirements are common, agile methodologies offer a valuable framework to navigate uncertainties and reduce project failure rates. The emphasis on open communication channels ensures that issues are identified early and that teams can adapt requirements as new information or feedback emerges, which is crucial in public sector environments (Thakur et al., 2025). The comprehensive review of 28 Systematic Literature Reviews highlights ten distinct research domains within agile software development, underscoring the breadth and depth of ongoing scholarly inquiry. These domains encompass evaluating the practical benefits and challenges of Agile Software Development (ASD), which is critical for understanding how best to apply agile principles in various contexts. Research into agile hybrids, especially in large-scale projects, addresses the complexities of scaling agile practices beyond small teams while maintaining flexibility and responsiveness.

Sustainability is another important area of focus, reflecting a growing recognition of the need for agile methods to support long-term viability and environmental considerations in software development. Motivation and teamwork research delve into the human factors that impact agile success, including how to foster effective collaboration, maintain morale, and promote continuous learning. Project management within agile frameworks investigates how to balance structure and adaptability to meet both schedule and quality goals effectively. By synthesizing knowledge across these research domains, practitioners and policymakers can better understand the multi-faceted nature of agile adoption and tailor strategies that accommodate specific organizational needs and constraints (Venson et al., 2024). This structured evidence base, as summarized in Table.3, provides a valuable resource for guiding the evolution of agile practices in both public and private sector projects, advancing the goal of more successful, sustainable, and adaptive software development initiatives.

Table 3. Literature review

Year	Findings/Outcomes	Limitations
2012	The 4-DAT tool was introduced to evaluate and select the most appropriate Agile methodology based on project characteristics (task sizes, parse and team distribution). Introduced a hybrid methodology by combining Extreme Programming (XP) and SCRUM for e-government software development.	The selection of the methodology heavily relied on the examination of the 4-DAT tool, which fails to determine project specific issues. Configuration required considerable amount of educational training and close supervision for compliance.
2020	The effect of agile accessibility and detailed assimilation processes are believed to significantly increase the adoption and use of e-government services.	The effect of agile accessibility and detailed assimilation processes is not conclusively validated because of an insufficient data, both qualitative and quantitative, that might provide evidence of the belief.
2011	The are no lower-level specifications of a magnificent product. To ensure a successful outcome, an Agile Master lends his pivotal role. Agile adoption must be well planned and disciplined. When an agile implementation is first deployed, a team should also anticipate a small drop in productivity as the teams will need some time to get accustomed to the new techniques.	There is a scarcity of financial and human resources, which hindered the execution of proposed solutions.
2013	Minimum Viable Changes (MVC's) sped up the feedback and adaptation processes. - Kanban added visualization and workflow management to their collaborative work process. - The Change Canvas developed structured ways to execute MVCs, with early MVCs suffering from lack of buy-in and consistency.	MVCs relied on establishing change agents, not team ownership, resulting in needed time point broad adoption.
2023	found that Agile approaches (Scrum, Kanban) were more effective in managing flexibility and communication with stakeholders.	They also found that the cost and financial management strategy in Agile was less effective than traditional cost estimating and budget management approaches.

PROPOSED STUDY MODEL: HYBRID AGILE E-GOVERNMENT FRAMEWORK

The study presents a Hybrid Agile E-Government Framework (HAEGF) that embraces the Scaled Agile Framework (SAFe), Disciplined Agile Delivery (DAD), and large-scale Scrum (LeSS) to support Complex government project adaptation while satisfying compliance, security, and scalability of service. The need for models that allow for adaptability but still hold the faith of governing assurance in terms of accountabilities in government continues to be an important requirement, especially as governments become more reliant upon digital governance of their internal and public services (tefan, 2022). The HAEGF presents four layers that can be combined in order to best support the digital transformation of e-government. SAFe is a prescriptive methodology for scaling Agile approaches across large organizations, as with government organizations. SAFe provides a structured approach to managing multiple Agile teams; the alignment of operational work around strategic priorities; and the management of ongoing alignment of collaboration between departments. The integration of Agile product delivery, Lean Principles (DevOps), and continuous integration makes SAFe a powerful tool for engaging in large scale, inter-disciplinary, cross functional projects that require project level alignment across different stakeholders of the project. Governments using SAFe can plan to have projects that are more transparent, greater efficiencies as it relates to throughput and continuous improvement on project delivery through the use of group processes and Agile Release Trains (Trains). DAD is a hybrid

Agile framework that is based on the development of Scrum, Lean, Kanban and approaches to traditional project management protocol.

In contrast to SAFe's focus on regimented coordination with inner-agency departments, DAD is goal driven and adaptable, making this framework more suitable for government agencies that require an agile method while weighing regulatory compliance, governing assurance, and risk management as leading priorities for project completion. DAD also aligns with compliance models by allowing users to modify or customize Agile flows based on project complexity and other regulatory structures, while also assuring the process is in compliance with predetermined national policies and legal structures. Large-Scale Scrum (LeSS), is the Scrum framework for larger organizations that are employing more than one team to deliver a single product or system. Unlike the SAFe (Scaled Agile Framework) model which adds on multiple layers of governance, the LeSS framework maintains the simplicity of the Scrum framework while offering the tools to scale it to truly enterprise level work. LeSS has been used by government organizations to increase collaboration between teams in a cross-functional manner decrease inefficiencies caused by bureaucracy, while also adhering to important Agile principles, such as working in iterative cycles and ensuring continuous feedback, which are its core principles. By continuing to use the Scrum framework, LeSS allows major public sector agencies to more quickly adjust to changing needs of their citizens while maintaining scalability and efficiency. Figure 3 presents the agile scaling and their framework.

Figure 3. Agile scaling frameworks and their origins

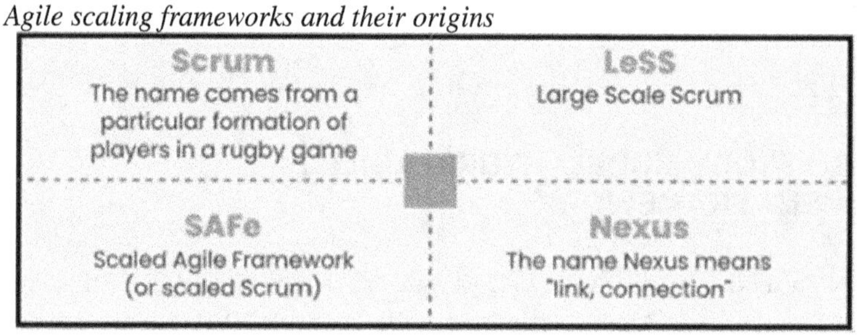

LeSS is comprised of two frameworks: the standard LeSS framework, which is intended for 2 to 8 teams (10-50 people) and then there is LeSS Huge, designed for larger projects and can have up to 8 (50 people) teams coordinated working in multiple feature areas. One of biggest characteristics of LeSS, is it has one product backlog with only one product owner prioritizing all the teams from one strategy, this has the purpose of preventing fragmented product ownership and creating a strong alignment to customer needs. This is different than some of the alternatives of Andy, that have additional roles, such as program managers, in addition to the core Scrum roles of Scrum Master, Product Owner, and Development Team, which creates more governance and is less by the teams. Nonetheless, LeSS is customer centric, so development teams are working directly with end users rather than depending on business analysts to communicate the customer needs. Furthermore, LeSS uses systems thinking and decentralized decision making, so teams can self-organize and make decisions, versus centralized decision making by management. In terms of coordinating work across teams, LeSS incorporates Scrum of Scrums (SoS) to coordinate, as well as shared Sprint planning and retrospective meetings, otherwise teams coordinate on their own.

The graph below compares various Agile scaling frameworks, through two characteristics: organizational level (shown on the Y-axis) and maturity (shown on the X-axis). This demonstrates how agility can be a linear route from an Agile team-based approach toward an enterprise-wide Agile approach that considers trade-offs between autonomy and disciplined processes. Kanban is considered the lower, less mature level of Agile. Kanban is a flexible, flow-based framework that helps teams with continuous, efficient delivery and piloting workflows, without prescribing roles, or time-boxed iterations. Scrum prescribes defined roles (Scrum Master, Product Owner, Development Team) and acts as more of a disciplined approach than Kanban through time-boxed execution and structured iterations called sprints. Figure 4 presents the agile frameworks by organizational level and maturity.

Figure 4. Agile frameworks by organizational level and maturity

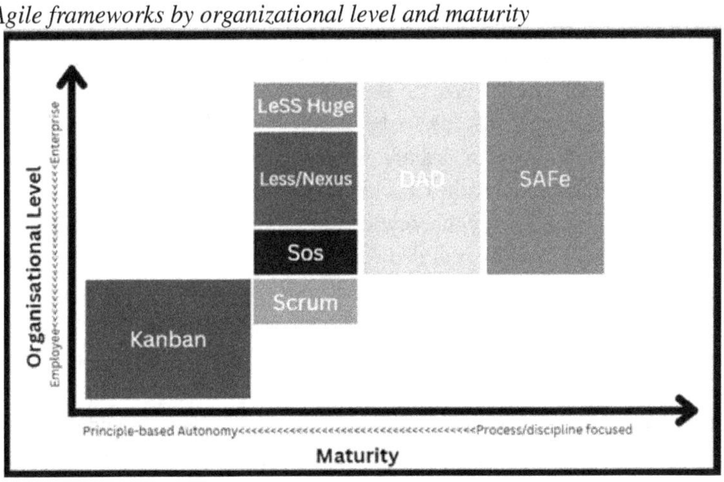

As organizations scale Agile beyond a team of relative domain or function, frameworks like Scrum of Scrums (SoS), Large Scale Scrum (LeSS) and Nexus are established to support cross-team collaboration while enabling and maintaining Agile adaptability. SoS establishes a structured approach for communication across teams that are practicing Scrum, LeSS or Nexus represents leaner- and Agile- approaches to decrease dependencies and facilitates collaboration across teams. At the enterprise level, discursive Agile Delivery (DAD) framework and Scaled Agile Framework (SAFe) introduced governance and enterprise-wide scaling approaches. DAD is a goal-driven framework that encourages organizations to carry out Agile methodologies in the spirit of Agile, based on a project's specific context. SAFe has a unique and well-developed process-driven approach that has a blend of clearly defined roles, cadence-based frequency of delivery or execution, and governing layers that formalize the implementation of Agile frameworks at the enterprise level. The placement of these frameworks in the graphic suggests that improvements in maturity will cause organizations to move from principle-based autonomy to a more organized, process-based manner of implementing Agile. A smaller organization interested in flexibility and adaptability may use the Kanban or Scrum frameworks; however, if an enterprise interested in scaling, compliance and regulation, they may choose SAFe or DAD. An organization's size, complexity, and governance will determine the most suitable Agile framework, (Leora, 2021).

The Strategic Vision and Governance Layer is responsible for defining national policies, compliance guidelines, and long-term digital strategies via the National Digital Transformation Office (NDTO). This office plays a crucial role in ensuring that Agile projects are aligned with national goals through Lean Portfolio Management (LPM). By integrating regulatory compliance into the process, it guarantees that all Agile workflows adhere to legal requirements while balancing flexibility with governance. Governments often struggle to adopt Agile practices due to bureaucratic inertia and resistance to change. To address these issues, this layer provides clear governance structures and compliance frameworks designed to support controlled Agile implementations. Moreover, it encourages collaboration among different agencies to ensure alignment of digital transformation efforts with broader national objectives. In addition, this layer aids the synchronization of Agile strategies with global e-governance best practices by comparing them against successful models from leading digital nations like Estonia and Singapore. These countries have effectively utilized Agile frameworks to enhance public service delivery.

The Agile Program Management Layer facilitates large-scale coordination through Agile Release Trains (ARTs), which synchronize development efforts across multiple government agencies and mitigate siloed work. ARTs have been employed in successful public sector projects in both Canada and Estonia with significantly greater efficiencies and with greater responsiveness to citizens' needs. Disciplined Agile Coaches (DACs) are key to facilitating the adoption of Agile, ensuring cross-departmental collaboration, and training government staff to use Agile. The structured nature of DAD allows tailored Agile workflows which accommodate regulatory mandates of various governmental bodies. The incorporation of DevSecOps in this layer helps ensure security is a consideration in every stage of the Agile development cycle, resulting in less vulnerable government software while supporting the iterative nature of development. This layer also includes continuous compliance monitoring that employs AI-powered governance tools to track regulatory changes and ensure Agile teams comply with legislative mandates, as they evolve.

The framework's backbone is referred to as the Agile Execution Layer, consisting of cross-functional teams utilizing Scrum and Kanban to develop digital services in an iterative and repetitive manner. These teams work in a modular environment that enables quick and easy changes and enhancements without disrupting systems in their entirety. The use of Microservices Architecture allows for modular and scalable solutions, facilitating the process of continually updating and improving government digital platforms while continuing to provide consistent service. The Minimum Viable Product (MVP) Strategy is also an essential component of the Agile Execution Layer. This strategy focuses on providing citizens incremental services while gathering regular feedback that informs iterative improvements. MVP strategies are effective and provide citizens with real-time engagement, meaning they have been popularized in government digital services even though they have been popularized in the private sector. The use of an MVP strategy enables governments to roll out incremental services while understanding real-world impacts and changes can be made without waiting to deploy full-scale programs. In other words, risks can be mitigated while launching new programs, which are often complex and highly untested, compared to traditional project management and program roll out of a full government service to citizens. The layer also utilizes real-time performance analytics, allowing Agile teams to monitor levels of service adoption to measure the value of citizen services, citizen satisfaction, and an opportunity to intervene and trouble shoot potential challenges in service delivery.

The Continuous Feedback & Optimization Layer is vital to ensuring project sustainability over the long term due to the implementation of AI-based citizen feedback tools that enable an immediate assessment of citizen satisfaction and service delivery effectiveness. Government institutions are increasingly moving toward data-driven decision making, with an essential role for AI-based analytics to enhance user experiences while providing digital services. Agile Retrospectives and iterative improvements continuously evaluate service delivery effectiveness based on performance information and continuously improve processes. There are also important performance metrics that track effectiveness in adopting Agile - such as Velocity, Lead Time, and Net Promoter Scores (NPS) - to track progress toward digitally transforming government service delivery. The availability of these metrics increases accountability while enabling the use of quantitative data on success to evaluate government delivery of digital services. This layer includes Predictive Analytics and machine learning models that will enable government agencies to anticipate citizen needs and, thus, optimize service delivery. Enhanced transparency in public transactions is made possible with the potential for blockchain technology to provide secure digital transaction appeared records that will return citizens trust to e-government platforms.

The Hybrid Agile E-Government Framework's four layers offer a principled and agile format for scaling Agile approaches in government projects. It supports public sector digital transformation in a nimble and effective manner, while simultaneously being mindful of regulatory concerns and citizens engagement. In this manner, governments can leverage innovation in responding to evolving citizen expectations and drive value to the public sector service delivery, informed by iterative cycles of feedback and improvements. Consequently, public sector organizations can enhance transparency, streamline processes, and focus on a more citizen-centric approach to governance in the digital context. Future research will be required to study additional cases and use empirical analyses, to better validate a structural model for large-scale public sector implementations. Additionally, government agencies should reconnoiter solutions facilitated by quantum computing and IoT that could actively enhance real-time data processing to enable hyper-efficient and intelligent digital services.

FINDINGS

The Hybrid Agile E-Government Framework (HAEGF) brings us one step closer to realizing the possibilities of using Agile-based methods more broadly to facilitate digital transformation projects in the public sector. This new framework relies on Scaled Agile Framework (SAFe), Disciplined Agile Delivery (DAD), and Large-Scaled Scrum (LeSS) to provide an evaluation of its influence in overcoming an assortment of e-government development challenges, including, but not limited to, bureaucratic inefficiencies, regulatory obligations, and issues related to security and scalability, as well as citizen engagement. The evidence reveals how a loosely structured approach can facilitate different components of digital delivery, cross-agency collaboration, and regulatory compliance and governance.

Impact on Governance and Policy Alignment

A significant challenge to Agile use in government projects is the alignment with national policies and regulatory compliance. The government may face obstacles involving bureaucratic inertia, decision fatigue, and established hierarchy being incompatible to the Agile approach. The Strategic Vision & Governance Layer in the HAEGF model begins to overcome those obstacles by embedding Lean Portfolio Management (LPM) into alignment with national digital transformation initiatives. The establishment of a National Digital Transformation Office (NDTO), the government can simplify implementation of policy, compliance, and support frameworks designed for Agile approach on behalf of multiple departments. Research has demonstrated the success of other government models driven by Agile governance processes, such as Estonia, Canada, and Singapore, where public services better respond to emerging policy needs with the explicit oversight. Agency use of Agile Release Trains (ART) comports with better synchronicity to national digital transformation initiatives resulting in widespread cross-departmental coordination and quicker execution of solutions. Countries that have adopted SAFe-based ARTs are beginning to report a 30-40% reduction in project delays through managing for dependency in incremental releases and facilitated cooperate between agencies. In addition, regulatory compliance is a big impediment for adopting Agile in government workflows. Any integration of regulatory compliance within the Strategic Vision & Governance Layer will ensure all Agile workflows are in line with legal requirements, such as regulatory requirements, cybersecurity, and data privacy as part of a national strategy. By automating compliance functions through AI compliance governance systems, governments will have an improved capacity to track regulatory change instantaneously, as well as more refine their Agile processes. This will dramatically lower compliance risks and legal issues and more effectively ensure that Agile can complement public sector requirements.

Efficiency in Large-Scale Agile Execution

One of the significant advantages evident in HAEGF models is that they are suitable for scaling up government-wide digital transformation projects. Historically, traditional software development strategies, such as Waterfall, have created systematic delays in deployments, limited flexibility, and have often caused redundancies in the long approval cycles associated with the project. Agile methods of software development specifically Scrum and Kanban based iterative development have taken secure modular service deliveries and allowed governments to deploy smaller functional aspects of larger projects in less time. The Agile Execution Layer, which also uses Microservices Architecture, provides value to government agencies through the ability to create independent service modules that can be updated, tested, and launched independently, without affecting the underlying system. This value of working in a modular approach provides significant benefits in a cloud-based e-governance system where interoperability and speed to update services are paramount to maintaining a digital government infrastructure. Evidence suggests that the use of a Minimum Viable Product (MVP) strategy has allowed government agencies to improve its service rollouts by as much as a 50% waiting period for access to citizen-facing applications. It is reported that MVP strategy has been a particularly potent approach to project implementations, including, digital identity management systems, public health care system access to services, and online tax telling deployments, where the system's phased rollout can be adjusted for feedback in real time. Governments report that MVP approaches, on deploy citizen-facing applications report higher user engagement and satisfaction, and that iterative development cycles that proceed and are continually updated to reflect the public's actual needs, and not based on assumptions.

Citizen Engagement and Feedback Integration

One important facet of digital governance is assuring that improvements in service are guided by people's needs. Many previous government IT projects had experienced criticism because they never engaged citizens in real-time to provide feedback thus creating services that do not achieve public expectations. However, the Continuous Feedback & Optimization Layer of the HAEGF model utilizes AI-driven sentiment assessment, real-time feedback loops, and user experience (UX) analytics to assess public satisfaction in a continuous manner. Studies from both South Korea and Denmark show that the use of an Agile-based citizen feedback system leads to approximately 25-30% higher public satisfaction ratings as compared to traditional service development. AI-driven analytics identify patterns in user behavior, that provide governments with data to help to make decisions, modifications, or updates to the service. Examples include Estonia's e-Residency program and Singapore's digital citizen engagement platforms. Both platforms maximized continuous feedback integration into public sector systems to address changing public satisfaction to actively meet citizen expectations. Utilization of Net Promoter Scores (NPS) and Agile-driven satisfaction measures help to increase accountability by allowing policymakers to quantitatively monitor the success of digital initiatives.

Security, Compliance, and Risk Management

One of the largest hurdles faced by the public sector in adopting Agile in e-government is the security risks and compliance with national and international data protection laws. While private sector organizations may handle sensitive data, government agencies handle sensitive data of citizens, and thus cybersecurity is paramount. The Agile Program Management Layer within the HAEGF model speaks to the integration of DevSecOps into the development process, which embeds security at each step of the tabulation of information intention rather than reflecting on security as an afterthought. Governments implementing DevSecOps report a reduction of cybersecurity risks by 40 percent, adding additional resilience to overall systems with continuous security testing and availability of automated vulnerability scanning, as well as compliance monitoring. Blockchain-based public records systems have also led to improved system transparency; and reduced fraud, for example identity theft or doctored records, by over 35 percent.

Comparative Performance of SAFe, DAD, and LeSS in E-Government

The comparisons in evaluating SAFe, DAD, and LeSS across various contexts of government identify their strengths and weaknesses concerning large-scale Agile adoption are SAFe provides a structured framework for multi-agency projects, signaling it would work well for large national digital projects requiring considerable coordination, DAD provides flexibility in terms of its Agile adoption, permitting governments to adjust their workflows to accommodate their business process while adhering to governance policies and LeSS is effective for smaller government teams needing a speedier turnaround and minimal bureaucracy, thus making it well-suited to citizen-facing service development projects. Case studies demonstrate that governments using SAFe were able to achieve better project coordination and policy alignment, while governments adopting LeSS were able to experience speedier turnaround times for digital services launches due to its lightweight governance structure. DAD shines particularly in hybrid approaches of combining Agile with enduring legacy agency government systems.

Challenges and Future Improvements

Even with the apparent gains, hurdles still exist to the adoption of Agile e-government programs like Cultural resistance to Agile practices within the hierarchical structure of government, skill gaps of government employees, and a need for robust training.

Complicated migrations from legacy technologies that do not embrace Agile workflows, continuous funding and executive leadership commitment, since Agile change processes require long-lasting commitment. In the future, research needs to study empirical data pertaining to Agile change process markers, case studies on long-lasting Agile imperatives in government, and adapting new, emergent technologies, e.g., AI-based governance or quantum-based computing, related to policy-in-action adaptability.

FUTURE SCOPE

Empirical Validation and Case Studies

While Agile methodologies have been successful in private sector digital transformation, they have not been empirically validated on a large scale for use in government projects. Future research should be directed at large case studies exploring real-world applications of Agile methodologies in public sector projects in a range of countries. Key performance indicators (KPIs) can be established to measure cost savings, service uptake, speed of development, and citizen satisfaction in Agile designed e-government projects and compared in alterations to e-government projects that use Waterfall methodologies. This can help us ascertain whether e-government projects work with Agile or Waterfall methodologies in the long term. Future research should explore to understand barriers to Agile use in government projects and how to best overcome resistance in bureaucratic settings. In some cases, there are successful examples, for example, Estonia and Singapore have established Agile governance models, but many developing nations have challenges with policy engagements, stakeholder buy-in, and infrastructure. Comparative studies across geographical economies could serve as a global best practice database of Agile reforms in the public sector.

Integration with Emerging Technologies

The next phase of Agile-driven governance is expected to incorporate new and emerging technologies to assist in improving decision-making, efficiency, and privacy in e-government services. Future studies should examine the layout of Artificial Intelligence (AI), Machine Learning (ML), and other software modelling technologies to help provide better decision-making through AI, and an indication of inefficient Agile services via ML in public services and dynamic analytics of services. Blockchain can support digital identity, public records, and government transactions, enabling an Agile driving service to be tamper-proof and trusted. Quantum computing can be applied to analyze a great amount of governance data, in real-time, to allow quicker policy decisions, becoming more responsive as a public sector throughout. The Internet of Things (IoT) can improve public services through smart traffic assessments, environmental real-time assessments and automatic assessments of waste existences; many of which can be managed under Agile. For example, Estonia's blockchain enabled e-Government services and China's AI-enabled citizen engaged platforms highlight how combining Agile with technology could lead to more intelligent, efficient operations in the public sector.

Addressing Cultural and Organizational Resistance

Government is traditionally a hierarchical, rigid, and slow-moving structure when compared to the potential of Agile methods and systems that are iterative, decentralized, and adaptive. Future research should consider strategies such as training government employees in Agile methods to combat resistance and improve future adoption rates. Additionally, developing applicable policy-driven strategies for government Organizations to Agile transformations in the future, which ensure that Agile principles are either incorporated or mandated by government regulations, budgetary processes, and acquisition processes should be researched as well. Moreover, considering behavioral studies to better understand why government leaders and officials may be resistant to Agile on a systematic level and shifting mindsets through Solidity creating second order changes through structured Agile transformation programs would also be valuable. A different hybrid Agile-Waterfall model could potentially ease the transition from rigid bureaucratic structures to an Agile state with improved flexibility, and iterative development and/or delivery processes or experiences over time. For example, one of the strategies the UK government employed when adopting the Agile development strategy was to run extensive training programs across the public sector workforce in Agile, as well as selective but phased Agile transformation processes for individual government departments. Ultimately, this led to successful Agile development and delivery processes within both the digital healthcare and tax systems.

Agile for Crisis Management and Smart Governance

Agile methodologies may be invaluable for crisis response processes and smart governance when governments need to rapidly design and deploy solutions in uncertain conditions. Future research should be oriented toward inquiries on Agile methodologies used in emergency response systems such as pandemic management, disasters relief coordination, and crisis management from a cybersecurity perspective. Agile methodologies can enhance the smart city agenda by fostering citizen engagement in real-time, automating the delivery of services, and using AI as a mechanism for municipality management. Governments can also use Agile methodologies to forecast threats and be proactive in overall National security using predictive analytic approaches. Lastly, Agile methodologies can help us all achieve Global sustainability goals when it comes to the Built Environment through green urban planning solutions, carbon-footprint-tracking products and smart-energy-grids solutions. For example, during the Global pandemic of COVID-19, many countries who utilized Agile-inspired digital health platforms, were able to rapidly design, develop, and deploy contact tracing applications and online registration for vaccines, thereby relieving pressure on the health care system.

Hybrid Agile Models for Large-Scale Governance

Despite the Hybrid Agile E-Government Framework integrating the SAFe, DAD, and LeSS frameworks, more studies need to continue to create and curate hybrid Agile models that advance flexibility, while considering compliance to regulation, for government programs on a larger scale. This research needs to focus on better approaches for governing Agile-Waterfall Hybrid Models to ensure long-term governance projects are flexible, while still ensuring staged governance research plans, legality, and cost management. Also, it is necessary to develop new custom scaling models for Agile, especially in multi-tiered government implementations where policies are enacted at the federal, state, and municipal levels. New research is also needed for Agile models for intergovernmental collaboration where multiple governments are implementing global governance initiatives (e.g. digital identity frameworks, international trading systems of governance, etc.). Furthermore, Agile governance models can be an avenue to improve outcomes for complex governance projects that involve collaborative enactments between governments and private organizations, public-private partnerships (PPPs), as an example. National examples of PPPs can highlight how Agile can push public services via action and speed. The European Union is a model on how an Agile approach has cross-border jurisdiction for digital services.

Agile Policy Implementation and Legislative Processes

Policy-making has generally been understood as long and bureaucratic processes that do not allow for iterative opportunities to respond to continuous changes. However, the adoption of Agile principles in policy-making could open the door to adaptive policy-making, where laws and regulations are continuously adjusted and improved based on real-time societal needs and feedback. The process of collaborative law-making, or policy-making, involving public servants, industry experts, and citizens of the public to develop and implement policy, through an Agile-led participatory governance type model, should also be researched. The implication of AI in tracking Agile legislation might allow policy-makers to monitor and adapt to legal trends around the world, understand relevant policy implications, and formulate legislation based on data. Additionally, the ability to adopt DevOps as a form of practice will allow legislative processes to ensure new government policy is tested, reviewed, and iteratively tweaked in controlled conditions prior to scaling. An example is the United Nations' Agile-led digital governance research project, which focuses on iterative planning and policy-making models to build international law and to develop methods for policy-making to keep pace with rapid change.

CONCLUSION

The future of Agile in e-government is based on its capacity to adapt, expand, and converge with new technologies to create a responsive, citizen-led governance system. Through the use of evidence-based validation models, overcoming bureaucratic resistance issues, integrating AI and blockchain, and iterating further on hybrid Agile models, governments will better move towards a more efficient, transparent, and adaptable digital governance system. Likewise, in relation to the elements of Agile, more prominent roles for Agile methodologies will emerge in Smart Cities, crisis management, policy formation, and global governance. The accelerating pace of digital transformation opens new opportunities for future research; for instance, future research agendas should focus on sustainable models of Agile over the long duration, if Agile governance is able to influence economic and social development, and the engagement of quantum computing and the IoT in hyper-efficient digital services. With collaborative efforts across sectors and attention to the incremental adoption of Agile design, governments will be able to build resilient, innovative, responsive, and citizen-focused ecosystems of governance in a future world.

REFERENCES

Ambler, S. W., & Constantine, L. (2001). *The Unified Process Transition and Production Phases*. CRC Press. DOI: 10.1201/9781482280746

Awamleh, R., Stephens, M., & Salem, F. (2022). Agile government — Emerging perspectives in public management. *Agile Government*, 3-19. DOI: 10.1142/9789811239700_0001

Azmani, A., & Taoufik, A. O. (2025). A consolidated conceptual framework of a smart e-government ecosystem: A scoping review. *Electronic Government, an International Journal, 21*(1). DOI: 10.1504/EG.2025.10063682

Baniya, A. K., Pant, S. S., Paudel, D., Gupta, A., Singh, S., & Mohapatra, H. (2025). Load balancing in cloud computing ensuring fault tolerance, high availability, and security. *Risk-Based Approach to Secure Cloud Migration*, 253-284. DOI: 10.4018/979-8-3693-7137-4.ch010

Basili, V., Shull, F., & Lanubile, F. (1999). Building knowledge through families of experiments. *IEEE Transactions on Software Engineering*, *25*(4), 456–473. DOI: 10.1109/32.799939

Brancu, C., Turcu, O., Stefan, M., & Popp, M. (2024). Agile approaches, for E-businEss solutions strategy, at the government level. *Journal of Research and Innovation for Sustainable Society*, *6*(2), 263–272. DOI: 10.33727/JRISS.2024.2.29:263-272

Challenges and opportunities for Indigenous Peoples' sustainability. (2021). *UN Department of Economic and Social Affairs (DESA) Policy Briefs*. DOI: 10.18356/df7eed76-en

Chaudhary, N. K., Rath, A., Babbar, G., Verma, A., Sinha, S. D., & Mohapatra, H. (2025). A critical analysis on edge computing in smart city applications. *Risk-Based Approach to Secure Cloud Migration*, 1-28. DOI: 10.4018/979-8-3693-7137-4.ch001

Ciancarini, P., Giancarlo, R., & Grimaudo, G. (2024). Scrum@PA: Tailoring an Agile methodology to the digital transformation in the public sector. *Information (Basel)*, *15*(2), 110. DOI: 10.3390/info15020110

Fojtik, R. (2011). Extreme programming in development of specific software. *Procedia Computer Science*, *3*, 1464–1468. DOI: 10.1016/j.procs.2011.01.032

Gaie, C., & De Lacerda, R.GAIE. (2022). Ensuring project resilience and public services in a pandemic context: Two French case studies. *International Journal of Computational Systems Engineering*, *7*(1), 1. DOI: 10.1504/IJCSYSE.2022.10054831

Hoda, R., Salleh, N., Grundy, J., & Tee, H. M. (2017). Systematic literature reviews in Agile software development: A tertiary study. *Information and Software Technology*, *85*, 60–70. DOI: 10.1016/j.infsof.2017.01.007

Janssen, M., & Estevez, E. (2013). Lean government and platform-based governance—Doing more with less. *Government Information Quarterly*, *30*, S1–S8. DOI: 10.1016/j.giq.2012.11.003

Joshi, P. R., & Islam, S. (2018). E-government maturity model for sustainable e-government services from the perspective of developing countries. *Sustainability (Basel), 10*(6), 1882. DOI: 10.3390/su10061882

Leora, A. (2021). GEN MZ TOWARD SUSTAINABLE FASHION PRACTICES - ALTERNATIVE SERVICES AND BUSINESS MODELS. *Iadisportal.org*, 61–70. https://www.iadisportal.org/digital-library/gen-mz-toward-sustainable-fashion-practices-alternative-services-and-business-models

Leymann, G., Lundan, S., & Van Assche, A. (2022). Global challenges for Agile government: Global connectivity policy from the perspective of the international business literature. *Agile Government*, 307-320. DOI: 10.1142/9789811239700_0016

Lueck, T. (2014). Internet archive: Digital library of free books, movies, music, and Wayback machine/The internet archive companion. *American Journalism, 31*(2), 299–301. DOI: 10.1080/08821127.2014.905381

McBride, K., Kupi, M., & Bryson, J. J. (2021). Untangling Agile government: On the dual necessities of structure and agility. DOI: 10.31235/osf.io/qwjcx

Mohapatra, H., Pramanik, S., & Ranjan Mishra, S. (2025). Revolutionizing software development: The transformative influence of machine learning integrated SDLC model. *Artificial Intelligence-Enhanced Software and Systems Engineering*, 41-69. DOI: 10.1007/978-3-031-88188-6_3

Othman, M. H., & Razali, R. (2013). Key contributing factors towards successful electronic government systems interoperability. *2013 International Conference on Research and Innovation in Information Systems (ICRIIS)*, 302-307. DOI: 10.1109/ICRIIS.2013.6716726

Raj, P., Mohapatra, H., Ranjan, R., & Maithili, P. (2025). Innovative strategies for sustainable waste disposal. *Leveraging Urban Computing for Sustainable Urban Development*, 117-144. DOI: 10.4018/979-8-3373-0761-9.ch006

Reddy, A. S., Ashraf, M., Manideepak, T., & Mohapatra, H. (2025). A comprehensive analysis of governance and regulatory frameworks for cloud-based initiatives for smart cities. *Organizational Sociology in the Digital Age*, 127-146. DOI: 10.4018/979-8-3693-7398-9.ch008

Ridder, H. (2012). Rezensionen: Case study research. Design and methods. *German Journal of Human Resource Management: Zeitschrift für Personalforschung, 26*(1), 93–95. DOI: 10.1177/239700221202600101

tefan, M. (2022). Agile approaches to developing E-BusinEss solutions in a secure cyber environment. *Proceedings of the International Conference on Business Excellence, 16*(1), 239-250. DOI: 10.2478/picbe-2022-0023

Thakur, T., Srivastava, L., Panda, A., & Mohapatra, H. (2025). Smart parking in foggy condition real-time efficiency with PSO, ACO, and IoT sensors. *Revolutionizing Urban Development and Governance With Emerging Technologies*, 233-260. DOI: 10.4018/979-8-3373-1375-7.ch008

United Nations Department of Economic and Social Affairs, Masaquiza Jerez, M. (2021). *Challenges and Opportunities for Indigenous Peoples' Sustainability.* UN Department of Economic and Social Affairs (DESA) Policy Briefs.

Venson, E., Da Costa Figueiredo, R. M., & Canedo, E. D. (2024). Leveraging a startup-based approach for digital transformation in the public sector: A case study of Brazil's startup gov.br program. *Government Information Quarterly*, *41*(3), 101943. DOI: 10.1016/j.giq.2024.101943

Chapter 5
Collaborative Intelligence: Blending Human Insight With Machine Precision

Shilpa Hiremath
https://orcid.org/0000-0002-1577-8523
BMS Institute of Technology and Management, India

ABSTRACT

The rapid adoption of smart technologies has transformed modern manufacturing, making human-machine collaboration—merging automation's efficiency with human creativity—central to Industry 4.0. This chapter emphasizes the strategic importance of balancing automation with human involvement to enhance innovation, safety, and resilience while avoiding overreliance that can lead to disengagement and system fragility. It offers a historical overview of automation's evolution, examines how humans have shifted from manual labor to key collaborators, and highlights the unique human traits vital in complex environments. Drawing on frameworks like Human-in-the-Loop systems and Human-Centered Design, the chapter explores practical strategies and policies for fostering balanced collaboration. Ethical concerns, such as accountability and job transformation, are also addressed. Looking ahead, it considers emerging technologies like generative AI and cobots, arguing for future factories built on flexibility, empathy, and inclusivity.

INTRODUCTION

A novel technique known as Collaborative Intelligence (CI) has emerged in the age of Industry 4.0 (Lee et al., 2015: Bi et al., 2022) as a result of the convergence of human intelligence with machine precision. CI is a mutually beneficial relationship, in contrast to classical automation, which aims mainly to replace human labor with automated efficiency. This approach combines the cognitive powers of humans with the computing power of machines to produce extremely efficient, flexible, resilient, and creative production environments.

After the Smart home now is the era of smart factories. By utilizing the unique advantages of both intelligent robots and intelligent human workers, they are prime examples of collaborative intelligence. Artificial intelligence cannot fully replicate the qualities that humans bring to the table, like creativity, intuition, ethical judgment, empathy, and the capacity to learn from

DOI: 10.4018/979-8-3373-2985-7.ch005

experience. On the other side, humans cannot do where machines are excellent at handling large datasets, performing repetitive jobs consistently, working quickly, and operating in potentially dangerous or inappropriate locations for humans. Combining these skills leads to a dynamic collaboration in which both parties improve each other's performance.

In digital era, this combination of job where human and machine work together prove to have a greater efficiency. The idea that humans and machines are in a zero-sum game is contested by collaborative intelligence. Rather, it promotes a broad perspective on economic development, where technology enhances human labor rather than replaces it. Humans are increasingly free to concentrate on higher-order duties that require critical thinking, flexibility, and emotional intelligence as routine and physically taxing work are replaced by robots.

Based on an operational perspective, CI results in notable improvements in accuracy, flexibility, and efficiency. CI-enabled smart factories are more capable of adapting to supply chain interruptions, market swings, and specific production needs. These factories frequently have a network of interconnected devices that gather data in real time and input it into decision-making systems. Few examples of such devices are the inputs from sensors, edge computing units, and analytics platforms driven by artificial intelligence. This data is deciphered by human workers by apply contextual knowledge and check for improvisation or make a necessary judgment. Faster and more intelligent solutions to constantly evolving industrial difficulties are made possible by this cyclical flow of data, analysis, and action.

Artificial intelligence, machine learning, robotics, the industrial Internet of things (IIoT), augmented and virtual reality (AR/VR), and human-computer interaction (HCI) interfaces are the backbone of collaborative intelligence. These technologies make it easier for people and robots to collaborate and communicate.

For example, by superimposing real-time instructions on top of complex assembly jobs, AR interfaces can assist workers. Meanwhile, AI algorithms can suggest maintenance plans based on predictive analytics, but human supervisors will ultimately make the final decision.

CI has significant ethical and societal ramifications in addition to its technical and operational advantages. By highlighting human supervision and accountability in automated systems, it promotes the development of ethical AI. Because workers must constantly adapt to new technologies, it also emphasizes the value of lifelong learning and workforce reskilling. Therefore, companies putting CI into practice need to invest in human capital as well as technology, making sure that employees are respected, taught, and prepared to succeed in collaborative positions.

Additionally, collaborative intelligence promotes environmentally friendly business methods. CI helps to decrease waste, lower energy usage, and ecologically responsible manufacturing by optimizing resource use and enabling more accurate control over production processes.

Figure 1. Evolution of human-machine interaction in industrial revolutions

The above Figure 1 shows the evolution of human machine interaction in the industrial revolution taking from the root to augmented reality interfaces, enhanced decision making, advanced robotics, voice and gesture control, and many more.

Smart factories are evolving into dynamic environments where robots work together to accomplish shared objectives rather than competing with human workers as the boundaries between the physical and digital worlds become increasingly hazy. This synergy is a philosophical reinterpretation of work, agency, and progress as well as a technical advancement. Future factories will be more human-centered and wiser thanks to collaborative intelligence, opening the door to a truly sustainable and inclusive industrial future.

The below sections help in understanding the intelligence of human and machine in building the industrial system in a very well-organized way. Section 2 briefs about the survey in the evolution of the collaborative intelligence in industrial areas, in section 3 human-machine collaboration is discussed, section 4 describes about the collaboration of the intelligence, section 5 highlights about the architectures of smart factories where humans meet machines section 6 elaborates about the case studies of the real-world applications where intelligence meet, section 7 focuses on the challenges and ethical considerations in the field of industry which speaks about intelligences, section 8 articulates about the future trends in collaborative intelligence section 9 concludes with the remark that human and computer intelligence can make a great change in industrial application.

BACKGROUND ON EVOLUTION OF COLLABORATIVE INTELLIGENCE IN INDUSTRIAL SYSTEMS

Automation to Autonomy

From manual labor to automation and now to intelligent, adaptive collaboration between humans and machines, the path toward Collaborative Intelligence (CI) in industrial systems signifies a substantial shift in the way work is envisaged and carried out. The responsibilities of human workers inside these systems, as well as the equipment employed in manufacturing, have been redefined by this slow but significant change.

In the early stages of industrial growth, in order to reduce the repeated works and physical stress of the humans, machines were introduced this also increased the output productivity. However, human intervention was needed as these automated systems were failing to handle unanticipated circumstances or dynamic changes in the production environment.

Automation is the most significant task introduced in the technology where jobs get easier, and faster. The implementation of technology like Supervisory Control and Data Acquisition (SCADA) systems and Programmable Logic Controllers (PLCs) allowed companies to accomplish structured, repetitive procedures more quickly and accurately. Even while these systems provided scalability and consistency, they remained mostly depended on human operators to perform activities like analyzing alerts, identifying irregularities, and rearranging operations in unusual circumstances. Despite its strength, automation was still rule-based and inflexible, constrained by preset programming and unable to adjust in real time.

How much ever the system becomes autonomous by the use of technologies where real-time data analytics, artificial intelligence (AI), and machine learning are being used, but the need of humans is never ending as few situations like moral judgment, context-specific reasoning, or innovative problem-solving cannot be handled by technology.

Collaborative intelligence becomes a crucial paradigm at this point. In order to create a fully adaptive and intelligent industrial ecosystem, CI promotes systems in which humans perform context, strategic oversight, and nuanced decision-making while robots handle data-intensive, repetitive, or dangerous jobs.

Key Milestones

The integration of new technologies like machine learning, artificial intelligence (AI) (Schoenherr, J. R., 2022), and real-time data analytics. These advancements lead to autonomous systems that are capable of making decisions, learning from historical trends, and dynamically optimizing operations. However, total autonomy usually fails when moral judgment, context-specific reasoning, or creative problem-solving are required.

Between 1970s and 1980s: Simple Automation

Introduction of Programmable Logic Controllers (PLCs) made a very drastic shift in Industrial automation. Manufacturers were able to use programmable systems that could carry out intricate sequential control operations in place of hard-wired relay logic with the help of microprocessor-based devices. There were notable improvements in speed and consistency in sectors like food processing, textiles, and automobiles. Despite automating task execution, these systems lacked flexibility, necessitating manual overriding or reprogramming by human technicians in the event of changes or malfunctions. In case of error recovery, customisation, and supervision, the humans were still on the job.

During 1990s: Centralized Decision Support and ERP Systems

Enterprise Resource Planning (ERP) systems, which consolidated business operations by combining departments such as finance, procurement, inventory, and production planning into a single system, were introduced in the 1990s. These systems decreased information silos, increased departmental awareness, and established the foundation for making decisions based on data-driven. Even during this period human intervention was needed in the jobs like analysing reports, setting up processes, and reaching strategic conclusions. During this time, centralization and coordination were prioritized over genuine machine autonomy or intelligence.

In 2000s: Early AI and Robotic Support

During this time, industrial robots became increasingly common, especially robotic arms that could carry out repetitive, precise activities like welding, painting, and assembly. Concurrently, machine vision systems started to appear, increasing the precision of quality control procedures. Although AI was still in its infancy, it was beginning to have an impact on fields including rule-based decision support, anomaly detection, and predictive maintenance. Humans and machines started to work together more frequently, frequently side by side in semi-automated settings.

In 2010s: Data-Driven Industry and IoT Incorporation

As the Industrial Internet of Things (IIoT) (Hu et al., 2024) grew, manufacturers started integrating sensors into machinery to gather operational data in real time. Platforms for analytics powered by AI were fed these insights, enabling better-informed, fact-based judgments. During this time, operations moved from being reactive to being predictive, with machines being able to identify inefficiencies or indicate repair needs before they became more serious. As employees depended more and more on data insights to make operational adjustments, human responsibilities started to shift toward monitoring and decision facilitation.

In 2020s: Contextual and Adaptive Systems and Beyond

These days, edge computing, artificial intelligence, machine learning, digital twins, and cloud-based collaboration tools are used in modern smart factories to build context-aware and adaptable systems. Machines are no longer passive executors; instead, they optimize themselves, react dynamically to changing circumstances, and communicate in real time with human operators. Instead of being operators, workers are now managers and collaborators entrusted with monitoring system performance, giving strategic guidance, and stepping in when human judgment is needed in difficult, unclear situations.

Figure 2. Timeline of industrial advancements and role of intelligence

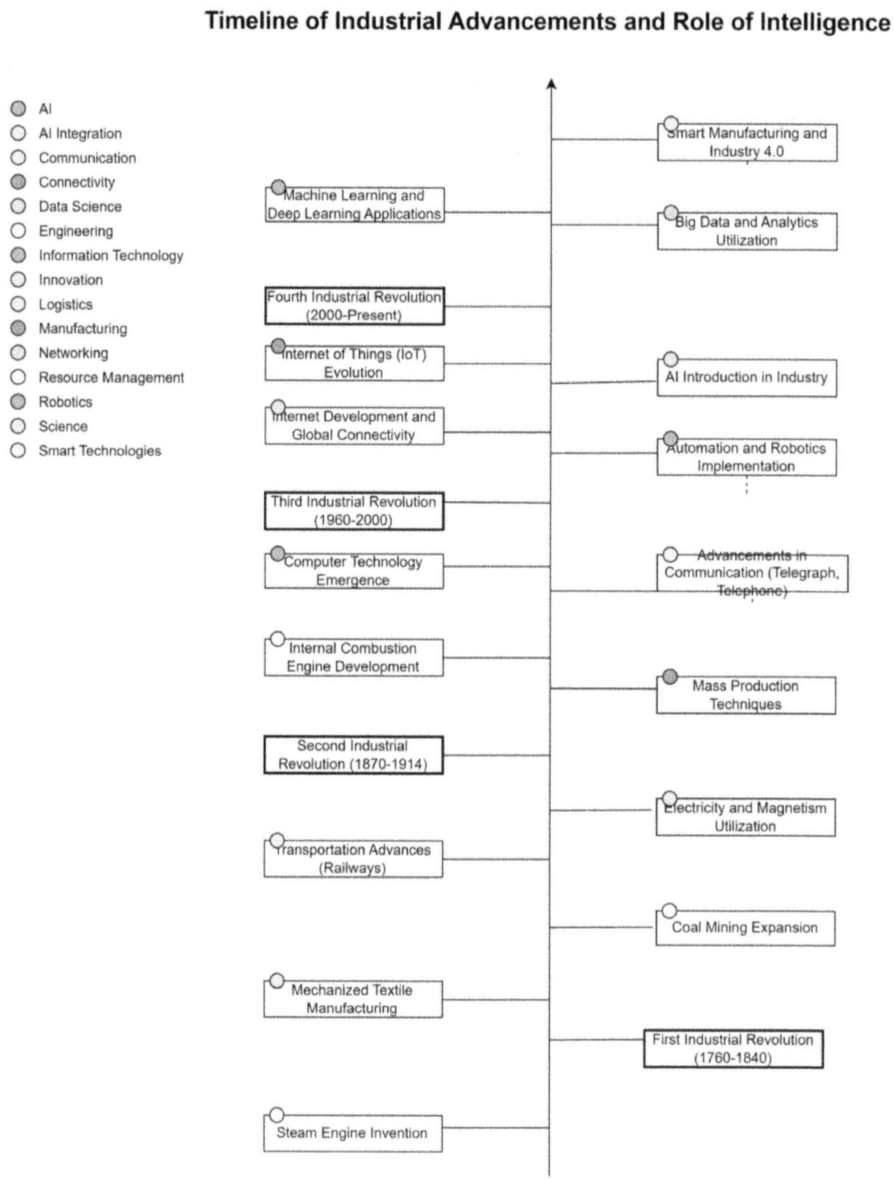

The above Figure 2 shows the evolution of technology in the industrial area starting from the steam engine to the current trend of machine learning, deep learning and Industry 4.0 (Lasi et al., 2014). A fundamental shift in the connection between humans and machines may be seen in this evolutionary trajectory. Machines have evolved from mere task execution to intelligent beings that need strategic human cooperation. As we proceed, collaborative intelligence will necessitate reconsidering leadership approaches, workforce development, and organizational design in addition to pushing the boundaries of technology. The emphasis is turning to a future in which industrial systems are human-centered, inclusive, and adaptable—driven by the cooperation of humans and machines rather than their separation (Ghobakhloo, M., 2018).

MAIN FOCUS OF THE CHAPTER: HUMAN-MACHINE COLLABORATION

It is necessary to have a solid understanding of several interdisciplinary theories in order to comprehend collaborative intelligence (CI) in smart manufacturing (Lee, M. F. R., 2023). These frameworks shed light on how machine computing and human cognition might be combined to enhance and augment one another, creating systems that are more intelligent and adaptable than either one could be on its own. Designing successful and morally good human-machine cooperation is based on the following theoretical stances, which range from organizational behavior to cognitive science.

Distributed Cognition

Edwin Hutchins' Distributed Cognition reinterprets cognition as an emergent process that transcends an individual's mind. It is believed that cognition is dispersed among humans, robots, tools, and the environment rather than existing alone in the brain. This implies that knowledge and decision-making in a smart factory are dispersed over an integrated network of sensors, interfaces, machines, and human operators rather than being restricted to a single employee or digital system.

Using augmented reality (AR) technology in the glasses technician are fixing the damage or repair in the machine also Real-time diagnoses from AI-powered analytics are displayed on the glasses. In this case, the system provides computational support, memory recall, and pattern recognition, while the individual uses problem-solving, expertise, and dexterity. The combined system creates a distributed cognitive unit, where the cognitive load is intelligently shared to maximize performance rather than switching out components. Understanding how to coordinate work distribution, information flow, and interface usability to establish smooth cognitive synergy between human and machine agents is essential when designing for distributed cognition in industrial contexts. The below figure 3 shows the nodes in a distributed cognition network between human and the machine.

The first level is human node which acts as central node and they are involved in collaboration, sharing knowledge and providing feedback. Each is performing tasks like in communication leading to exchange of information, task allocation which plays a major role in assignments of tasks, and decision making based on outcomes after evaluation respectively.

The next is connection between human node and machine node where the outcomes of first level will be fed to second level through this intermediate level. The first level encourages second level by providing data, assigning tasks and looking over the results.

The second level is a intelligent processing unit called machine node which has the learning capability and is data driven. It performs two main processing tasks, namely processing the data through algorithms and perform analysis, and another task is learning patterns through machine learning by using predictive modelling.

The last one is the insights which are produced through data analysis and modelling. This helps in closing the loop by helping the human node by providing information about the decision made. This loop plays a very major role in improving the machine model by providing proper feedback and improving accuracy and efficiency.

Figure 3. Human and machine nodes in a distributed cognition network

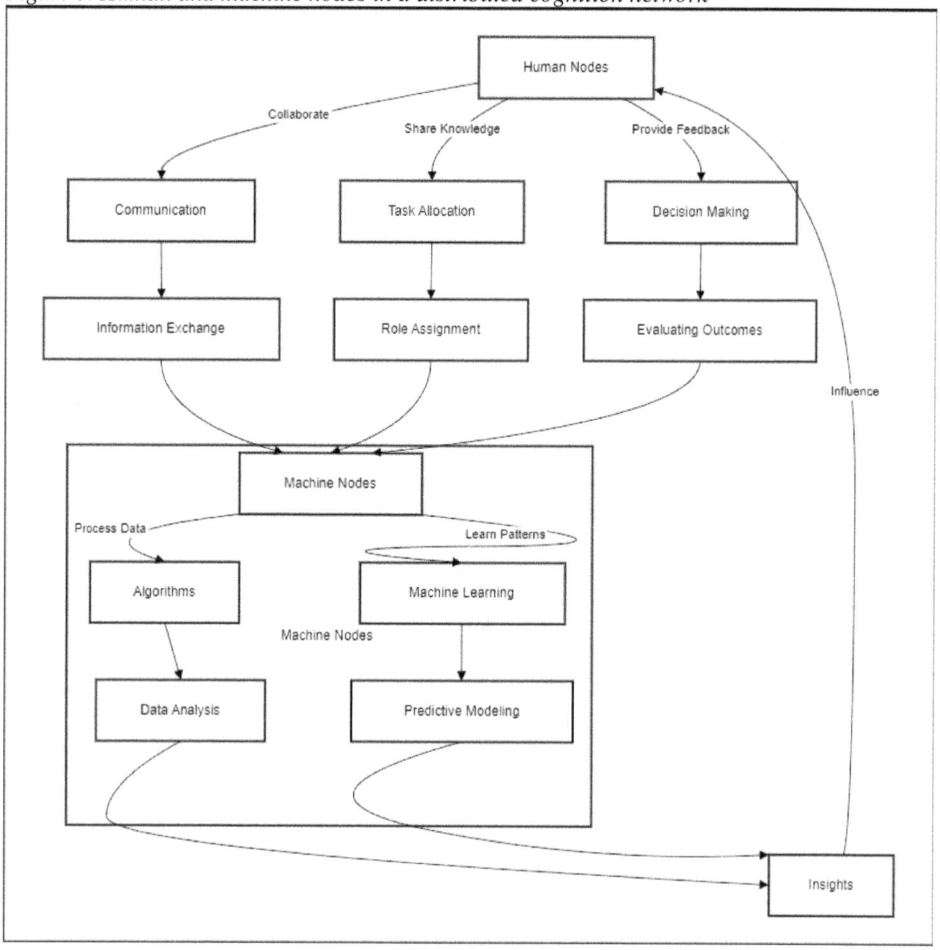

Joint Cognitive Systems (JCS)

Authors David Woods and Erik Hollnagel's Joint Cognitive Systems (JCS) produced a hypothesis in 2005 which emphasizes on mutual adaptability between human and machine entities, extending it to the idea of shared cognition. According to the authors, human-machine interaction is a partnership in which both parties actively engage in cognitive processes rather than a one-way help paradigm. Because it co-evolves—humans adjust to the limitations and capacities of computers, while intelligent systems change dynamically based on user behavior and environmental context—the system is referred to as "joint."

This could entail a collaborative robotic workstation in a smart factory environment, where a robot (Wei et al., 2023) and an operator work together to complete an assembly operation. While the human keeps an eye on output and makes contextual decisions, the robot suggests workflow optimizations based on real-time data. Crucially, the system is built to learn from human inputs; for example, it records when a person overrides a robotic action for later pattern recognition and behavior adjustment. Instead of complete automation, JCS emphasizes shared control and resilience. This concept is particularly applicable in high-stakes or complex industrial settings where trust, transparency, and error tolerance are essential.

Human-computer interfaces that facilitate bidirectional communication are necessary when designing for JCS. This allows humans to intuitively step in and reroute processes while allowing machines to "explain" their recommendations.

Levels of Autonomy and Collaboration

The interaction of human and machine is based on the autonomy and collaboration where the interaction can be very little or very high. System designers can decide how much decision-making power should be given to humans and robots and how to structure interactions for efficiency, safety, and trust by having a better understanding of this continuum. There are various levels based on the jobs assigned, namely:

- Level 1 Coexistence: Humans and machines share a workspace but function separately. This is common in early automation, when human responsibilities are maintenance or supervisory and robots follow set patterns.
- Level 2 Coordination: Human operators give machines predetermined commands. Machine behavior changes are triggered outside, and communication is unidirectional.
- Level 3 Cooperation: Although machines are capable of interpreting sensor data and making real-time adjustments to their operations, they still rely on human judgment. It is a semi-autonomous system with bidirectional information flow.
- Level 4 Collaboration: Real-time mutual adaptation and goal-sharing are practices shared by humans and machines. Both sides dynamically negotiate control during the interactive decision-making process.
- Level 5 Co-agency: The human-machine team functions as a collective intelligence with blurred boundaries. The system demonstrates great levels of autonomy while maintaining human ethical control, and roles are flexible.

This is a very significant work in defining the jobs roles as it decides the levels of control, accountability, and transparency in an organized manner—all of which are essential for fostering trust in cooperative systems.

Human-Centred AI Design

The entire goal of designing human centred AI is to create intelligent systems that enhance, empower, and support human users but not to replace them. The ground truth is to work in line with humans in all required situations. This results in systems and interfaces in manufacturing settings that are clear, easy to use, and flexible enough to accommodate varying user skill levels.

Explainability is a feature of a well-designed HCAI that enables consumers to comprehend the reasoning behind an AI system's recommendations. Additionally, it gives priority to feedback loops, which allow users to contribute ideas or override recommendations, facilitating a circle of reciprocal learning. For jobs like predictive maintenance in smart factories, where an AI tool may recommend downtime scheduling based on pattern recognition, these systems are essential. However, the human operator can make the final decision based on operational context.

Based on the training given to AI on the different backgrounds or abilities, ensures the proper interface with the user. When AI systems were shaped by keeping in mind about the user, they stand-in a collaborative culture, increase trust, and make people feel in charge of their technology rather than isolated by it. In essence, HCAI shifts the focus from automation to augmentation, recognizing the value of human judgment and creativity in industrial operations.

Socio-Technical Systems (STS)

According to the Socio-Technical Systems (STS) theory, a system's social context and technical infrastructure must be integrated for it to be successful. This means that, in the context of smart factories, technology advancements like artificial intelligence (AI), the Internet of Things (IoT), and robotics must be implemented in conjunction with careful evaluations of workplace culture, communication procedures, staff training, and ethical standards.

According to the author Appelbaum, S. H., (1997), for a consultant or an executive it's a difficult job to merge the inventions in technology and the organizational development into one system.

Even a technically sound system can go wrong if it doesn't take user behavior into consideration. When autonomous scheduling software is introduced without worker input, for instance, it may be met with opposition, mistrust, or abuse. In order to ensure that solutions are in line with workflows and restrictions found in the real world, STS promotes co-design approaches, in which staff members provide their domain expertise and tacit knowledge during the design phase.

Additionally, STS emphasizes the significance of diversity, openness, and governance. Are there avenues for employees to voice concerns? Can decisions made by AI be audited? Does the organization have the capacity to upskill its employees?

According to this approach, success is determined by the general health and adaptability of the socio-technical ecosystem rather than just ROI or productivity. When properly applied, STS guarantees that human-machine cooperation results in systems that are not only effective but also moral, long-lasting, and empowering.

COLLABORATIVE INTELLIGENCE

A collection of interconnected technologies that establish an ecosystem in which humans and machines may coexist peacefully are essential to the implementation of Collaborative Intelligence (CI) in smart factories. Effective human-machine collaboration is powered by these technologies, which also act as the cognitive engine and infrastructure. In the era of Industry 4.0 (Xu et al., 2018), these enablers revolutionize the design, execution, and optimization of work by facilitating real-time data interchange, context-aware computing, and adaptive interaction.

Figure 4. Technology enablers of collaborative intelligence

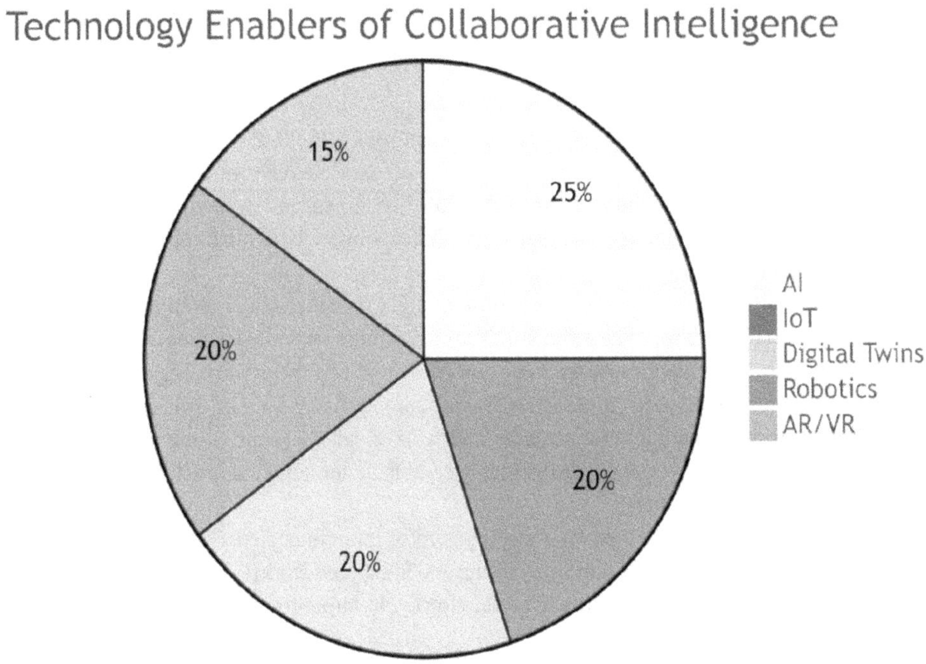

The above Figure 4 indicates the collaboration of the technology with the intelligence where artificial intelligence is about 25%, internet of things holds 20%, digital twining and robotics are as same as IoT i.e. 20% each, and augmented reality and virtual reality is 15%.

Artificial Intelligence and Machine Learning

The intellectual foundation of CI systems is made up of machine learning (ML) and artificial intelligence (AI) (Mohanan, 2025). They enable machines to perform tasks that were previously limited to human cognition, including as interpreting complicated datasets, identifying patterns, and making judgments. Applications like visual inspection, defect categorization, and anomaly detection are powered by supervised and unsupervised learning approaches, which allow machines to continuously improve through feedback loops driven by data.

Autonomy and adaptability are introduced into dynamic situations through reinforcement learning, particularly in robotics and real-time production optimization. For instance, robots that have received reinforcement learning training are able to adjust to changes in the industrial layout while learning the most effective routes and workflows.

By enabling machines to comprehend written or spoken instructions, Natural Language Processing (NLP) makes technology accessible to non-technical users. This facilitates intelligent chatbots, voice-activated interfaces, and multilingual shop floor help systems (Shah, 2024).

All things considered, AI and ML not only improve machine capabilities but also facilitate responsive and easy collaboration with human counterparts. These systems' capacity to learn from human input and modify their behaviour guarantees that they stay in line with human objectives and the operating environment.

Internet of Things (IoT)

The sensing and communication layer of smart factories is the Internet of Things (IoT) (Dou et al., 2023). IoT allows for continuous monitoring and data interchange across all production layers by integrating sensors and actuators into tools, machines, and even goods. Temperature, vibration, torque, and energy usage are just a few of the many variables that these sensors gather and provide the raw data for AI-driven insights.

In order to facilitate context-aware decision-making, IoT is essential. IoT-enabled alarms, for example, can alert maintenance algorithms and human operators when a machine exhibits unusual vibration patterns, allowing for prompt, well-informed replies. These warnings can be handled locally in almost real-time when combined with edge computing, reducing downtime and latency.

IoT also improves operational awareness and worker safety. While proximity sensors can warn people and machines to avoid collisions, wearable technology can track worker vitals or exposure to dangerous situations.

Beyond a single device, IoT builds an ecosystem of interconnected systems that opens the door to predictive maintenance, smart scheduling, and integrated workflows. It ensures visibility, traceability, and reactivity throughout the production cycle, supporting both technical efficiency and human-centric operations.

Augmented Reality (AR) and Virtual Reality (VR)

By overcoming the gap between digital and physical reality, augmented reality (AR) and virtual reality (VR) are transforming how people view and engage with industrial systems. By superimposing contextual digital data on top of actual surroundings, augmented reality (AR) gives employees real-time guidance, alerts, and diagnostics without taking them away from their work. When performing maintenance, for instance, a technician with AR glasses can view overlayed visual guidelines, greatly increasing speed and accuracy.

According to the authors Ma, Gausemeier, Fan, and Grafe (2011) in their book titled Virtual reality & augmented reality in industry mentioned that AR & VR are the valuable tools to increase the product production and faster development of the process.

Conversely, virtual reality (VR) makes it possible for operators to practice complex or high-risk processes in a safe and controlled environment through immersive training environments. These simulations shorten training times, improve muscle memory, and get employees ready for unforeseen circumstances.

Remote cooperation is supported by both AR and VR. Through AR interfaces, experts can provide real-time visual comments to on-site workers. This is particularly helpful in situations when specialized involvement is needed but physical presence is impractical, or in operations that are spread out geographically.

By fusing human perception with data-rich interfaces, these technologies enhance worker potential, decrease human error, and improve knowledge retention. AR and VR play a crucial role in collaborative intelligence by ensuring that machine-generated knowledge complements human understanding rather than replaces it.

Robotics and Collaborative Robots (Cobots)

Rigid, segregated automation has given way to flexible, intelligent systems that can interact closely with humans in modern robotics. Cobots, or collaborative robots, are made especially to work alongside human operators without the need for physical barriers or safety cages. They can recognize human proximity, gestures, and force feedback thanks to their sensors and adaptive algorithms, which guarantees smooth and secure contact.

As mentioned by the authors Sherwani, Asad, and Ibrahim (2020) we can achieve high productivity and production cost by collaborating humans and robos i.e. using the decision-making capability of human repeatability and strength of robots".

By taking over tedious taxing, or hazardous jobs, cobots increase human productivity by allowing employees to concentrate on higher-value duties like inspection or decision-making. For instance, a human operator may supervise quality and coordination while a cobot manages component placement.

This capability is further expanded by mobile robots, which can deliver parts, inspect conditions, and navigate production areas on their own. In order to make context-aware decisions and react quickly to changes in their environment, these robots frequently rely on artificial intelligence.

Robots can do more than just perform predetermined jobs when AI is integrated with them. They can contribute to common objectives, adjust to operator behaviour, and learn from demonstration. Their capacity to work with humans, rather than merely coexisting with them, represents a paradigm change in industrial automation. Robots are collaborators in production as well as instruments in the context of continuous integration.

Digital Twins

A digital twin (Ranawaka et al., 2024) is an actual digital duplicate of a system, process, or physical object. It offers a real-time, interactive model that replicates manufacturing activities and is continuously fed by data from IoT sensors. Digital twins are essential to continuous integration (CI) in complex contexts because they allow for real-time monitoring, diagnostics, simulation, and prediction (Tao et al., 2022).

A digital twin (Tao et al., 2018) of a production line, for instance, enables engineers and operators to model modifications—like changing temperature thresholds or rearranging workflows—before implementing them in practice. This feature facilitates proactive decision-making, risk assessment, and what-if analysis.

By simulating equipment wear and tear based on past trends, digital twins aid in predictive maintenance by seeing possible faults before they happen. This maximizes asset use and reduces downtime (Hui, 2024).

Digital twins also provide a common interface for human-machine cooperation. Through user-friendly dashboards, engineers can engage with the virtual model, and AI agents are constantly updating simulations using real-time data. A coherent decision-making environment is produced by this real-time feedback loop, which makes use of both machine accuracy and human intuition.

Edge and Cloud Computing

The infrastructure needed for scalable, responsive, and intelligent CI systems is supported by the two pillars of cloud and edge computing. Centralized data storage, model training, and enterprise-wide integration are made possible by cloud computing. Massive datasets, multi-user cooperation, and ongoing AI model refinement are all supported (Bajic et al., 2019).

Conversely, edge computing positions processing power on devices, sensors, or local gateways that are closer to the data source. This significantly lowers latency, enabling real-time decision-making in applications that require quick responses, such as emergency notifications or robotic control.

A balanced strategy is made possible by the synergy between cloud and edge architectures, where the edge takes care of urgent operational requirements while the cloud handles large-scale analysis and long-term learning. To prevent equipment failure within milliseconds, for example, an edge node may examine sensor data. The cloud platform then aggregates this data across factories to improve predictive models.

Collaborative intelligence systems prove to be clever, responsive, dependable, and robust by providing a architecture which interacts with human and machine in a smooth manner.

Cybersecurity and Data Governance

Data governance and cybersecurity are essential for maintaining trust, compliance, and long-term viability in an interconnected industrial ecosystem where data is continuously transferred between people, machines, and cloud platforms. Data integrity, privacy, and secure communication are critical components of collaborative intelligence systems.

In order to mitigate the risks associated with outdated perimeter-based models, Zero Trust Architecture (ZTA) is being used more and more to validate all requests, regardless of where they come from. Blockchain technology (Van Rijmenam, M., 2019) is being investigated in the interim to improve traceability, safeguard transaction records, and guarantee tamper-proof data exchanges.

When dealing with consumer who are sensitive to their information then Data governance plays a major role as it is a set of rules or policies which speaks of accessibility given to control data, usage of this data, and also speaks of legal and ethical requirements.

Smart factories' operational continuity is safeguarded, and user confidence is increased through efficient cybersecurity and governance. In the face of escalating cyberthreats, they guarantee that collaborative intelligence is not just potent but also reliable, moral, and robust.

ARCHITECTURES OF SMART FACTORIES: WHERE HUMANS MEET MACHINES

The physical, digital, and cyber-physical configuration of systems that facilitate human-machine cooperation is embodied in the architecture of a smart factory. Agility, interoperability, and co-adaptability are key considerations in the design of these architectures, which enable real-time decision-making and collaboration across roles and technology.

Core Architectural Components

The following interrelated layers are commonly found in smart factories:

Machines, sensors, actuators, and human operators operating in close proximity make up the physical layer. It includes cobots, robots, and production lines.

Network Layer

Provides smooth connectivity via IoT protocols, 5G, Wi-Fi 6, and industrial Ethernet.

Data Layer

Manages the gathering, combining, cleaning, and preparation of data. Cloud data lakes, edge nodes, and IoT gateways fuel it. AI models, dashboards, manufacturing execution systems, enterprise resource planning, and simulation tools are all part of the application layer.

Interface Layer

Facilitates communication between humans and machines via voice commands, control panels, augmented reality displays, and mobile applications.

Figure 5. Layered architecture of a smart factory

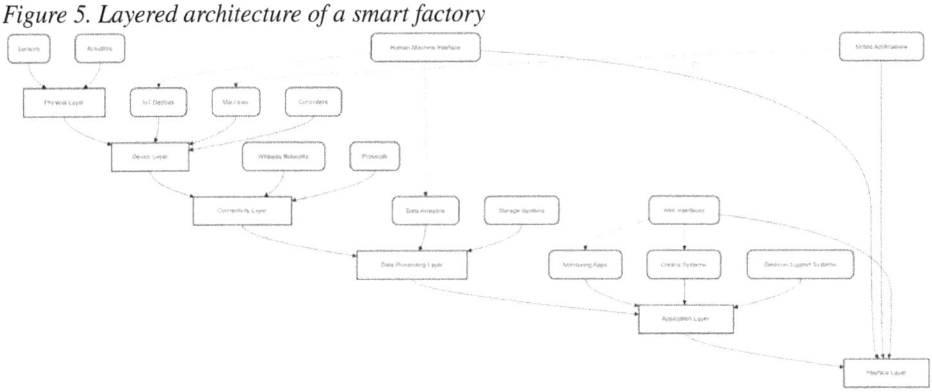

The above Figure 5 shows the various layer architecture of an IoT enabled human machine system of collaboration of a smart factory or industrial automation and industry 4.0 environments. There is various layer namely physical, device, connectivity, data processing, application and interface layers. The explanation for each of these layers is given below:

Physical layer has sensors and actuators as the components and their role is to collect real-world information in terms of temperature, motion and pressure, and perform physical actions like opening a valve or moving a robotic arm etc. respectively. The main function of this layer is to form a strong groundwork by associating the physical world with the digital system.

Device layer has IoT devices and machines, IoT device perform the capturing, preprocessing or processing of the information and any automation in the production/operation is taken care by machines. There main function is to perform the interface between physical components and digital communication systems.

Connectivity layer has three components wireless networks, controllers and protocols each of which has a special role to ensure that the devices communicate with each other and also with higher layers by using the standard communication protocols like MQTT, OPC-UA, etc. the processing of the systems can happen reliably by transfer od data from devices.

Data Analytics and Storage Systems are the components of Data processing layer where data analytics perform the processing of raw data and show into the use of artificial intelligence or machine learning, or statistical models and storage systems performs analysis and compliance based on the retained data. The main function of this layer is to transform data into actionable information.

The components of application layer are Monitoring Apps, Control Systems, Decision Support Systems, Web Interfaces each of which help in displaying the real time information, adjust operations by enabling the operators or algorithms, in order to optimise the performance, provide recommendations, and system functionality is provided by browser-based access, over all this layer supports user interaction and also help in operational decision making.

Interface layer has two components namely Human–Machine Interface (HMI), and mobile applications where HMI performs the task of operator consoles or dashboards for interacting with machines and processes, and mobile applications enable monitors and controls remotely. The function of this layer is to provide the user to interact with the system in real-time.

The flow of the interaction can be highlighted as follows:

- **Data Acquisition:** From Physical layer to Device layer.
- **Data Transmission:** From Device layer to Connectivity layer.
- **Data Processing:** From Connectivity layer to Data processing layer.
- **Application Execution:** From Processed data to Application layer.
- **User Interaction:** From Application layer to Interface layer (HMI / Mobile Apps).
- **Control Feedback:** From interface/application back to physical layer via actuators.

Workstations for Collaboration

The purpose of collaborative workstations is to enable human-machine task execution. Cobots, AR interfaces, voice-activated AI assistants, and adaptable tooling systems are a few examples of these configurations. To accommodate a variety of product lines, they are frequently movable and adaptable.

A worker completing intricate assembly under the guidance of AR overlays and supported by a cobot holding components are two examples.

An operator looking over predictive maintenance notifications produced by an AI dashboard that combines sensor information with past patterns.

Human workers may concentrate on decision-making and flexibility rather than physical effort thanks to these workstations, which place a high priority on ergonomics, safety, and intuitiveness.

The Continuum from Edge to Cloud

Smart factories are able to strike a compromise between the requirements for long-term data analytics and real-time processing which is provided by the architectural movement from edge to cloud.

The jobs taken by edge devices are process control, quality assurance, and safety which are time-sensitive jobs. Supply chain coordination, worldwide operations, model training, and historical data collection are the jobs taken by Cloud systems. Cloud-based orchestration combined with edge AI guarantees rapid and consistent cooperation across time zones and locales.

HITL and HOTL (human-in-the-loop and human-on-the-loop)

HITL or HOTL models are commonly used in collaborative architectures:

HITL

A human actively engages in decision-making and task execution, including overriding machine recommendations as can be seen in Figure 6.

HOTL

A human oversees automated processes and steps in only when required.

Interfaces that support trust calibration, quick input when needed, and comprehension of machine logic (explainable AI) are necessary for both approaches as can be seen in Figure 7.

Figure 6. HITL architecture

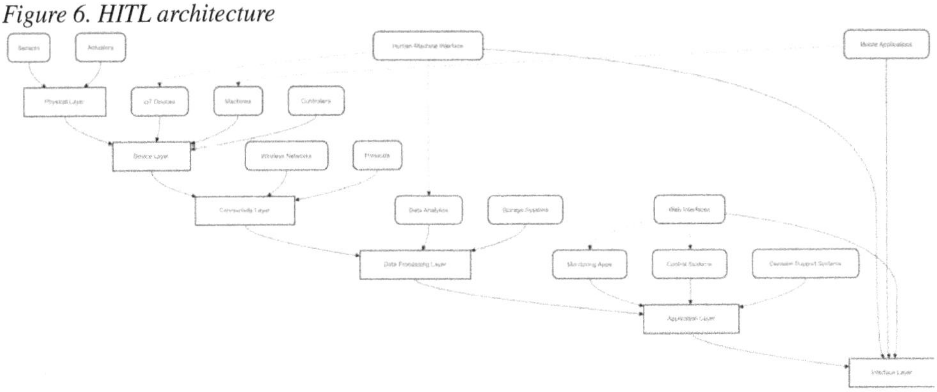

Figure 7. HOTL architecture

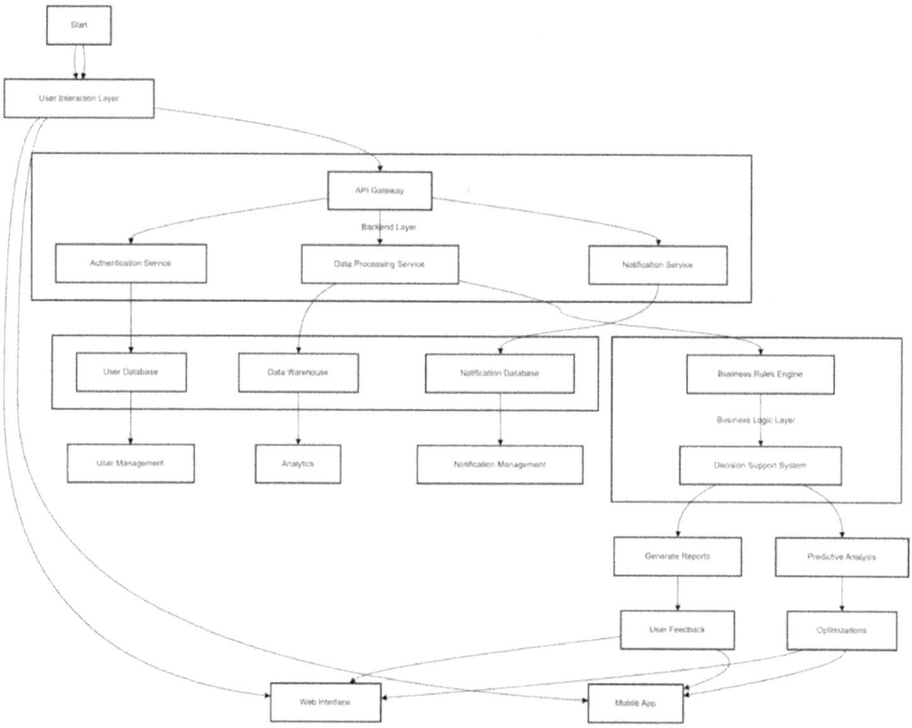

Cyber Physical Systems (CPS)

CPS are systems in which networking, computation, and physical processes are closely intertwined. CPS enables real-time physical process monitoring and control through virtual representations in smart factories.

AI-driven optimization engines, real-time feedback loops between machinery and MES, and digital twins of production lines are a few examples.

Additionally, CPS facilitate situational awareness, allowing for proactive modifications in response to operator input or changes in the surroundings.

Modularity and Interoperability

Architectures must support the following in order to handle ongoing evolution and changing needs: Plug-and-play parts that can be improved or replaced without causing significant problems are known as modularity.

Interoperability

Using open standards (like OPC UA and MQTT) to guarantee smooth communication across various suppliers and systems.

In order to scale and modify the production environment without sacrificing performance or safety, this flexibility is crucial.

Layers of Safety and Ethics

Smart industrial designs incorporate ethical and safety considerations:

safety procedures that make use of proximity sensors and AI-based vision systems.

Fair task distribution, less prejudice in AI decision-making, and consideration for operator preferences are all ensured by ethical logic layers.

To have a long-term relationship between humans and robots, there has to be feedback system which accepts and adjusts based on the input given by humans on bias, irregularities, or discomfort happened in the process.

SOLUTIONS OF CASE STUDIES: REAL-WORLD APPLICATIONS

The potential for human-machine collaboration to revolutionize industrial operations is demonstrated by real-world applications of collaborative intelligence in smart factories. Selected case examples from various industries are presented in this part to highlight the variety of uses, advantages, and lessons discovered through the integration of collaborative systems.

BMW Group: Joint Production Lines

The new trending technologies are considered by the company i.e. collaborative robots, or cobots, so that humans get help from the robots during the assembly processes. The best-known example is the use of Cobots in door assembly at the Leipzig facility.

Application

Human workers carry out complex assembly activities while cobots grasp and position heavy door components. Utilized technologies include force-feedback sensors, ergonomic augmented reality displays, and AI-controlled cobots.

Impact

Less physical stress on employees, increased speed and accuracy of assembly, hard workload leads to increased job satisfaction.

Takeaway

This setup highlights human adaptability combined with machine strength and consistency.

Siemens Amberg Electronics Plant: Digital Twin from Birth until End

The use of printed circuit board (PCB) during the production as an end-to-end digital twin is well known at Siemens' Amberg facility.

Application

Simulation and optimization are made possible by real-time digital twins of production lines and machinery. IoT sensors, cloud analytics, machine learning, and simulation software were among the technologies used.

Impact

Predictive maintenance reduced downtime, smooth cooperation between AI stems, operators, and engineers, enhanced agility and throughput.

Takeaway

Digital twins enable seamless human-machine collaboration by providing transparency, predictive insights, and decision support across the entire product lifecycle.

The below Figure 8 shows the digital twin workflow followed in Siemens, this innovation is a closed loop which helps in improved development, production and operation.

Figure 8. Siemens digital twin workflow (https://cambashi.com/siemens-goes-all-in-on-digital-twin/)

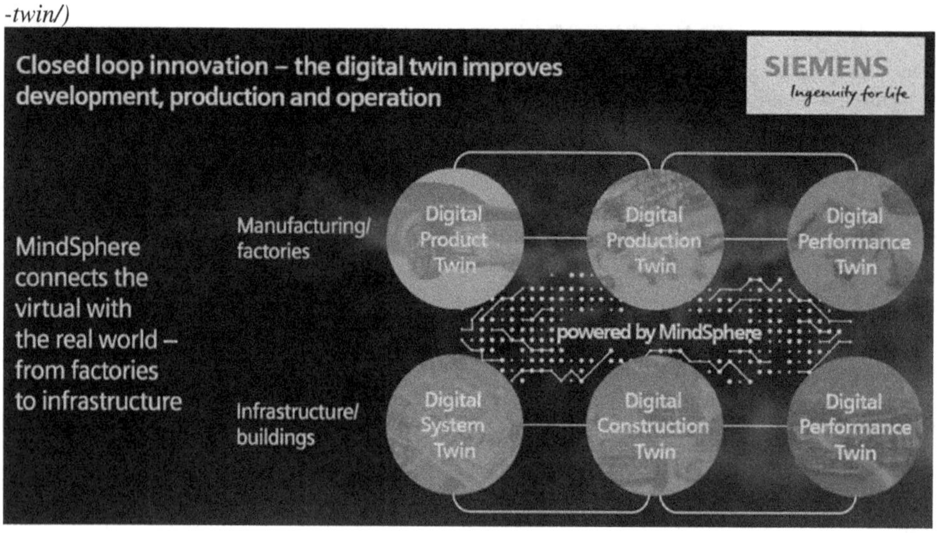

Foxconn: Quality Inspection Driven by AI

During the quick assembly lines, any flaws can be inspected by the use of AI algorithms is the technique that is been used by Foxconn, a multinational electronics manufacturing.

Application

Real-time surface defect and anomaly detection is achieved by AI algorithms analyzing camera feeds. Utilized technologies include computer vision, edge computing, and deep learning models.

Impact

Missed flaws can be reduced by up to 95%. The focus of human inspectors shifted to system training and edge cases, constant enhancement via feedback loops.

Takeaway

This example demonstrates how CI technologies enhance human perception and judgment.

Procter & Gamble (P&G): Maintenance Using Augmented Reality

AR-assisted maintenance practices have been implemented at several P&G production facilities.

Application

Field technicians can access remote assistance and interactive maintenance instructions by using AR glasses as in figure 9, utilized technologies include digital maintenance logs, IoT integration, and augmented reality headsets.

Impact

Less downtime for the machines, quicker upskilling and onboarding of technicians, enhanced adherence to routine operating procedures.

Takeaway

Augmented reality enhances human capabilities by delivering real-time information and expert support, reducing errors and downtime while accelerating skill development.

Figure 9. Technician using AR glasses for maintenance support

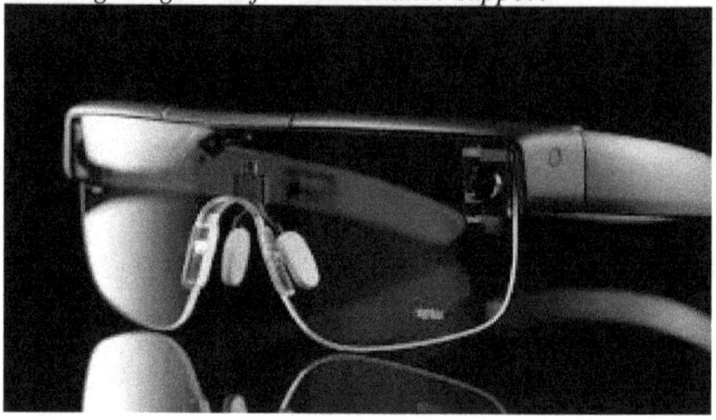

Airbus: Aerospace Human-Robot Cooperation

Cobots are used by Airbus to help with precision drilling operations on airplane fuselages, which call for a high degree of accuracy.

Application

Human workers oversee quality control as cobots drill thousands of holes each aircraft, Cobots, AI alignment systems, and AR overlays for verification are some of the technologies used.

Impact

Enhanced structural integrity and precision, reduced rates of mistake, kept employees safe while performing high-risk and repetitive duties.

Takeaway

This illustrates collaborative precision engineering, in which machine uniformity and human oversight coexist.

Toyota Production System – Lean Manufacturing with Human Judgment

Toyota blends automation with a strong emphasis on human insight through the *Andon* system, which allows workers to stop the production line if a fault is detected.

Collaborative Intelligence

Automation is paired with human authority and responsibility, ensuring quality through shared control.

Impact

Quality assurance, worker empowerment, and minimized waste—all critical to Toyota's global success.

Takeaway

Empowering human workers within automated systems enhances quality, accountability, and continuous improvement by placing trust in human judgment at critical moments.

GE Aviation – Predictive Maintenance with Human Analysis

GE uses both machine and human intelligence i.e it considers AI for predictive maintenance on aircraft engines and the expertise of an engineers to verify and interpret the model's outputs.

Collaborative Intelligence

AI detects anomalies; human experts investigate contextual causes and make final decisions.

Impact

Enhanced safety, reduced downtime, and effective risk mitigation via a human-in-the-loop approach.

Takeaway

Human expertise adds critical judgment, contextual awareness, and accountability to AI-driven diagnostics, ensuring reliability and trust in high-stakes systems.

Bosch Rexroth – Smart Workstations with Adaptive Assistance

The AI driven sensors input and the ergonomic data both are considered by the Bosch who developed an intelligent workstation that adapt to individual worker needs.

Collaborative Intelligence

The system guides workers through assembly tasks while adjusting table height, lighting, and tool suggestions based on worker behavior and experience level. Adaptive workstations that adjust ergonomics, display guidance, and tools based on worker behavior and skill level.

Impact

Reduced training time, fewer errors, increased worker satisfaction, and improved accessibility for diverse labor forces.

Takeaway

Intelligent environments can enhance human performance and well-being by adapting to individuals, not replacing them.

TESLA – HUMAN SUPERVISION IN AUTOMATED GIGAFACTORIES

Tesla's highly automated Gigafactories still rely heavily on human technicians to calibrate robots, handle non-standard tasks, and ensure smooth line functioning.

Collaborative Intelligence

AI handles production speed and precision; humans intervene for quality control, problem-solving, and system adjustments. Highly automated production lines where humans oversee, troubleshoot, and fine-tune robotic systems.

Impact

Maintains high output while keeping systems flexible in response to design or model changes.

Takeaway

Human oversight ensures flexibility and resilience in large-scale automation, especially when unexpected variability arises.

HAIER GROUP – USER-DRIVEN MASS CUSTOMIZATION WITH HUMAN-AI TEAMS

This company uses the combination of skills of design engineers and AI to create consumer appliance which are customized based on the needs of customers i.e COSMOPlat in the industrial platform.

Collaborative Intelligence

AI analyzes user feedback and configures production, while humans validate the feasibility and fine-tune design features. Co-creation of custom appliances via AI design and human validation on the COSMOPlat platform.

Impact

Enables large-scale customization while maintaining efficiency and consumer responsiveness.

Takeaway

Blending AI and human creativity enables mass customization without sacrificing efficiency or product quality.

SCHNEIDER ELECTRIC – AUGMENTED OPERATORS IN ENERGY-INTELLIGENT FACTORIES

At Schneider Electric's Le Vaudreuil smart factory, workers use augmented reality (AR) and AI analytics to manage energy efficiency and machine health.

Collaborative Intelligence

AR provides real-time data overlays; workers interpret results and act based on context and experience. AR and AI tools empower factory workers to monitor energy usage, diagnose faults, and make data-informed decisions.

Impact

Increased energy efficiency, reduced machine downtime, and enhanced situational awareness on the shop floor.

Takeaway

Augmenting humans with digital tools transform them into proactive decision-makers, boosting sustainability and uptime.

FANUC – MACHINE LEARNING WITH REMOTE HUMAN GUIDANCE

FANUC uses AI to optimize robotic motion paths, while remote engineers monitor and guide the learning process to ensure relevance and safety.

Collaborative Intelligence

AI learns from process data, but humans validate, tune, and apply solutions in real-time. AI robots learn motion paths while human experts validate safety and effectiveness remotely.

Impact

Improved learning speed, fewer accidents, and safer AI deployment in robotic systems.

Takeaway

Even self-learning systems benefit from human context-awareness and ethical oversight, ensuring safer AI deployment.

ABB – COLLABORATIVE ROBOTICS IN ELECTRONICS MANUFACTURING

ABB's YuMi cobot is used in electronics assembly where intricate components require fine manipulation and adaptability.

Collaborative Intelligence

YuMi handles precision tasks, while humans manage part feeding, inspection, and design variation handling. ABB's YuMi cobots work side-by-side with humans in precision electronics tasks requiring dexterity and adaptability.

Impact

Combines robot precision with human adaptability in small-batch, high-variability manufacturing environments.

Takeaway

Cobots extend human capabilities in fine-detail work, allowing scalable collaboration in high-variability environments.

PHILIPS – LIGHTS-OUT MANUFACTURING WITH HUMAN REMOTE MONITORING

Philips' lighting plant in the Netherlands operates with minimal onsite staff, but humans remotely monitor processes, resolve anomalies, and maintain oversight.

Collaborative Intelligence

Automation runs routine tasks; humans intervene only for strategic or unexpected decisions. Nearly fully automated factory with remote human teams managing anomalies, strategic changes, and interventions.

Impact

Maximizes automation efficiency without compromising operational safety or adaptability.

Takeaway

Automation can operate autonomously, but human-in-the-loop monitoring ensures adaptability, safety, and strategic control.

CHALLENGES AND ETHICAL CONSIDERATIONS

Despite the promising potential of collaborative intelligence in smart factories, several challenges and ethical dilemmas must be addressed to ensure sustainable, inclusive, and responsible implementation. These challenges span technical, organizational, social, and moral dimensions.

Complexity and Technological Limitations

Diverse technologies, each with unique limitations, are integrated into collaborative intelligence systems:

Generalization of AI Models

In new product lines or unexpected situations, machine learning models may perform poorly.

Integration of Systems

It is difficult to provide smooth interoperability between new platforms, heterogeneous devices, and old systems.

Reliability and Maintenance

Cyber-physical coordination, sensor calibration, and frequent updates are necessary for autonomous systems.

Technical staff resistance, inefficiencies, and downtime might result from these technological obstacles.

Skill Gaps and Workforce Readiness

Human workers' roles are shifting toward adaptation, decision-making, and supervision as computers take on increasingly complex jobs.

Upskilling Requirements

Employees must receive instruction in collaborative robotics, digital literacy, and AI interpretation.

Opposition to Change

Adoption may be hampered by suspicion of automation or fear of job displacement.

Cognitive Overload

Juggling supervision duties across several intelligent systems might cause focus and poor decision-making.

Figure 10. Survey on workforce sentiment towards collaborative technologies

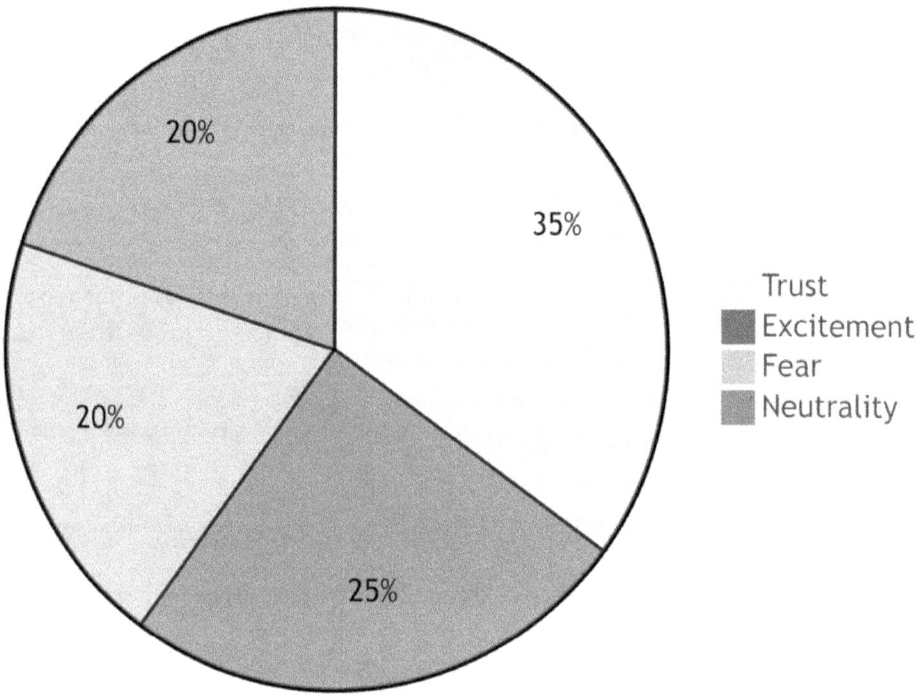

The above Figure 10 shows the pie graph of the sentiments considered in the collaborative technology in terms of percentage, where trust has the highest weightage of 35%, excitement has 25% of weightage, fear and neutrality are 20% each.

Privacy and Surveillance Issues with Data

Large volumes of data are gathered by smart manufacturing environments (Tiwari et al., 2023), including:

- Location and behavior of workers and machine performance
- Biometric or verbal inputs (e.g., voice instructions, signs of weariness)

This poses important queries:

- The data produced by human-machine interaction belongs to whom?
- What are the boundaries of performance monitoring surveillance?
- How is personal information safeguarded by international laws such as the CCPA or GDPR?

To foster confidence and guarantee the ethical use of data, clear policies and transparency are essential.

Fairness and Algorithmic Bias

Biases in training data or design may be reflected in AI systems utilized in smart factories as in Figure 11:

Task Allocation Bias

Inadvertently, algorithms may give preference to particular roles, shifts, or profiles.

Interpretability Problems

Decisions made by black-box models may be hard for people to challenge or confirm.

Fairness in algorithmic judgments is crucial, particularly when it comes to hiring, assigning tasks, or evaluating performance.

Figure 11. Sources of bias in industrial AI pipelines

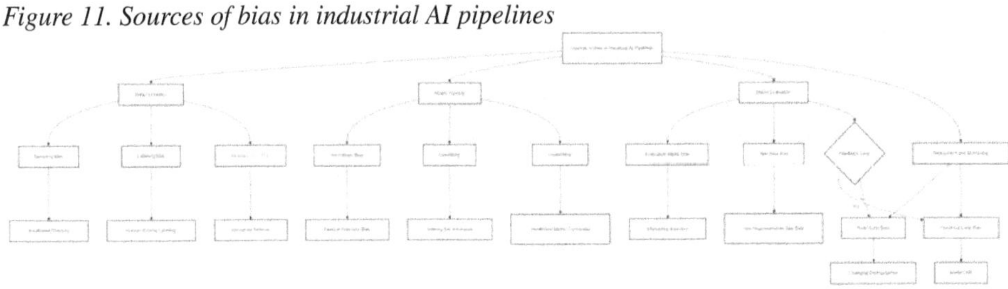

Ethical Machine Decision-Making

Ethical quandaries emerge as machines assume increasingly independent roles:

- Should a robot put production efficiency (such as pace or task difficulty) ahead of worker well-being?
- How should an AI reconcile contradicting commands from several operators?

To guarantee alignment with company and societal values, human-in-the-loop governance mechanisms and ethical design frameworks must be in place.

Liability, Accountability, and Trust

In systems of shared autonomy, it is essential to establish distinct lines of accountability:

- Who is in charge if an AI system breaks down—executives, vendors, engineers, or operators?
- How can one balance confidence between machine recommendations and human judgment?

Organizational, contractual, and legal structures must change to answer these issues.

Social and Economic Effects

Collaborative intelligence may change labor markets on a larger scale:

Job Displacement

Economic inequality may result from the automation of routine tasks.

Creation of New Jobs

Human-machine training, system design, and AI supervision will all see an increase in roles.

Global Disparities

There may be a growing adoption gap between developed and developing nations.
To manage these changes fairly, policy, education, and inclusive design are required.

FUTURE RESEARCH DIRECTIONS IN COLLABORATIVE INTELLIGENCE

The next ten years will see the rise of revolutionary trends that will reshape human-machine cooperation as smart factories continue to develop. Technological developments, changing socioeconomic paradigms, and an increasing focus on sustainable and ethical innovation will all influence these trends (Benbya et al., 2020).

Designing AI with Humans in Mind

- Human values, emotions, and context will be given more importance in future collaborative systems:

Affective AI refers to systems that can recognize and react to human emotions, resulting in more sympathetic user interfaces.

- Context-Aware Systems: Devices that modify their actions in response to real-time perceptions of human intent, weariness, or skill levels.
- Cognitive Digital Twins: Tailored digital models that mimic human abilities, inclinations, and performance trends in order to maximize teamwork.

More natural, intuitive, and encouraging interactions will be supported by these platforms.

The Emergence of Autonomous Co-Creation and Generative AI

In industrial settings, generative AI technologies such as GPT and DALL·E are opening the door for innovative co-design:

- Collaborative Design: AI tools and human engineers work together to jointly develop maintenance schedules, industrial layouts, and product designs.

AI agents that can learn from operators and create machine code or do robotic tasks on their own are known as autonomous programmers.

This tendency lessens the cognitive load on human workers and democratizes creativity.

Highly Tailored Processes

Workflows will be customized to each person's preferences and talents with the use of AI analytics and more detailed data collection:

- biometric feedback-based dynamic workload balancing.
- AR/VR-based customized training routes.
- Role adaptability according to mental state and skill level.
- These developments will maximize performance while improving job satisfaction and human well-being.

Multi-Agent Systems and Collaborative Swarms

Future factories will use coordinated swarms of robots in addition to individual systems or robots:

Small autonomous robots working together on big projects under human supervision is known as swarm robotics.

- Multi-Agent AI: Dispersed AI systems that collaborate with humans and one another to negotiate, divide, and carry out tasks.

Figure 12. Framework for multi-agent collaboration

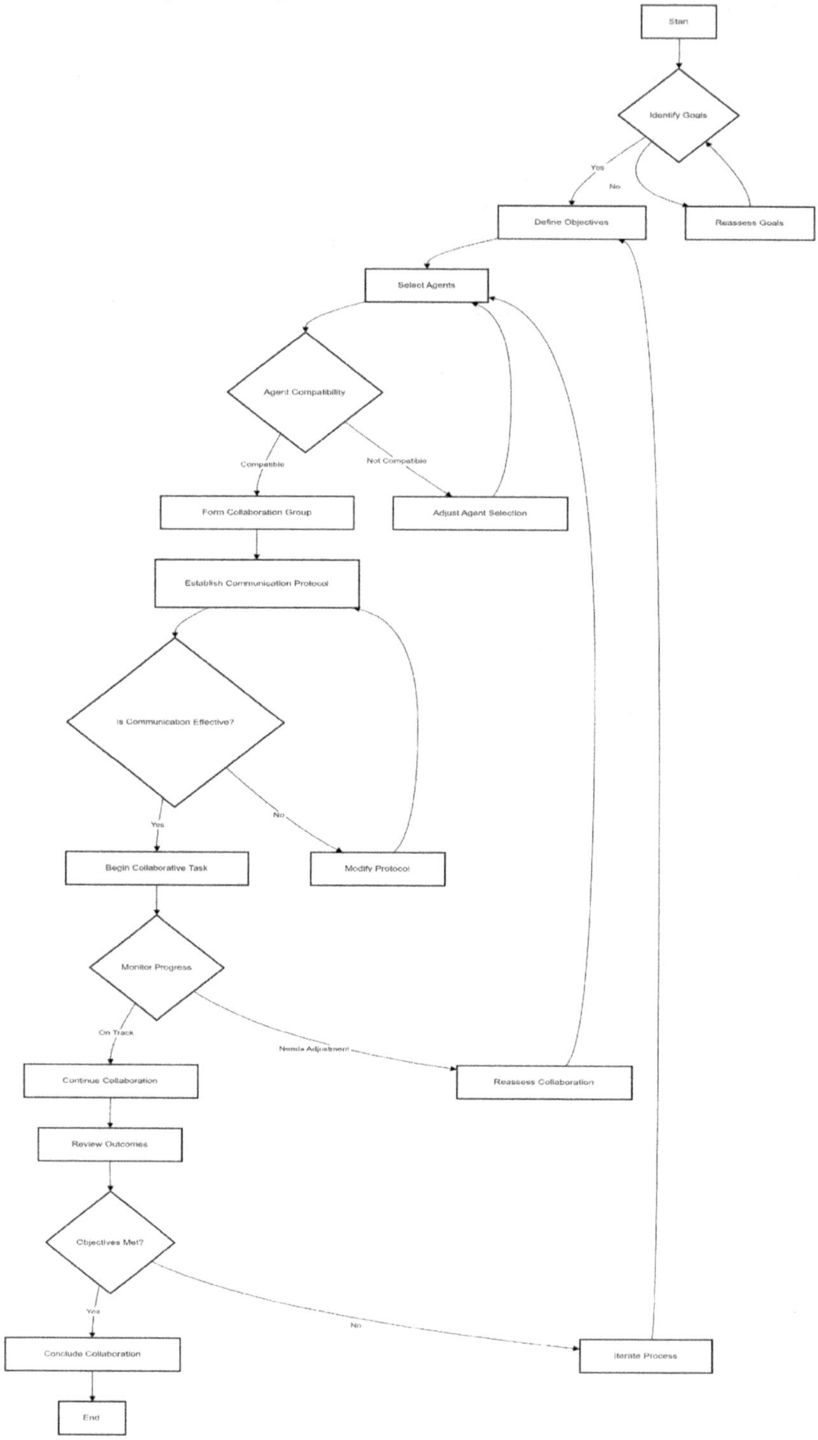

These structures improve factory operations scalability, resilience, and adaptability. The flowchart in Figure 12 shows that the goals have to be identified, if not then reassess the goals until its identified, once it is identified then define the objectives, based on the objectives select the agents and check for the compatibility for framing collaboration. establish the protocol and check for effective communication for performing the task. Continuously monitor the progress and adjust the collaboration, accordingly, review the outcomes, if the defined objectives are met then conclude else reiterate the process by considering the objectives.

Combining Neuromorphic and Quantum Computing

Real-time decision-making will be extended by advanced computing paradigms:

- Quantum AI: Exceptionally quick scheduling, material flow, and logistics optimization.
- Neuromorphic Systems: Chips designed to resemble brains that allow for effective real-time learning and edge pattern recognition.

This will make it possible to collaborate with increased energy efficiency almost instantly.

Federated Learning and Zero Trust for Safe Cooperation

Data sovereignty and security will be important facilitators rather than barriers:

- Federated Learning: Enables the training of AI models in several locations without exchanging private information.
- Zero Trust Architecture: Constant identity checks for all interactions between devices, users, and systems.

These frameworks facilitate decentralized, secure cooperation throughout international operations.

Interdisciplinary and Cross-Sector Fusion

The use of collaborative intelligence will not be limited to the manufacturing floor.

- Cross-sector Collaboration: Sharing AI models and best practices with the healthcare, logistics, and agricultural sectors.
- Humanities and Design Thinking: Including sociologists, designers, and ethicists in the creation of systems.

This all-encompassing strategy guarantees that collaborative systems are resilient, inclusive, and sensitive to context.

In the Direction of Sensitive Factors

The long-term goal calls for factories that exhibit sentient behaviour in addition to intelligence have a Maturity Model of Collaborative Intelligence in Industry as in Figure 13.

- Self-Awareness: Systems and machines keep an eye on their own conditions, anticipate malfunctions, and optimize themselves.

Adaptive governance refers to factory systems that modify operating guidelines in response to ethics and performance.

These factories will function as living ecosystems where machine accuracy and human intuition interact dynamically.

Figure 13. Maturity model of collaborative intelligence in industry

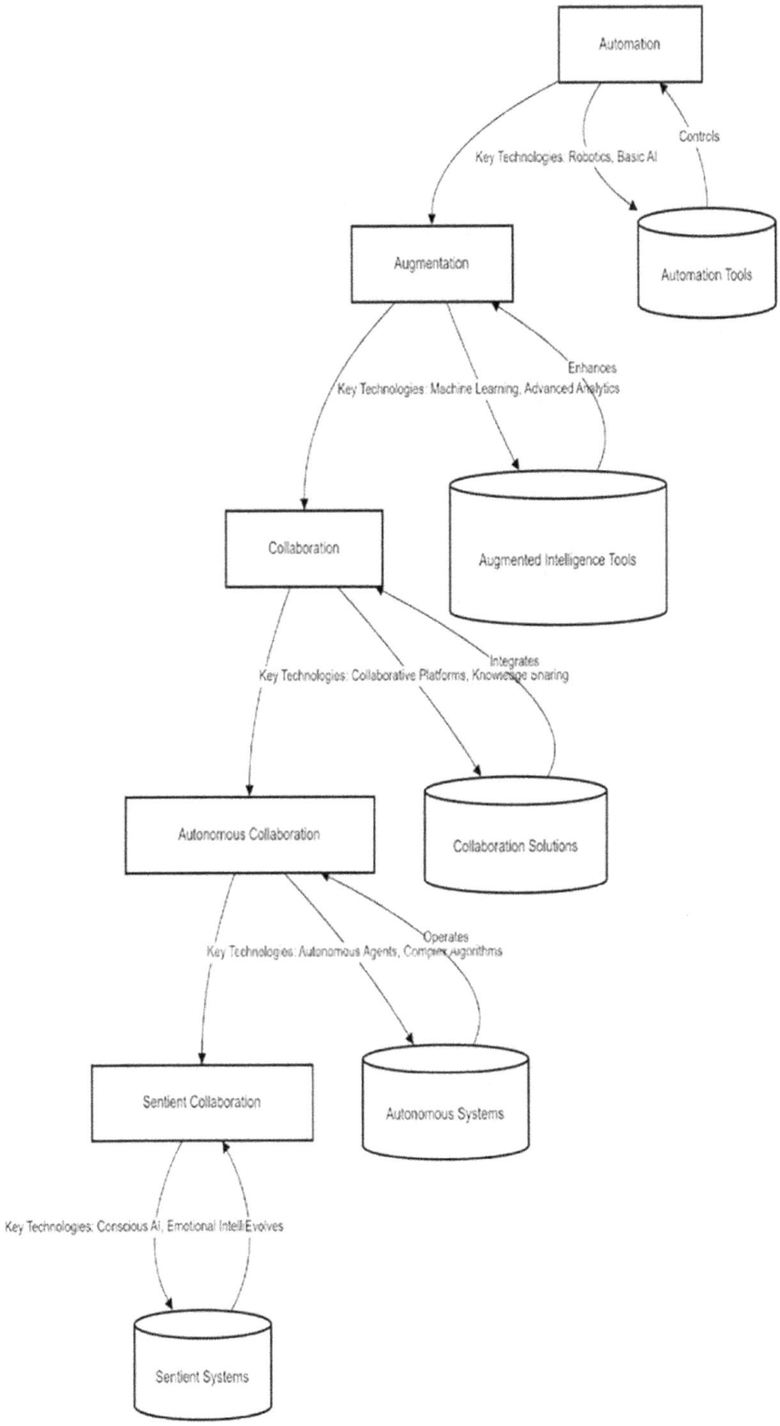

CONCLUSION: THE PATH FORWARD FOR COLLABORATIVE INTELLIGENCE IN SMART FACTORIES

Collaborative Intelligence (CI) is a significant change in how humans and robots will collaborate in smart factories as we approach the dawn of a new age in industrial operations. This evolution involves not just the smooth integration of cutting-edge technology like robotics, AI, and the Internet of Things, but also a fundamental rethinking of the roles that humans play in these systems. The concept of collaborative intelligence highlights how human judgment, creativity, and insight complement the accuracy, speed, and scalability of robots. In manufacturing settings, the combination of human and machine skills is fostering a cooperative ecosystem that presents previously unheard-of chances for productivity, creativity, and problem-solving.

This concept of combining the intelligence of human expertise and machine expertise has solved many unsolved problems. Also, came up with new creative solutions.

The evolution of CI from the automation of manual activities to increasingly complex, autonomous systems that need human participation for ethical monitoring, decision-making, and adaptability has been discussed throughout this chapter. Collaborative intelligence will make it possible to implement self-learning systems, cognitive robots, and adaptive decision-support tools that complement human capabilities rather than take their place as smart factory design develops. The new paradigm focuses on co-evolution, in which mechanical accuracy and human intelligence continuously improve one another, benefiting workers, companies, and society as a whole.

But as we move forward, we need to be cautious about integrating collaborative intelligence. The difficulties in managing employment displacement, addressing ethical issues, guaranteeing fair access to this technology, and preserving human control over AI systems are crucial elements that will determine how well this transition goes. Human workers must continue to play an essential role in the process as sectors rely more on automation and autonomous decision-making. They must do so not only as operators but also as moral stewards, innovators, and decision-makers.

Looking ahead, collaborative intelligence in smart factories will succeed if it can consistently adjust to changing socioeconomic conditions and technological advancements. CI's future will be influenced by the emergence of emotional computing, which allows machines to comprehend and react to human emotions, swarm robotics, which allows for more decentralized control, and quantum computing, which has the ability to completely transform data processing. Digital twins and generative design will enable real-time simulations and streamlined procedures, giving employees the means to accomplish more, more quickly, and more precisely.

In summary, although there are many chances and difficulties along the way, the future of smart manufacturing will ultimately depend on how well human insight and machine accuracy are combined. Collaborative intelligence is more than just a new technology; it is a paradigm shift that will change how industries operate, how work is structured, and how we engage with the digital world. Building systems that put sustainability, equity, and collaboration first will be essential as we advance to make sure that the development of smart factories improves everyone's quality of life.

REFERENCES

Appelbaum, S. H. (1997). Socio-technical systems theory: An intervention strategy for organizational development. *Management Decision, 35*(6), 452–463. DOI: 10.1108/00251749710173823

Bajic, B., Cosic, I., Katalinic, B., Moraca, S., Lazarevic, M., & Rikalovic, A. (2019). EDGE COMPUTING VS. CLOUD COMPUTING: CHALLENGES AND OPPORTUNITIES IN INDUSTRY 4.0. *Annals of DAAAM & Proceedings, 30.*

Benbya, H., Davenport, T. H., & Pachidi, S. (2020). Artificial intelligence in organizations: Current state and future opportunities. *MIS Quarterly Executive, 19*(4).

Bi, Z. M., Chen, B., Xu, L., Wu, C., Malott, C., Chamberlin, M., & Enterline, T. (2022). Security and safety assurance of collaborative manufacturing in industry 4.0. *Enterprise Information Systems, 16*(12), 2008512. DOI: 10.1080/17517575.2021.2008512

Dou, F., Ye, J., Yuan, G., Lu, Q., Niu, W., Sun, H., . . . Song, W. (2023). Towards artificial general intelligence (agi) in the internet of things (iot): Opportunities and challenges. *arXiv preprint arXiv:2309.07438.*

Ghobakhloo, M. (2018). The future of manufacturing industry: A strategic roadmap toward Industry 4.0. *Journal of Manufacturing Technology Management, 29*(6), 910–936. DOI: 10.1108/JMTM-02-2018-0057

Hollnagel, E., & Woods, D. D. (2005). *Joint cognitive systems: Foundations of cognitive systems engineering*. CRC press. DOI: 10.1201/9781420038194

Hu, Y., Jia, Q., Yao, Y., Lee, Y., Lee, M., Wang, C., Zhou, X., Xie, R., & Yu, F. R. (2024). Industrial internet of things intelligence empowering smart manufacturing: A literature review. *IEEE Internet of Things Journal, 11*(11), 19143–19167. DOI: 10.1109/JIOT.2024.3367692

Hui, Y. (2024). *Using Artificial Intelligence Solutions Towards Asset Performance Management System in Smart Manufacturing: A Case Study with Capgemini on Service Integration for Operation and Maintenance Purposes* (Master's thesis, UIS).

Lasi, H., Fettke, P., Kemper, H. G., Feld, T., & Hoffmann, M. (2014). Industry 4.0. *Business & Information Systems Engineering, 6*(4), 239–242. DOI: 10.1007/s12599-014-0334-4

Lee, J., Bagheri, B., & Kao, H. A. (2015). A cyber-physical systems architecture for industry 4.0-based manufacturing systems. *Manufacturing Letters, 3*, 18–23. DOI: 10.1016/j.mfglet.2014.12.001

Lee, M. F. R. (2023). A review on intelligent control theory and applications in process optimization and smart manufacturing. *Processes (Basel, Switzerland), 11*(11), 3171. DOI: 10.3390/pr11113171

Ma, D., Gausemeier, J., Fan, X., & Grafe, M. (Eds.). (2011). *Virtual reality & augmented reality in industry*. Springer Berlin Heidelberg. DOI: 10.1007/978-3-642-17376-9

Mohanan, P. V. (2025). Artificial Intelligence and Biological Sciences.

Ranawaka, A., Alahakoon, D., Sun, Y., & Hewapathirana, K. (2024). Leveraging the Synergy of Digital Twins and Artificial Intelligence for Sustainable Power Grids: A Scoping Review. *Energies, 17*(21), 5342. DOI: 10.3390/en17215342

Schoenherr, J. R. (2022). *Ethical artificial intelligence from popular to cognitive science: trust in the age of entanglement*. Routledge. DOI: 10.4324/9781003143284

Shah, W. (2024). Human-AI Collaboration: Designing Intelligent Applications for Enhanced Productivity. *Journal of AI Range, 1*(3), 37–48.

Sherwani, F., Asad, M. M., & Ibrahim, B. S. K. K. (2020, March). Collaborative robots and industrial revolution 4.0 (ir 4.0). In *2020 international conference on emerging trends in smart technologies (ICETST)* (pp. 1-5). IEEE.

Tao, F., Xiao, B., Qi, Q., Cheng, J., & Ji, P. (2022). Digital twin modeling. *Journal of Manufacturing Systems, 64*, 372–389. DOI: 10.1016/j.jmsy.2022.06.015

Tao, F., Zhang, H., Liu, A., & Nee, A. Y. (2018). Digital twin in industry: State-of-the-art. *IEEE Transactions on Industrial Informatics, 15*(4), 2405–2415. DOI: 10.1109/TII.2018.2873186

Tiwari, S., Bahuguna, P. C., & Srivastava, R. (2023). Smart manufacturing and sustainability: A bibliometric analysis. *Benchmarking, 30*(9), 3281–3301. DOI: 10.1108/BIJ-04-2022-0238

Van Rijmenam, M. (2019). *The Organisation of Tomorrow: How AI, blockchain and analytics turn your business into a data organisation*. Routledge. DOI: 10.4324/9780429279973

Wei, H. H., Zhang, Y., Sun, X., Chen, J., & Li, S. (2023). Intelligent robots and human-robot collaboration in the construction industry: A review. *Journal of Intelligent Construction, 1*(1), 1–12. DOI: 10.26599/JIC.2023.9180002

Xu, M., David, J. M., & Kim, S. H. (2018). The fourth industrial revolution: Opportunities and challenges. *International journal of financial research, 9*(2), 90-95.

KEY TERM AND DEFINITIONS

Artificial Intelligence (AI): It is the process of providing intelligence to the system developed in an artificial way so as to perform various jobs like learning, identifying, making decisions.

Augmented Reality (AR): It is the technology that uses devices like smart glasses or smartphone and combines real world environment with technology so as to enhance the view.

Collaborative Intelligence: It is method where the intelligence of an individual and machine are combined to produce a solution to problems, take decisions in a critical situation and provide a better result.

Human-Computer Interaction (HCI): It is a process where the individuals interact with the machine i.e a digital system which focuses on the usability, user experience, and the effectiveness of communication between humans and machines.

Industrial Internet of Things (IIoT): In order to improve the efficiency in the industrial application by analysing the data in real time, providing safety to manufacturing and production units we need the interconnection of the internet connected devices like sensors, this is called as Industrial IoT.

Industry 4.0: It is a trend in manufacturing technologies where automation and information sharing happen, this includes cloud computing, AI, and IoT to build intelligent and networked factories.

Supervisory Control and Data Acquisition (SCADA): It is a system of control which is employed in industries to remotely track and handle machinery and procedures, gathering data in real time from sensors and devices to assure secure and effective operations.

Virtual Reality (VR): A realistic digitally produced environment that mimics the real world, this is frequently accessed through headsets that completely stimulate the individual's senses while blocking off the outside world.

Chapter 6

Excellence in New Era of Human–Machine Collaboration and Interaction for Smart Factories:

Artificial Intelligence and Machine Learning for Industry 5.0 Transformation

Cristina Raluca Gh. Popescu
https://orcid.org/0000-0002-5876-0550
University of Bucharest, Romania & The Bucharest University of Economic Studies, Romania

Sandeep Garg
Motilal Nehru College (Eve), University of Delhi, India

Jyoti Kalita
Assam Down Town University, India

ABSTRACT

New paradigm shift set in motion by Industry 5.0 complements proficiently existing Industry 4.0 by specifically positioning critical thinking, knowledge, innovation, and research at core of transitioning to sustainable, healthy, human-centric, inclusive, and resilient industry. The research draws on collaborative technologies reshaping human-machine relationships in modern factories, addresses key drivers in undergoing economic and societal transition, and provides a vision of flourishing industry that goes beyond efficiency and productivity. On the one hand, the study emphasizes todays' challenges for society and, on the other hand, offers solutions for climate change, preservation of resources, and social stability. Findings show artificial intelligence and machine learning robustly support Industry of the Future, actively attract best talents, and

DOI: 10.4018/979-8-3373-2985-7.ch006

vigorously address evolving skills and training of individuals. Likewise, communities of practice are crucial in sharing steps in implementing good principles as well as knowledge and innovation ecosystems for new collaborations.

INTRODUCTION

In this day and age, centering on excellence in the new era of human-machine collaboration and interaction for smart factories, while focusing, in particular, on artificial intelligence and machine learning for Industry 5.0 transformation is a fascinating and timely area of research. On the one hand, based on specialists worldwide, artificial intelligence and machine learning are at the heart of what makes Industry 5.0 both revolutionary and human-centric. On the other hand, likewise, explaining and understanding the differences between Industry 5.0 and Industry 4.0 is crucial in today's complex and multidisciplinary business environment. In this given context, some pressing questions requiring in-depth analysis and valuable examples are, as follows: Can specialists explain the common points and the differences between Industry 5.0 and Industry 4.0? Additionally, what are some real-world examples of artificial intelligence transforming Industry 5.0? How will the future look like in the next ten or twenty years if the major target is acquiring excellence in the new era of human-machine collaboration and interaction for smart factories, while addressing, in particular, the aspects that surround the very hot topic of artificial intelligence and machine learning for Industry 5.0 transformation?

In the lines below some compelling directions concerning centering on excellence in the new era of human-machine collaboration and interaction for smart factories, while focusing, in particular, on artificial intelligence and machine learning for Industry 5.0 transformation have been mentioned:

(A.) Human-Centric Artificial Intelligence in Industry 5.0;
(B.) Ethical and Transparent Artificial Intelligence for Sustainable Manufacturing;
(C.) Machine Learning for Hyper Personalization and Customization at Scale;
(D.) Digital Twins, Artificial Intelligence, and Real Time Optimization;
(E.) Artificial Intelligence-Driven Workforce Up-Skilling.

The elements highlighted above generate a lot of thought-provoking ideas for each of the topics and the aspects depicted in the lines below are prefect for research, innovation proposals, and strategic planning, as follows:

Human-Centric Artificial Intelligence in Industry 5.0

While addressing human-centric artificial intelligence in Industry 5.0, it is of outmost importance to explore how artificial intelligence is being designed to augment, rather than replace human labor. There are some vital examples that could be taken into consideration here, namely: the collaborative robots (co-bots); the adaptive learning systems; and, also, among others, the artificial intelligence systems trained to support human decision-making in smart factories. Specialists highlight the fact that artificial intelligence is increasingly designed so that it can amplify human capabilities rather than replace humans, especially in complex industrial settings like smart factories (Chaparro-Banegas et al., 2023). In this matter, this particular shift is regarded as crucial in Industry 5.0, where the focus is placed on the synergies between humans and machines, rather than pure automation (Xu et al., 2024).

Noteworthy, collaborative robots (co-bots) are built to work safely along humans, hence having different characteristics when compared with traditional industrial robots that are known to work in isolation (Xu et al., 2022). One of the notable advantages of using collaborative robots (co-bots) in industry today is that human workers are, consequently, able to center on more complex, creative, or supervisory roles (Zyoud, 2023). Hence, collaborative robots are the ones handling repetitive and physically demanding tasks, adapting to human movements, being designed to learn from interactions, and even assist in precision tasks, such as, for example, assembling different parts or becoming a valuable instrument in inspecting the processes that take place (Zhang & Wang, 2023).

Interestingly, adaptive learning systems which are artificial intelligence-driven are rapidly transforming workforce training these days. These platforms provide tailored content to individual learning style and places, providing tremendous help to workers to up-skill more efficiently (Valentine et al., 2016). While taking into consideration the case of the use of adaptive learning systems which are artificial intelligence-driven in factories, it can be mentioned that employees can, on the one hand, quickly learn to operate new machinery and, on the other hand, interpret data from artificial intelligence systems. However, human experience is to be found at the center of innovation (Vega & Frenz, 2013).

Moving on to the next type of opportunity offered by the use of artificial intelligence-driven tools these days the focal point can be represented by artificial intelligence for decision support in smart factories (UNESCO, 2025). In this case it needs to be remembered that smart factories are leveraging artificial intelligence to enhance and not to override human-decision making processes (van Driel et al., 2024). In this case there are several pivotal examples that could be mentioned, namely: firstly, predictive maintenance leverages artificial intelligence to anticipate equipment failure, hence reducing downtime which is regarded as a valuable accomplishment in industrial settings; secondly, artificial intelligence-based quality control systems ensure that total quality management characteristics are respected inspecting very rapidly the products, while centering on catching defects with almost perfect accuracy; and, thirdly, predictive scheduling optimizes production planning by improving resource allocation and reducing waste (Zhang et al., 2025).

Essentially, human-centric artificial intelligence in Industry 5.0 depends on the power of collaboration in reshaping the workforce and in supporting the labor market by providing on-going support to individuals to become more skilled and more dedicated to improving their knowledge, creativity, innovative potential, and creative thinking competences (Van Tulder et al., 2021). Besides all these, studies show that human-centric artificial intelligence in Industry 5.0 supports the Sustainable Development Goals, putting a particular accent on workers well-being and health (United Nations Environment Program (UNEP). Finance Initiative, 2025). With artificial intelligence handling monotonous tasks or dangerous activities, job satisfaction increased significantly in the last years, while work injuries dropped significantly due to the fact that the tasks were taken by artificial intelligence-driven tools and instruments (Tikadar et al., 2026). It needs to be stressed that workers are now more engaged in roles that require more creativity, like system optimization and strategic planning (United Nations Environment Program (UNEP), 2025b).

Ethical and Transparent Artificial Intelligence for Sustainable Manufacturing

Nowadays ethical and transparent artificial intelligence for sustainable manufacturing is paramount (United Nations Environment Program (UNEP), 2025a). It is vital to delve into the way in which artificial intelligence can be explained and utilized so that Industry 5.0 would support sustainability, safety, and ethics. Likewise, it is of great essence to highlight the way in which responsible machine learning models could shape Industry 5.0 in order to foster sustainability, safety, and ethics in the benefit of the planet and individuals all around the globe (Zhao & Kumar, 2022). Very important case studies in this context could be analyzed and, among them, green artificial intelligence to reduce energy consumption and waste could represent a consistent source of interest (Zhao et al., 2022).

Machine Learning for Hyper Personalization and Customization at Scale

In this day and age everything around is changing rapidly and the challenges are becoming more and more complex and, in like manner, overwhelming. Specialists promote the vital idea that Industry 5.0 is mass customization. Hence, machine learning algorithms are responsible for enabling tailored products and services. Moreover, machine learning algorithms are known to foster predictive analytics, especially in sectors, such as: automotive; customer electronics, and healthcare devices (Zhang, 2024).

Digital Twins, Artificial Intelligence, and Real Time Optimization

Tackling with Industry 5.0 aims is essential these days and adaptability and resilience are pivotal in this case. What is more, the complex process optimization that aligns with the Industry 5.0 goals takes into account the following pivotal issues that express the convergence of artificial intelligence, digital twins and real time optimization. Furthermore, the outcome is to lead all these processes to smarter simulations, maintenance forecasting, and process optimization (Zhang et al., 2025).

Artificial Intelligence-Driven Workforce Up-Skilling

Interestingly, the artificial intelligence-driven workforce up-skilling is a major topic these days. To begin with, several questions come to support that artificial intelligence-driven workforce up-skilling is paramount these days (Bergman et al., 2018). The first question could be: What is the role of artificial intelligence in reshaping the workforce? The second question could be: Which are the ways in which continuous re-skilling and human-machine collaboration can be fostered and which solutions can be found to empower them? And, the third question could be: What role do adaptive learning platforms, artificial intelligence mentors, and skill-gaps analysis systems play in reshaping the workforce based on artificial intelligence?

LITERATURE REVIEW

In the lines below the literature sections delves into the most relevant aspects that embrace the advantages of artificial intelligence, hence promoting the in-depth analysis of some directions that are targeting excellence in the new era of human-machine collaboration and interaction for smart factories, while addressing, in particular, artificial intelligence and machine learning for Industry 5.0 transformation:

Human-Centric Artificial Intelligence in Industry 5.0

According to reputed specialists these days, human-centric artificial intelligence in Industry 5.0 is of utmost importance. Here the discussion delves mainly in to major directions, namely the collaborative robots (co-bots) and the adaptive learning in vocational training.

Collaborative robots (co-bots) are seen as extremely valuable across a range of sectors, each with its own set of characteristics that make it absolutely perfect for collaboration. To begin with, in automotive industry, collaborative robots assist with precision tasks like component assembling, painting, and welding. Being very flexible, they are able to switch between models on the production line, without the necessity of major reprograming (Zhan & Santos-Paulino, 2021). Continuing, in the electronics sector, in those environments where delicate handling is pivotal, the collaborative robots are known to excel in several areas among which could be mentioned: assembling smart components, testing circuits, and even packaging (United Nations Environment Program (UNEP). Department of Economic and Social Affairs Sustainable Development, 2025). Moving on, in healthcare sectors the collaborative robots are used for tasks ranging from preparing medication, assisting in surgeries, and handling the laboratory samplings, which, on the one hand, significantly reduces contamination risks and, on the other hand, enables the medical staff to have more time for patient care (Ullah et al., 2021). The food and beverage sectors rely significantly on the collaborative robots, since they help maintain hygiene standards while boosting thorough outcomes, being centered among others on packaging and quality inspection (Yip et al., 2023). In the end, logistics and warehousing sectors are highly representative because the collaborative robots work alongside to pick, sort, and transport goods especially in current challenging times and in the continuously changing e-commerce environments.

All the above mentioned ideas show the incredible nature of the collaborative robots (co-bots) which are not only replacing repetitive or dangerous manual labor but, in turn, are designed to successfully augment human roles. All these issues are imperative to allow workers to focus on oversight, customization, and problem-solving (Wang et al., 2023).

Adaptive learning in vocational training provides new perspectives in terms of vocational education and sheds a new light on artificial intelligence-powered adaptive learning in reshaping the way in which people gain hands-on skills (Ugwu et al., 2025). The current directions are: personalized learning paths, learning nuggets, and hand-on collaborative robots training. The personalized learning paths rely on platforms that provide tailored content depending on the learners' pace, background, and goals. Likewise, these specific personalized learning paths are extremely useful in vocational training where learners come from extremely diverse educational and professional environments (Chen et al., 2022; Chuong et al., 2025). When analyzing the learning nuggets it is of great importance to show-case that different educational and training program developers are combining digital content with physical exercises to create hybrid learning environments. In the same time, these experienced educational and training program developers prepare for their students modular and bite-sized lessons and, also, these learning nuggets adapt in real time to the learners' progress so that the knowledge will be enabled. It is worth to mention that hands-on collaborative robots training are based on enabling novices to program and to learn how to operate collaborative robots by using intuitive and low-code interfaces. Additionally, studies show that even beginners could significantly reduce the time in which they are able to solve their tasks (Wang et al., 2021).

It ought to be emphasized that the adaptive learning in vocational training approach is intended to build confidence and enables adaptability, while facilitating an elaborate teaching environment focused on highlighting the technical skills. Consequently, human-centric artificial intelligence in Industry 5.0 creates notable connections between: collaborative artificial intelligence systems; emotionally intelligent interfaces; adaptive workflows; and, inclusive design frameworks. In this regard, collaborative artificial intelligence systems are keen on designing artificial intelligence that augments human decision-making rather than replacing it. A valuable example in this case is Operator 5.0 where humans and machines are involved in a powerful process of collaboration and creation, co-creating pivotal solutions, hence constantly improving the processes. Continuing with the next point above, emotionally intelligent interfaces are of great essence to human-centric artificial intelligence in Industry 5.0. In essence, the emotionally intelligent interfaces develop artificial intelligence systems that can interpret to human feelings and respond to human emotions in industrial settings to improve the employees' safety and the enhance workers' morale (Yamin et al., 2019). Moving on to the next aspect represented by adaptive workflows, specialists have shown that the use of artificial intelligence in the complex process of dynamically and on-going adjusting manufacturing processes is based, among others, on human feedback and on ergonomic data. Lastly, inclusive design frameworks are known to ensure that the artificial intelligence systems are co-designed with workers to reflect diverse needs and multiple capabilities (Wang & Zhao, 2021).

Ethical and Transparent Artificial Intelligence for Sustainable Manufacturing

Even though the topics related to the importance and role of artificial intelligence are rather new, there already are numerous valuable studies on ethical and transparent artificial intelligence for sustainable manufacturing (Chawla et al., 2025). Consequently, the ethical and transparent artificial intelligence for sustainable manufacturing combines ideas from different themes and topics, while the interdisciplinary approach and vision are crucial, as follows: firstly, the artificial intelligence ethics auditing tools are specially designed to create frameworks intended to continuously audit artificial intelligence models for bias, fairness, and environmental impact; secondly, explainable artificial intelligence for supply chain targets the implementation of transparent artificial intelligence systems that are able to identify and trace sustainability data across the supply chain; thirdly, the green artificial intelligence algorithms are the one optimizing artificial intelligence models for energy efficiency and minimal carbon footprint, during different processes, namely, training and deployment; and, fourthly, the artificial intelligence for circular economy is based on the use of artificial intelligence to identify those opportunities concerning reuse and recycling in real time, having as major goal centering on reducing resource consumption and minimizing waste (Brodny & Tutak, 2021).

Machine Learning for Hyper Personalization and Customization at Scale

Interestingly, machine learning for hyper personalization and customization at scale is crucial nowadays, when there are numerous preoccupations regarding: real-time personalization engines; generative artificial intelligence for custom design; behavior clustering; and, feedback loops for continuous learning. To begin with, the real-time personalization engines center themselves on developing machine learning models that are keen on adapting the products' configurations and settings instantly, depending on the users' behavior and the people's preferences. In continuation, generative artificial intelligence (GenAI) for custom design uses GenAI to co-create product designs together with clients, being centered on mass customization by making a major priority in satisfying customers' needs and desires (Calderon-Monge & Ribeiro-Soriano, 2023). Likewise, behavior clustering applies unsupervised learning to segment users starting from nuanced behavior patterns, making a priority in centering on ensuring the satisfaction of the customers' needs and desires, rather than targeting demographics (Ul Rehman, 2025). Interestingly, the feedback loops for continuous learning help the integration in complex processes, such as the machine learning pipelines, of the users' feedbacks, hence aiming to refine the personalization of products and services over time (Bibri et al, 2023).

Digital Twins, Artificial Intelligence, and Real Time Optimization

Nowadays, researchers are highly drawn to digital twins, artificial intelligence, and real time optimization. Of great interest, there are numerous aspects due to be considered here: self-combing digital twins; predictive maintenance ecosystems; scenario simulations for inclusivity, resilience, and robustness; and, edge artificial intelligence integration. Given the fact that all the aforementioned issues are of great importance, in the lines below all these elements are analyzed and characterized accordingly. Firstly, self-combing digital twins represent a combination of artificial intelligence and digital twins, to enable autonomous process optimization in different areas and domains, such as, for example, in manufacturing (Zhu & Yang, 2025). Secondly, predictive maintenance ecosystems are the ones making use of artificial intelligence-enhanced twins to predict equipment failures and program maintenance with the minimal disruption of the processes, so that continuity is ensured (Zhuge et al., 2023). Thirdly, scenario simulations for inclusivity, resilience, and robustness use artificial intelligence to simulate "what-if" scenarios in digital twins, for production planning, businesses processes, and supply chain. Lastly, edge artificial intelligence integration deploys lightweight artificial intelligence models at the edge in order to foster real-time decision-making in decentralized environments. Besides all these, there are many other aspects surrounding digital twins, artificial intelligence, and real time optimization, but the focus in this particular situation was intended to be the issues promoted above (Bai et al., 2024).

Artificial Intelligence-Driven Workforce Up-Skilling

The turbulent times in which humanity lives led to the importance of dedicating energy and time regarding artificial intelligence-driven workforce up-skilling, since this a very complex process that involves concerted and serious work from academics, researchers, specialists, governmental officials, and policymakers. There is a great concern that human beings will be less and less used in industry, while more and more instruments, tools, and techniques driven by artificial intelligence will come to occupy important places in different areas and domains, putting the workforce in danger (Verdejo Espinosa et al., 2021). Many powerful international voices came with the idea that humanity is little by little losing its identity and will to decide, in particular, due to artificial intelligence-driven systems. Hence, the artificial intelligence-driven workforce up-skilling refers to: artificial intelligence learning platforms; digital mentorship robots (bots); skill forecasting models; and, "gamified reskilling". In the beginning, while targeting artificial intelligence learning platforms, it needs to be stressed that creating adaptive learning systems that are able to tailor content to individual skill gaps and learning styles are very important today (Yuan et al., 2024). According to the United Nation's Sustainable Development Goals, no one should be left behind and everyone should be included in the educational and training processes; hence, all individuals should receive support in order to constantly improve their individual skill gaps and make use in the best way possible to achieve new knowledge depending on their learning styles (Aziz et al., 2024). The digital mentorship bots deploy conversational artificial intelligence agents that have the power to guide employees to become more and more skilled and have more competences in their educational and training journey, benefiting by real-time support. When addressing the aspects dedicated to skill forecasting models there is a lot to be said. For example, the use of artificial intelligence to predict future skills and competences demands is pivotal today. Also, likewise, for instance, the use of artificial intelligence can implicate the recommendation of personalized learning paths in the case of individuals, so that they would benefit accordingly from life-long learning programs (Azim et al., 2025). The "gamified reskilling" is the process in which the game mechanisms are integrated into artificial intelligence-driven training (Agrawal et al., 2025). The goal is to boost employees' engagement and retention, ultimately determining people to perfect their competences and skills in an interactive and pleasant environment (Acharya et al., 2018).

RESEARCH METHODS AND METHODOLOGY

The new era of human-machine collaboration and interaction for smart factories receives plenty of attention these days. Specialists are drawn to this theme, since it is a fascinating and timely topic and, additionally, the smart factories evolve under the influence of the Industry 5.0 principles. In the lines below, the authors propose a breakdown of some powerful and valuable key research methods and methodological approaches. These major research methods and methodological approaches are thoroughly being explored in the new era of human-machine collaboration: the research focus areas; the common research methods; and, the enabling technologies.

The research focus areas are very complex. For example, for the human-centric digital twins the focus of these fascinating virtual replicas is on integrating real-time data to simulate and optimize human-machine interactions, placing human needs at the center of different processes, such as, for instance, the manufacturing processes (Darwish, 2024). Also, among the research focus areas could be considered the collaborative decision-making models (Deng et al., 2025). In this matter, for instance, connecting human cognitive strengths with machine computation represents a vital research topic today and, also, has a scope to improve decision-making under uncertainty (Dhaoui, 2022). Moving on to another theme, trust and transparency in artificial intelligence systems are of great importance today (de Andrade et al., 2021). Specialists are centered in exploring how to build trust in artificial intelligence and artificial intelligence systems by making use of several instruments, namely: adaptive interfaces; the capacity to explain its advantages and offer support to the users; and, also, user control (Duháček Šebestová & Popescu, 2025).

What is more, there are numerous common research methods that could fit very nicely with this current theme and which are of great interest these days. One of these research methods is simulation and experimental studies. Noteworthy, especially in hybrid human-artificial intelligence teams, simulations are the ones testing collaborative scenarios in controlled environments, empowering these forms of collaborative work and enabling humans to receive support from artificial intelligence-driven instruments and robots (Duraimutharasan et al., 2025). Moving on, the case studies and the grounded theory are seen as common research methods these days in which these qualitative methods are being used to test real-world implementations and, also, drive theory from practice (Elavarasan et al., 2022). Also, the systematic literature reviews are pivotal in the cases in which attention is placed on offering a better understanding on existing knowledge (Elzein, 2024). In this matter, researchers synthetize information to identify gaps and future directions, using different frameworks, such as, for example, the PRISMA ones (Fang & Cao, 2023; Filho et al., 2024; Frichi & Jawab, 2025).

Furthermore, enabling technologies – among which could be highlighted the IoT and sensor networks, the artificial intelligence, and the machine learning, the augmented reality (AR) and wearables, are essential today (Gill & Mathur, 2024; Goalla et al., 2024). Going more in-depth, regarding the internet of things (IoT) and sensor networks it is noteworthy to state that these are very useful for real-time data collections and feedback loops (Guo et al., 2023). In continuation, when addressing the artificial intelligence and the machine learning the major concern is to adapt systems dynamically to human behavior (Guo et al., 2022). In the end, the augmented reality (AR) and wearables, which are of great interest today, the major preoccupation is to human-machine interfaces in factories and not only (Guo et al., 2022).

There are some intriguing research questions that this current book chapter on "Excellence in New Era of Human-Machine Collaboration and Interaction for Smart Factories: Artificial Intelligence and Machine Learning for Industry 5.0 Transformation" addresses and focuses on. In essence, these interesting questions are keen to explore the evolving landscape of human-machine collaboration and interaction in smart factories. The general structure on which these interesting and inquisitive questions diving into the evolving landscape of human-machine collaboration and interaction in smart factories are based is presented in the lines below:

Human-Machine Interface (HMI) and Interaction Design

For the human-machine interface (HMI) and the interaction design the key questions are:

- How can intuitive interface designs improve employees' engagement and reduce errors in smart factories (He et al., 2024)?
- What is are the importance and role played by augmented reality (AR) in enhancing human-machine interactions in factories (Helldén et al., 2022)?
- How can specialists design collaborative instruments, tools, and robots (co-bots) that adapt dynamically to human behavior and workflow changes (Huang et al., 2024; Irajifar et al., 2023)?

Trust, Ethics, and Decision-Making

For the aspects dedicated to trust, ethics, and decision-making there are some vital research questions that ought to be considered, as follows:

- How does the level of transparency in artificial intelligence decision-making influence employees' trust and adoption (Kaur, 2024)?
- What are the ethical frameworks needed to enable and to promote safe and trustworthy shared decision-making between human and machines (Kenwright et al., 2025)?
- How can specialists in artificial intelligence ensure that automated systems support and not replace human expertise in critical operations (Khan et al., 2025; Kieny et al., 2017)?

Training, Learning, and Adaption

In this day and age, it became crucial to empower workers with the rigorous knowledge and training on artificial intelligence, so that businesses will grow and be more and more prosperous (Kumar et al., 2024). That is the reason why asking mind-provoking questions on training, learning, and adaption is essential today:

- What are those training methods which are more effective for empowering and up-skilling workers to collaborate with intelligent systems (Kumar et al., 2024)?
- How do machines learn from human actions and, also, how do humans learn from machines in complex, collaborative, and shared work environments (Leal Filho et al., 2022)?
- Can learning be effectively and safely applied in real-time human-machine collaborations (Liang et al., 2023)?

Performance, Safety, Health, and Well-Being

Some research questions on performance, safety, health, and well-being that are asked and which should be answered are, as follows:

- How can smart factories balance growth and productivity gains with workers' safety, health, and well-being, especially in a world profoundly governed by the Sustainable Development Goals (Liu et al., 2024; Luo & Liu, 2024)?
- What are those metrics that can evaluate the effectiveness of human-machine collaboration beyond just efficiency and just transition (Luo et al., 2022)?
- Do collaborative systems react to people's stress and how do these collaborative systems recognize and respond to human beings fatigue, problems, and stress (John Joseph et al., 2026)?

Scalability and Integration

Today, scalability and integration research questions come to support the studies on the complex interaction and relationship between human and machine learning activities and processes. Some of the thought-provoking research questions may be encountered below, as follows:

- What are the barriers to scaling human-machine collaboration (MacPherson et al., 2022)?
- What are the solutions to successfully overcome the barriers to scaling human-machine collaboration across divers settings, such as, for instance, manufacturing domain (Matsui et al., 2022)?
- How do legacy systems and infrastructures impact the integration and the functionality of intelligent collaborative systems, tools, and technologies (Mehta et al., 2024)?

DELVING INTO THE WORLD OF ARTIFICIAL INTELLIGENCE AND MACHINE LEARNING FOR INDUSTRY 5.0 TRANSFORMATION

The section on delving into the world of artificial intelligence and machine learning for Industry 5.0 transformation has its roots into the previously highlighted ideas and intends to cover during the analysis the thought-provoking research questions depicted above. What is more, this particular section addresses some compelling insights as well as some valuable case studies that explore how artificial intelligence and machine learning are driving the transformation towards Industry 5.0.

Some of the major ideas, insights, and cases studies refer to: the human-centric manufacturing with artificial intelligence; the smart cities and predictive maintenance; the artificial intelligence-driven customization in production; the ethical artificial intelligence and sustainable innovation; and, the healthcare system and artificial intelligence within the Industry 5.0 paradigm.

Human-Centric Manufacturing with Artificial Intelligence

To begin with, the human-centric manufacturing with artificial intelligence shows the way in which businesses can be transformed in the digital age, highlighting how different organizations are integrating the use of artificial intelligence not only for automation processes but, also, to enhance people's capabilities and skills (Nilashi et al., 2023). An interesting example is the one showing how manufacturers are using artificial intelligence powered co-bots (collaborative robots) that are known to adapt to employees pace, desires, and preferences (Mohammed & Ghebreyesus, 2018). In this particular example, the advantages and benefits of using artificial intelligence powered co-bots (collaborative robots) may be encountered in improving both productivity and worker satisfaction (Murshed, 2024).

Smart Cities and Predictive Maintenance

The smart cities and predictive maintenance represent very interesting case studies and offer valuable examples of how powerful the influence of artificial intelligence is these days (Ozturk et al., 2024). There are already important situations in which specialists explored how generative artificial intelligence and large language models are being used in smart cities infrastructure (Qi et al., 2023). Hence, researchers showed that artificial intelligence and large language models are very important technologies which today help predict equipment failures (Palau-Pinyana et al., 2023). Also, studies emphasize that artificial intelligence and large language models are crucial in optimizing energy usage and in improving citizen services, having the great property of blending industrial efficiency and performance together with societal well-being and environmental health (Pandey et al., 2025; Pandey & Joseph, 2025).

The Artificial Intelligence-Driven Customization in Production

The artificial intelligence-driven customization in production is another potential case study that requires specific attention when centering on the topic of excellence in new era of human-machine collaboration and interaction for smart factories, while accentuating the aspects that enable artificial intelligence and machine learning to support the Industry 5.0 transformation (Popescu & Verma, 2025a). Hence, there are different studies that offer important details on how artificial intelligence and machine learning are enabling hyper-personalization manufacturing (Popescu & Verma, 2025b). Additionally, by analyzing customer data and production variables in real time it was presented how companies can produce customize goods at scale, which was something almost impossible to create during the previous industrial revolutions (Parikh et al., 2021).

Ethical Artificial Intelligence and Sustainable Innovation

While addressing the Global Goals, the ethical artificial intelligence and sustainable innovation become paramount (Popescu & Popescu, 2025). Moreover, the importance of ethical frameworks and sustainability in Industry 5.0 transformation is essential in today's continuously changing and challenging business environment (Peng, 2025). In consequence, researches show that artificial intelligence is being used with several, such as, for instance: to optimize operations (Ponnada, 2026); to reduce waste (Popescu, 2025); to lower the carbon dioxide emissions (Raygoza-L. et al., 2025); and, to ensure fair labor practices (Rahman et al., 2023). Noteworthy, the ultimate goal is to generate a significant and notable shift from profit-centric to purpose-driven innovation and industry, hence becoming more and more aware that the Sustainable Development Goals are the ones that generate the expected equilibrium between the humans, the environment, and, also, the businesses (Raman et al., 2025).

The Healthcare System and Artificial Intelligence within the Industry 5.0 Paradigm

There are many other fascinating case studies and research insights that showcase how artificial intelligence is transforming different domains by making use of Industry 5.0 paradigm. The healthcare system within the Industry 5.0 paradigm, for example, registers major changes due to artificial intelligence (Reddy et al., 2024). In this case the accent should be placed on: explainable artificial intelligence for patient-centric care; human-artificial intelligence collaboration in clinical settings; and, also, smart healthcare technologies with artificial intelligence, the internet of medical things (IoMT), and blockchain. Today it is pivotal to integrate smart healthcare technologies with artificial intelligence, the internet of medical things (IoMT), and blockchain in order to generate secure healthcare ecosystems and, also, to create intelligent and responsible healthcare ecosystems (Ranjbari et al., 2021). Some of the case studies include remote patient monitoring, predictive analytics for early disease detection and, also, virtual reality (VR) and augmented reality (AR) that are simulated imaging tools used in surgery (Rehman Khan et al., 2022).

In the end it is worth mentioning that Industry 5.0 is truly a fascinating leap forward. The primary aim is to position people back at the center of industrial innovation, where technology acts as an enhancer and enabler of knowledge and advantages for humanity and the planet, rather than a replacement in the industrial settings and not only there (Ruth et al., 2024). In this given context, artificial intelligence and machine learning are – according to prominent specialists, far more than just tools for efficiency and instruments for performance (Saluja & Mongia, 2025). They become key enablers to a more inclusive, personalized, responsible, robust, and sustainable industrial ecosystem. When compared to Industry 4.0 – mainly characterized by automation, connectivity, and smart systems, Industry 5.0 successfully adds the human layer (Santarius et al., 2023). The vision brought by artificial intelligence and machine learning in the Industry 5.0 complex setting is far more complex and much more interesting and fascinating than all the previous scenarios.

CONCLUSION, LIMITATIONS, AND RECOMMENDATIONS

These days, the new vision embraced by artificial intelligence and machine learning is impressive. The targeted issues are, among others, human-artificial intelligence collaboration, customization at scale, sustainability intelligence, and resilience and flexibility.

Firstly, the human-artificial intelligence collaboration addresses the importance of smart systems assisting human beings rather than replacing them, having in mind very important examples such as: the collaborative robots (co-bots) which are not intended to replace workers and which are designed to support them in their tasks; and, also, in addition, the predictive systems and personalized interfaces which adapt to human behavior and which are created to ensure the well-being of the workers and to offer them more time to innovate and rely on creative thinking rather than tasks that are repetitive (Sayyed et al., 2024).

Secondly, customization at scale is pivotal these days and Industry 5.0 is extremely powerful in this matter (Šebestová & Popescu, 2025). There are some crucial aspects that require immediate attention here: one noteworthy situation is the one regarding machine learning models which can dynamically adjust production lines and which are keen on hyper-personalization and ultra-personalization of goods and services (Šebestová & Popescu, 2025); also, this goes hand-in-hand with the scenario of "batch size one" manufacturing which brings a new meaning to the idea of customization at scale and care for consumers' needs and desires (Smaniotto et al., 2020).

Thirdly, sustainability intelligence makes everything revolve around the Global Goals and the sustainable development principles (Sreenivasan et al., 2023). That is the reason why the human-artificial intelligence collaboration was brought to a new level, where artificial intelligence has as aims to reduce waste, optimize energy usage, and, even more than all these, support the shift towards the circular economy by analyzing environmental impact in real time (Sundaram et al., 2024).

Fourthly, resilience and flexibility are those characteristics that have the capacity to keep industries agile amid uncertainty (Şimşek, 2025). Moreover, resilience and flexibility characterize environments that rely on advanced machine learning, supply chains that can on the one hand, self-heal and, on the other hand, predict disruptions. Furthermore, these environments profoundly influenced by artificial intelligence and machine learning enable supply chains to adapt their operations so that the focus will not solely be on deploying cutting-edge technology. Besides all these, the employees should thrive alongside intelligent machines being under the influence of sustainability and sustainable development which ensures the need for ethics, transparency, and up-skilling.

There are many aspects that ought to be noted concerning artificial intelligence and machine learning. To begin with, artificial intelligence and machine learning are evolving rapidly and they impose rapid transformation in all sectors. Besides all these, the labor market is changing and the ambitions to support in a better and more complex way human resources worldwide display a great vision that relies on tremendous responsibilities.

There are many limitations to this current book chapter, even though the research done in this case is highly complex and addresses numerous challenges and valuable suggestions of a better and safer future for all. Some key limitations are presented in the below lines, as follows:

Human-Artificial Intelligence Collaboration

There is still a visible struggle for artificial intelligence ecosystems to offer a better and more in-depth understanding to different complex scenarios, since there are several contexts and nuances that these instruments and techniques based on smart technologies are not yet able to grasp, which practically shows that empathy is specific to human beings and is irreplaceable. In this case, a limitation of this research paper addresses the fact that the analysis on human-artificial intelligence collaboration could focus in more on the arguments according to which human beings and is irreplaceable. Also, in terms of future research new studies could approach topics such as: the disadvantages of relying on human-artificial intelligence collaboration in scenarios that require a lot of empathy and where certain nuances could affect the way in which interpretations are being made by artificial intelligence systems. In this case, companies that rely on human-artificial intelligence collaboration ought to be aware of those situations in which workers could end up by stating that they have misplaced their trust in automated systems when some errors occur.

Customization of Scale

When addressing this situation, the current research has some limitations in offering a more in-depth understanding of how the personalization algorithms function. Hence, additional future research needs to delve into the world of personalization algorithms centering on massive datasets, while being very keen to understand aspects, such as: the importance of data privacy, the algorithm bias, the data ownership, and the discriminatory patterns.

All in all, there are several recommendations for future research on the theme of artificial intelligence and machine learning for a responsible and sustainable path forward, as follows: the human-centered design; the ethical customization; the green artificial intelligence patterns; the robustness by design; and, the policy and regulations.

REFERENCES

Acharya, S., Lin, V., & Dhingra, N. (2018). The role of health in achieving the sustainable development goals. *Bulletin of the World Health Organization, 96*(9), 591–591A. DOI: 10.2471/BLT.18.221432 PMID: 30262936

Agrawal, R., Priyadarshinee, P., Kumar, A., Luthra, S., Garza-Reyes, J. A., & Kadyan, S. (2025). Are emerging technologies unlocking the potential of sustainable practices in the context of a net-zero economy? An analysis of driving forces. *Environmental Science and Pollution Research International, 32*(12), 7130–7148. DOI: 10.1007/s11356-023-26434-2 PMID: 36934193

Azim, H., Johnston, A. L., Nixon, M., Woodliffe, J. L., Theunissen, R., Suresh, R., Sivapalan, S., Bobo, J., & Licence, P. (2025). Collaborating for Impact: Navigating Partnerships and Overcoming Challenges across the Sustainable Development Goals. *ACS Sustainable Chemistry & Engineering, 13*(3), 1164–1173. DOI: 10.1021/acssuschemeng.4c10171 PMID: 39886474

Aziz, G., Strielkowski, W., Sarwar, S., & Tiwari, A. K. (2024). Implications of circular economy, digitalization and technological innovation to achieve sustainable environmental goal: Pre and post-vision 2030. *Heliyon, 10*(10), e30978. DOI: 10.1016/j.heliyon.2024.e30978 PMID: 38770279

Bai, L., Wang, X., Zhang, L., Feng, J., Liao, J., Chen, B., Wang, P., & Zhang, X. (2024). A systematic study of interactions between sustainable development goals (SDGs) in Hainan Island. *Scientific Reports, 14*(1), 26613. DOI: 10.1038/s41598-024-77984-5 PMID: 39496804

Bergman, Z., Bergman, M. M., Fernandes, K., Grossrieder, D., & Schneider, L. (2018). The Contribution of UNESCO Chairs toward Achieving the UN Sustainable Development Goals. *Sustainability (Basel), 10*(12), 4471. DOI: 10.3390/su10124471

Bibri, S. E., Alexandre, A., Sharifi, A., & Krogstie, J. (2023). Environmentally sustainable smart cities and their converging AI, IoT, and big data technologies and solutions: An integrated approach to an extensive literature review. *Energy Informatics, 6*(1), 9. DOI: 10.1186/s42162-023-00259-2 PMID: 37032812

Brodny, J., & Tutak, M. (2021). Assessing the level of digitalization and robotization in the enterprises of the European Union Member States. *PLoS One, 16*(7), e0254993. DOI: 10.1371/journal.pone.0254993 PMID: 34293023

Calderon-Monge, E., & Ribeiro-Soriano, D. (2023). The role of digitalization in business and management: A systematic literature review. *Review of Managerial Science*, •••, 1–43. Advance online publication. DOI: 10.1007/s11846-023-00647-8

Chaparro-Banegas, N., Ibañez Escribano, A. M., Mas-Tur, A., & Roig-Tierno, N. (2023). Innovation facilitators and sustainable development: A country comparative approach. *Environment, Development and Sustainability, 26*(4), 1–29. Advance online publication. DOI: 10.1007/s10668-023-03055-w PMID: 37363019

Chawla, M., Sharma, P., Gautam, S., & Popescu, C. R. (2025). Sustainability in Human Capital Management: Shaping the Future With Technology. In Kapoor, T., Singh, A., Sahay, K., Kaur, H., & Verma, R. (Eds.), *Sustainable Management Practices for Employee Retention and Recruitment* (pp. 265–290). IGI Global Scientific Publishing., DOI: 10.4018/979-8-3693-8679-8.ch013

Chen, M., Chen, L., Li, Y., & Xian, Y. (2022). Developing computable sustainable urbanization science: Interdisciplinary perspective. *Computational Urban Science*, *2*(1), 17. DOI: 10.1007/s43762-022-00048-9 PMID: 35755716

Chuong, H. N., Uyen, V. T. P., Ngan, N. D. P., Tram, N. T. B., Han, N. D. M., & Duyen, P. H. K. (2025). The impact of globalization, renewable energy, and labor on sustainable development: A cross-country analysis. *PLoS One*, *20*(2), e0315273. DOI: 10.1371/journal.pone.0315273 PMID: 39982918

Darwish, D. (2024). Human-AI Collaboration in Industry 5. In Kumar, N., Pal, S., Agarwal, P., Rosak-Szyrocka, J., & Jain, V. (Eds.), *Human-Machine Collaboration and Emotional Intelligence in Industry 5.0* (pp. 44–70). IGI Global Scientific Publishing., DOI: 10.4018/979-8-3693-6806-0.ch003

de Andrade, L. C., Borges-Pedro, J. P., Gomes, M. C. R. L., Tregidgo, D. J., do Nascimento, A. C. S., Paim, F. P., Marmontel, M., Benitz, T., Hercos, A. P., & do Amaral, J. V. (2021). The sustainable development goals in two sustainable development reserves in central amazon: Achievements and challenges. *Discover Sustainability*, *2*(1), 54. DOI: 10.1007/s43621-021-00065-4 PMID: 35425916

Deng, D., Zhang, J., Wang, J., & Zong, X. (2025). Sustainable performance evaluation of pharmaceutical companies: Sustainable balanced scorecard and hybrid MCDM approach. *Frontiers in Public Health*, *12*, 1495156. DOI: 10.3389/fpubh.2024.1495156 PMID: 39845657

Dhaoui, I. (2022). E-Government for Sustainable Development: Evidence from MENA Countries. *Journal of the Knowledge Economy*, *13*(3), 2070–2099. DOI: 10.1007/s13132-021-00791-0 PMID: 40477231

Duháček Šebestová, J., & Popescu, C. (Eds.). (2025). *Building a Green Future Through Essential Decision-Making Competencies*. IGI Global., DOI: 10.4018/979-8-3373-0189-1

Duraimutharasan, N., Deepan, A., Swadhi, R., Velmurugan, P. R., & Varshney, K. R. (2025). Enhancing Control Engineering Through Human-Machine Collaboration: AI for Improved Efficiency and Decision-Making. In Mellal, M. (Ed.), *Harnessing AI for Control Engineering* (pp. 155–176). IGI Global Scientific Publishing., DOI: 10.4018/979-8-3693-7812-0.ch008

Elavarasan, R. M., Pugazhendhi, R., Shafiullah, G. M., Kumar, N. M., Arif, M. T., Jamal, T., Chopra, S. S., & Dyduch, J. (2022). Impacts of COVID-19 on Sustainable Development Goals and effective approaches to maneuver them in the post-pandemic environment. *Environmental Science and Pollution Research International*, *29*(23), 33957–33987. DOI: 10.1007/s11356-021-17793-9 PMID: 35032263

Elzein, B. (2024). Nano Revolution: "Tiny tech, big impact: How nanotechnology is driving SDGs progress". *Heliyon*, *10*(10), e31393. DOI: 10.1016/j.heliyon.2024.e31393 PMID: 38818162

Fang, X., & Cao, Y. (2023). Spatial Association Network Evolution and Variance Decomposition of Economic Sustainability Development Efficiency in China. *International Journal of Environmental Research and Public Health, 20*(4), 2966. DOI: 10.3390/ijerph20042966 PMID: 36833661

Filho, W. L., Sierra, J., Price, E., Eustachio, J. H. P. P., Novikau, A., Kirrane, M., Dinis, M. A. P., & Salvia, A. L. (2024). The role of universities in accelerating the sustainable development goals in Europe. *Scientific Reports, 14*(1), 15464. DOI: 10.1038/s41598-024-65820-9 PMID: 38965303

Frichi, Y., & Jawab, F. (Eds.). (2025). *Multi-Stakeholder Collaboration for Sustainable Supply Chain.* IGI Global Scientific Publishing., DOI: 10.4018/979-8-3693-8925-6

Gill, A., & Mathur, A. (2024). Emotional Intelligence in the Age of AI: Enhancing Workforce Development for Human-Machine Collaboration. In Sharma, A., Moses, O., Sharma, R., & Gupta, S. (Eds.), *Sustainable Innovation for Industry 6.0* (pp. 271–294). IGI Global Scientific Publishing., DOI: 10.4018/979-8-3693-3140-8.ch014

Goalla, A., Shefali, P., & Sushmitha, K. (2024). Human-AI Collaboration in Environmental Decision Making: Ethical Considerations. In Kannan, H., Rodriguez, R., Paprika, Z., & Ade-Ibijola, A. (Eds.), *Exploring Ethical Dimensions of Environmental Sustainability and Use of AI* (pp. 330–345). IGI Global Scientific Publishing., DOI: 10.4018/979-8-3693-0892-9.ch016

Guo, H., Chen, F., Tang, Y., Ding, Y., Chen, M., Zhou, W., Zhu, M., Gao, S., Yang, R., Zheng, W., Fang, C., Lin, H., Roders, A. P., Cigna, F., Tapete, D., & Xu, B. (2023). Progress toward the sustainable development of world cultural heritage sites facing land-cover changes. *Innovation (Cambridge (Mass.)), 4*(5), 100496. DOI: 10.1016/j.xinn.2023.100496 PMID: 37663934

Guo, H., Huang, L., & Liang, D. (2022). Further promotion of sustainable development goals using science, technology, and innovation. *Innovation (Cambridge (Mass.)), 3*(6), 100325. DOI: 10.1016/j.xinn.2022.100325 PMID: 36193207

Guo, S., Diao, Y., & Du, J. (2022). Coupling Coordination Measurement and Evaluation of Urban Digitalization and Green Development in China. *International Journal of Environmental Research and Public Health, 19*(22), 15379. DOI: 10.3390/ijerph192215379 PMID: 36430095

He, W. W., He, S. L., & Hou, H. L. (2024). Digital economy, technological innovation, and sustainable development. *PLoS One, 19*(7), e0305520. DOI: 10.1371/journal.pone.0305520 PMID: 39042638

Helldén, D., Weitz, N., Nilsson, M., & Alfvén, T. (2022). Situating Health Within the 2030 Agenda-A Practical Application of the Sustainable Development Goals Synergies Approach. *Public Health Reviews, 43*, 1604350. DOI: 10.3389/phrs.2022.1604350 PMID: 35465141

Huang, L., Swain, R. B., Jeppesen, E., Cheng, H., Zhai, P., Gu, B., Barceló, D., Lu, J., Wei, K., Luo, L., Wang, F., Wang, H., Zeng, J., & Guo, H. (2024). Harnessing science, technology, and innovation to drive synergy between climate goals and the SDGs. *Innovation (Cambridge (Mass.)), 5*(6), 100693. DOI: 10.1016/j.xinn.2024.100693 PMID: 39399226

Irajifar, L., Chen, H., Lak, A., Sharifi, A., & Cheshmehzangi, A. (2023). The nexus between digitalization and sustainability: A scientometrics analysis. *Heliyon, 9*(5), e15172. DOI: 10.1016/j.heliyon.2023.e15172 PMID: 37153424

John Joseph, F., Chinnusamy, K., Jeganathan, J., & Obaid, J. A., & Rajest, S. (Eds.). (2026). *Machine Learning, Predictive Analytics, and Optimization in Complex Systems*. IGI Global Scientific Publishing. https://doi.org/DOI: 10.4018/979-8-3373-5203-9

Kaur, J. (2024). The Nexus of Care: Human-Machine Collaboration Redefining Healthcare Delivery. In T. Murugan, J. W., & V. P. (Eds.), *Technologies for Sustainable Healthcare Development* (pp. 230-253). IGI Global Scientific Publishing. https://doi.org/DOI: 10.4018/979-8-3693-2901-6.ch013

Kenwright, B. (Ed.). (2025). *Cutting-Edge Techniques for Effective Data Mining and Machine Learning*. IGI Global Scientific Publishing., DOI: 10.4018/979-8-3693-1606-1

Khan, K. A., Ahmad, W., Zubair, A. O., Subhan, M., & Shah, M. I. (2025). From invention to progress: Energy technology innovation and sustainable development in OECD economies. *PLoS One*, *20*(2), e0310104. DOI: 10.1371/journal.pone.0310104 PMID: 39946336

Kieny, M. P., Bekedam, H., Dovlo, D., Fitzgerald, J., Habicht, J., Harrison, G., Kluge, H., Lin, V., Menabde, N., Mirza, Z., Siddiqi, S., & Travis, P. (2017). Strengthening health systems for universal health coverage and sustainable development. *Bulletin of the World Health Organization*, *95*(7), 537–539. DOI: 10.2471/BLT.16.187476 PMID: 28670019

Kumar, N., Pal, S. K., Agarwal, P., Rosak-Szyrocka, J., & Jain, V. (Eds.). (2024). *Human-Machine Collaboration and Emotional Intelligence in Industry 5.0*. IGI Global Scientific Publishing., DOI: 10.4018/979-8-3693-6806-0

Kumar, N., Pal, S. K., Agarwal, P., Rosak-Szyrocka, J., & Jain, V. (Eds.). (2024). *Harnessing Artificial Emotional Intelligence for Improved Human-Computer Interactions*. IGI Global Scientific Publishing., DOI: 10.4018/979-8-3693-2794-4

Leal Filho, W., Vidal, D. G., Chen, C., Petrova, M., Dinis, M. A. P., Yang, P., Rogers, S., Álvarez-Castañón, L., Djekic, I., Sharifi, A., & Neiva, S. (2022). An assessment of requirements in investments, new technologies, and infrastructures to achieve the SDGs. *Environmental Sciences Europe*, *34*(1), 58. DOI: 10.1186/s12302-022-00629-9 PMID: 37128271

Liang, D., Guo, H., Nativi, S., Kulmala, M., Shirazi, Z., Chen, F., Kalonji, G., Yan, D., Li, J., Duerler, R., Luo, L., Han, Q., Deng, S., Wang, Y., Kong, L., & Jelinek, T. (2023). A future for digital public goods for monitoring SDG indicators. *Scientific Data*, *10*(1), 875. DOI: 10.1038/s41597-023-02803-x PMID: 38062062

Liu, Y., Yang, Y., Zhang, X., & Yang, Y. (2024). The impact of technological innovation on the green digital economy and development strategies. *PLoS One*, *19*(4), e0301051. DOI: 10.1371/journal.pone.0301051 PMID: 38662690

Luo, L., Wang, X., & Guo, H. (2022). Contribution of UNESCO designated sites to the achievement of Sustainable Development Goals. *Innovation (Cambridge (Mass.))*, *3*(3), 100227. DOI: 10.1016/j.xinn.2022.100227 PMID: 35373170

Luo, S., & Liu, J. (2024). Enterprise service-oriented transformation and sustainable development driven by digital technology. *Scientific Reports*, *14*(1), 10047. DOI: 10.1038/s41598-024-60922-w PMID: 38698122

MacPherson, J., Voglhuber-Slavinsky, A., Olbrisch, M., Schöbel, P., Dönitz, E., Mouratiadou, I., & Helming, K. (2022). Future agricultural systems and the role of digitalization for achieving sustainability goals. A review. *Agronomy for Sustainable Development, 42*(4), 70. DOI: 10.1007/s13593-022-00792-6 PMID: 35818482

Matsui, T., Suzuki, K., Ando, K., Kitai, Y., Haga, C., Masuhara, N., & Kawakubo, S. (2022). A natural language processing model for supporting sustainable development goals: Translating semantics, visualizing nexus, and connecting stakeholders. *Sustainability Science, 17*(3), 969–985. DOI: 10.1007/s11625-022-01093-3 PMID: 35136451

Mehta, S., Gupta, S. K., Aljohani, A. A., & Khayyat, M. (Eds.). (2024). *Impact and Potential of Machine Learning in the Metaverse*. IGI Global Scientific Publishing., DOI: 10.4018/979-8-3693-5762-0

Mohammed, A. J., & Ghebreyesus, T. A. (2018). Healthy living, well-being and the sustainable development goals. *Bulletin of the World Health Organization, 96*(9), 590–590A. DOI: 10.2471/BLT.18.222042 PMID: 30262935

Murshed, M. (2024). Is digitalization essential for abating carbon emission growth in South Asia? *Heliyon, 10*(20), e39012. DOI: 10.1016/j.heliyon.2024.e39012 PMID: 39497997

Nilashi, M., Ali Abumalloh, R., Mohd, S., Nurlaili Farhana Syed Azhar, S., Samad, S., Hang Thi, H., Alghamdi, O. A., & Alghamdi, A. (2023). COVID-19 and sustainable development goals: A bibliometric analysis and SWOT analysis in Malaysian context. *Telematics and Informatics, 76*, 101923. DOI: 10.1016/j.tele.2022.101923 PMID: 36510580

Ozturk, I., Alqassimi, O., & Ullah, S. (2024). Digitalization and SMEs development in the context of sustainable development: A China perspective. *Heliyon, 10*(6), e27936. DOI: 10.1016/j.heliyon.2024.e27936 PMID: 38510016

Palau-Pinyana, E., Llach, J., & Bagur-Femenías, L. (2023). Mapping enablers for SDG implementation in the private sector: a systematic literature review and research agenda. *Management Review Quarterly*, 1–30. Advance online publication. https://doi.org/.DOI: 10.1007/s11301-023-00341-9

Pandey, A., & Joseph, J. (2025). Human-Machine Interaction: Ethical and Legal Considerations. In Chakraborty, S. (Ed.), *Exploring AI Implications on Law, Governance, and Industry* (pp. 177–200). IGI Global Scientific Publishing., DOI: 10.4018/979-8-3373-3384-7.ch007

Pandey, D., Muniandi, B., Pandey, B. K., & George, A. (Eds.). (2025). *Interdisciplinary Approaches to AI, Internet of Everything, and Machine Learning*. IGI Global Scientific Publishing., DOI: 10.4018/979-8-3373-1032-9

Parikh, P., Diep, L., Hofmann, P., Tomei, J., Campos, L. C., Teh, T. H., Mulugetta, Y., Milligan, B., & Lakhanpaul, M. (2021). Synergies and trade-offs between sanitation and the sustainable development goals. *UCL Open Environment, 3*, e016. DOI: 10.14324/111.444/ucloe.000016 PMID: 37228800

Peng, W. (2025). Shaping sustainable perceptions: The role of metaphors in Olympic news discourse. *PLoS One, 20*(1), e0317380. DOI: 10.1371/journal.pone.0317380 PMID: 39804915

Ponnada, V. T. (Ed.). (2026). *Innovative Machine Learning Applications in the Aerospace Industry*. IGI Global Scientific Publishing., DOI: 10.4018/979-8-3693-7525-9

Popescu, C. (Ed.). (2025). *Building Inclusive Global Knowledge Societies for Sustainable Development*. IGI Global., DOI: 10.4018/979-8-3693-6298-3

Popescu, C. R., & Popescu, G. N. (2025). The Power of Data Analysis in Transforming Businesses Performance: Improved Decision-Making, Real-World Applications, and the Sustainable Development Goals. In Popescu, C. (Ed.), *Building Inclusive Global Knowledge Societies for Sustainable Development* (pp. 169–208). IGI Global Scientific Publishing., DOI: 10.4018/979-8-3693-6298-3.ch008

Popescu, C. R., & Verma, R. (2025). Green Future Ideas and Scalable Technologies Essential for Innovators, Scientists, Researchers, and Engineers: Cutting-Edge Initiatives and Commercially-Viable Solutions for a Fairer Future. In Duháček Šebestová, J., & Popescu, C. (Eds.), *Building a Green Future Through Essential Decision-Making Competencies* (pp. 63–88). IGI Global Scientific Publishing., DOI: 10.4018/979-8-3373-0189-1.ch003

Popescu, C. R., & Verma, R. R. (2025). Quality Education to Meet the Sustainable Development Goals: Tremendous Role of Sustainable Management Practices for Employee Retention and Recruitment. In Kapoor, T., Singh, A., Sahay, K., Kaur, H., & Verma, R. (Eds.), *Sustainable Management Practices for Employee Retention and Recruitment* (pp. 233–242). IGI Global Scientific Publishing., DOI: 10.4018/979-8-3693-8679-8.ch011

Qi, S., Huang, D., Ma, Q., & Zhou, M. (2023). Factors Influencing Sustainable Development Literacy among Engineering Undergraduates in China: Based on the College Impact Model. *International Journal of Environmental Research and Public Health*, *20*(2), 1249. DOI: 10.3390/ijerph20021249 PMID: 36673998

Rahman, M. M., Hasan, M. J., Chandra Deb, B., Rahman, M. S., & Kabir, A. S. (2023). The effect of social media entrepreneurship on sustainable development: Evidence from online clothing shops in Bangladesh. *Heliyon*, *9*(9), e19397. DOI: 10.1016/j.heliyon.2023.e19397 PMID: 37662716

Raman, R., Sreenivasan, A., Kulkarni, N. V., Suresh, M., & Nedungadi, P. (2025). Analyzing the contributions of biofuels, biomass, and bioenergy to sustainable development goals. *iScience*, *28*(4), 112157. DOI: 10.1016/j.isci.2025.112157 PMID: 40230525

Ranjbari, M., Shams Esfandabadi, Z., Zanetti, M. C., Scagnelli, S. D., Siebers, P. O., Aghbashlo, M., Peng, W., Quatraro, F., & Tabatabaei, M. (2021). Three pillars of sustainability in the wake of COVID-19: A systematic review and future research agenda for sustainable development. *Journal of Cleaner Production*, *297*, 126660. DOI: 10.1016/j.jclepro.2021.126660 PMID: 34785869

Raygoza-L., M. E., Orduño-Osuna, J. H., Mercado-Herrera, A., Jimenez-Sanchez, R., & Murrieta-Rico, F. N. (Eds.). (2025). *Exploring Psychology, Social Innovation and Advanced Applications of Machine Learning*. IGI Global Scientific Publishing., DOI: 10.4018/979-8-3693-6910-4

Reddy, C. K. K., Reddy, A. V., Doss, S., & Priyanka, K. (2024). Human-Machine Collaboration and Emotional Intelligence in Industry 6.0: Concepts, Challenges, and Future Directions. In Nijalingappa, P., Gururaj, T., Goyal, S., Shukla, V., & Bruno, A. (Eds.), *Examining the Metaverse in Healthcare: Opportunities, Challenges, and Future Directions* (pp. 221–246). IGI Global Scientific Publishing., DOI: 10.4018/979-8-3693-1515-6.ch009

Rehman Khan, S. A., Ahmad, Z., Sheikh, A. A., & Yu, Z. (2022). Digital transformation, smart technologies, and eco-innovation are paving the way toward sustainable supply chain performance. *Science Progress*, *105*(4), 368504221145648. DOI: 10.1177/00368504221145648 PMID: 36573795

Ruth, J. A., Mahesh, V. G., Visalakshi, P., Uma, R., & Meenakshi, A. (Eds.). (2024). *Machine Learning and Cryptographic Solutions for Data Protection and Network Security*. IGI Global Scientific Publishing., DOI: 10.4018/979-8-3693-4159-9

Saluja, A., & Mongia, A. (2025). Human-Machine Collaboration: Augmenting Human Abilities With Robotic Assistance in the Workplace. In Kaur, J. (Ed.), *Technological Enhancements for Improving Employee Performance, Safety, and Well-Being* (pp. 145–170). IGI Global Scientific Publishing., DOI: 10.4018/979-8-3693-9631-5.ch007

Santarius, T., Bieser, J. C. T., Frick, V., Höjer, M., Gossen, M., Hilty, L. M., Kern, E., Pohl, J., Rohde, F., & Lange, S. (2023). Digital sufficiency: Conceptual considerations for ICTs on a finite planet. *Annales des Télécommunications*, *78*(5-6), 277–295. DOI: 10.1007/s12243-022-00914-x PMID: 37593439

Sayyed, M., Jadhav, B. R., Barnabas, V., & Gupta, S. K. (2024). Human-Machine Interaction in the Metaverse: A Comprehensive Review and Proposed Framework. In Mehta, S., Gupta, S., Aljohani, A., & Khayyat, M. (Eds.), *Impact and Potential of Machine Learning in the Metaverse* (pp. 1–28). IGI Global Scientific Publishing., DOI: 10.4018/979-8-3693-5762-0.ch001

Šebestová, J. D., & Popescu, C. R. (2025). Basic Competency Frameworks Within Sustainable Business Model. In Duháček Šebestová, J., & Popescu, C. (Eds.), *Building a Green Future Through Essential Decision-Making Competencies* (pp. 1–26). IGI Global Scientific Publishing., DOI: 10.4018/979-8-3373-0189-1.ch001

Šebestová, J. D., & Popescu, C. R. (2025). Basic Competency Frameworks Within Sustainable Business Model. In Duháček Šebestová, J., & Popescu, C. (Eds.), *Building a Green Future Through Essential Decision-Making Competencies* (pp. 1–26). IGI Global Scientific Publishing., DOI: 10.4018/979-8-3373-0189-1.ch001

Şimşek, T. (2025). Human-Machine Collaboration: How Do Employees Interact With Artificial Intelligence? In Demircioğlu, A. (Ed.), *AI Use in Social Sciences* (pp. 157–180). IGI Global Scientific Publishing., DOI: 10.4018/979-8-3373-2612-2.ch006

Smaniotto, C., Battistella, C., Brunelli, L., Ruscio, E., Agodi, A., Auxilia, F., Baccolini, V., Gelatti, U., Odone, A., Prato, R., Tardivo, S., Voglino, G., Valent, F., Brusaferro, S., Balzarini, F., Barchitta, M., Carli, A., Castelli, F., Coppola, C., & Sisi, S. (2020). Sustainable Development Goals and 2030 Agenda: Awareness, Knowledge and Attitudes in Nine Italian Universities, 2019. *International Journal of Environmental Research and Public Health*, *17*(23), 8968. DOI: 10.3390/ijerph17238968 PMID: 33276530

Sreenivasan, A., Suresh, M., Nedungadi, P., & R, R. R. (2023). Mapping analytical hierarchy process research to sustainable development goals: Bibliometric and social network analysis. *Heliyon*, *9*(8), e19077. DOI: 10.1016/j.heliyon.2023.e19077 PMID: 37636443

Sundaram, A. & L. V., N. (2024). Collaborate and Energize: Human-Machine Synergy in Advanced Energy Industries (AEI) Within the Framework of Industry 5.0. In N. Kumar, S. Pal, P. Agarwal, J. Rosak-Szyrocka, & V. Jain (Eds.), *Human-Machine Collaboration and Emotional Intelligence in Industry 5.0* (pp. 71-89). IGI Global Scientific Publishing. https://doi.org/.DOI: 10.4018/979-8-3693-6806-0.ch004

Tikadar, S., Rodrigues, J. J., Shaw, R., Hassan, A., & Patra, B. K. (Eds.). (2026). *Driving Smarter Human-Computer Interaction With Multidisciplinary Personalized Systems*. IGI Global Scientific Publishing., DOI: 10.4018/979-8-3373-1444-0

Ugwu, C. N., Ugwu, O. P., Alum, E. U., Eze, V. H. U., Basajja, M., Ugwu, J. N., Ogenyi, F. C., Ejemot-Nwadiaro, R. I., Okon, M. B., Egba, S. I., & Uti, D. E. (2025). Sustainable development goals (SDGs) and resilient healthcare systems: Addressing medicine and public health challenges in conflict zones. *Medicine*, *104*(7), e41535. DOI: 10.1097/MD.0000000000041535 PMID: 39960902

Ul Rehman, S. (Ed.). (2025). *Challenges and Solutions for Cybersecurity and Adversarial Machine Learning*. IGI Global Scientific Publishing., DOI: 10.4018/979-8-3373-2200-1

Ullah, A., Pinglu, C., Ullah, S., & Hashmi, S. H. (2021). Nexus of regional integration, socioeconomic determinants and sustainable development in belt and road initiative countries. *PLoS One*, *16*(7), e0254298. DOI: 10.1371/journal.pone.0254298 PMID: 34242342

UNESCO. (2025). UNESCO and Sustainable Development Goals. Retrieved https://www.unesco.org/en/sdgs[REMOVED HYPERLINK FIELD]. Accessed on the 15th of June 2025.

United Nations Environment Program (UNEP). Finance Initiative (2025). SDGs and Impact. Holistic impact analysis and management to close the SDG financing gap. Retrieved https://www.unepfi.org/impact/positive-impact/[REMOVED HYPERLINK FIELD]. Accessed on the 15th of June 2025.

United Nations Environment Program (UNEP). (2025a). UNEP and the Sustainable Development Goals. Retrieved https://www.unep.org/explore-topics/sustainable-development-goals. Accessed on the 15th of June 2025.

United Nations Environment Program (UNEP). (2025b). Sustainable Development Goals. Understand how the environmental SDGs are measured, the progress being made and the current gaps that exist. Retrieved https://sdgs.unep.org/[REMOVED HYPERLINK FIELD]. Accessed on the 15th of June 2025.

United Nations Environment Program (UNEP). Department of Economic and Social Affairs Sustainable Development (2025). United Nations Environment Program (UNEP). Retrieved https://sdgs.un.org/un-system-sdg-implementation/united-nations-environment-programme-unep-34570[REMOVED HYPERLINK FIELD]. Accessed on the 15th of June 2025.

Valentine, N. B., Koller, T. S., & Hosseinpoor, A. R. (2016). Monitoring health determinants with an equity focus: A key role in addressing social determinants, universal health coverage, and advancing the 2030 sustainable development agenda. *Global Health Action*, *9*(1), 34247. DOI: 10.3402/gha.v9.34247 PMID: 27989275

van Driel, M., Biermann, F., Kim, R. E., & Vijge, M. J. (2024). The impact of the Sustainable Development Goals on global policies on sustainable consumption and production. *Globalizations*, *21*(8), 1366–1382. DOI: 10.1080/14747731.2024.2351301 PMID: 39634121

Van Tulder, R., Rodrigues, S. B., Mirza, H., & Sexsmith, K. (2021). The UN's Sustainable Development Goals: Can multinational enterprises lead the Decade of Action? *Journal of International Business Policy*, *4*(1), 1–21. DOI: 10.1057/s42214-020-00095-1

Vega, J., & Frenz, P. (2013). Integrating social determinants of health in the universal health coverage monitoring framework. *Revista panamericana de salud publica = Pan. American Journal of Public Health*, *34*(6), 468–472. PMID: 24569977

Verdejo Espinosa, Á., Lopez Ruiz, J., Mata Mata, F., & Estevez, M. E. (2021). Application of IoT in Healthcare: Keys to Implementation of the Sustainable Development Goals. *Sensors (Basel)*, *21*(7), 2330. DOI: 10.3390/s21072330 PMID: 33810606

Wang, F., Harindintwali, J. D., Yuan, Z., Wang, M., Wang, F., Li, S., Yin, Z., Huang, L., Fu, Y., Li, L., Chang, S. X., Zhang, L., Rinklebe, J., Yuan, Z., Zhu, Q., Xiang, L., Tsang, D. C. W., Xu, L., Jiang, X., & Chen, J. M. (2021). Technologies and perspectives for achieving carbon neutrality. *Innovation (Cambridge (Mass.))*, *2*(4), 100180. DOI: 10.1016/j.xinn.2021.100180 PMID: 34877561

Wang, M., Zhu, X., & Yin, S. (2023). Spatial-temporal coupling coordination and interaction between digitalization and traditional industrial upgrading: A case study of the Yellow River Basin. *Scientific Reports*, *13*(1), 17997. DOI: 10.1038/s41598-023-44995-7 PMID: 37865702

Wang, Z., & Zhao, Y. (2021). Energy Strategy for Sustainable Development of Rural Areas Based on the Analysis of Sustainable Digital Economy. *Frontiers in Psychology*, *12*, 788026. DOI: 10.3389/fpsyg.2021.788026 PMID: 34987452

Xu, J., She, S., & Liu, W. (2022). Role of digitalization in environment, social and governance, and sustainability: Review-based study for implications. *Frontiers in Psychology*, *13*, 961057. DOI: 10.3389/fpsyg.2022.961057 PMID: 36533022

Xu, X., Wang, S., Li, J., & Qiao, T. (2024). Environmental regulatory intensity, green finance and corporate green sustainable development performance. *Heliyon, 10*(9), e30114. DOI: 10.1016/j.heliyon.2024.e30114 PMID: 38694050

Yamin, A. E., & Mason, E. (2019). Why accountability matters for universal health coverage and meeting the SDGs. *Lancet, 393*(10176), 1082–1084. DOI: 10.1016/S0140-6736(19)30434-9 PMID: 30894258

Yip, W. S., Zhou, H., & To, S. (2023). A critical analysis on the triple bottom line of sustainable manufacturing: Key findings and implications. *Environmental Science and Pollution Research International, 30*(14), 41388–41404. DOI: 10.1007/s11356-022-25122-x PMID: 36631618

Yuan, Q., Zhang, G., Wang, R., Ma, X., & Niu, J. (2024). Does technological innovation in National Sustainable Development Agenda Innovation Demonstration Zones promote green development?-the case from Chengde City, China. *PLoS One, 19*(5), e0300315. DOI: 10.1371/journal.pone.0300315 PMID: 38805430

Zhan, J. X., & Santos-Paulino, A. U. (2021). Investing in the Sustainable Development Goals: Mobilization, channeling, and impact. *Journal of International Business Policy, 4*(1), 166–183. DOI: 10.1057/s42214-020-00093-3

Zhang, J. (2024). Research on the impact of digitalization on green development: An empirical analysis from the low-carbon strategy perspective. *PLoS One, 19*(3), e0300288. DOI: 10.1371/journal.pone.0300288 PMID: 38457442

Zhang, Q., Dai, W., Chen, J., Gu, Y., & Zhao, Y. (2025). The ‘side effects’ of digitalization: A study on role overload and job burnout of employees. *PLoS One, 20*(4), e0322112. DOI: 10.1371/journal.pone.0322112 PMID: 40305484

Zhang, Q., & Wang, Q. (2023). Digitalization, Electricity Consumption and Carbon Emissions-Evidence from Manufacturing Industries in China. *International Journal of Environmental Research and Public Health, 20*(5), 3938. DOI: 10.3390/ijerph20053938 PMID: 36900949

Zhang, Y., Lin, Y., Zheng, G., Liu, Y., Sukiennik, N., Xu, F., Xu, Y., Lu, F., Wang, Q., Lai, Y., Tian, L., Li, N., Fang, D., Wang, F., Zhou, T., Li, Y., Zheng, Y., Wu, Z., & Guo, H. (2025). MetaCity: Data-driven sustainable development of complex cities. *Innovation (Cambridge (Mass.)), 6*(2), 100775. DOI: 10.1016/j.xinn.2024.100775 PMID: 39991486

Zhao, J., & Kumar, V. (Eds.). (2022). *Handbook of Research on Technologies and Systems for E-Collaboration During Global Crises*. IGI Global Scientific Publishing., DOI: 10.4018/978-1-7998-9640-1

Zhao, W., Yin, C., Hua, T., Meadows, M. E., Li, Y., Liu, Y., Cherubini, F., Pereira, P., & Fu, B. (2022). Achieving the Sustainable Development Goals in the post-pandemic era. *Humanities & Social Sciences Communications, 9*(1), 258. DOI: 10.1057/s41599-022-01283-5 PMID: 35967482

Zhu, W., & Yang, G. (2025). Research on the balanced, coordinated and sustainable development of China manufacturing industry. *PLoS One, 20*(4), e0322400. DOI: 10.1371/journal.pone.0322400 PMID: 40294056

Zhuge, K., Lin, W., Yuan, Y., He, H., & Zhang, Y. (2023). Does Digital Capability Promote Sustainable Development of New Ventures? The Dual Impact of Green Knowledge Creation and Green Pressure. *International Journal of Environmental Research and Public Health*, *20*(3), 2274. DOI: 10.3390/ijerph20032274 PMID: 36767640

Zyoud, S. H. (2023). Analyzing and visualizing global research trends on COVID-19 linked to sustainable development goals. *Environment, Development and Sustainability*, *25*(6), 5459–5493. DOI: 10.1007/s10668-022-02275-w PMID: 35340818

Chapter 7
Harnessing Artificial Intelligence in Marketing and Communication for Smart Factories:
Regulatory and Policy Landscape for Sustainable Development Goals

Cristina Raluca Gh. Popescu
https://orcid.org/0000-0002-5876-0550
University of Bucharest, Romania & The Bucharest University of Economic Studies, Romania

Karthick Natarajan
https://orcid.org/0009-0007-8287-3989
University of Madras, India

X. Vaishali
https://orcid.org/0009-0003-4760-7918
University of Petroleum and Energy Studies, Dehradun, India

ABSTRACT

Mastering artificial intelligence (AI) in marketing and communication based on digital strategies and analytics creates sustainable competitive advantage for smart factories and enables marketing teams boost their consumer relationship management programs. The research accentuates how AI shapes marketing today, highlighting advantages and disadvantages, while centering on happiness management, sustainability, and well-being. Additionally, the study sheds a new light on AI regulatory and policy landscape for Sustainable Development Goals, exploring the multifaceted impact of AI on innovations, research, conversations, trends, and consumer behavior. Outcomes point up AI's capabilities in enhancing marketing excellence, while focusing on achieving organic growth with the aid of best marketing practices, marketing ecosystem, marketing agility, strategic

DOI: 10.4018/979-8-3373-2985-7.ch007

thinking, and leadership. Findings emphasize that building an effective marketing AI strategy depends on hyper-personalized marketing campaigns and tailored communications that offers engagement, precision, and creativity.

INTRODUCTION

Nowadays, harnessing artificial intelligence in marketing and communication for smart factories is crucial for the sustainability and sustainable development of a healthy and competitive business environment (Xu, 2009). Hence, specialists believe there is a cutting-edge intersection between artificial intelligence, marketing, and next generation communication (Yaqoob *et al.*, 2023).

In this current book chapter several pivotal aspects concerning how artificial intelligence can be harnessed in marketing and communication, especially within the context of a sustainable business environment and smart factories. Among these aspects could be highlighted: the intelligent customer engagement (Zhu *et al.*, 2023); predictive marketing (Zhou, 2022); automated content creation (Thanigaivelan *et al.*, 2022); enhanced internal communication (Teng *et al.*, 2021); marketing sentiment analysis (Talele *et al.*, 2024); and digital twin and communication strategy (Weir *et al.*, 2023).

To begin with, some of the main questions that this introductory section intends to answer are displayed in the lines below. Firstly, how can artificial intelligence be harnessed in marketing and communication? Secondly, how predictive marketing can be explained in an in-depth way, in the case of smart factories? Thirdly, how does artificial intelligence marketing in smart factories compare with traditional marketing? And, fourthly, how are digital twins used beyond marketing in smart factories?

In this particular section answers to the first aforementioned question will be presented: Firstly, how can artificial intelligence be harnessed in marketing and communication?

These days, one of the vital ways that show how artificial intelligence can be harnessed in marketing and communication is represented by intelligent customer engagement (Zhang *et al.*, 2023). Researchers point out the role of artificial intelligence-driven tools like chatbots and virtual assistance that are known to enable manufacturers to engage directly with customers, hence providing real-time updates general key issues, such as: the order status; the delivery timeliness; and the product customization options (Weir *et al.*, 2023; Webb *et al.*, 2023). One highly important aspect in this matter is that the natural language processing (NLP) supports smooth and context-aware communication in a highly competitive business environment (Wang *et al.*, 2025).

In continuation, predictive marketing strongly characterizes smart factories equipped with the Internet of Things (Iot) and artificial intelligence which can analyze product trends and center on offering a better understanding to customer demand, in these cases predicting what products will be needed and when (Wang *et al.*, 2019). What is more, predictive marketing is vital in allowing and supporting just-in-time marketing campaigns which ought to target specific customer segments with hyper-personalized messaging known to be of tremendous interest and extremely desired by potential clients (Waheed, 2023).

Additionally, the automated content creation artificial intelligence can generate product descriptions, promotional content, and technical documentation by leveraging data from design and manufacturing processes (Vázquez *et al.*, 2019). In this way, previous research shows that the contents were maintained consistently, up-to-date, and tailored to different audiences, including specialists from different areas, engineers, buyers, and customers (Shawkat & Thivagar, 2024).

Besides all these, among other ways that foster how artificial intelligence can be harnessed in marketing and communication, especially within the context of a sustainable business environment and smart factories could successfully be mentioned enhanced internal communication (Sharma *et al.*, 2025). Under these circumstances, artificial intelligence can streamline internal communication which could generate translating technical data into understandable insights for sales, marketing, and executive teams (Seeniappan *et al.*, 2023). Great opportunities arise, in these cases, through cross-functional collaboration and shorten decision-making cycles (Šebestová & Popescu, 2022).

Moving on to marketing sentiment analysis it could be emphasized that by processing data from on-line sources, such as, for example, social media, forums, and reviews, artificial intelligence can greatly help factories understand how their products are perceived on the marketplace (Simbi *et al.*, 2024). Also, with the aid of marketing sentiment analysis the messaging and other additional features can be adjusted accordingly (Sari *et al.*, 2024).

In the end, digital twins and communication strategy are essential in this day and age. Digital twins (which are represented by virtual replicas of physical manufacturing systems), powered by artificial intelligence, are used with numerous purposes among which could be stressed the following ones: the operational optimization (Rasheed & Jianhua, 2023); the visualization processes for stakeholders (Sajil & Nair, 2020); and, the creation of marketing presentations (Ramakuri *et al.*, 2019). All these solutions are making marketing systems more digestible and engaging (Rajaram & Kumar, 2021).

In this particular part of the introduction, answers to the second aforementioned question will be presented: Secondly, how predictive marketing can be explained in an in-depth way, in the case of smart factories?

Moving on to predictive marketing in smart factories it needs to be stressed that by using data and artificial intelligence both customer needs and market dynamics can be anticipated. In this manner, things can be anticipated successfully before they happen. When taking a deep dive into all these matters, several aspects ought to be tackled: firstly, what powers predictive marketing in smart factories (Popescu & Popescu, 2019); and, secondly, how does this fuel predictive marketing (Popescu, 2019).

When willing to offer an in-depth understanding of what powers predictive marketing in smart factories there are at least three key aspects that come to mind: the real time data collection; the artificial intelligence-powered analytics; and the integrated customer insights (Paramarthalingam *et al.*, 2025; Pei, 2022; Popescu, 2025a). Hence, real time data collection refers to creating the right environment for relevant data collection due to empowering future marketing strategies (Popescu, 2025b). Smart factories today rely on sensors and IoT devices that continuously generate data, putting a specific accent on key issues, namely: customer orders, production rates, inventory levels, supply chain flows, and machine health (Popescu, 2020).

When addressing the artificial intelligence-powered analytics the following details are worth highlighting: the ocean of data is being processed and analyzed by artificial intelligence models in order to identify patterns and trends; also, these artificial intelligence models are of paramount importance since they might detect a seasonal spike in demand when targeting the trends analysis of certain products or they might forecast that supply shortages will impact one way or another the delivery timelines (Qu, 2016; Qu, 2018).

Likewise, the integrated customer insights are vital to the Customer Relationship Management (CRM) and the online behavior. In these situations, for instance, by creating a combination between production data, CRM, and online behavior – characterized by web traffic and engagement on digital channels, the artificial intelligence can build detailed customer profiles and, also, support the purchase likelihood models (Niu & Fan, 2024).

There are numerous ways in which the elements described synthetically above productive marketing, some of them being: the dynamic pricing; the hyper-targeted campaigns; the personalized offers; and the proactive content strategy (Niggli *et al.*, 2023).

The dynamic pricing is extremely important in today's business environment which promotes inclusivity, resilience, and robustness (Neufeld *et al.*, 2023). In those moments when the inventory is projected to be tight or in those situations in which costs are anticipated to rise, the marketing and the sales teams have an excellent opportunity to implement real-time pricing strategies so that the margin or real stock can be maximized accordingly at the right time (Ngonghala *et al.*, 2024).

When addressing the hyper-targeted campaigns, there are several characteristics that can be brought into analysis, as follows: factories can time their marketing efforts perfectly, which implicate that they can consistently focus on customers' needs and desires in a timely manner; additionally, when specific problems arise, such as, for instance, if artificial intelligence predicts a surge in demand for a product due to market factors or seasonal determinants marketers will be inclined to launch the marketing campaigns just at the right moment when customers are starting to look for the product (Nazir *et al.*, 2023; National Academy of Medicine; National Academies of Sciences, Engineering, and Medicine; Health and Medicine Division; Board on Health Sciences Policy; Committee on Creating a Framework for Emerging Science, Technology, and Innovation in Health and Medicine, 2023).

Personalized offers are essential in extremely dynamic business environments. In these situations, the artificial intelligence tools can tailor marketing messages and promotions down to the individual level (National Academy of Medicine; National Academies of Sciences, Engineering, and Medicine; Health and Medicine Division; Board on Health Sciences Policy; Committee on Creating a Framework for Emerging Science, Technology, and Innovation in Health and Medicine, 2023). A very powerful example in this case is the one in which clients are receiving personalized e-mails concerning new features of certain products, as a result of previously purchased products, right before they are likely to consider new acquisitions of upgraded goods and services (Nasir *et al.*, 2020).

Going more in-depth with the characteristics of the proactive content strategy, the aspects that are crucial here are: the product cycles and the upcoming product releases are the key trigger for the complex and diverse activities of the marketing teams; hence, based on the product cycles and the upcoming product releases, the marketing teams are expected to prepare relevant content for blog posts, videos, and webinars, long before customers start asking for information regarding their products and services of interest (Nagarsheth & Sharma, 2021).

When focusing on the issues depicted above it can clearly be noticed that harnessing artificial intelligence in marketing and communication for smart factories and, in particular, predictive marketing in smart factories flips traditional marketing on its head, making a notable shift from reactive to proactive, which is especially valuable in business to business (B2B) industrial and manufacturing markets and scenarios (Mubuli *et al.*, 2021).

LITERATURE REVIEW

Today it is paramount to consider the benefits and advantages of predictive in smart factories, especially when addressing the importance and the role played by the Sustainable Development Goals (SDGs) for humanity and society as a whole (Roundtable on Environmental Health Sciences, Research, and Medicine; Board on Population Health and Public Health Practice; Institute of Medicine, 2013). In addition, there are numerous real-world-inspired use cases that show how predictive marketing can work in a smart factory context (Minos *et al.*, 2016). Noteworthy, some valuable examples could be offered at any moment when bringing into discussion and when shedding new light on the global leaders in industrial automation (Adewuyi & Awodumi, 2021).

An illustrative hypothetical example can be chosen in order to demonstrate how predictive marketing can work in a smart factory context, especially when willing to focus on the case of a global leader in industrial automation that employs smart manufacturing and that relies on artificial intelligence. Hence, as use case the focal point could be the launching of a new line servo drivers, which are highly specialized electronic devices that control the motion of electronic motors, in particular, servo motors with high precision. It needs to be acknowledged that servo drivers are responsible for the smooth and accurate movement in robotics, to them being credited: the Computer Numerical Control (CNC) machines that automate machining processes like cutting, drilling, and milling; the conveyor systems which are mechanical systems that move materials from one point to another, being used anywhere efficiency is key, such as, for instance, in warehouses, airports, and factories; and, the camera gimbals, which are stabilizing mounts that keep the cameras steady while moving. In this case, the smart factory is assiduously preparing to launch a next generation line of servo drivers centers on gathering real-time production data, supplier trends, and customer usage insights through IoT sensors and machine learning platforms (Abid *et al.*, 2022; Ali *et al.*, 2022). Several steps are considered, namely: step 1, predictive insights from smart manufacturing; step 2, artificial intelligence-driven customer segmentation; step 3, targeted marketing campaigns; and, lastly, step 4, feedback loop and continuous refinement. In the lines below, all of these four important steps and described and analyzed in order to offer more complex details on the benefits and advantages of predictive in smart factories.

Step 1: Predictive Insights from Smart Manufacturing

The first step of the use case addresses predictive insights from smart manufacturing. This particular step centers on major elements, such as: the artificial intelligence instruments and models analyze usage patterns and warranty data from the already existing drives in the field; additionally, the IoT data signals that many customers' current drives will reach very soon (namely, for instance, in 6-12 months) the end-of-life; and, also, external data reveals increased automation adoption in regions such as, for example, Southeast Asia, South America, and so on and so forth.

In this step it is shown that predictive insights mean using technologies like machine learning, IoT sensors, and real-time analytics to forecast what is going to happen before it happens so that the most important aspects are acknowledged and covered accordingly. In addition, it needs to be mentioned that this can be done in a few keyways that would enhance the role and the importance of smart manufacturing, such as:

- **Predictive Maintenance:** In order to save time and money, among others, machines are the ones that are able to notify and provide the symptoms that they are about to break down before they will actually be in that unfortunate and troubling situation (Agustina *et al.*, 2019; Ahmad & Jabeen, 2023).
- **Demand Forecasting:** In order to meet demand more precisely, manufacturers can optimize inventory and production schedules by centering on market trends analysis and historical sales data interpretations (Ahmed *et al.*, 2019; Akkalkot *et al.*, 2025).
- **Quality Control:** Reducing waste and improving product quality are of paramount importance today (Azar *et al.*, 2021; Aziz & Waheed, 2023). Hence, it becomes crucial to rely on advanced algorithms that can detect patterns in production data that have the purpose of signaling potential defects and errors (Chichaibelu *et al.*, 2023).
- **Energy Efficiency:** Energy supplies are vital in this day and age, being an impressive resource that calls for great responsibility when being used, starting from the aspects surrounding the Global Goals and continuing with the desire to support and foster resilient and inclusive communities worldwide (Choi *et al.*, 2019). Studies show that it is crucial that systems are able to predict when energy usage peak and adjust operations accordingly to lower the costs and, also, the environmental impact (Chelladurai *et al.*, 2025; Badía *et al.*, 2019; Baena *et al.*, 2023).
- **Supply Chain Resilience:** Nowadays, it is expected from companies to constantly adapt and be highly proactive in response to the challenges and changes in the business environment (Canales & Fears, 2023). Moreover, being proactive can only be done by anticipating disruptions starting, for example, with parts shortages and, ending with, for instance, logistics delays (Chen *et al.*, 2020; Bai *et al.*, 2012; Bashir *et al.*, 2022).

Step 2: Artificial Intelligence-Driven Customer Segmentation

The second step has as focal point the artificial intelligence-driven customer segmentation (Colombo *et al.*, 2016; Dawane & Malwatkar, 2025). In this case combining CRM data with predictive models is of great essence, ultimately leading to several high-priority segments, among which the following ones could be mentioned:

- **Already existing customers who are very close to product upgrading cycles:** This is a wonderful opportunity for upselling and cross-selling, according to research (Diao *et al.*, 2023). Additionally, continuous focus on this category of customers can increase revenue and strengthen customer loyalty. There are several strategies that refer to existing customers nearing their product upgrade cycle, among which could be stressed the following ones: the targeted messages about new features; showcasing the benefits of upgrading and, also, offering different types of incentives and tempting benefits for early adoption.
- **New customers known to make investments in automation projects:** When centering on the new customers known to make investments in automation projects it ought to be noted that this is the case in which customers are most often driven by the desire to build cutting-edge facilities from the ground up (Dubé *et al.*, 2014). The scope is to create highly optimized automation projects characterized by efficiency and sustainability in a world that prioritizes sustainable development in competitive and turbulent times (Elastic *et al.*, 2020; Elhosseini, 2020; Fears & Canales, 2023).

- **System integrators with upcoming proposals for large installations:** Handling large installations implicates the existence of well-positioned integrators having expertise in industries, such as the audiovisual systems; the automation; and the energy (Ebadinezhad *et al.*, 2019; Fertő & Harangozó, 2025).

Step 3: Targeted Marketing Campaigns

Moving on, the third step centers on the targeted marketing campaigns looking into the following major elements, namely:

- First of all, based on each segment of clients, the marketing team launches several artificial intelligence-personalized campaigns. It is always very important to consider the desires of the consumers as well as their expectations so that the newly launched artificial intelligence-personalized campaigns turn out to be successful.
- Second of all, the content of the targeted marketing campaigns contains specific content that includes, among others:

  (a) An interactive return on investment (ROI) calculator allowing the businesses to estimate the potential ROI of their marketing efforts before, during, and after the targeted marketing campaigns, which typically implicates having access to campaign specific data, such as, for example, the target audience, the marketing budget and, also, the estimated conversion rates (Hamad *et al.*, 2025). The scope is to project the potential revenue, the costs and, additionally, the ROI (Geetha *et al.*, 2025).
  (b) Digital twin demons of the new product for marketing prioritize the future of the customer experience being characterized by photorealistic accuracy and creative anonymity (Gupta *et al.*, 2021). The purpose is to drive sales by centering on the way in which consumers depend on the product images and videos. Another element to be considered in this particular case is the one represented by interactive experiences having in mind that digital twin of a consumer (DToC) in marketing is an emerging technology that marketers ought to strategically incorporate into their long-run plans.
  (c) Additionally, the webinars featuring technical questions and answers, which turn out to be extremely useful today and which are of essence when willing to offer all the needed support to the clients (Hatamlou, 2018; Hao *et al.*, 2021).

- Third of all, besides the aspects brought to the attention in the above lines, campaigned timing is of major essence and depends on the predicted customer buying windows which turn out to be pivotal at present and, also, for the future, in capturing the customers' behavior, interests, and motivations (Gh. Popescu & Ban a, 2019).

Step 4: Feedback Loop and Continuous Refinement

In the end, the last step in the use case in which the smart factory is assiduously preparing to launch a next generation line of servo drivers centers on gathering real-time production data, supplier trends, and customer usage insights through IoT sensors and machine learning platforms, is represented by the feedback loop and continuous refinement.

There are two major points of interest in this situation, namely:

- Firstly, the engagement metrics feedback into the predictive model, where the engagement metrics that are considered are some of the following ones: the engagement metrics; the webinar attendance; and, also, the queries referring to pricing (Herrero *et al.*, 2023; Hodson de Jaramillo *et al.*, 2023).
- Secondly, it is noteworthy to mention that the campaigns rhythm, the messaging and, also, the follow-up actions that take place in real time support the way in which the system dynamically adjusts (Hordofa, 2023; Hye *et al.*, 2023).

All in all, the outcomes derived from this use case are increasingly positive. Based on the support of each and every step imagined as being taken and which were carefully, diligently, and thoroughly described in the lines above the sales enquiries increased. For this particular situation, they increased by 45%, with a higher-than-expected early adoption rate in new markets. It is paramount to emphasize that companies have other advantages by adopting the steps described above which enables them to minimize marketing waste (Islam *et al.*, 2024). Another immense benefit of the aforementioned steps is that the companies are able to minimize timed outreach perfectly with customer readiness, which represents, in essence, the sweet spot of predictive marketing (Khalid, 2025; Korstanje, 2017).

KEY RESEARCH AREAS AND METHODS

This section delves into the major key research areas and methods that fit the general theme of navigating human-machine collaboration in smart factories and, particularly, in this current situation the specific topic of harnessing artificial intelligence in marketing and communication for smart factories, while emphasizing the importance and role of the regulatory and policy landscape for SDGs.

There are numerous issues on which the focus ought to be placed, as follows (Milić, 2024): the policy analysis; the stakeholder engagement; the impact assessment; the systemic analysis; the data analysis and visualization; and, also, the scenario planning and foresight.

When addressing policy analysis, there are several elements that ought to be mentioned with priority (Laborde & Torero, 2023). Policy analysis involves thoroughly examining the effectiveness of existing policies and regulations having as pivotal outcome promoting the SDGs' progress. Also, there are some crucial aspects that are focal points in smart factories in order to ensure the effectiveness of existing policies and regulations which make reference to the following aspects, namely: the comparative policy analysis and the legal reviews (Li *et al.*, 2023).

The stakeholder engagement refers to understating the perspectives and contributions of various stakeholders – such as, for example, the governments, the businesses, the civil societies, and the citizens, which becomes crucial in these days. In the case of research papers, the research methods include surveys; interviews; focus groups; and, also, participatory workshops (Lina, 2024).

The impact assessment is a vital stage of research areas and methods. Starting from the major idea that evaluating the social, environmental, and economic impacts of policies and interventions related to the Global Goals is extremely important. Some of the methods that can be successfully used include, among others, the life cycle assessment, the cost-benefit analysis and, importantly, the social return on investment (SROI).

Systemic analysis shows that making connections is of great importance these days. When specialists recognize the interconnectedness of the Global Goals, they make use of research methods that can analyze complex systems. These very important research methods that can analyze complex systems are the agent-based modeling, causal loop diaphragm, and the network analysis (Kavitha *et al.*, 2023).

Data analysis and visualization are pivotal for research papers that depict the major topic of harnessing artificial intelligence in marketing and communication for smart factories, while emphasizing the importance and role of the regulatory and policy landscape for SDGs. The purpose of using data analysis and visualization methods is to track progress, identify trends, and communicate findings effectively, while the goal of using data analysis and visualization methods in research is crucial (Lee *et al.*, 2020).

The scenario planning and foresight implicates exploring potential future scenarios and developing strategies to navigate uncertainties which are of great importance for long-term Global Goals planning. Among the methods that can be used here are the Delphi studies and the scenario workshops (Leape *et al.*, 2023).

FINDINGS AND OUTCOMES

This section tackles the findings and the outcomes that address the general theme of navigating human-machine collaboration in smart factories and, particularly, in this current situation, the specific topic of harnessing artificial intelligence in marketing and communication for smart factories, while emphasizing the importance and role of the regulatory and policy landscape for SDGs (Mahajan *et al.*, 2024).

To begin with, when making connections with the major key research areas and methods aforementioned, there are several examples of research applications that fit the profile, as follows:

- Analyzing the role of the public – private partnerships in financing projects related to the SDGs: it is a major point of interest these days, the public – private partnerships being extremely valuable for the success of the Global Goals (Mi *et al.*, 2019);
- Assessing the impact of carbon pricing mechanisms on achieving the SDG 13 (Climate Action): is strongly related to the actions that citizens, organizations, policymakers, and leaders ought to take for the success of the Global Goals in order to foster well-being and health for all, while considering the environment (Iwasokun & Alimi, 2022);
- Evaluating the effectiveness of the policies promoting SDG 5 (Gender Equality) in specific sectors, regions, and countries: gender equality is of essence for the SDGs and has profound connections with the current research (Jumbri *et al.*, 2019);
- Examining the social impacts of urbanization on achieving SDG 11 (Sustainable Cities and Communities): urbanization is one of the factors that influence the well-being and health of individuals and, also, one of the issues that influence the environment, especially in high-urbanized areas, where pollution is very high (Jalil *et al.*, 2021);

- Investigating the effectiveness of the education programs in fostering sustainable consumption patterns (SDG 12): the success of the education programs in promoting the SDGs is paramount these days (Kabir *et al.*, 2024).

There are several challenges and considerations when addressing the aspects surrounding the general theme of navigating human-machine collaboration in smart factories and, particularly, in this current situation, the specific topic of harnessing artificial intelligence in marketing and communication for smart factories, while emphasizing the importance and role of the regulatory and policy landscape for SDGs:

Firstly, Data Availability and Quality

In this given case reliable and comparable data on the SDGs progress can be acknowledged as being limited, requiring innovative data collection and diverse analysis methods.

Secondly, Complexity, Relying on Links, and Relationships

It needs to be stressed that the interconnectedness of the Global Goals makes isolating the impact of specific interventions and policies extremely challenging, while the inter-linkages are highly complex and continuously growing and evolving.

Thirdly, Political and Institutional Barriers

These days, when analyzing the references for the creation of this current research it was noticed that implementing policies that support the SDGs can face resistance from lack of political will or strong interests.

Fourthly, Measuring Progress

It is of great importance today to address the necessity of developing robust indicators and methodologies that measure the continuous progress of the SDGs. This happens particularly when addressing complex social issues and major environmental problems, on-going challenges can be encountered.

Fifthly, Integrating Sustainability into Research

When analyzing recent data and reports it was noticed that researchers need to consider the way in which smart factories address the sustainability implications and, also, the way in which these smart factories promote sustainable practices among their employers and not only.

The acceleration of the progress of the Global Goals and, additionally, making a notable contribution to a more sustainable and equitable future can be done by employing wide range of actions and addressing accordingly the challenges derived from the current interventions, practices, and policies.

CONCLUSION

All in all, the regulatory and policy landscape for the SDGs involves a very complex interplay of international agreements, national policies, local initiatives, and international actions that have as paramount scope to achieve the 2030 Agenda for Sustainable Development. Among the successful research methods that were described at length in the section dedicated to the key research areas and methods could be highlighted: the policy analysis; the stakeholder engagement; the impact assessment; the systemic analysis; the data analysis and visualization; and, also, the scenario planning and foresight.

What is more, the current research addresses the importance of digital twins for marketing. It was noted that the digital twins for marketing promote photorealistic accuracy and creative autonomy, which relies on the product images, videos, and interactive experiences that enable sales.

Furthermore, the potential of ROI is emphasized, and several questions are answered: What is an interactive ROI calculator? Why should an interactive ROI calculator be used? What is the common use of an interactive ROI calculator?

Besides all this, this section presents the limitations and future research, as follows:

- **Limitations:** An interesting limitation of this research is represented by the fact that more success examples could have been offered as well as the utility of certain artificial intelligence-driven instruments, tools, and techniques.
- **Future research:** In continuation, the future research could center on offering answers to some major pressing questions, namely: How can data related to customers' needs and preferences increase sales and cross-sales? How can smart factories strengthen the relationship with the customers?

REFERENCES

Abid, N., Ceci, F., & Ikram, M. (2022). Green growth and sustainable development: Dynamic linkage between technological innovation, ISO 14001, and environmental challenges. *Environmental Science and Pollution Research International, 29*(17), 25428–25447. DOI: 10.1007/s11356-021-17518-y PMID: 34843051

Adewuyi, A. O., & Awodumi, O. B. (2021). Environmental pollution, energy import, and economic growth: Evidence of sustainable growth in South Africa and Nigeria. *Environmental Science and Pollution Research International, 28*(12), 14434–14468. DOI: 10.1007/s11356-020-11446-z PMID: 33215283

Agustina, R., Dartanto, T., Sitompul, R., Susiloretni, K. A., Suparmi, , Achadi, E. L., Taher, A., Wirawan, F., Sungkar, S., Sudarmono, P., Shankar, A. H., Thabrany, H., Agustina, R., Dartanto, T., Sitompul, R., Susiloretni, K. A., Suparmi, , Achadi, E. L., Taher, A., & Khusun, H. (2019). Universal health coverage in Indonesia: Concept, progress, and challenges. *Lancet, 393*(10166), 75–102. DOI: 10.1016/S0140-6736(18)31647-7 PMID: 30579611

Ahmad, M., & Jabeen, G. (2023). Do economic development and tourism heterogeneously influence ecological sustainability? Implications for sustainable development. *Environmental Science and Pollution Research International, 30*(37), 87158–87184. DOI: 10.1007/s11356-023-28543-4 PMID: 37420157

Ahmed, H. G. M.-D., Sajjad, M., Li, M., Azmat, M. A., Rizwan, M., Maqsood, R. H., & Khan, S. H. (2019). Selection Criteria for Drought-Tolerant Bread Wheat Genotypes at Seedling Stage. *Sustainability (Basel), 11*(9), 2584. DOI: 10.3390/su11092584

Akkalkot, A. I., Mahawar, S., Modak, K. C., Srivastava, P., Shavkatov, N., & Dhanraj, J. A. (2025). Brain-Computer Interface-Based Real-Time Movement Control for Virtual Learning Environments. In Darwish, D., & Pandey, D. (Eds.), *Concepts and Applications of Brain-Computer Interfaces* (pp. 417–432). IGI Global Scientific Publishing., DOI: 10.4018/979-8-3693-7427-6.ch025

Ali, M., Kirikkaleli, D., Sharma, R., & Altuntaş, M. (2022). The nexus between remittances, natural resources, technological innovation, economic growth, and environmental sustainability in Pakistan. *Environmental Science and Pollution Research International, 29*(50), 75822–75840. DOI: 10.1007/s11356-022-21228-4 PMID: 35661303

Azar, A. T., Serrano, F. E., & Ahmad Kamal, N. (2021). Synchronization and Anti-Synchronization of Unidirectional and Bidirectional Coupled Chaotic Systems by Terminal Sliding Mode Control. In Azar, A., & Kamal, N. (Eds.), *Handbook of Research on Modeling, Analysis, and Control of Complex Systems* (pp. 399–433). IGI Global Scientific Publishing., DOI: 10.4018/978-1-7998-5788-4.ch016

Aziz, G., & Waheed, R. (2023, August 18). (2023). Relevance of oil price, military expenditure and exports to achieve the Saudi Arabian sustainable economic and environmental objectives: A non-linear ARDL approach. *Environmental Science and Pollution Research International.* Advance online publication. DOI: 10.1007/s11356-023-28979-8 PMID: 37592070

Badía, G., Pina, V., & Torres, L. (2019). Financial Performance of Government Bond Portfolios Based on Environmental, Social and Governance Criteria. *Sustainability (Basel), 11*(9), 2514. DOI: 10.3390/su11092514

Baena, P. A., Brunel, A., & Fernández-de-Larrinoa, Y.. (2023). In Brief: The White/Wiphala Paper on Indigenous Peoples' Food Systems. 2023 Jan 2. In von Braun, J., Afsana, K., & Fresco, L. O. (Eds.), *Science and Innovations for Food Systems Transformation* [Internet]. Springer., Available from https://www.ncbi.nlm.nih.gov/books/NBK599602/, DOI: 10.1007/978-3-031-15703-5_13

Bai, X., Chen, J., & Shi, P. (2012). Landscape urbanization and economic growth in China: Positive feedbacks and sustainability dilemmas. *Environmental Science & Technology, 46*(1), 132–139. DOI: 10.1021/es202329f PMID: 22103244

Bashir, M. F., Sadiq, M., Talbi, B., Shahzad, L., & Adnan Bashir, M. (2022). An outlook on the development of renewable energy, policy measures to reshape the current energy mix, and how to achieve sustainable economic growth in the post COVID-19 era. *Environmental Science and Pollution Research International, 29*(29), 43636–43647. DOI: 10.1007/s11356-022-20010-w PMID: 35416580

Belgasmia, M. (2021). *Numerical Methods for Solving Dynamic Equilibrium Equations*. IGI Global Scientific Publishing., DOI: 10.4018/978-1-7998-4399-3.ch007

Canales, C., & Fears, R. (2023). The Role of Science, Technology, and Innovation for Transforming Food Systems in Europe. 2023 Jan 2. In von Braun, J., Afsana, K., & Fresco, L. O. (Eds.), *Science and Innovations for Food Systems Transformation* [Internet]. Springer., Available from https://www.ncbi.nlm.nih.gov/books/NBK599615/, DOI: 10.1007/978-3-031-15703-5_40

Chelladurai, S. P., Krishna, S. H., Sarkar, P., Lami, Z. F., Lourens, M., & Al-Shaikhli, T. R. (2025). Smartphones With Brain-Computer Interface Control for HR Secure Data Management. In B. Pandey, A. George, S. Tiwari, S. Albermany, & H. Hung (Eds.), *Brain-Computer Interfaces and Applications in Business* (pp. 215-232). IGI Global Scientific Publishing. https://doi.org/DOI: 10.4018/979-8-3373-5122-3.ch013

Chen, K., Huang, T., Liu, E., Lu, Y., Huo, Z., Mi, L., Zhang, W., & Frazer, I. (2020). The Innovation: A Journal to See the Unseen and Change the Unchanged. *Innovation (Cambridge (Mass.)), 1*(1), 100014. DOI: 10.1016/j.xinn.2020.04.014 PMID: 34557702

Chichaibelu, B. B., Bekchanov, M., & von Braun, J.. (2023). The Global Cost of Reaching a World Without Hunger: Investment Costs and Policy Action Opportunities. 2023 Jan 2. In von Braun, J., Afsana, K., & Fresco, L. O. (Eds.), *Science and Innovations for Food Systems Transformation* [Internet]. Springer., Available from https://www.ncbi.nlm.nih.gov/books/NBK599618/, DOI: 10.1007/978-3-031-15703-5_34

Choi, D., Chung, C. Y., & Young, J. (2019). An Economic Analysis of Corporate Social Responsibility in Korea. *Sustainability (Basel), 11*(9), 2676. DOI: 10.3390/su11092676

Colombo, F., García Goñi, M., & Schwierz, C. (2016). Addressing multimorbidity to improve healthcare and economic sustainability. *Journal of Comorbidity, 6*(1), 21–27. DOI: 10.15256/joc.2016.6.74 PMID: 29090168

Dawane, M. K., & Malwatkar, G. M. (2025). Comparative Study on Implementation of PID Control Architecture and Sliding Mode Control With IoT Integration on Inverted Pendulum Systems. In D. Pandey, B. Muniandi, B. Pandey, & A. George (Eds.), *Interdisciplinary Approaches to AI, Internet of Everything, and Machine Learning* (pp. 49-72). IGI Global Scientific Publishing. https://doi.org/DOI: 10.4018/979-8-3373-1032-9.ch003

Diao, X., Reardon, T., & Kennedy, A.. (2023). The Future of Small Farms: Innovations for Inclusive Transformation. 2023 Jan 2. In von Braun, J., Afsana, K., & Fresco, L. O. (Eds.), *Science and Innovations for Food Systems Transformation* [Internet]. Springer., Available from https://www.ncbi.nlm.nih.gov/books/NBK599622/, DOI: 10.1007/978-3-031-15703-5_10

Dubé, L., Jha, S., Faber, A., Struben, J., London, T., Mohapatra, A., Drager, N., Lannon, C., Joshi, P. K., & McDermott, J. (2014). Convergent innovation for sustainable economic growth and affordable universal health care: Innovating the way we innovate. *Annals of the New York Academy of Sciences*, *1331*(1), 119–141. DOI: 10.1111/nyas.12548 PMID: 25294668

Ebadinezhad, S., Dereboylu, Z., & Ever, E. (2019). Clustering-Based Modified Ant Colony Optimizer for Internet of Vehicles (CACOIOV). *Sustainability (Basel)*, *11*(9), 2624. DOI: 10.3390/su11092624

Elastic, P., Bakhtiar, T., & Jaharuddin. (2020). An Optimal Control Problem of Knowledge Dissemination. In E. Idemudia (Ed.), *Handbook of Research on Social and Organizational Dynamics in the Digital Era* (pp. 461-482). IGI Global Scientific Publishing. https://doi.org/.DOI: 10.4018/978-1-5225-8933-4.ch022

Elhosseini, M. A. (2020). Using Fuzzy Logic to Control Combined Cycle Gas Turbine During Ambient Computing Environment. [IJACI]. *International Journal of Ambient Computing and Intelligence*, *11*(3), 106–130. DOI: 10.4018/IJACI.2020070106

Fears, R., & Canales, C. (2023). The Role of Science, Technology and Innovation in Transforming Food Systems Globally. 2023 Jan 2. In von Braun, J., Afsana, K., & Fresco, L. O. (Eds.), *Science and Innovations for Food Systems Transformation* [Internet]. Springer., Available from https://www.ncbi.nlm.nih.gov/books/NBK599656/, DOI: 10.1007/978-3-031-15703-5_44

Fertő, I., & Harangozó, G. (2025). Trade-offs and synergies when balancing economic growth and globalization for sustainable development goals achievement. *Scientific Reports*, *15*(1), 8634. DOI: 10.1038/s41598-025-93360-3 PMID: 40082545

Geetha, B. T., Tripathi, S. P., Saxena, C. R., Chouhan, K., Shavkatov, N., & Dadheech, P. (2025). Brain-Computer Interface-Controlled Smartphones for Streamlined Organizational Data Management. In B. Pandey, A. George, S. Tiwari, S. Albermany, & H. Hung (Eds.), *Brain-Computer Interfaces and Applications in Business* (pp. 183-198). IGI Global Scientific Publishing. https://doi.org/DOI: 10.4018/979-8-3373-5122-3.ch011

Gh. Popescu, C. R., & Ban a, V. C. (2019). Performance Evaluation of the Implementation of the 2013/34/EU Directive in Romania on the Basis of Corporate Social Responsibility Reports. *Sustainability*, *11*(9), 2531. DOI: 10.3390/su11092531

Gupta, A., Bharadwaj, N., & Rastogi, V. (2021). Computational Framework of Various Semi-Active Control Strategies for Road Vehicles Thorough Bondgraphs. [IJSDA]. *International Journal of System Dynamics Applications, 10*(4), 1–29. DOI: 10.4018/IJSDA.20211001.oa9

Hamad, A. A., Jha, S., & Al-Badri, K. (Eds.). (2025). *Integrating Intelligent Control Systems With Sensor Technologies*. IGI Global Scientific Publishing., DOI: 10.4018/979-8-3373-0330-7

Hao, W., Rasul, F., Bhatti, Z., Hassan, M. S., Ahmed, I., & Asghar, N. (2021). A technological innovation and economic progress enhancement: An assessment of sustainable economic and environmental management. *Environmental Science and Pollution Research International, 28*(22), 28585–28597. DOI: 10.1007/s11356-021-12559-9 PMID: 33544345

Hatamlou, A. (2018). Numerical Optimization Using the Heart Algorithm. [IJAEC]. *International Journal of Applied Evolutionary Computation, 9*(2), 33–37. DOI: 10.4018/IJAEC.2018040103

Hendriks, S., Soussana, J. F., & Cole, M.. (2023). Ensuring Access to Safe and Nutritious Food for All Through the Transformation of Food Systems. 2023 Jan 2. In von Braun, J., Afsana, K., & Fresco, L. O. (Eds.), *Science and Innovations for Food Systems Transformation* [Internet]. Springer., Available from https://www.ncbi.nlm.nih.gov/books/NBK599649/, DOI: 10.1007/978-3-031-15703-5_4

Herrero, M., Hugas, M., & Lele, U.. (2023). A Shift to Healthy and Sustainable Consumption Patterns. 2023 Jan 2. In von Braun, J., Afsana, K., & Fresco, L. O. (Eds.), *Science and Innovations for Food Systems Transformation* [Internet]. Springer., Available from https://www.ncbi.nlm.nih.gov/books/NBK599638/, DOI: 10.1007/978-3-031-15703-5_5

Hodson de Jaramillo, E., Niggli, U., & Kitajima, K.. (2023). Boost Nature-Positive Production. 2023 Jan 2. In von Braun, J., Afsana, K., & Fresco, L. O. (Eds.), *Science and Innovations for Food Systems Transformation* [Internet]. Springer., Available from https://www.ncbi.nlm.nih.gov/books/NBK599628/, DOI: 10.1007/978-3-031-15703-5_17

Hordofa, D. F. (2023). Impacts of external factors on Ethiopia's economic growth: Insights on foreign direct investment, remittances, exchange rates, and imports. *Heliyon, 9*(12), e22847. DOI: 10.1016/j.heliyon.2023.e22847 PMID: 38058435

Hye, Q. M. A., Ul-Haq, J., Visas, H., & Rehan, R. (2023). The role of eco-innovation, renewable energy consumption, economic risks, globalization, and economic growth in achieving sustainable environment in emerging market economies. *Environmental Science and Pollution Research International, 30*(40), 92469–92481. DOI: 10.1007/s11356-023-28945-4 PMID: 37491494

Islam, M. Z., Rahaman, S. H., & Chen, F. (2024). How do R&D and remittances affect economic growth? Evidence from middle-income countries. *Heliyon, 10*(9), e30160. DOI: 10.1016/j.heliyon.2024.e30160 PMID: 38694034

Iwasokun, G. B., & Alimi, S. A. (2022). Genetic Algorithm Model for Stock Management and Control. [IJSDS]. *International Journal of Strategic Decision Sciences, 13*(1), 1–20. DOI: 10.4018/IJSDS.309119

Jalil, A., Rauf, A., Sikander, W., Yonghong, Z., & Tiebang, W. (2021). Energy consumption, economic growth, and environmental sustainability challenges for Belt and Road countries: A fresh insight from "Chinese Going Global Strategy". *Environmental Science and Pollution Research International, 28*(46), 65987–65999. DOI: 10.1007/s11356-021-15549-z PMID: 34324149

Jumbri, I. A., Ikeda, S., Jimichi, M., Saka, C., & Managi, S. (2019). Inequality of health stock and the relation to national wealth. *International Journal for Equity in Health, 18*(1), 188. DOI: 10.1186/s12939-019-1096-x PMID: 31791346

Kabir, M. A., Nandi, S. K., Suma, A. Y., Abdul Kari, Z., Mohamad Sukri, S. A., Wei, L. S., Al Mamun, A., Seguin, P., Herault, M., Khoo, M. I., & Téllez-Isaías, G. (2024). The Potential of Fish Protein Hydrolysate Supplementation in Nile Tilapia Diets: Effects on Growth and Health Performance, Disease Resistance, and Farm Economic Analysis. *Applied Biochemistry and Biotechnology, 196*(10), 7145–7167. DOI: 10.1007/s12010-024-04913-7 PMID: 38489116

Kavitha, C. R., Varalatchoumy, M., Mithuna, H. R., Bharathi, K., Geethalakshmi, N. M., & Boopathi, S. (2023). Energy Monitoring and Control in the Smart Grid: Integrated Intelligent IoT and ANFIS. In Arshad, M. (Ed.), *Applications of Synthetic Biology in Health, Energy, and Environment* (pp. 290–316). IGI Global Scientific Publishing., DOI: 10.4018/978-1-6684-6577-6.ch014

Khalid, S. (2025). Advanced Electric Propulsion Systems for Hybrid Electric Vehicles: Design, Control, and Performance Optimization. In Khalid, S., Ali, J., & Yadav, D. (Eds.), *Hybrid Electric Vehicles and Distributed Renewable Energy Conversion: Control and Vibration Analysis* (pp. 1–54). IGI Global Scientific Publishing., DOI: 10.4018/979-8-3693-5797-2.ch001

Korstanje, M. E. (Ed.). (2017). *Threat Mitigation and Detection of Cyber Warfare and Terrorism Activities*. IGI Global Scientific Publishing., DOI: 10.4018/978-1-5225-1938-6

Laborde, D., & Torero, M. (2023). Modeling Actions for Transforming Agrifood Systems. 2023 Jan 2. In von Braun, J., Afsana, K., & Fresco, L. O. (Eds.), *Science and Innovations for Food Systems Transformation* [Internet]. Springer., Available from https://www.ncbi.nlm.nih.gov/books/NBK599630/, DOI: 10.1007/978-3-031-15703-5_7

Leape, J., Micheli, F., & Tigchelaar, M.. (2023). The Vital Roles of Blue Foods in the Global Food System. 2023 Jan 2. In von Braun, J., Afsana, K., & Fresco, L. O. (Eds.), *Science and Innovations for Food Systems Transformation* [Internet]. Springer., Available from https://www.ncbi.nlm.nih.gov/books/NBK599605/, DOI: 10.1007/978-3-031-15703-5_21

Lee, K. H., Noh, J., & Khim, J. S. (2020). The Blue Economy and the United Nations' sustainable development goals: Challenges and opportunities. *Environment International, 137*, 105528. DOI: 10.1016/j.envint.2020.105528 PMID: 32014791

Li, D., Bae, J. H., & Rishi, M. (2023). Sustainable Development and SDG-7 in Sub-Saharan Africa: Balancing Energy Access, Economic Growth, and Carbon Emissions. *European Journal of Development Research, 35*(1), 112–137. DOI: 10.1057/s41287-021-00502-0 PMID: 35106034

Liao, Z. (2023). Supply chain optimization for environmental sustainability and economic growth. *Environmental Science and Pollution Research International, 30*(58), 121599–121613. DOI: 10.1007/s11356-023-30521-9 PMID: 37957491

Lina, Z. (2024). Towards the sustainable financial management in green economic growth through big data technology innovation; Case study in China. *Heliyon, 10*(19), e35517. DOI: 10.1016/j.heliyon.2024.e35517 PMID: 39403525

Liu, L., Yu, B., & Wu, W. (2019). The Formation and Effects of Exploitative Dynamic Capabilities and Explorative Dynamic Capabilities: An Empirical Study. *Sustainability (Basel), 11*(9), 2581. DOI: 10.3390/su11092581

Lokshina, I. V., & Lanting, C. J. (2020). Analysis of Queueing Networks in Equilibrium: Numerical Steady-State Solutions of Markov Chains. [IJITN]. *International Journal of Interdisciplinary Telecommunications and Networking, 12*(4), 1–17. DOI: 10.4018/IJITN.2020100101

Ma, Q., Yang, R., & Lin, Y. (2024). Dynamic role of clean energy and sustainable economic growth in coastal region: Novel observations from China. *Environmental Research, 251*(Pt 2), 118659. DOI: 10.1016/j.envres.2024.118659 PMID: 38462089

Mahajan, S. L., Obiene, S., Ojwang, L., Olwero, N., Valdivia, A., Wosu, A., Adrid, E., Andradi-Brown, D. A., Andriamalala, G., Ban, N. C., Bennett, N. J., Blythe, J., Cheng, S. H., Darling, E., De Nardo, M., Drury O'Neill, E., Epstein, G., Fidler, R. Y., Fisher, K., Geldmann, J., … Ahmadia, G. (2024). Introducing Elinor for monitoring the governance and management of area-based conservation. *Conservation biology: the journal of the Society for Conservation Biology, 38*(2), e14213. https://doi.org/.DOI: 10.1111/cobi.14213

Martins, H. S., & Guerreiro, S. L. (2021). Access Control Challenges in Enterprise Ecosystems: Blockchain-Based Technologies as an Opportunity for Enhanced Access Control. In I. Management Association (Ed.), *Research Anthology on Blockchain Technology in Business, Healthcare, Education, and Government* (pp. 503-528). IGI Global Scientific Publishing. https://doi.org/DOI: 10.4018/978-1-7998-5351-0.ch029

Mezgár, I., & Pedone, G. (2021). Cloud-Based Manufacturing (CBM) Interoperability in Industry 4.0. In I. Management Association (Ed.), *Research Anthology on Cross-Industry Challenges of Industry 4.0* (pp. 540-567). IGI Global Scientific Publishing. https://doi.org/DOI: 10.4018/978-1-7998-8548-1.ch029

Mi, C., Chen, Y., Cheng, C.-S., Uwanyirigira, J. L., & Lin, C.-T. (2019). Exploring the Determinants of Hot Spring Tourism Customer Satisfaction: Causal Relationships Analysis Using ISM. *Sustainability (Basel), 11*(9), 2613. DOI: 10.3390/su11092613

Milić, J. V. (2024). Sustainable Development Goals in Chemistry in Switzerland. *Chimia, 78*(6), 369–371. DOI: 10.2533/chimia.2024.369 PMID: 38946407

Milousi, M., Souliotis, M., Arampatzis, G., & Papaefthimiou, S. (2019). Evaluating the Environmental Performance of Solar Energy Systems Through a Combined Life Cycle Assessment and Cost Analysis. *Sustainability (Basel), 11*(9), 2539. DOI: 10.3390/su11092539

Minos, D., Butzlaff, I., Demmler, K. M., & Rischke, R. (2016). Economic Growth, Climate Change, and Obesity. *Current Obesity Reports*, *5*(4), 441–448. DOI: 10.1007/s13679-016-0234-7 PMID: 27714641

Mubuli, A., Nigdeli, S. M., & Bekdaş, G. (2021). Control of Full-Size Frame Structures via Optimum Tuned Mass Dampers. [IJDIBE]. *International Journal of Digital Innovation in the Built Environment*, *10*(1), 47–61. DOI: 10.4018/IJDIBE.2021010104

Nagarsheth, S. H., & Sharma, S. N. (2021). The Impact of Fractional-Order Control on Blood Pressure Regulation. [IJEHMC]. *International Journal of E-Health and Medical Communications*, *12*(3), 38–54. DOI: 10.4018/IJEHMC.20210501.oa3

Nasir, A. W., Kasireddy, I., & Singh, A. K. (2020). Fractional-Order-Based Inferential Control: A Case Study for the Level Control of Three Interacting Tank Processes. In Matouk, A. (Ed.), *Advanced Applications of Fractional Differential Operators to Science and Technology* (pp. 322–341). IGI Global Scientific Publishing., DOI: 10.4018/978-1-7998-3122-8.ch014

National Academy of Medicine, & the National Academies of Sciences, Engineering, and Medicine. Health and Medicine Division; Board on Health Sciences Policy; Committee on Creating a Framework for Emerging Science, Technology, and Innovation in Health and Medicine (2023). Toward Equitable Innovation in Health and Medicine: A Framework. Washington (DC): National Academies Press (US); 2023 Aug 22. 3, The Innovation Life Cycle in Health and Medicine and the Challenge of Equity. Available from: https://www.ncbi.nlm.nih.gov/books/NBK599705/

National Academy of Medicine, & the National Academies of Sciences, Engineering, and Medicine. Health and Medicine Division; Board on Health Sciences Policy; Committee on Creating a Framework for Emerging Science, Technology, and Innovation in Health and Medicine (2023). Toward Equitable Innovation in Health and Medicine: A Framework. Washington (DC): National Academies Press (US); 2023 Aug 22. Available from: https://www.ncbi.nlm.nih.gov/books/NBK599708/ DOI: 10.17226/27184

Nazir, M., Zaman, K., Khan, S., Nassani, A. A., Khan, H. U. R., & Haffar, M. (2023). Economic growth and carbon emissions in Pakistan: The effects of China's Logistics Industry. *Environmental Science and Pollution Research International*, *30*(18), 53778–53795. DOI: 10.1007/s11356-023-26150-x PMID: 36867335

Neufeld, L. M., Huang, J., & Badiane, O.. (2023). Advance Equitable Livelihoods. 2023 Jan 2. In von Braun, J., Afsana, K., & Fresco, L. O. (Eds.), *Science and Innovations for Food Systems Transformation* [Internet]. Springer., Available from https://www.ncbi.nlm.nih.gov/books/NBK599631/, DOI: 10.1007/978-3-031-15703-5_8

Ngonghala, C. N., Enright, H., Prosper, O., & Zhao, R. (2024). Modeling the synergistic interplay between malaria dynamics and economic growth. *Mathematical Biosciences*, *372*, 109189. DOI: 10.1016/j.mbs.2024.109189 PMID: 38580079

Niggli, U., Sonnevelt, M., & Kummer, S. (2023). Pathways to Advance Agroecology for a Successful Transformation to Sustainable Food Systems. 2023 Jan 2. In von Braun, J., Afsana, K., & Fresco, L. O. (Eds.), *Science and Innovations for Food Systems Transformation* [Internet]. Springer., Available from https://www.ncbi.nlm.nih.gov/books/NBK599641/, DOI: 10.1007/978-3-031-15703-5_18

Niu, W., & Fan, M. (2024). Control and Research of Computer Virus by Multimedia Technology. [IJISSCM]. *International Journal of Information Systems and Supply Chain Management, 17*(1), 1–17. DOI: 10.4018/IJISSCM.333896

Paramarthalingam, A., Janarthanan, A., Arivunambi, A., Ariyangavu, S. S., Senthamaraikannan, H., & Ganapathy, R. (2025). AI-Powered Virtual Mouse Control Through Hand Gestures With Computer Vision. In Mobo, F. (Ed.), *Harnessing AI in Geospatial Technology for Environmental Monitoring and Management* (pp. 61–76). IGI Global Scientific Publishing., DOI: 10.4018/979-8-3693-8104-5.ch003

Pei, J. (2022). Approaches Toward Building the Digital Enterprise and Sustainable Economic Development: The Moderating Role of Sustainability. *Frontiers in Psychology, 13*, 835602. DOI: 10.3389/fpsyg.2022.835602 PMID: 35360590

Popescu, C. R. G. (2019). Corporate Social Responsibility, Corporate Governance and Business Performance: Limits and Challenges Imposed by the Implementation of Directive 2013/34/EU in Romania. *Sustainability (Basel), 11*(19), 5146. DOI: 10.3390/su11195146

Popescu, C. R. G. (2020). Sustainability Assessment: Does the OECD/G20 Inclusive Framework for BEPS (Base Erosion and Profit Shifting Project) Put an End to Disputes Over The Recognition and Measurement of Intellectual Capital? *Sustainability (Basel), 12*(23), 10004. DOI: 10.3390/su122310004

Popescu, C. R. G. (2025a). Editorial: Durable, Inclusive, Sustainable Economic Growth and Challenge. *Journal of Risk and Financial Management, 18*(5), 257. DOI: 10.3390/jrfm18050257

Popescu, C. R. G. (2025b). Preface to the Special Issue on "Recent Advances in Business and Industry: Mathematical Analysis, Sustainability Assessment Instruments and Methods". *Mathematics, 13*(7), 1207. DOI: 10.3390/math13071207

Popescu, C. R. G., & Popescu, G. N. (2019). An Exploratory Study Based on a Questionnaire Concerning Green and Sustainable Finance, Corporate Social Responsibility, and Performance: Evidence from the Romanian Business Environment. *Journal of Risk and Financial Management, 12*(4), 162. DOI: 10.3390/jrfm12040162

Qu, X. (2016). Journal of Materials Chemistry B top picks web collection: "seeing the unseen: advances in bioimaging and biosensors". *Journal of Materials Chemistry. B, 4*(26), 4500–4501. DOI: 10.1039/C6TB90078J PMID: 32263391

Qu, X. (2018). Journal of Materials Chemistry B Editor's choice web collection: "Seeing the unseen updated: advances in biosensing". *Journal of Materials Chemistry. B, 6*(19), 2922–2923. DOI: 10.1039/C8TB90055H PMID: 32254329

Rajaram, C., & Kumar, R. P. (2021). Numerical Modeling of Near Fault Seismic Ground Motion for Denali Fault. [IJGEE]. *International Journal of Geotechnical Earthquake Engineering*, *12*(2), 1–22. DOI: 10.4018/IJGEE.2021070104

Ramakuri, S. K., Chithaluru, P., & Kumar, S. (2019). Eyeblink Robot Control Using Brain-Computer Interface for Healthcare Applications. [IJMDWTFE]. *International Journal of Mobile Devices, Wearable Technology, and Flexible Electronics*, *10*(2), 38–50. DOI: 10.4018/IJMDWTFE.2019070103

Rasheed, M., & Jianhua, L. (2023). Unleashing the power of innovation promoters for sustainable economic growth: A global perspective. *Environmental Science and Pollution Research International*, *30*(45), 100979–100993. DOI: 10.1007/s11356-023-29313-y PMID: 37642909

Roundtable on Environmental Health Sciences, Research, and Medicine; Board on Population Health and Public Health Practice; Institute of Medicine (2013). Public Health Linkages with Sustainability: Workshop Summary. Washington (DC): National Academies Press (US); 2013 Jul 19. 2, Overview Of The Links Between Sustainability And Human Health. Available from: https://www.ncbi.nlm.nih.gov/books/NBK202304/

Sajil, C. K., & Nair, A. S. (2020). Role of Acoustic Properties in Biomedical Active Noise Control. [IJBCE]. *International Journal of Biomedical and Clinical Engineering*, *9*(1), 48–60. DOI: 10.4018/IJBCE.2020010104

Sari, F. T., Nigdeli, S. M., Bekdaş, G., & Geem, Z. W. (2024). A Review of Optimization of Structural Control Systems: Usage of Metaheuristics in Structural Control. In Ahmad, A., & Camp, C. (Eds.), *Advanced Optimization Applications in Engineering* (pp. 152–171). IGI Global Scientific Publishing., DOI: 10.4018/979-8-3693-2161-4.ch007

Šebestová, J. D., & Popescu, C. R. G. (2022). Factors Influencing Investments into Human Resources to Support Company Performance. *Journal of Risk and Financial Management*, *15*(1), 19. DOI: 10.3390/jrfm15010019

Seeniappan, K., Akram, C. M., Soundararajan, S., & Natrayan, L. (2023). Modelling and Development of Energy Systems Through Cyber Physical Systems With Optimising Interconnected With Control and Sensing Parameters. In T. R., S. Kaliappan, & C. Jegadheesan (Eds.), *Cyber-Physical Systems and Supporting Technologies for Industrial Automation* (pp. 280-295). IGI Global Scientific Publishing. https://doi.org/DOI: 10.4018/978-1-6684-9267-3.ch016

Sharma, V. K., Joshi, S. K., & Bora, A. (2025). Quantum Control of Networked Robotic Systems. In Gajbhiye, P., Kannan, H., Rodriguez, R., & Rojas-Méndez, J. (Eds.), *The Quantum AI Era of Neuromarketing* (pp. 357–368). IGI Global Scientific Publishing., DOI: 10.4018/979-8-3693-7673-7.ch015

Shawkat, S. A., & Thivagar, M. L. (2024). Fundamentals of Numerical. In Hamad, A., & Jha, S. (Eds.), *Coding Dimensions and the Power of Finite Element, Volume, and Difference Methods* (pp. 67–91). IGI Global Scientific Publishing., DOI: 10.4018/979-8-3693-3964-0.ch003

Simbi, C. H., Yao, F., Zhang, J., Tenaw, D., Sugira Murekezi, J., Magati, M. T., Hirwa, H., Al-Sakkaf, A. S., Kayiranga, A., & Peng, J. (2024). Balancing growth and preservation: Unravelling Africa's carbon-economic nexus through the environmental kuznets curve. *Heliyon, 10*(20), e39269. DOI: 10.1016/j.heliyon.2024.e39269 PMID: 39640728

Talele, M., Jain, R., & Mapari, S. (2024). Complex Face Emotion Recognition Using Computer Vision and Machine Learning. In Kumar, N., Pal, S., Agarwal, P., Rosak-Szyrocka, J., & Jain, V. (Eds.), *Harnessing Artificial Emotional Intelligence for Improved Human-Computer Interactions* (pp. 180–196). IGI Global Scientific Publishing., DOI: 10.4018/979-8-3693-2794-4.ch011

Teng, Y., Cox, A., & Chatziantoniou, I. (2021). Environmental degradation, economic growth and tourism development in Chinese regions. *Environmental Science and Pollution Research International, 28*(26), 33781–33793. DOI: 10.1007/s11356-021-12567-9 PMID: 33625700

Thanigaivelan, R., Rajan, N., & Argul, T. (Eds.). (2022). *Advanced Manufacturing Techniques for Engineering and Engineered Materials*. IGI Global Scientific Publishing., DOI: 10.4018/978-1-7998-9574-9

Vázquez, J. E., Martin-Ortiz, M., Olmos-Pineda, I., & Olvera-Lopez, A. (2019). Wheelchair Control Based on Facial Gesture Recognition. [IJITSA]. *International Journal of Information Technologies and Systems Approach, 12*(2), 104–122. DOI: 10.4018/IJITSA.2019070106

Waheed, R. (2023). Energy Challenges, Green Growth, Blue Indicators, and Sustainable Economic Growth: A Study of Saudi Arabia. *Evaluation Review, 47*(6), 983–1024. DOI: 10.1177/0193841X221134653 PMID: 36281482

Wang, C.-N., Tsai, T.-T., Hsu, H.-P., & Nguyen, L.-H. (2019). Performance Evaluation of Major Asian Airline Companies Using DEA Window Model and Grey Theory. *Sustainability (Basel), 11*(9), 2701. DOI: 10.3390/su11092701

Wang, W., Zhang, P., & He, T. (2025). Climate investment and financing, economic growth target constraints and urban green development efficiency. *Journal of Environmental Management, 382*, 125369. DOI: 10.1016/j.jenvman.2025.125369 PMID: 40254006

Webb, P., Flynn, D. J., & Kelly, N. M.. (2023). The Transition Steps Needed to Transform Our Food Systems. 2023 Jan 2. In von Braun, J., Afsana, K., & Fresco, L. O. (Eds.), *Science and Innovations for Food Systems Transformation* [Internet]. Springer., Available from https://www.ncbi.nlm.nih.gov/books/NBK599660/, DOI: 10.1007/978-3-031-15703-5_48

Weir, S., Arstein-Kerslake, A., Eadie, T., & McVilly, K. (2023). Realizing economic and social rights for children with communication and swallowing disability: Sustainable Development Goals 1, 8 and 10. *International Journal of Speech-Language Pathology, 25*(1), 37–41. DOI: 10.1080/17549507.2022.2153166 PMID: 36562595

Xu, X. (2009). *Integrating Advanced Computer-Aided Design, Manufacturing, and Numerical Control: Principles and Implementations*. IGI Global Scientific Publishing., DOI: 10.4018/978-1 59904-714-0

Yaqoob, N., Ali, S. A., Kannaiah, D., Khan, N., Shabbir, M. S., Bilal, K., & Tabash, M. I. (2023). The effects of Agriculture Productivity, Land Intensification, on Sustainable Economic Growth: A panel analysis from Bangladesh, India, and Pakistan Economies. *Environmental Science and Pollution Research International*, *30*(55), 116440–116448. DOI: 10.1007/s11356-021-18471-6 PMID: 35165846

Zhang, Q., Qu, Y., & Zhan, L. (2023). Great transition and new pattern: Agriculture and rural area green development and its coordinated relationship with economic growth in China. *Journal of Environmental Management*, *344*, 118563. DOI: 10.1016/j.jenvman.2023.118563 PMID: 37418914

Zhou, R. (2022). Sustainable Economic Development, Digital Payment, and Consumer Demand: Evidence from China. *International Journal of Environmental Research and Public Health*, *19*(14), 8819. DOI: 10.3390/ijerph19148819 PMID: 35886669

Zhu, P., Ahmed, Z., Pata, U. K., Khan, S., & Abbas, S. (2023). Analyzing economic growth, eco-innovation, and ecological quality nexus in E-7 countries: Accounting for non-linear impacts of urbanization by using a new measure of ecological quality. *Environmental Science and Pollution Research International*, *30*(41), 94242–94254. DOI: 10.1007/s11356-023-29017-3 PMID: 37531055

KEY TERMS AND DEFINITIONS

Digital Twin of a Customer (DToC): Marketers these days are keen on emerging technologies, such as, DToC, which they strive to incorporate in their long-term plans and predictions (Liu *et al.*, 2019; Lokshina & Lanting, 2020). A valuable digital twin example in the automotive is to create virtual replicas intended to test and try out different aerodynamics and materials of the new vehicle designs. Among the priorities of such powerful business activities could be mentioned the marketers' ability to choose the ideal design, which is expected in the end to improve performance, create efficient, and mind-blowing models (Ma *et al.*, 2024).

Interactive Return on Investment (ROI) Calculator: Is a tool represented, for instance, by a web-based application that allows users to input data and have access to the calculated ROI for the marketing campaign (Mezgár & Pedone, 2021). Additionally, the interactive ROI calculator relies on formulas and benchmarks that are relevant to marketing teams and which are highly interested in the cost per lead, customer acquisition costs and, also, sales (Martins & Guerreiro, 2021; M. Belgasmia, 2021).

Transforming Industries: The future of the customer experience is represented, according to specialists, on digital twins for commerce: since the role of digital twins in marketing and in product development mirror real-world systems in real time being centered on sophisticated simulations of consumer behavior and marketing channels (Hendriks *et al.*, 2023; Milousi *et al.*, 2019). Omni-channel digital marketing predicts marketing performance and helps businesses to make more informed decisions, improving the value of the marketing activity (Liao, 2023).

Chapter 8
Human Resistance to Automation and the Fear of Job Displacement

Nilüfer Rüzgar
https://orcid.org/0000-0002-9598-3390
Bursa Technical University, Turkey

ABSTRACT

Using a new and complex technological system is stressful in itself, and this paradigm shift may not be immediately internalized by employees. This can cause high levels of technostress in employees, which can manifest itself as anxiety, timidity and frustrationThe concept of technostress is evaluated in multiple dimensions. In particular, the concept can trap employees in a negative psychological state that has a permanent effect on their physical, emotional and mental well-being. Employees feel that they always have to be alert and connected and therefore they can feel that they have no control on anything including their lives.In this sense, the purpose of this study is first to scrutinize the concept of technostress, which is caused by exposure to technology and especially to automation in organizations. The fear of job displacement is going to be handled in detail, since nowadays the most important cause of this fear is technology.

INTRODUCTION

The industrial revolutions that have taken place from the beginning of industrialization to the present day are called the 1st, 2nd, 3rd and 4th Industrial Revolutions. While water and steam power were adapted to the mechanical production process in the 1st Industrial Revolution, mechanical production was standardized for mass production with the help of electrical division of labor in the 2nd Industrial Revolution. In the 3rd Industrial Revolution, information and communication technology (ICT) was the basis of production using electronics. All these revolutions indicate major changes in the organization of the production process, social practices and established cultural norms. While the development of previous industrial systems took a long time, these

DOI: 10.4018/979-8-3373-2985-7.ch008

changes have occurred much faster with the development of the internet and the influence of ICT technologies (Rainnie & Dean, 2020; Saraç, 2022).

Industry 4.0 has been released in 2011 and since then, organizations have benefited from the benefits of digital technologies. Today, almost all organizations aim to carry out their business functions in automated style which helps them to gain competitive advantage. Industry 4.0, includes technological processes such as the internet of things, artificial intelligence (AI), smart factory, cloud computing and automation (Avunduk & Kazan, 2019; Avunduk, 2023). In particular, the benefits provided by automation, data analytics, artificial intelligence and machine learning are becoming increasingly evident in daily life (Brynjolfsson & McAfee, 2014). These changes in business processes constitute cruciality. Although this change is extremely beneficial for businesses, it can cause problems in terms of employees' adaptation and acceptance of new technology. In this sense, businesses should take various measures for harmonization. Because, it is foreseen that if adaptation cannot be achieved with harmonization studies and measures, this situation may pose a risk to the success of new technologies and systems (Rossignoli, Mauro, & Rocco, 2016; Türen & Erdem, 2017; Akkoyun, 2021). This risk is mainly called technostress.

Using a new and complex technological system is stressful in itself, and this paradigm shift may not be immediately internalized by employees. This can cause high levels of technostress in employees, which can manifest itself as anxiety, timidity and frustration (Sasidharan, 2022). The concept of technostress is evaluated in multiple dimensions. In particular, the concept can trap employees in a negative psychological state that has a permanent effect on their physical, emotional and mental well - being (Salanova et al., 2013; Akman, 2023). Employees feel that they always have to be alert and connected and therefore they can feel that they have no control on anything including their lives (Küçükcivil, Gargalık & Koçyiğit, 2024).

In this sense, the purpose of this study is first to scrutinize the concept of technostress, which is caused by exposure to technology and especially to automation in organizations. The fear of job displacement is going to be handled in detail, since nowadays the most important cause of this fear is technology. Finally, the solutions and reccommendations for both individuals and organizations is going to be discussed.

BACKGROUND

The effects of the integration of the virtual and real worlds, the internet of things studies and issues such as artificial intelligence are crucial concepts that need to be examined by both academicians and business professionals. The fact that technology is perceived as a threat by employees, increase the gap in every business field between those who have technological abilities and those who do not. Therefore, all organizations need to prepare themselves and their human resources for this rapid technological change. In this sense, it is essential to examine both the professional and the social dimension of this situation (Akkoyun, 2021).

In this sense, job insecurity has become a phenomenon that is increasingly felt and gaining importance as a result of new organizational structures and flexibility practices. Views that draw attention to the effects of job security on the economy, employment and unemployment and emphasize the importance of flexibility in the competitiveness of businesses, argue that job insecurity should be accepted as an inevitable result of this flexibility. However, in a work environment where the human element, which plays a major role in the competitiveness of businesses, is uncertain, where wages and working conditions are deteriorating and where there is anxiety about when they will be fired, it becomes impossible for them to exhibit the expected work attitudes and behaviors (Çakır, 2007).

Researchers have various views on how automation will affect the labor market. In this context, there are two different views on automation and unemployment. The first of these is that it will create a destructive effect as workers will have to reallocate their labor as technology replaces labor. The second is the view that companies entering sectors with high productivity will lead to the expansion of employment in these sectors (Frey & Osborne, 2017; Saraç, 2022).

THEORETICAL BACKGROUND OF TECHNOLOGY AND AUTOMATION

When the historical development process of technology is considered, the relevant process dates back to prehistoric times when people first started using tools they obtained from various materials. With the help of time, technological developments have facilitated important changes in the social sense and have undertaken a very important task in changing the course of history (Şen, 2024). Between 1800 and 1950, a transition from a farming and feudal society to an industrial and capitalist society took place. After the World War II, the Western world entered a period of rapid industrial expansion. After 1980, it can be mentioned that the society has entered the post - industrial process and the transition to the service society, where the workforce is freed from physical activities and focuses more on creativity and innovation. Afterwards, while the de - industrialization process continued in developed countries, a large part of the labor - intensive industries were transferred to East Asia and a rapid industrialization process began in this region (Prisecaru, 2017). On the other hand, it can be stated that industrial revolutions do not only affect the workforce in a limited region, but also cause serious changes in the labor markets of all countries with globalization (Saraç, 2022).

The technological development process has gained momentum over the centuries. Milestones that have played an active and important role in the mentioned development process include the invention of the steam engine during the Industrial Revolution, the emergence of electricity, and the development of telecommunication systems in the late 19th and early 20th centuries. Important issues in this development process include the increase in the understanding of digitalization in recent years (Finniston, 1982). It can be said that these developments, which have manifested themselves with the influence of time and research, have affected the change of social life in various ways in terms of facilitating tasks, increasing productivity, expanding human capabilities, and connecting people at long distances. On the other hand, these have also led to social and cultural changes, creating new industries, organizations and business lines, and also increasing some concerns in the minds of employees (Jovanovic & Rousseau, 2005; Şen, 2024).

Industry 4.0 connect sensors, machines and information technology systems to each other throughout the value chain beyond a single business with new digital industrial technologies. These connected systems can interact with each other using internet - based protocols. As a result, it becomes possible to produce high - quality products at lower costs while a more flexible and efficient process is carried out. All these changes and transformations are expected to seriously affect the labor market (Rübmann et. Al., 2015). Although the production process becomes smarter with Industry 4.0, it can be observed that the requirements regarding the knowledge and skills demanded from the workforce changes (Dworschak & Zaiser, 2014).

In this sense, it carries cruciality to mention the historical background of automation. The invention of the sock knitting machine developed by Lee Williams in 1589 is considered a concrete beginning of automation. Lee presented the machine to Queen Elizabeth I, stating that the machine would free workers from hand knitting and thus save time and effort. Elizabeth I refused to grant a patent, stating the negative effects the machine would have on employment rather than being interested in the invention (Milojevic & Davidow, 2018). In the following years, resistance to mechanization continued in different ways. The main ones are the Luddite (machine breaking) actions that took place between 1811 and 1816. This uprising started when a group gathered to protest working conditions and broke 60 sock knitting machines (Saraç, 2022).

However, this resistance was broken and the machines spreaded to factories. With the spread of machines and the decrease in wages, the number of products incresed and developments were observed in trade. In the mid - 19th century, as a result of the increase in the superiority of the parliament over the crown in England, the propertied classes began to become effective in England. This situation played an important role in the decrease of resistance to mechanization in England, where industrialization began. The growth of organizations, increasing complexity and the geographical expansion of the market, created the need to employ more white - collar workers in addition to blue - collar workers (Frey & Osborne, 2017).

At the beginning of the 20th century, significant changes occurred in the structure of production organization. The production organization structure applied by Henry Ford in his automobile assembly plant, can be given as an example of this change. In the assembly system, a moving belt system, where workers remained stationary was implemented. Through the moving belt system, the work/product came in front of the responsible workers and the speed of the belt system could be controlled by them. Thus, mass production system was started to be implemented and it was aimed to minimize the loss of time spent in the production processes. The price of vehicles produced with mass production was kept sufficiently low. With this change in the organizational structure, serious growth occurred in some industries with large economies of scale (Mokyr & Strotz, 1998). On the other hand, as a result of mass production, production surplus emerged, the production - consumption balance was disrupted and subsequently, the Great Depression of 1929 occured and resulted in mass unemployment. Then, ways out of the crisis were sought with the implementation of Keynesian policies. In the following period, mass production continued and the use of electrical energy increased. Since the 1970's, production has been attempted to be further automated by using information technologies and electronic systems (Saraç, 2022).

To sum up, digitalization and automation are fundamental transformation processes focused on technology. The roots of the concepts are based on the rapid advancement of digital technologies and their increasing penetration into all areas of life (Stolterman & Fors, 2004). For example, when the period 2012 - 2021 is considered, the number of people employed as ICT (information and communication technologies) specialists in the European Union has increased by more than 50.0%. Over the course of 10 years, the number of ICT specialists in employment has increased continuously with an average annual growth rate of 4.6% (Eurostat, 2021). These growth trends reflect the dimensions of the digital transformation in the economy (Dragano & Lunau, 2020; Çini et al., 2023; Akman, 2023). In this sense, technology management in organizations comes to the fore. Technology management activities refer to all studies and practices related to the effective management of technologies in organizations. It includes the planning, implementation and control of technological resources and processes in order to increase performance and achieve strategic goals, which are of great importance in organizational management (Becker, 2009). As technological advances shape the way organizations work and compete, the concept of technology management has become one of the concepts that have been given much more importance over time. Considering the historical process, it can be said that technology management primarily focuses on technical aspects such as research and implementation of new technologies. However, with the impact of change and developments, technology has increasingly become integrated into all aspects of organizations, and the scope of technology management has expanded to include broader management functions and continues to expand today (Phaal, Farrukh & Probert 2006). Today, technology management activities are considered by many researchers and managers as a critical discipline that directly affects the competitiveness and success of an organization (Becker, 2009). This increasing expansion process has brought about the development of academic programs and professional certificates in technology management, as well as the establishment of professional managements and the increase in associations and research institutes dedicated to developing the field (Liao, 2005; Şen, 2024).

THEORETICAL BACKGROUND OF TECHNOSTRESS

Technostress is mostly used to describe the psychological and physiological negative reactions that some employees may experience when they use ICT intensively in their work (Arnetz & Clairy, 1997). Technostress is a specific type of stress related to the use of information and communication technologies, which mostly results from the high speed at which technological change occurs (Şahin & Çoklar, 2009). As digital technologies develop and their use becomes widespread, the technostress levels experienced by employees increase. For employees, the fear of not being able to keep up, especially in the internet age, causes many negative situations, especially stress. This situation requires being constantly connected to the internet and being online. In this context, the intensive and compulsive use of the internet and smartphones can cause an addictive state called techno - addiction, which leads to technostress (Young, 2015). In this context, some negative situations that arise with the development of information and communication technologies (techno - overload, inability to keep up with technological developments, fear of dismissal or stress due to inability to adapt to technology, etc.) are called technostress (Akman, 2023).

Brod (1984), one of the first authors to define the concept of technostress, defined the concept as a modern adaptation disease resulting from the inability to cope with new computer technologies in a healthy way. Tarafdar et al. (2007) defined technostress as the stress caused by the inability to adapt to Information Technology (IT) in a healthy way or to cope with IT. Although there are measures taken by the organizations against the changes created by new information technologies, it is widely observed that employees are individually affected by this change at different levels and that this situation can cause negative effects such as anxiety and stress on the individual due to the complexity, information overload and uncertainty it creates (Tarafdar et al., 2007). Since this type of stress is fed by the change in technologies, it is called techno - stress (Alam, 2016; Türen, Erdem & Kalkın, 2015). More specifically, technostress is a state of anxiety that occurs when employees perceive that they do not have the skills necessary to effectively use computers in their jobs, adapt to new ICTs, or cope with them, and the physiological correlates of this anxiety (Zhao et al., 2020). Although the causes of technostress vary depending on many factors, they can generally be listed as technical problems related to human - machine interaction, inadequate usability, and the new skills that are required by the organization (Çini et al., 2023).

According to Tarafdar et al. (2007), benefits from the socio - technical system theory while defining the concept of techno - stress and mentions that it is the stress created by information and computer technologies. The socio - technical system approach was first used during the Tavistock research, which was conducted on coal miners in England during World War II (Heller, 1997). According to the socio - technical system theory, organizations are socio - technical systems and consist of two important aspects. The first of these is the *social dimension* and deals with the skills, attitudes, values and roles that people have, in addition to the reward systems and authority structures. The second is the *technical and task - oriented dimension*, which is related to the actual work performed by individuals or the related processes and technologies. Roles within the organization are also divided into two groups in accordance with these dimensions. While the first group is classified as the roles that individuals undertake through the social system they are in, the second group is classified as the roles related to the tasks that individuals perform and the technical systems they interact with. While the social systems organize subordinate - superior relationships, who will work with whom, who will do what job with whom, technical systems organize hierarchy, coordination mechanisms and reporting systems, division, control and authority structures, standard rules, centralization and decentralization balances. These two related systems determine the roles of employees within the organization (Tarafdar et al. 2007; Türen, Erdem, & Kalkın, 2022).

Technostress is a phenomenon at the individual level. Individuals with technostress are included at the first level. At the second level, technostress is directly or indirectly caused by information and communication technology. The cause can take different forms depending on the usage environment. For example, a situation experienced in a workplace environment is an example of this. At the third level, technostress is a negative experience. It is usually accompanied by psychological, emotional, physical, and behavioral discomforts such as feeling exhausted (Weil & Rosen, 1997; Dahabiyeh et al., 2022; Akman, 2023).

To sum up, the anxiety and stress experienced by individuals due to technology prevent them from using technology more (Joo et al., 2015) and cause work efficiency to decrease (Marcoulides, 1989; Tarafdar et al., 2007). Individuals who have difficulty accepting computer technology, feel pressure to use this technology. This pressure, called technostress, is a modern adaptation disease that occurs as a result of not being able to cope with new computer technologies in a healthy way (Brod, 1984). Weil and Rosen (1997) defined technostress as the negative attitudes, thoughts and behaviors that technology directly or indirectly creates on people's psychology. Technostress is one of the results of an individual's efforts and struggles to cope with the ever - evolving information and communication technologies (ICT) and the changing cognitive and social requirements related to their use (Tarafdar et. al., 2007). Technostress has a negative impact on individual productivity by resulting in various problems such as dissatisfaction, fatigue, anxiety and overwork (Nelson & Kletke, 1990). Technostress is a reflection of the individual's dissatisfaction, fear, distress and anxiety that prevents him/her from learning or using computer technology at a more advanced level due to psychological and emotional reasons (Doğrular, 2019). Stress experienced in work environments is associated with technostress, which causes job burnout and techno burnout in individuals, and the basis of this is inadequacies in ICT (Maier, Laumer & Eckhardt, 2015; Gökbulut & Dindaş, 2022).

RESISTANCE TO AUTOMATION AS THE RESULT OF TECHNOSTRESS AND FEAR OF JOB DISPLACEMENT

Job insecurity covers situations that arise as a result of all kinds of legal or illegal organizational changes that will prevent the continuity of the current job, and that cause the employee to worry about losing his/her job based on the idea of uncertainty. Accordingly, the elements of job insecurity can be expressed as uncertainty, the perception of threat arising from the variability of working conditions, and anxiety. Especially uncertain situations, reduce the individuals' sense of control and increase their sense of helplessness (Wichert, 2001). The uncertainty of how long the job will continue will also cause the decisions the individual will make regarding his/her future in his/her job and his/her private life to become uncertain. The individual who is uncertain about whether he/she will be laid off, will not know what awaits him/her and will not be able to decide how to deal with the situation in question (Çakır, 2007).

Job insecurity, as an emotion, emerges depending on the evaluation and interpretation of various signs in the external and internal environment of the organization by individuals. Job insecurity, without a real job loss, emerges latently with the perception of indicators that threaten the job (Jacobson & Hartley, 1991). Accordingly, one of the elements of job insecurity is the perception of regulations and practices that threaten the continuity of the job. For example, practices that endanger the continuity of the job such as working hours and advancement opportunities cause job insecurity anxiety in employees (Ashford et al., 1989). In addition, organizational changes cause a loss of status for employees, reduce their authority, provide less autonomy, narrow career paths and provide less income, which causes the feeling of job insecurity to appear and increase (Holm & Hovland, 1999; Çakır, 2007).

As for technological/digital transformation (DT), it promotes the Industry 4.0 era in the structures of organizations (Hanelt et al., 2021). In this sense, they firstly facilitate "the combination and connectivity of innumerable, dispersed information, communication and computing technologies" (Bharadwaj et al., 2013). Secondly, they also facilitate far - reaching transformation (Guenzi & Nijssen, 2021; Kellogg et al., 2020) and connects the physical world with the digital world by promoting human - machine interactions (Pereira et al., 2023). Finally DT share with earlier automation the capacity to replace human roles (Faizal et al., 2022; Hanelt et al., 2021; Pereira et al., 2023).

Early automation required repetitive and physically demanding tasks. On the other hand, today DT have evolved and have begun to require cognitive and creative functions with human - like intelligence (Robert et al., 2020). In addition, these technologies are implemented at every organisational level and thus, all employees, regardless of their previous qualifications, must be equipped with the necessary skills and learn new digital behaviours (Faizal et al., 2022; Hanelt et al., 2021; Rožman et al., 2023). Consequently, DT are considered to be "game - changing" (Sebastian et al., 2020; Cieslak & Valor, 2025).

In this sense, it constitutes importance to mention the concept of "change". Change can be expressed in its most basic sense as "moving towards a new state". Change is to abandon a state due to the mission it carries and continue on its way with a random new structure. Since change is a spontaneous process, it does not require power, but innovation refers to a systematic, planned and purposeful process. Organizations, like individuals, have dynamic structures. Since organizations aim to create added value, produce and develop, they expect positive results from change in order to protect their existence. With this feature, technological and social developments, increasing competition and advances in creative thinking, make organizational change necessary and even compulsory (Saraçoğlu & Duran, 2009).

The structure of the digital society is trying to protect its existence by abandoning old boards and determining strategies suitable for today's business world, business management and business life. With the effect of the change in new management systems and organizational structures, organizations have turned to the most suitable state of factors such as technology, quality, speed and cost (Öğüt, 2007). Organizational change is a transformational development move that covers all phenomena such as development, creativity, growth, turning to the positive and being open to innovation (Dinçer, 2008; Özbalcı & Çevik, 2020).

Change management, on the other hand, is the whole of what the organization understands from change, the steps taken within this scope, the studies carried out and the strategies followed (Salisbury, 1996). The change process planned on an organizational scale requires a significant management capability (Senior, 2002) since change is perceived by employees as a process that means abandoning the familiar, is not easy to accept, contains threats within itself and needs to be fought (Genç, 2005). The main reason for this is the fear that the results of change will be negative. Employees tend to resist change with the thought that the change to be implemented within the organizational structure will have negative effects or that their habits will be broken. Even if the effects of change are large or small, it is natural to develop a behavior against this situation (Yeniçeri, 2002). In this context, it is important to mention the concept of resistance. Resistance has several dimensions such as *affective, cognitive* and *behavioral* (Erwin & Garman, 2010; Oreg, 2006; Peiperl, 2005) and the emotions that are felt because of these dimensions are experinced in both individual and in organizational level (Niedenthal & Brauer, 2012). In this sense, *affective dimension* carries cruciality the most in the frame of organizational life when employees face with change. Affective dimension underlines the feelings that an individual experience when he/she perceives a threat (Brief & Weiss, 2002). These threats mostly include new working habits that employees have to adopt because of technological changes (Cieslak & Valor, 2025). In addition, the fear of being incapable of using new technologies makes the employees to feel the fear of job displacement.

Here, technological unemployment concept comes to the fore. The concept of technological unemployment states that significant technological developments, especially those involving automation, can lead to job displacement and as a result, unemployment due to the takeover of human tasks by machines and automated processes. The concept of technological unemployment, based on the proposition that with the progressive evolution of technology, machines can take over roles previously performed by humans, indicates the potential decrease in human labor demand. As organizations increasingly turn to technology, the overall effect is the obsolescence of some human roles (Rawashdeh, 2023). Technological unemployment anxiety, which has emerged as a result of disruptive technological developments in the world, is a feeling of insecurity in employees regarding their jobs and the anxiety of technological unemployment causes increased stress in individuals, reduces self - motivation, reduces job commitment and job performance, and leads to less participation of employees in decision - making processes (Civelek & Pehlivanoğlu, 2020).

Concerns about technological unemployment are not a new phenomenon. Indeed, in recent years, digital technologies have been replacing many jobs, including those of accountants, cashiers and telephone operators. Recently, the poor performance of labor markets in advanced economies has intensified the debate among economists about technological unemployment (Akçakanat, 2024).

There are many studies in the literature that prove that computerization has reduced employment in routine - intensive occupations in terms of labor market outcomes (Frey & Osborne, 2017). For example, Felten, Raj and Seamans (2023) researched job categories and found out that 16 of 20 jobs are affected by AI models and thus, they are cetagorized as "high skill jobs".

Acemoglu & Restrepo (2020), researched AI - related job vacancies between 2010 and 2018 in the United States of America (USA). According to the findings, demand for AI - related skills increased through this period. In addition, it was also revealed that non - AI jobs were decreased through this period. This is called "skill - biased technological change" (SBTC) (Levy & Murnane 1992).

According to SBTC, jobs are cetagorized as high - skill, middle - skill and low - skill. High - skill jobs consists engineers, lawyers and doctors, in the sense that these jobs require creativity and complex decision making. Middle - skill and low - skill jobs, on the other hand, coonsists routine jobs that are amenable to automation such as assembly - line workers, cashiers, sales representatives and clerical employees. While high - skill employees have access to resources, own the necessary tools and of course required level of expertise and knowledge, these middle - skill and low - skill jobs may not always require formal education and/or a degree. However, they also have to adopt technological change (Dahlin, 2024).

To sum up, automation technologies, namely digital technologies, have a crucial impact on organizations and organizational development. This situation is considered as a new industrial revolution. Automation has also a crucial impact on employment and unfortunately job displacement. In this sense, it is thought that the main problem is the fact that the emotional situations of employees, such as their concerns and fears, are ignored. On the other hand, the main aim of digital development is to not only to provide organizational growth but also consider the emotional well - being of the employees' (Chen, 2024). The adoption of new technologies is a complex process that necessitates a deep understanding of various factors, including employee reactions (Frick et al., 2021; Vial, 2019). In fact, approximately 70% to 95% of digital transformation efforts fail (Bonnet, 2022). The failure rates of digital transformation projects are higher than those of other change initiatives (Wade & Shan, 2020), and one significant factor that contributes to these failures is employee resistance since they have the fear of job displacement (Cieslak & Valor, 2025).

Block Quotes

> *"Life is ten percent what you experience and ninety percent how you respond to it." (Dorothy M. Neddermeyer)*
>
> *"Adopting the right attitude can convert a negative stress into a positive one." (Hans Selye)*
>
> *"Any sufficiently advanced technology is equivalent to magic." (Arthur C. Clarke)*
>
> *"It has become appallingly obvious that our technology has exceeded our humanity." (Albert Einstein)*
>
> *"One machine can do the work of fifty ordinary men. No machine can do the work of one extraordinary man." (Elbert Hubbard)*
>
> *"Automation is driving the decline of banal and repetitive tasks." (Amber Rudd)*
>
> *"You cannot endow even the best machine with initiative." (Walter Lippmann)*
>
> *"The first rule of any technology used in a business is that automation applied to an efficient operation will magnify the efficiency. The second is that automation applied to an inefficient operation will magnify the inefficiency." (Bill Gates)*

"I was not depressed when they got me out. I have always taken my dismissals as part of the game." (Frank Woolley)
"Believe you can and you're halfway there." (Theodore Roosevelt)
"We cannot solve our problems with the same thinking we used when we created them." (Albert Einstein)

SOLUTIONS AND RECOMMENDATIONS

Technology is of vital importance for organizations in order to be able to provide sustainable competitive advantage and achieve above - average returns in the long term. Organizations are making efforts to utilize technology more effectively every day to achieve their goals. Of course, one of the ways to do this is to understand the components of technology well, especially in the field of management science. The connection between technology and science and its effective applications in other scientific disciplines have increased significantly with the developments experienced in the industrial age and afterwards. However, the idea that technology is a controllable factor consisting of administrative elements, has started to attract much more attention after the second half of the 20th century (Şen, 2024).

Artificial intelligence (AI), mainly robots, are transforming the nature of human work. Although human - robot collaborations can create new jobs and increase productivity, pundits often warn about how robots might replace humans at work and create mass unemployment (Yam, et. al, 2022). In this context, this situation creates anxiety and Wang and Wang (2022) classified AI anxiety into the following dimensions: *Job change anxiety*, which refers to the fear of the negative effects of AI on work life, *sociotechnical blindness,* which refers to the anxiety caused by the lack of full understanding of AI's dependence on humans, *AI configuration anxiety*, which refers to the fear of human - like AI and *AI learning anxiety*, which refers to the anxiety related to learning AI technologies.

Job change may play a role in AI anxiety, especially for individuals who are directly affected by the automation of tasks and job loss due to the adoption of AI technologies. Individuals who are displaced or at risk of displacement due to automation may experience anxiety and uncertainty about future job prospects. Research has shown that layoffs can lead to a number of negative outcomes, such as reduced job satisfaction and life satisfaction and poor physical health. However, not all individuals are affected in the same way by being displaced from their jobs due to the adoption of AI. Some individuals may be able to transition into new roles created as a result of the adoption of AI technologies, while others may require education and training to acquire the necessary skills. Some research suggests that anxiety about displacement can be reduced through policies and programs that support individuals affected by automation (Hopcan et al., 2023; Akçakanat, 2024).

Numerous researchers continue to make efforts to systematize technology management and create a managerial framework (Akkoyun, 2016). Technology management can be defined as the process of planning, implementing and controlling technological activities of organizations. This field of science helps organizations use technological competencies effectively in order to maintain their competitive advantage and achieve their strategic goals. In this way, organizations can adapt to new technologies more easily, encourage innovation and gain an advantage in the competitive order. Technology management is also considered as a talent because it includes the ability of businesses to adapt to constantly changing technological environments. Today, successful multinational companies and institutions of developed countries manage planning, implementation and control processes using technology management to achieve their strategic goals (Ünsal, 2009; Şen, 2024).

FUTURE RESEARCH DIRECTIONS

Today technology is an inevitable part of both individuals and organizations. On the other hand, employees especially who are used to use traditional methods in their tasks, can feel stressed when they have to adopt technological processes. In addition, they can think that their organizations will not need them anymore since automated systems and robots complete all the tasks properly and with minimum faults. The difficulty in adopting new technologies and the ambiguity of their future in the organization, make them feel fear of job displacement. Technology management and harmonization processes are among solutions for dismissing this fear. Nevertheless, organizations still do not properly develop effective techniques in this context. In this sense, it constitutes cruciality for the researchers to focus on technostress and its causes and effects so as to draw attention to the cruciality of the issue. Moreover, it is also crucial to focus on the fear of job displacement with its causes and effects. In this frame, both managers and employees would be able to draw lessons regarding their sectors and develop solutions in order to prevent the fear of job displacement and technostress. In addition, it would also help to create a strong organizational culture.

CONCLUSION

As mentioned in the literature review, using new and complex technological systems is stressful and thus, this situation may not be immediately adopted by employees. In this sense, this causes technostress, which is known as anxiety, timidity and frustration towards technology (Sasidharan, 2022). The concept of technostress is evaluated multidimensionally. In particular, the concept can trap employees in a negative psychological state that has a permanent effect on their physical, emotional, and mental well - being (Salanova et al., 2013). In addition, technostress is often attributed to inadequate training and lack of post - implementation information - support activities (La Torre et al., 2019; Akman, 2023). Although there are measures taken by the organization against the change created by new technologies, it is inevitable for employees to be affected individually by technostress at different levels (Tarafdar et al., 2007). In this context, it is possible for organizations to determine first the stress level of their employees and then the technostress levels of them and to take measures to manage this stress and keep it at reasonable levels (Türen, Erdem & Kalkın, 2015).

According to the current research, it is obvious that new technological advances in organizational life affects employees in a negative way since they are afraid of adopting new Technologies and see themselves as incompetent. Therefore, this technostress makes them feel they will löse their current jobs. In this sense, organizations should take necessary precautions in order to prevent technostress. In addition, the management teams should also create solutions for existing and potential problems that arise and likely arise from technostress.

REFERENCES

Acemoglu, D., & Restrepo, P. (2020). Robots and jobs: Evidence from US labor markets. *Journal of Political Economy, 128*(6), 2188–2244. DOI: 10.1086/705716

Akçakanat, Ö. (2024). Yapay zekâ kaygısının teknoloji kaynaklı işsizlik endişesi üzerine etkisi: Muhasebe meslek mensupları üzerine bir araştırma. *Kocatepe İİBF Dergisi, 26*(special issue), 53–67. DOI: 10.33707/akuiibfd.1458358

Akkoyun, B. (2016). Çalışanların teknoloji yönetimi algılarının demografik değişkenlere göre farklılaşması üzerine bir araştırma. *Elektronik Sosyal Bilimler Dergisi, 15*(56), 70–82. DOI: 10.17755/esosder.13489

Akkoyun, B. (2021). Bilgi ve teknoloji becerileri ile sistem kullanılabilirliklerinin, işletmelerin yeni teknolojiyi kabul düzeyi üzerinden verimliliklerine etkisi, *İşletme Araştırmaları Dergisi, 13* (1), 739 - 759.

Akman, A. Z. (2023). Türkiye'de "teknostres" konusunda gerçekleştirilen çalışmalara yönelik bir inceleme. *International Journal of Entrepreneurship and Management Inquiries, 2*(special issue), 71–86. DOI: 10.55775/ijemi.1326272

Alam, M. A. (2016). Techno - stress and productivity: Survey evidence from the aviation industry. *Journal of Air Transport Management, 50*, 62–70. DOI: 10.1016/j.jairtraman.2015.10.003

Arnetz, B. B., & Clairy, W. (1997). Technological stress: Psychophysiological symptoms in modern offices. *Journal of Psychosomatic Research, 43*(1), 35–42. DOI: 10.1016/S0022-3999(97)00083-4

Ashford, S. J., Lee, C., & Bobko, P. (1989). Content, causes, and consequences of job insecurity: A theory - based measure and sunstantive test. *Academy of Management Journal, 32*(4), 803–829. DOI: 10.2307/256569

Avunduk, Z.B. (2023). Endüstri 4.0 ve inovasyon etkileşiminin uluslararası literatürdeki yansımaları: Bibliyometrik bir analiz. *Girişimcilik ve İnovasyon Yönetimi Dergisi/Journal of Entrepreneurship and Innovation Management, 12* (2), 59 - 80.

Avunduk, Z. B., & Kazan, H. (2019). *New trends and ındustrial changes in production technology.* in *Sosyal Bilimlerde Yeni Araştırmalar,* (Ed.s, Doç. Dr. Emel İslamoğlu & Doç. Dr. Sinem Yıldırımalp). Ankara: Berikan Yayınevi, 453 - 465.

Becker, P. R. (2009). Technology management degree programs: Meeting the needs of employers. (In) *PICMET'09 - 2009 Portland International Conference on Management of Engineering & Technology*, 2171 - 2183.

Bharadwaj, A., El Sawy, O. A., Pavlou, P. A., & Venkatraman, N. (2013). Digital business strategy: Toward a next generation of insights. *Management Information Systems Quarterly, 37*(2), 471–482. DOI: 10.25300/MISQ/2013/37:2.3

Bonnet, D. (2022). 3 stages of a successful digital transformation. *Harvard Business Review.* https://hbr.org/2022/0 9/3 - *stages - of - a - successful - digital - transformation.*

Brief, A. P., & Weiss, H. M. (2002). Organizational behavior: Affect in the workplace. *Annual Review of Psychology, 53*(1), 279–307. DOI: 10.1146/annurev.psych.53.100901.135156

Brod, K. (1984). On the uniqueness of solution for all wavenumbers in acoustic radiation. *The Journal of the Acoustical Society of America, 76*(4), 1238–1243. DOI: 10.1121/1.391418

Brynjolfsson, E., & McAfee, A. (2014). *The second machine age: Work, progress, and prosperity in a time of brilliant technologies*. W.W. Norton & Co.

Çakır, Ö. (2007). *İşini kaybetme kaygısı: İş güvencesizliği*. Çalışma ve Toplum.

Chen, F., & Li, R. (2024). Improvement and replacement: The dual ımpact of automation on employees' job satisfaction. *Systems, 12*(2), 46. DOI: 10.3390/systems12020046

Cieslak, V., & Valor, C. (2025). Moving beyond conventional resistance and resistors: An integrative review of employee resistance to digital transformation. *Cogent Business & Management, 12*(1), 2442550. Advance online publication. DOI: 10.1080/23311975.2024.2442550

Çini, M. A., Erdirençelebi, M., & Akman, A. Z. (2023). The effect of organisation employees' perspective on digital transformation on their technostress levels and performance: The public institutions example. *Central European Business Review, 12*(4), 1–25. DOI: 10.18267/j.cebr.331

Civelek, M. E., & Pehlivanoğlu, M. Ç. (2020). Technological unemployment anxiety scale development. *Eurasian Business & Economics Journal, 22*, 64–76. DOI: 10.17740/eas.econ.2020.V22 - 05

Dahabiyeh, L., Najjar, M. S., & Wang, G. (2022). Online teaching during COVID - 19 crisis: The role of technostress and emotional dissonance on online teaching exhaustion and teaching staff productivity. *The International Journal of Information and Learning Technology, 39*(2), 97–121. DOI: 10.1108/IJILT-09-2021-0147

Dahlin, E. (2024). Who says artificial intelligence is stealing our jobs? *Socius: Sociological Research for a Dynamic World, 10*, 1–13. srd.sagepub.com. DOI: 10.1177/23780231241259672

Dinçer, Ö. (2008). *Örgüt geliştirme teori, uygulama ve teknikleri* (2 b.). İstanbul: Alfa Yayınları.

Doğrular, M. (2019). *Teknostresin verimlilik üzerine etkisi* (Published PhD dissertation). Marmara University, İstanbul.

Dragano, N., & Lunau, T. (2020). Technostress at work and mental health: Concepts and research results. *Current Opinion in Psychiatry, 33*(4), 407–413. DOI: 10.1097/YCO.0000000000000613

Dworschak, B., & Zaiser, H. (2014). Competences for cyber - physical systems in manufacturing - first findings and scenarios. *Procedia CIRP, 25*, 345–350. DOI: 10.1016/j.procir.2014.10.048

Erwin, D. G., & Garman, A. N. (2010). Resistance to organizational change: Linking research and practice. *Leadership and Organization Development Journal, 31*(1), 39–56. DOI: 10.1108/01437731011010371

Eurostat (2021). ICT specialists in employment, (https://ec.europa.eu/eurostat/statisticsexplained/index.php?title=ICT_specialists_in_employment#Relative_share_of_ICT_specialists_i n_the_total_workforce, (Retrieved: 3.04.2025).

Faizal, S., Jaffar, N., & Mohd Nor, A. S. (2022). Integrate the adoption and readiness of digital technologies amongst accounting professionals towards the fourth industrial revolution. *Cogent Business & Management*, *9*(1), 1–14. DOI: 10.1080/23311975.2022.2122160

Felten, E., Manav, R., & Robert, S. (2023). How will language modelers like ChatGPT affect occupations and industries? *arXiv*. (Retrieved: 4.04.2025). https://arxiv.org/ abs/2303.01157

Finniston, M. (1982). Technology: The future determinant of society. *Technology in Society*, *4*(1), 1–13. DOI: 10.1016/0160-791X(82)90010-0

Frey, C. B., & Osborne, M. A. (2017). The future of employment: How susceptible are jobs to computerisation? *Technological Forecasting and Social Change*, *114*, 254–280. DOI: 10.1016/j.techfore.2016.08.019

Frick, N. R. J., Mirbabaie, M., Stieglitz, S. & Salomon, J. (2021). Maneuvering through the stormy seas of digital transformation: The impact of empowering leadership on the AI readiness of enterprises. *Journal of Decision Systems, 30* (2 - 3), 235 - 258. .DOI: 10.1080/12460125.2020.1870065

Genç, N. (2005). *Yönetim ve organizasyon*. Seçkin Yayınları.

Gökbulut, B., & Dindaş, S. (2022). Öğretmenlerin mesleki tükenmişlik ve teknostres düzeylerinin incelenmesi. [IJOESS]. *International Journal of Eurasia Social Sciences*, *13*(47), 42–59. DOI: 10.35826/ijoess.3096

Guenzi, P., & Nijssen, E. J. (2021). The impact of digital transformation on salespeople: An empirical investigation using the JD - R model. *Journal of Personal Selling & Sales Management*, *41*(2), 130–149. DOI: 10.1080/08853134.2021.1918005

Hanelt, A., Bohnsack, R., Marz, D., & Antunes Marante, C. (2021). A systematic review of the literature on digital transformation: Insights and implications for strategy and organizational change. *Journal of Management Studies*, *58*(5), 1159–1197. DOI: 10.1111/joms.12639

Heller, F. (1997). Sociotechnology and the environment. *Human Relations*, *50*(5), 605–624. DOI: 10.1177/001872679705000507

Holm, S., & Hovland, J. (1999). Waiting for the other shoe to drop: Help for the job - insecure employee. *Journal of Employment Counseling*, *36*(December), 36, 156–164. DOI: 10.1002/j.2161-1920.1999.tb01018.x

Hopcan, S., Türkmen, G., & Polat, E. (2023). Exploring the artificial intelligence anxiety and machine learning attitudes of teacher candidates. *Education and Information Technologies*. Advance online publication. DOI: 10.1007/s10639-023-12086-9

Jacobson, D., & Hartley, J. (1991). *Mapping the context*, in *Job ınsecurity coping with jobs at risk*, (Ed.s: J. Hartley, D. Jacobson, B. Klandermans & T. V. Vuuren), London: Sage Publications, 1 - 22.

Joo, Y. J., Lim, K. Y., & Kim, N. H. (2015). The effects of secondary teachers' technostress on the intention to use technology in South Korea. *Computers & Education, 95*, 114–122. DOI: 10.1016/j.compedu.2015.12.004

Jovanovic, B., & Rousseau, P. L. (2005) *General purpose technologies*, in *Handbook of economic growth*, (Ed.s: Philippe A. & Steven N. D. 1181 - 1224, London: Elsevier DOI: 10.3386/w11093

Kellogg, K. C., Valentine, M. A., & Christin, A. (2020). Algorithms at work: The new contested terrain of control. *The Academy of Management Annals, 14*(1), 366–410. DOI: 10.5465/annals.2018.0174

Küçükcivil, B., Gargalık, E., & Koçyiğit, M. (2024). Üniversitelerde dijital eğitim - öğretim faaliyetleri ve teknostres. *Erciyes İletişim Dergisi, 11*(1), 105–134. DOI: 10.17680/erciyesiletisim.1352212

La Torre, G., Esposito, A., Sciarra, I., & Chiappetta, M. (2019). Definition, symptoms and risk of techno - stress: A systematic review. *International Archives of Occupational and Environmental Health, 92*(1), 13–35. DOI: 10.1007/s00420-018-1352-1

Levy, F., & Murnane, R. J. (1992). US earnings levels and earnings inequality: A review of recent trends and proposed explanations. *Journal of Economic Literature, 30*(3), 1333–1381.

Liao, S. H. (2005). Technology management methodologies and applications: A literature review from 1995 to 2003. *Technovation, 25*(4), 381–393. DOI: 10.1016/j.technovation.2003.08.002

Maier, C., Laumer, S., & Eckhardt, A. (2015). Information technology as daily stressor: Pinning down the causes of burnout. *Journal of Business Economics, 85*(4), 349–387. DOI: 10.1007/s11573-014-0759-8

Marcoulides, G. A. (1989). Measuring computer anxiety: The computer anxiety scale. *Educational and Psychological Measurement, 49*(3), 733–739. DOI: 10.1177/0013164489049003028

Milojevic, I., & Davidow, S. (2018). The futures of work. *TEXT Specia, 52*, 1–12.

Mokyr, J. & Strotz, RH (1998). The Second industrial revolution, 1870 - 1914. *Storia dell'ekonomia Mondiale,* 1 - 16.

Nelson, D. L., & Kletke, M. G. (1990). Individual adjustment during technological innovation: A research framework. *Behaviour & Information Technology, 9*(4), 257–271. DOI: 10.1080/01449299008924242

Niedenthal, P. M., & Brauer, M. (2012). Social functionality of human emotion. *Annual Review of Psychology, 63*(1), 259–285. DOI: 10.1146/annurev.psych.121208.131605

Öğüt, A. (2007). *Bilgi çağında yönetim.* Çizgi Kitapevi.

Oreg, S. (2006). Personality, context, and resistance to organizational change. *European Journal of Work and Organizational Psychology, 15*(1), 73–101. DOI: 10.1080/13594320500451247

Özbalcı, A. A., & Çevik, O. (2020). *Sağlık çalışanlarının bilişim sistemlerindeki değişimlere karşı gösterdikleri direnç ile demografik özellikleri arasındaki ilişki,* in *Current Debates on Social Sciences 5, Human Studies.* (Ed.s: Salih Çeçen, Zeynel Karacagil, Ali Osman Tiro, Ömer Kahya & Şafak Bozgun). Bilgin Kültür Sanat Yayın Dağıtım Pazarlama Ltd. Şti.

Peiperl, M. (2005). *Resistance to change,* in *The Blackwell encyclopedia of management. Organizational behaviour,* (Ed.s: N. Nicholson, P. G. Audia, and M. M. Pillutula), 348 - 349. Blackwell.

Pereira, V., Hadjielias, E., Christofi, M., & Vrontis, D. (2023). A systematic literature review on the impact of artificial intelligence on workplace outcomes: A multi - process perspective. *Human Resource Management Review, 33*(1), 100857. DOI: 10.1016/j.hrmr.2021.100857

Phaal, R., Farrukh, C., & Probert, D. (2006). Technology management tools: Concept, development and application. *Technovation, 26*(3), 336–344. DOI: 10.1016/j.technovation.2005.02.001

Prisecaru, P. (2017). The challenges of the industry 4.0. *Global Economic Observer, 5*(1), 66–72.

Rainnie, A. & Dean, M. (2020). Industry 4.0 and the future of quality work in the global digital economy. *Labour & Industry: A journal of the social and economic relations of work, 30* (1), 16 - 33.

Rawashdeh, A. (2023). The consequences of artificial intelligence: An investigation into the impact of AI on job displacement in accounting. *Journal of Science and Technology Policy Management.* .DOI: 10.1108/JSTPM-02-2023-0030

Robert, L. P., Pierce, C., Marquis, L., Kim, S. & Alahmad, R. (2020). Designing fair AI for managing employees in organizations: A review, critique, and design agenda. *Human - Computer Interaction, 35* (5 - 6), 545 - 575. .DOI: 10.1080/07370024.2020.1735391

Rossignoli, C., Mauro, G., & Rocco, A. (2016). Introducing and discussing information and technology management for organizational innovation and change. *Springer International Publishing., 13,* 1–7. DOI: 10.1007/978-3-319-22921-8_1

Rožman, M., Tominc, P., & Milfelner, B. (2023). Maximizing employee engagement through artificial intelligent organizational culture in the context of leadership and training of employees: Testing linear and non - linear relationships. *Cogent Business & Management, 10*(2), 1–30. DOI: 10.1080/23311975.2023.2248732

Rübmann, M., Lorenz, M., Gerbert, P., Waldner, M., Justus, J., Engel, P., & Harnisch, M. (2015). Industry 4.0: The future of productivity and growth in manufacturing industries. *Boston Consulting Group, 9* (1), 54 - 89.

Şahin, Y. L., & Çoklar, A. N. (2009). Social networking users' views on technology and the determination of technostress levels. *Procedia: Social and Behavioral Sciences, 1*(1), 1437–1442. DOI: 10.1016/j.sbspro.2009.01.253

Salanova, M., Llorens, S., & Cifre, E. (2013). The dark side of technologies: Technostress among users of information and communication technologies. *International Journal of Psychology, 48*(3), 422–436. DOI: 10.1080/00207594.2012.680460

Salisbury, D. F. (1996). *Five technologies for educational change.* Educational Technology Publications.

Saraç, H. (2022). Otomasyonun yükselişi. Kitlesel işsizlik mi? Yeni istihdam mı? İşgücü piyasasına etkisine dair bir değerlendirme. *Çalışma İlişkileri Dergisi, 13* (2). 55 - 76.

Saraçoğlu, M. & Duran, C. (2009). Yaratıcı girişimcinin gelişiminde çevrenin rolü. *Elektronik Sosyal Bilimler Dergisi,* 131 - 151.

Sasidharan, S. (2022). Technostress in the workplace: A social network perspective. *Information Technology & People, 35*(4), 1219–1238. DOI: 10.1108/ITP-09-2020-0649

Sebastian, I. M., Ross, J. W., Beath, C., Mocker, M., Moloney, K. G., & Fonstad, N. O. (2020). How big old companies navigate digital transformation. In *Strategic information management* (pp. 133–150). Routledge. DOI: 10.4324/9780429286797-6

Şen, A. T. (2024). Kamu çalışanlarının teknoloji yönetimi algılarının demografik değişkenler esasında analiz edilmesi: Kastamonu örneği. [Manisa Celal Bayar University Journal of Social Sciences]. *Manisa Celal Bayar Üniversitesi Sosyal Bilimler Dergisi, 22*(4), 182–200. DOI: 10.18026/cbayarsos.1514391

Senior, B. (2002). *Organisational change.* Prentice Hall.

Stolterman, E., & Fors, A. (2004). *Information technology and the good life*, in *Information systems research. IFIP ınternational federation for ınformation processing*, (Ed.s: B. Kaplan, D. P. Truex, D. Wastell, A. T. Wood - Harper, & J. I. DeGross), 687 - 692. Springer. DOI: 10.1007/1-4020-8095-6_45

Tarafdar, M., Tu, Q., Ragu-Nathan, B. S., & Ragu-Nathan, T. S. (2007). The impact of technostress on role stress and productivity. *Journal of Management Information Systems, 24*(1), 301–328. DOI: 10.2753/MIS0742-1222240109

Türen, U., & Erdem, H. (2017). Örgütsel adalet algısının işgörenlerin bilgi teknolojileri verimliliği üzerindeki etkisi: Havacılık sektöründe bir araştırma. *Uluslararası İktisadi ve İdari İnceleme Dergisi, 1*(1), 89–110.

Türen, U., Erdem H. & Kalkın, G. (2015). İşyerinde tekno - stres ölçeği: Havacılık ve bankacılık sektöründe bir araştırma techno - stress at work scale: A research in aviation and banking sectors. *Çalışma İlişkileri Dergisi,* 6 (1), 1 - 19.

Ünsal, E. (2009). Dinamik bir yetenek olarak teknoloji yönetimi: Teknoloji yönetimi yeteneği. *Savunma Bilimleri Dergisi, 8*(2), 167–189.

Vial, G. (2019). Understanding digital transformation: A review and a research agenda. *The Journal of Strategic Information Systems, 28*(2), 118–144. DOI: 10.1016/j.jsis.2019.01.003

Wade, M., & Shan, J. (2020). Covid - 19 has accelerated digital transformation, but may have made it harder not easier. *MIS Quarterly Executive, 19*(3), 213–220. DOI: 10.17705/2msqe.00034

Wang, Y. Y., & Wang, Y. S. (2022). Development and validation of an artificial intelligence anxiety scale: An initial application in predicting motivated learning behavior. *Interactive Learning Environments*, *30*(4), 619–634. DOI: 10.1080/10494820.2019.1674887

Weil, M., & Rosen, L. (1997). *Technostress: Coping with technology at work at home at play.* John Wiley.

Wichert, I. (2001). *Job insecurity and work intensification, The effects of an health and well - being*, in *Job Insecurity & Work Intensification.* (Ed.s: Brendan Burchell, David Ladipo & Frank Wilkinson). London: Routledge.

Yam, K. C., Tang, P. M., Jackson, J. C., Su, R., & Gray, K. (2022). The rise of robots increases job ınsecurity and maladaptive workplace behaviors: Multimethod evidence. *The Journal of Applied Psychology*. Advance online publication. DOI: 10.1037/apl0001045

Yeniçeri, Ö. (2002). Örgütleri etkinleştirme aracı olarak bilgi ve bilgi yönetimi. *Standard Ekonomik ve Teknik Dergi,* 71 - 79.

Young, K. (2015). *The evolution of internet addiction disorder,* in *Internet addiction: Neuroscientific approaches and therapeutical ınterventions,* (Ed.s: C. Montag & M. Reuter), 3 - 17. Springer Science + Business Media. DOI: 10.1007/978-3-319-07242-5_1

Zhao, X., Xia, Q., & Huang, W. (2020). Impact of technostress on productivity from the theoretical perspective of appraisal and coping processes. *Information & Management*, *57*(8), 1–11. DOI: 10.1016/j.im.2020.103265

KEY TERMS AND DEFINITIONS

Technology: The application of scientific knowledge for practical purposes, especially in industry.

Automation: The use or introduction of automatic equipment in a manufacturing or other process or facility.

Stress: A state of mental or emotional strain or tension resulting from adverse or demanding circumstances.

Technostress: Stress or psychosomatic illness caused by working with computer technology on a daily basis.

Job Displacement: The involuntary loss of employment due to various factors such as economic downturns, technological advancements, company restructuring, or offshoring.

Chapter 9
Human-Machine Collaboration in Industry 4.0:
Balancing Automation, Innovation, and the Human Factor

Manoj Govindaraj
https://orcid.org/0000-0003-2830-7875
VelTech Rangarajan Dr. Sagunthala R&D Institute of Science and Technology, India

P. Indira
V.S.B. Engineering College, Karur, India

P. Anitha Kumari
Vels Institute of Science, Technology, and Advanced Studies, India

Jayakranth Rapoori
https://orcid.org/0000-0002-6002-2645
Siddharth Institute of Engineering and Technology, India

Jenifer Lawrence
https://orcid.org/0000-0002-4115-1521
Woldia University, Ethiopia

ABSTRACT

Industry 4.0 marks a new era of intelligent manufacturing, where advanced technologies such as Artificial Intelligence (AI), the Internet of Things (IoT), and robotics are transforming industrial processes. Central to this transformation is the evolving relationship between humans and machines. Rather than replacing the human workforce, the future of Industry 4.0 lies in fostering effective human-machine collaboration. This article examines how organizations can balance automation and innovation with the irreplaceable human qualities of creativity, critical thinking, and ethical judgment. By analysing case studies and emerging trends, the paper highlights strat-

DOI: 10.4018/979-8-3373-2985-7.ch009

egies for designing collaborative work environments that leverage technological advancements while preserving the human factor. It also addresses challenges such as workforce displacement, skill development, and organizational change in the face of rapid digital transformation.

INTRODUCTION

The industrial world globally experiences substantial business modifications due to the emergence of Industry 4.0 which people also call the Fourth Industrial Revolution. The industrial period named Industry 4.0 combines Artificial Intelligence (AI) with Internet of Things (IoT) together with robotics and cloud computing systems along with big data analytics platforms into contemporary production systems. These technologies let machines function independently to handle communication and analysis duties while making decisions. Industry 4.0 operates on two essential pillars: automation and innovation yet the main emergence comes from human-machine collaborative approaches.

Machines started as human-operated tools during the earlier industrial revolutions to boost industrial production rates until now. Machines in Industry 4.0 detect independent thinking which transforms the existing work structure. Machines progress from basic tools to cooperative systems with independent operational capacity which learns from data as well as generates predictions. Concerns about machine automation have emerged because of the industry transition. Various organizations together with subject matter experts predict that future work strategies should emphasize teamwork between human operators and machines. Human-machine collaboration bridges the capabilities of machines and people through machine speed and data processing strength and human creative excellence and judgment and emotional intelligence and problem-solving abilities.

Smart factories benefit most from this developing partnership because automated systems share tasks with personnel. The system combines human-led quality evaluation and problem resolution tasks with robotic systems performing repetitive work. When these two resources collaborate the result leads to elevated productivity together with better innovation capabilities alongside enhanced workplace conditions. AI systems that analyze broad data collections enable workers to improve their decision-making abilities by concentrating on strategic work activities.

The move toward human-machine collaboration faces serious implementation difficulties that have developed alongside its positive advantages. A major drawback of human-machine cooperation involves job replacement because numerous positions face automation since they require standard manual skills and cognitive abilities. Employee instability together with job loss fears has emerged as a result. Jobs currently in development require both state-of-the-art technical abilities as well as modern interpersonal competencies. The workforce needs mandatory training to prepare staff for digital-based positions in a newly emerging work environment. Another challenge is organizational change. Entrepreneurial organizations need to revise organizational frameworks and business procedures and workplace environments for human-robot collaboration. The implementation calls for fresh workplace systems and the incorporation of design methods focused on human needs together with protective protocols between humans and machines. The newly developed systems need to solve problems regarding data privacy and cybersecurity while addressing ethical AI usage to establish trust among users.

Human-machine collaboration depends on the development of technologies which place the human core at the centre of design decisions. Achieving success depends on comprehending human worker capabilities and limitations and developing technology that boosts their capabilities instead of obstructing their work processes. Human-cantered design emphasizes the production of systems whose operational capabilities match natural human logic and cater to diverse user groups and respect human ethical principles. Within this framework machines operate best when they help human workers rather than governing their actions.

Multiple organizations together with different industries now operate successful methods that unite human operators with machines. Cobots (collaborative robots) are now used in automotive production lines because they cooperate safely alongside human workers. AI systems support medical diagnostics to generate precise and rapid disease identifications through human-doctor teamwork because doctors maintain patient care and emotional expertise that machines lack. Rightful human-machine partnerships designed with purpose lead to enhanced results that benefit organizations and their workforce members.

People are core elements within Industry 4.0 despite its primary focus on machines together with technology. This industrial revolution will reach its utmost potential through effective human-machine collaboration. The collaborative approach between industries enables human workers to unite with machines in order to resolve complex problems while generating innovative solutions and developing sustainable workplaces that welcome all employees. The successful adoption of Industry 4.0 demands strategic planning and ethical practices and human development to guarantee equal access to its advantages for society.

BACKGROUND OF THE STUDY

Industry 4.0 uses human-machine collaboration by connecting robots to workers to achieve safety goals and personalized cooperation methods which allows better production results. The balanced approach improves innovation methods in addition to managing social concerns as well as the effectiveness of robot-human interactions at common workstations (Galin & Meshcheryakov, 2019).

The document demonstrates how workplaces benefit from human-machine synchrony by implementing smart technology and teamwork robots which enable operators to concentrate on essential responsibilities as robots execute routine hazardous operations thus redefining job profiles and work scheduling (Saluja & Mongia, 2024).

The Operator 4.0 initiative adopts product-friendly concepts within Industry 4.0 settings to develop robot-human working relationships. The collaboration improves stability across socio-technical systems by integrating human-machine interface tools which maintain innovation levels alongside essential human components in automated environments (Singh & Tretten, 2020). The paper demonstrates its focus on human-oriented approaches of Industry 4.0 by recommending machines to support human capabilities instead of replacing them while enabling collaborative processes which improve production effectiveness in industrial environments through the design of cyber-physical systems (Zarte et al., 2020).

The paper analyses human-machine operational cooperation within Industry 4.0 by studying autonomous cyber-physical systems. This research explains how industrial cooperation has evolved while maintaining its restrictions especially within maintenance activities but requires machine-human coordination for successful partnerships (Gely et al., 2020).

The partnership of humans and machines within Industry 4.0 provides flexibility because it unites robotic automation with human capabilities. The partnership creates conditions for innovation while keeping human elements prominent which leads to improved productivity and better manufacturing adjustments for evolving production requirements (Pullagura et al., 2024). This paper demonstrates that human control plays an essential role in Industry 4.0 human-machine collaborations by establishing automation and innovation balance which supports human-based dynamic environment task management and overall well-being (Nissoul et al., n.d.).

Within Industry 4.0 self-organized production systems the paper establishes a discussion about human-machine cooperation through implementation of cognitive work analysis to achieve effective human-machine system collaboration by balancing automation and innovation alongside human factor considerations (Berdal et al., 2018).

The paper stresses that Industry 4.0 depends on humans working with machines because robots help boost performance levels yet people maintain their core position. Intelligent sensors and artificial intelligence systems enable safe human-machine interactions by making technology flexible to human requirements together with the flexibility of automation throughout innovation processes (Tatarinov et al., n.d.).

The paper suggests that human-machine collaboration stands as a fundamental requirement in Industry 4.0 because workers now assume strategic functions that adaptive assistance systems support to maximize worker abilities when dealing with complex challenges in cyber-physical production systems (Gorecky et al., 2014). The paper explores human-robot collaboration (HRC) approaches within Industry 4.0 while explaining implementation barriers in manufacturing processes accompanied by analysis of one-way and two-way collaboration approaches and the requirement to achieve innovative results through automation and human engagement (Inkulu et al., 2021).

Research investigates human-robot collaboration in industrial environments by presenting how robotics and AI and IoT work together and demonstrates specific configuration methods with interaction types which help achieve harmonious human-robot cooperation for future factories (Michalos et al., 2022). The paper makes HRI the top priority for balancing Industry 4.0 automation against human involvement to produce higher productivity along with cost reductions through robot collaboration with human decision-making in shared work zones (Sherwani et al., 2020).

The paper discusses human-centred design strategies for Industry 4.0 systems by showing how advanced robotics implement collaboration with human operators. The system combines deep learning together with hybrid optimization to improve both user experience and productivity and safety in computer human interactions throughout different fields (Chinthamu et al., 2025).

The analysis stresses that Industry 4.0 should focus on human solutions through interfaces which are easy to use and training and workplace support systems which create balance between automated systems and innovation improvements while helping workers develop skills to use advanced technologies effectively (Krishnaveni & Schwartz, 2024).

The paper examines important human elements affecting factory automation teamwork alongside effective implementation techniques which help generate innovation by maintaining appropriate robot-human interactions specifically within different production stages of industrial development (Charalambous et al., 2013).

The research centre's its focus on efficient approaches to human-robot interaction during Industry 4.0 operations and explains how robotic systems and AI systems improve collaboration through the resolution of automation versus innovation versus human factor dilemmas in industrial operations (Amalia, 2023). The research shows how Industry 4.0 needs cognitive-enabled collaborative robots (CoCoBots) for improved flexibility and productivity through human-robot partnerships but insists that humans should retain all emergency decision authority (Sandini et al., 2024).

The writing stresses that human elements remain vital despite higher automation in Industry 4.0. The paper supports worker education regarding effective assistance system interaction alongside preserving human involvement throughout production methods (Galaske et al., 2017). The paper evaluates Human-Robot Collaboration within Industry 4.0 because sophisticated collision protection systems need development together with cloud-based computational abilities to balance robotic automation against human factory workers (Nagy et al., 2019).

RESEARCH OBJECTIVES

1. To explore the role of human-machine collaboration in enhancing productivity and innovation in industry 4.o environments.
2. To identify strategies that organizations can adopt a balance automation with human-centred values such as creativity and ethical decision-making.
3. To assess the challenges related to workforce displacement and skills development due to increasing automation.
4. To analyse case studies of successful implementation of human-machine collaboration in smart manufacturing settings.

RESEARCH QUESTION

1. How does human-machine collaboration contribute to productivity and innovation in smart factories under industry 4.0?
2. What strategies can organizations implement to ensure a balance between automation and essential human factories like creativity and ethical judgement?
3. What are the major challenges organizations face in addressing workforce displacement and reskilling in the age of digital manufacturing?
4. Which industries or companies have successfully implemented human-machine collaboration and what lessons can be learned from their experiences?

METHODOLOGY

The research questions receive analysis through a qualitative approach which includes reviewing both academic literature and industrial reports and real-world case studies of human-machine collaboration in Industry 4.0. The study examines winning implementation cases from multiple sectors which help understand current and forthcoming trends as well as implementation barriers and successful methods. Modern research focuses on human-centred methods combined with workforce evolution and the social aspects when implementing intelligent systems in contemporary industrial infrastructure.

SYNERGISTIC CAPABILITIES OF AI AND BLOCKCHAIN IN GOVERNANCE SYSTEMS

Human-machine collaboration stands as the primary driver for production gains and innovation success in environments based on Industry 4.0. Row-based manufacturing optimization and reactive production system development occurs when organizations unite machine precision and speed and data management skills with human adaptiveness and decision-making characteristics.

The main value HMC brings to productivity improvement stems from its implementation of collaborative robots (cobots) that optimize production routines. Human operators join forces with these robots to handle monotonous or strenuous tasks thus enhancing product quality and operation speed without risking employee safety (Galin & Meshcheryakov, 2019). Cobots improve human practical abilities instead of taking their place because they let workers handle complex and thought-demanding tasks (Saluja & Mongia, 2024). The World Economic Forum (2023) confirms that HMC and automation techniques will transform 85 million jobs instead of eliminating them to enhance productivity levels.

Through its implementation HMC promotes innovation by building agile manufacturing environments. The adaptation systems of these platforms perform changes in production specifications or customized needs which traditional automated devices struggle to handle. Human-machine teams perform better in unexpected situations since they combine human instinct with artificial intelligence-driven real-time data processing systems according to Charalambous et al. (2013). The process of developing new products becomes faster as a result of this innovation technique thus reducing product delivery periods.

Through combination with intelligent interfaces and wearable technologies human workers receive live data as well as procedure direction and operational reviews which enhance their capabilities. Workers benefit from augmented reality (AR) technology because it uses guidance to help them perform complex assembly duties which shortens learning durations and enhances operational precision (Inkulu et al., 2021). Per Ong et al. (2021) smart exoskeletons serve to minimize physical strain in workers' bodies and extend their strength capabilities which simultaneously enhances product safety and operational efficiency.

Cyber-Physical Systems now play a principal role in smart factories through which they enable neighbourhood-based decision functionality and automated manufacturing adjustments for production alterations. Resilient autonomous process operations become possible through human supervision, and these systems contribute positively to productivity levels (Gely et al., 2020). Through digital twin technologies humans gain real-time simulation abilities to view possible scenarios before implementing physical updates thus speeding up innovation periods (Lee, Bagheri, & Kao, 2015).

Importantly, HMC fosters a culture of continuous improvement. Work environments that promote collaboration between employees lead to the exchange of insights along with new improvement proposals. The application of kaizen by Toyota in digital platforms reveals how workers at the frontlines generate innovations through smart system access as reported by Shao et al. (2022). Simulations enable workers to test new ideas in virtual spaces which cuts down operational risks and enhances operational speed (Matits, 2022).

AI together with IoT technologies enable predictive maintenance because of human-machine collaborative efforts. The proactive maintenance approach demonstrates great significance for industries such as automotive and aerospace due to their high costs from unanticipated downtime events. Human-centred design principles support system development by integrating human factors which create automated tools that enhance human performance instead of presenting barriers to work (Zarte et al., 2020). Through these strategies the organization achieves better performance results alongside minimizing worker resistance and enhancing job satisfaction according to Brougham & Haar (2018).

HMC brings forth the new paradigm of Operator 4.0 which demonstrates how technological augmentation transforms human workers in manufacturing operations. Operator 4.0 uses smart wearable technology combined with interfaces to enhance operator capabilities while changing their role into knowledge work and problem-resolution (Wuest et al., 2017). The worker of the future presents as a "cyber-physical human agent" which renders their duties closer to both digital operations and physical service delivery according to Bogue (2018).

The real-world experiences from different settings demonstrate these positive outcomes. HMC systems deployed by BMW and Bosch unite workers with cobot assemblies to enhance product quality and decrease production duration according to Michalos et al. (2022). Human-machine co-creation takes place in Siemens' Amberg factory through the implementation of AI assistants which enable both human and machine value generation (Kang et al., 2021).

HMC technology enhances both worker satisfaction as well as employee retention rates. Work becomes more meaningful to employees when they do not face repetitive dangerous assignments according to Ghanghorkar & Pillai (2024). Eurofound (2020) reports that this job enriched strategies decrease employee turnover together with preventing absenteeism.

The joint work between humans and factory automation systems increases smart factory operational efficiency by distributing work effectively and responding quickly while providing workers with innovative tools and promoting ongoing development across the workforce. For Industry 4.0 to advance successfully organizations need to design systems which merge automated tasks with essential human work involvement.

ENHANCING TRANSPARENCY AND ACCOUNTABILITY THROUGH AI-BLOCKCHAIN INTEGRATION

Enterprises advancing through Industry 4.0 require a strategic plan to preserve human attributes which include creativity together with ethical sensibility and adaptability in modern workplaces. Organizations need to develop a comprehensive plan that combines human core competencies with automation system achievements. Organizations need to establish integrated systems which allow machines to handle repetitive data-driven operations while humans maintain responsibility for critical thinking-based tasks as well as innovative work and ethical sensitivity (Charalambous et al., 2013).

Organizations achieve the most significant results through the implementation of human-centred design principles. Human operators' needs along with their skills and constraints serve as the main factors for technology development under this approach. Zarte et al. (2020) explain that system designs which focus on human abilities result in improved automated system usability and both safety performance and user acceptance. People-centred approaches let workers stay involved in intellectual tasks by having machines handle physical work together with measurements.

The organization focuses on developing continuous learning and skill enhancement programs as an essential strategic approach. Modelling technological evolution requires workforce members to develop their expertise in parallel. Companies need to develop training programs that allow their personnel to move into positions which work well with automated systems through skill improvement. Galaske et al. (2017) explained that workforce management systems need to change their assessment approach for digital manufacturing readiness through appropriate training program adjustments aligned with modern technology. The curricula must teach practical abilities together with educational content that focuses on both creative problem-solving and ethical problem selection.

Collaborative robotics systems known as cobots represent an important method to implement. The design of cobots prioritizes human safety when working alongside people thus they support workers instead of taking their positions. The research by Sandini, Sciutti, and Morasso (2024) discusses robots that use cognitive abilities to mimic human actions during shared industrial work in flexible manufacturing environments. The integrated system allows organizations to boost productivity by maintaining human supervision throughout every stage of operation.

Development and deployment of automation technologies need proper ethical frameworks and governance systems to incorporate. The system aims to produce automated choices which show ethical awareness together with human-assisted execution of significant value-based decisions. The implementation of human-oriented principles within AI algorithms serves as crucial for AI systems to show moral judgment and social value alignment according to Chinthamu et al. (2025).

Employee participation during the system development process through participatory design results in improved matching of tools with genuine working needs. Gorecky et al. (2014) established that early system development engagement with human workers creates better trust relations along with high usability that optimizes teamwork performance. Such employee involvement creates psychological safety environments according to Edmondson & Lei (2014) which enables workers to share ideas and voice concerns fearlessly.

The use of digital twins together with simulation models assists organizations in finding equilibrium between operations. Organizations use these digital modelling systems to evaluate human-machine workflows ahead of implementation since they find system friction areas and enhance system efficiency and user comfort. The tools described by Ghanghorkar and Pillai (2024) facilitate planning ahead through continuous iterations which are made possible by obtaining immediate feedback data.

Hybrid intelligence represents a new method that joins artificial intelligence with human input to resolve complex issues at a superior level than separate systems can reach. Dellermann et al. (2019) suggest that hybrid intelligence builds socio-technical systems which distribute decision-making through the combination of human emotional and ethical understanding with AI data processing abilities.

The equilibrium between human employees and organizational culture depends significantly on workplace culture. An organization with this type of culture fosters human-technology synergy through its values of both human capabilities and technological acceptance. Strong leadership should demonstrate automation value clearly along with showing methods to prevent employment loss and present how technological advancement will boost personal development.

According to Krishnaveni and Schwartz (2024) organizations must develop a team-oriented approach which emphasizes the unique abilities of people and equipment for achieving successful Industry 4.0 implementation.

Value-sensitive design (VSD) principles must be part of ethical organizational design because they incorporate ethical values including autonomy alongside privacy and justice in technology development (Friedman et al., 2006). The implementation of technology needs to maintain human dignity while supporting meaningful work beside economic advantages.

A combination of engineers and ethicists along with frontline team members creates multidisciplinary groups that facilitate complete technology adoption. Organizational values become an integral part of automation systems through collaborative teamwork which ensures technological efficiency and social responsibility (Amalia 2023). Organizations need to establish ongoing socio-technical alignment systems to enable technology and policy development alongside new business requirements (Sommerville et al., 2012).

The implementation of automation in Industry 4.0 demands diverse strategies including human-focused design combined with ongoing training along with mixtures of artificial and human intelligence and team-based design processes and an environment that supports all team members. Organizations use these strategies to utilize technological benefits while safeguarding human judgment capabilities and creative instincts.

SECTORAL INNOVATION IN AI – BLOCKCHAIN GOVERNANCE APPLICATIONS

Managers must adapt their production lines and facilities to advanced automation technologies which arose from Industry 4.0. The Industrial Revolution creates significant obstacles for factories to address because it results in job elimination alongside the requirement to retrain workers continually. Intelligent machines along with AI systems continue to replace repetitive tasks together with physically demanding work which leads to quick changes in human labour requirements. The social and technical aspects of digitization need thorough evaluation by organizations to achieve successful digital transformation that includes all stakeholders.

The primary challenge emerges from technological employment because automated machines and robots force employees out of work. Modern smart systems maintain a superior ability to complete tasks faster and more precisely than humans so numerous occupations now face obsolescence. Galin and Meshcheryakov (2019) demonstrated that collaborative robots in production lines (cobots) have decreased the need for unskilled workers especially in assembly, packaging and logistics roles. The positive effect of automation on productivity produces negative consequences mostly for unskilled employees who lack digital skills and formal training. Acemoglu and Restrepo (2020) discovered that automation produces productivity growth but simultaneously generates substantial job division by substituting standard operations and developing more professionally demanding positions.

A parallel issue has developed while skills mismatches continue to escalate. New emerging roles involving data analysis, robot coordination and maintenance engineering need skilled workers but there are inadequate qualified applicants available to take these positions. Several organizations face challenges locating workers who possess the necessary technical along with cognitive and emotional competencies needed for digital integration according to Galaske et al.

(2017). Training standards from the past are unable to match the quick speed required by technological innovations. Hecklau et al. (2016) state that businesses need to redesign the educational format of vocational programs to prioritize problem-solving abilities as well as adaptability and ICT competency development.

The prevention of change resistance emerges as another major implementation hurdle involving both staff members and highest organizational leaders. Employee job insecurity produces anxiety with consequent effects on morale and productivity. Large-scale training programs receive no quick returns which causes managers to hold back from investing in them. Workers and managers according to Singh and Tretten (2020) work in many organizations with outdated approaches where workforce development expenses get treated as operational expenses instead of strategic investments. Bonekamp and Sure (2015) argue that organizations need cultural transformations to create learning hotspots through which workers can innovate safely because failure is a critical part of growth.

The implementation of reskilling initiatives manages to encounter challenges at both the structural level as well as with logistical requirements. The entire process demands substantial time investment and resources together with organizational dedication. Organizations should develop customized training programs that meet the requirements of their diverse employees regarding different learning approaches and digital knowledge capacities. Chinthamu et al. (2025) emphasize the need for training methods focused on people to combine technological teaching aids along with practical training and mental wellness support for staff relocating to tech-assisted positions. The authors Dombrowski and Wagner (2014) point to lean along with agile methodologies as key methods for workforce development that deliver flexibility to evolving industry skill demands.

Several gaps remain between different generations during their acceptance of Industry 4.0 technological solutions. Youth workers tend to adapt to new systems better than older employees who find digital interfaces challenging to handle. The inclusive digital transformation needs specific support aimed at various demographic segments to limit marginalization as well as age-based exclusion according to Nissoul, Pacaux-Lemoine, and Chaabane (n.d.). The research by Morrar, Arman, and Mousa (2017) shows that programs designed to link different generations enable cross-generational learning which enables parallel advancement of a cohesive transformation process.

Regional differences as well as economic standing affect the ability for different areas to adopt and be prepared for adoption of new systems. Little reskilling infrastructure exists to support learners in developing nations and economically stressed communities that includes vocational centres and e-learning access together with industrial-academic partnerships.

The development of equitable online environments for education and employment possibilities requires joint work between governments and industry partners according to Inkulu et al. (2021). According to Frey and Osborne (2017) the economic gap might expand between rural areas and underdeveloped locations because of exclusion from digital workforce plans.

Organizations resolve these difficulties through new workforce formats which bring innovative solutions. Wuest and his co-authors from Wuest, Romero, and Stahre (2017) advocate for Operator 4.0 which aims to empower workers through augmented technologies like AI-based decision support systems and wearables as well as augmented reality systems. Under this method employees become primary participants who interact with cyber-physical production systems while transcending mere automation observation status. The research by Wuest, Romero, &

Stahre (2017) demonstrates the capability of this augmentation to enhance three key areas: safety, productivity and cognitive engagement at the same time.

The successful approach to lifelong learning development has been tested through research. The partnership between companies and universities along with tech providers creates continuous professional development systems according to the research by Michalos et al. (2022). Through these educational programs organizations manage to keep their workforce while elevating their professional competence and maintaining their interest. The OECD (2019) demonstrates that organizations implementing learning culture throughout their workplace become successful digital economy performers.

The principles of human-centric design function as essential components for this situation. Zarte Pechmann and Nunes (2020) maintain that smart factory environments should include ergonomics and user-friendly interfaces in combination with participatory design. By considering human abilities and requirements in system design the learning process becomes shorter and employees experience higher workplace satisfaction. Lu and Weng (2018) present inclusive innovation as an essential factor to guarantee technological adoption reaches all groups including low-skilled workers and persons with disabilities.

The implementation of Industrial 4.0 involves multiple complex issues which affect workers facing displacement and require retraining efforts. Digital transformation requires organizations to adopt an entire system of progressive communication methods and training inclusion plus psychological support protocols which surpass mere technological advances. Through this approach organizations lower the probabilities of employment loss and depreciated skills while enabling employees to excel within a digitalized environment.

INDUSTRY-WIDE ADOPTION OF HUMAN-MACHINE COLLABORATION: INSIGHTS AND BEST PRACTICE

Industry 4.0 environments succeed through human-machine collaboration (HMC) which represents their key innovation and competitive advantage. The automotive and healthcare along with electronic industries embrace HMC strategies to deliver enhancements in three areas which include production performance and workplace safety and worker commitment levels. The programs demonstrate essential insights which show technology operates better as an enhancement beyond basic human skills.

BMW Group has become the leading adopter of collaborative robots (cobots) throughout its manufacturing production lines. The German automotive manufacturer selects connectors to perform strenuous body-positioned work and component-manipulation tasks that human employees find difficult. The system allows employees to handle precise tasks requiring making decisions and to minimize physical stress according to Zarte, Pechmann, and Nunes (2020). The lesson emerges from this example to demonstrate why human-centred design should be used with automation because it helps machines support human abilities instead of competing with them.

The diagnostic imaging department of Philips Healthcare together with its surgical assistance functions have successfully integrated AI and robotic systems into their practice. AI-powered radiology systems provide assistance to radiologists in scanning analysis which results in lessened diagnostic mistakes and quicker report generation. The capability to make ultimate choices rests with human experts because machines do not possess ethical oversight nor do they possess emotional intelligence (Chinthamu et al., 2025). The case illuminates how joint work of machines with professionals enhances judgment abilities above replacing them in fields requiring empathy-based critical thinking and ethical oversight.

Foxconn stands as a prominent example in the electronics manufacturing sector because it produces products for Apple and Sony among other clients. Foxconn received first-round criticism about its labour-intensive approach until the company began implementing automation measures and cobotic robots on assembly lines. The company found out that complete automation remained unattainable because the tasks performed many different ways and were very complex. Human participation across quality control tasks and product customization in collaborative automation has boosted operational performance while making workers more content according to Inkulu et al. (2021).

The logistics leader DHL implements robotic technology alongside artificial intelligence at warehouses for inventory management purposes and delivers automated picking and packing services. The implementation of Sawyer and other co-bot robots provides DHL workers with the ability to operate next to robots which produces significant improvements in both speed and precision of operations. DHL provided extensive training for employees to turn them into supervisors who work with robots as part of their skill transformation (Galin & Meshcheryakov, 2019). Successful HMC implementation requires both the active involvement of staff members and programs to develop new skills according to a fundamental lesson.

General Electric (GE) demonstrates another commendable demonstration of HMC implementation through Brilliant Manufacturing which unites Industrial Internet of Things (IIoT) with AI and machine learning capabilities in manufacturing operations. The staff at GE Durham's North Carolina factory employs tablets together with augmented reality technology to observe manufacturing equipment output data while making process adjustments via their augmented reality system. GE began their process by including workforce participation for system design so personnel could develop an acceptance and ease of use for these systems (Galaske et al., 2017). Participatory design with inclusive digital transformation proves to be essential because of its revealed importance.

John Deere achieved a remarkable HMC success by introducing autonomous tractors combined with AI-based crop monitoring systems in the farming industry. These technological systems enable farmers to implement better output strategies as well as decrease resource loss while making decisions based on collected data. The human farmer continues to maintain a crucial role in strategy development along with handling exceptions during farm operations.

These test cases exhibit several fundamental learnings which become apparent.

1. The necessity of human-centric design stands as the foundation for technology development since it ensures acceptance alongside efficiency (Zarte et al., 2020).
2. The objective of innovation should focus on improving human working capabilities without aiming to replace human employees (Chinthamu et al., 2025).
3. Organizations that excel in business practices maintain continuing education and retraining programs that match technology development patterns (Galin & Meshcheryakov, 2019).
4. Labor employees must get involved with technology implementation from start to finish because their participation leads to better tool fit and acceptance (Galaske et al., 2017).
5. The sector of healthcare among others demands human supervision because ethical decisions require empathy from humans instead of machines (Inkulu et al., 2021).

Organizations succeeding in HMC implementation use a socio-technical method which connects technological infrastructure with structural organization elements and human requirements. Human talent forms their strategic asset and organizations adopt inclusive practices because they view their employees as valuable resources for sustainable digital changes. Organizations that manage human-machine collaboration successfully serve as examples for handling the intricate situations of Industry 4.0. These companies establish future-proof operations through their commitment to human contributions alongside technological inclusivity and workforce development which creates ethical guidelines that other companies should adopt.

CASE STUDY EXAMPLES

Medical organizations together with policymakers and technical developers need thorough consideration of data security issues alongside interoperability challenges when they adopt Artificial Intelligence (AI) in healthcare. The reliability and trustworthiness and effectiveness of clinical AI applications undergo direct negative effects because of these challenges. Multiple healthcare scenarios illustrate the necessity of implementing AI systems through safe interfaces in medical practice.

Singapore Health Data Breach (2018)

The healthcare system of Singapore encountered one of its biggest data privacy incidents during 2018 after intruders accessed and stole personal data from 1.5 million patients including Prime Minister Lee Hsien Loong. SingHealth the biggest healthcare institution group in the nation suffered a data breach through network vulnerabilities which attackers targeted using advanced persistent threats (APT). Attackers managed to breach hospital systems unsuccessfully because network infrastructure required enhanced partitioning and active threat detection was absent according to Tan et al. (2020). The stolen medical information included personal identities together with domestic locations and medical documents and doctor-prescribed medications. AI systems in healthcare encounter serious operational and ethical risks from poor cyber security practices which affect their ability to use large volumes of sensitive patient data for learning and operational decision-making. This incident revealed essential cybersecurity necessities such as encryption protocols that are strong, security architectures with multiple layers, threats detection systems and routine cybersecurity audit procedures. Staff education about security presented itself as an important element for building a security-focused organizational culture. AI-powered healthcare depends on security protocols because such measure protects these systems from infiltration attempts while sustaining trust in the system.

Interoperability Challenges in NHS England's AI Pilot Programs

English National Health Service (NHS) conducted several AI application tests by developing both cancer diagnostic technologies and hospital admission forecasting systems. The expansion of these pilot programs faces important barriers because different EHR systems in use today fail to communicate with each other. The research done by Greenhalgh et al. (2021) showed NHS trusts and their departments operated with unconnected IT systems which made unified patient data processing for AI models challenging. The barriers to spread AI throughout the network came from inconsistent data formats combined with coding language differences along with varied documentation practices. The examined case exposed the absolute necessity of implementing workable standards such as HL7 FHIR while building centralized data storing facilities. Healthcare organizations need to allocate funds for data integration systems as well as technologies that optimize workflow management between departments. Advanced AI tools fail to work properly in complicated healthcare networks unless healthcare providers address the interoperability issue.

CHALLENGES IN HUMAN-MACHINE COLLABORATION IN INDUSTRY 4.0

The industrial revolution called Industry 4.0 brings forward three main technological advancements: artificial intelligence (AI), robotics and Internet of Things (IoT) that continue to enter manufacturing sectors and services. The real implementation of human-machine collaboration faces multiple major difficulties which organizations need to properly address. The main obstacles stem from three factors: workers need to adapt to changes; different systems must work together effectively and organizations need time to prepare for these changes.

Workforce Displacement and Skill Gap

Workforce displacement from automation represents the major barrier organizations encounter when they implement HMC. Job security requires constant attention because intelligent machines increasingly take over tasks which require strict compliance and repetition especially affecting workers with basic skills. Galin and Meshcheryakov (2019) declare that economic instability might occur when workforce alterations occur but proper transition plans help minimize negative effects. The market transition has created an immediate lack of qualified talent because new hybrid work positions now combine technical abilities with analytics skills and collaboration competencies. Galaske et al. (2017) emphasize the need to launch reskilling initiatives which match the developing technological environment because these programs form the bridge to digital transformation readiness.

System Integration and Interoperability

Difficulties exist in making collaborative technology compatible with current industrial system architectures. Modern organizations continue using outdated infrastructure systems that lack the capability to conduct real-time data sharing and AI automation. HMC faces performance constraints due to non-uniform communication protocols that link automated systems to software networks and human operator interfaces according to Wuest et al. (2017). Old data management systems with uncoupled sources together with different digital user interfaces create obstacles for fluid operator-machine communications. Organization-wide integration depends on accepting open communication protocols combined with interoperable system platforms as well as adaptable system architecture. Unified frameworks with Michalos et al. (2022) which facilitate human-robot-digital communication systems serve as essential components for enabling safe usable efficient operations.

Ethical and Psychological Resistance to Automation

One main obstacle to human-machine collaboration implementation occurs because employees show resistance through psychological and ethical objections about job replacement and decreased autonomy. Workers usually view technology machines as competitors rather than partners which causes a decrease in company spirit alongside worker rejection of new processes. When AI systems perform surveillance duties together with the oversight of employee performance and task allocation workers typically raise ethical issues about monitoring and machine-based decisions. The development of trusted Human-Machine Collaborative (HMC) systems happens when organizations use open communication and human-focused design practices and follow ethical guidelines to protect dignity plus secure both consent and continuous involvement from users during system deployment stages according to Singh and Tretten (2020).

Improving these obstacles goes beyond upgrading systems through adding better technology because organizational cultures need ethical reforms along with continuous learning opportunities. Organizations can attain maximum human-machine collaboration potential to build a sustainable Industry 4.0 future by actively addressing these barriers.

DISCUSSION

AI and blockchain technologies create a strong operating partnership that unifies AI decision-making systems with blockchain transparent record maintenance when integrated together. Dynamic analyses work together with automated responses through AI and blockchain achieves both data trustworthiness and regulatory compliance. The main obstacle for this partnership involves making both systems work together effectively as well as solving energy usage problems and limitations in operational performance. The analysis demonstrates the requirement for an AI and blockchain co-design framework which employs blockchain to verify AI output results. Implementing development with technical experts together with policy makers remains crucial because it establishes ethical standards and enables efficient public governance deployments.

The implementation of artificial intelligence with blockchain throughout governance systems produces quantifiable enhancements of transparency as well as auditability and accountability capabilities. Data efficiency through AI integration allows blockchain technology to establish unalterable records that keep decision execution processes both traceable and easily explainable. Research evidence demonstrates that public sector performance includes lower instances of fraud next to quicker services and more confident citizens. Essential to the success of AI implementations is both equal access to systems along with unbiased algorithm operation. The discussion shows how essential organizations must establish strong governance systems and constantly monitor their systems so fairness persists. Public service implementation of these technologies will succeed through proper ethical deployment methods in partnership with stakeholders.

The chapter evaluates the critical problems organizations experience due to workforce loss caused by automation and digital manufacturing methods. Robotics and AI seeking to enhance operational performance simultaneously generate skill shortages causing employees to lose their routine jobs. Organizations experience challenges in rapidly upgrading their workforce abilities because they face opposition from both worker and management teams.

The chapter establishes that organizations need human-centred training initiatives together with policies that embrace inclusion along with educational institution partnerships. The difficulties in workforce reskilling become more challenging because of diversity in digital proficiency across generations and regions. Social technical and ethical elements need to be consistently integrated in proactive workforce transition strategies according to the final section in the discussion.

The chapter demonstrates human-machine partnership success by showing various real-life applications in automotive manufacturing combined with healthcare facilities and logistics operations and agricultural processes. The automotive sector represents various successful examples of human-machine collaboration through BMW to manufacturing with DHL and Philips and John Deere demonstrating this approach.

These examples emphasize that organizations must focus on designing friendly systems alongside providing inclusive training alongside engaging their personnel in development. Organizational success necessitates continuous spending and interdisciplinary teamwork along with ethical standards which prevent any workforce discrimination. The discussion reveals the critical need for constant adaptation together with socio-technical alignment which provides organizations with implementation guidelines for ethical Industry 4.0 adoption.

MAIN FINDINGS

The combination of Artificial Intelligence technology with blockchain operations enables users to speed up complex data analysis and automatic decision processes that require extensive data. Blockchain provides secure digital records which stay transparent and cannot be altered. Their integration in governance creates automated trustworthy decisions between data integrity and intelligent processing that represents a crucial need for efficient and fair as well as transparent administrative and policy-making activities across sectors.

AI technology makes blockchain more efficient by recognizing patterns and by enhancing operational processes as well as improving system speed. Blockchain security enables the verification and detection of AI data points because it makes the system completely tamper-proof and operational records traceable. This mutual operation strengthens automated platform trust which allows stakeholders to develop transparent administrative frameworks with accountable systems.

The combination of AI and blockchain technology allows organizations to track all actions and decisions through automatically recorded and real-time monitored uneditable records. The system logs and verifies all transactions which promotes substantial transparency in the process. The technology allows both citizens and institutions to analyze governance processes which decreases fraud and corruption in public together with private sector systems.

Using AI alongside blockchain logs automates auditing procedures by detecting irregularities which leads to increased responsibility tracking for systems or personnel. Digital trails allow officials and stakeholders to receive constant secure monitoring because they generate measurable improvements that create both reliable service oversight and delivery.

The current rise of automation along with AI in digital manufacturing sectors leads to the elimination of workers who hold low-skilled positions. Organizations experience difficulties managing this transition because their current workforce elements lack the needed capabilities to work in technologically advanced Industry 4.0 environment roles.

Most organizations have inadequate reskilling strategies together with insufficient financial support for employee upskilling programs and strategic direction for workforce development. Organizations face three key barriers to digital education access which combine with cultural reluctance to change and training content discrepancies with industrial requirements. These gaps create barriers that block workers from learning new digital and analytical and technical abilities which upcoming roles require.

The automotive logistics healthcare fields have applied robots and AI approaches with success for their production processes. The combination of BMW and DHL represents how machines excel at physical labor and repetitive work so employees free themselves to supervise creatively with better productivity and operational efficiency and enhanced worker satisfaction.

Organizations which achieve success allocate funds to ethical implementation together with employee instruction and system design inclusivity. The learned lessons reveal that organizations must prioritize human-oriented practices and must include stakeholders at the beginning and keep developing employee skills to integrate new technology in an efficient manner that deals with the concerns about displaced workforces and automation of human activities in collaborative work systems.

SOLUTIONS AND RECOMMENDATIONS

Promote integrated framework: The integration of AI with blockchain systems requires a unified governance structure that both governments and technical developers should create. Guide AI operations with blockchain transparency in order to create automated systems that protect data reliability. The combination of sectors through collaborative efforts ensures development of systems which deliver improved services while preventing unauthorized activity or poor effectiveness and wrong usage of data.

Standardize Interoperability Protocols

The integration process requires the creation of technical standards together with interoperability protocols for smooth connection. AI systems should maintain standards which secure reliable and protected interpretation of blockchain-stored data. A standardized framework will provide government entities with one deployment strategy to build interoperable digital systems that strengthen public administration trust as well as governance systems.

Implement Audit-Ready Blockchain Systems

The implementation of blockchain-ledgers by governments should serve to record every decision made through AI systems for immediate auditing purposes. These unalterable recorded actions improve accountability since they enable remote reviews of system execution. The visible record system discourages misbehaviour while giving citizens alongside monitoring agencies sufficient tools to hold public officers responsible through trackable digital evidence.

Develop Citizen Focused Interfaces

Citizens need adaptable screen-based dashboards to monitor AI-run governmental actions which get automatically documented through blockchain technology. Interactive decision systems that reveal management processes will establish trust between citizens while stimulating civic input and provide citizens tools to examine policy developments and distribution of public resources.

Launch Continuous Learning Programs

Organizations need to establish methodical continuous training initiatives that deliver digital abilities particularly in AI operation alongside data analysis and robotics. The programs should transform according to current industrial requirements because their accessibility and adaptable structure must support personnel transitions from traditional positions to modern technical work.

Establish Workforce Redeployment Strategies

Businesses need to produce predictive employee transition strategies that show how to move former employees into developing positions. Companies need to offer career guidance and skill development services together with mentor-based assistance. Strategic planning in advance will reduce job termination effects while maintaining employee participation throughout digital company transformations.

Adopt Human Centric Design Approaches

Organizations which achieve success integrate machines by putting ergonomics together with safety and user-friendly systems at the forefront. Workers who use applications daily should join the process of designing collaborative systems to guarantee application relevance and acceptance by the workforce. The involvement of frontline workers in system design helps produce improved cooperation along with easier transfers which produces enhanced accuracy through human-machine collaboration.

Facilitate Cross Functional Teams

Manufacturing facilities should create teams composed of engineers along with machine operators and management personnel to direct human-machine system integration. When businesses foster combined collaboration, they resolve problems instantly and encourage fresh ideas which spawn shared performance goals. Leading organizations demonstrate that inclusively managed responsibility models lead to optimal teamwork results and decrease resistance among employees.

FUTURE RESEARCH DIRECTIONS

Smarter and More Trustworthy Public Services

When artificial intelligence operates with blockchain's transparent mechanisms the result becomes automated public service systems which are decentralized and self-executable. Experienced citizens receive discreet real-time services from transparent systems which fortify public trust as well as improve administrative fairness and accountability.

Redefinition of Decision-Making Authority

AI-blockchain technology allows public policy enforcement and regulatory compliance automation which may cause human administrators to supervise these intelligent systems rather than remaining active decision-makers. Such changes in governance power dynamics would create evidence-based policy systems yet prompt fresh inquiries about system ethics and responsibility in accountability.

Institutionalization of Real-Time Oversight

Real-time oversight can be established through integrated AI-blockchain governance because independent bodies as well as the public can perform continuous monitoring of decisions and transactions. The system would cut down on corruption significantly which would result in an unceasing accountable environment particularly in public procurement routines along with land records management and social service distribution.

Emergence of Decentralized Citizen Auditing Models

An accessible data framework will eventually develop into an auditing system that makes citizens active watchdogs. Through blockchain dashboards together with AI-alerting technologies communities should be able to track government actions immediately to foster participatory governance and create an era where citizens maintain open control of public and ethical monitoring standards.

Rise of Hybrid Workforce

Organizations which adopt smart manufacturing technologies will create workforces consisting of humans who use machine precision while maintaining creativity and adaptability. A complete shift in professional roles exists alongside adaptive learning systems which are essential competencies needed for industries to employ future workers.

National-Level Workforce Policy Reforms

Such workforce reskilling deserves nationwide attention from government bodies to implement reform plans. The increasing scale of displacement requires government-funded education along with public-private partnership training initiatives because these measures will provide equitable digital economy access and decrease socioeconomic gaps produced by automation.

Benchmarking Collaboration Model Access Sectors

Modifying human-machine interaction successfully within industries transforms them into industry-leading models which establish best practices for work environment design and robot-human system integration and worker welfare management. Modern collaboration systems have the power to determine worldwide sectoral practices which might lead healthcare and logistics sectors to implement similar safety and efficiency frameworks.

Expansion of AI Enhanced Job Roles

Modern collaborative human-AI systems should bring about brand-new positions that blend human capabilities with AI tools. The emerging job categories of "AI Process Coordinator" together with "Robotic Integration Specialist" will likely require new educational programs for teaching the skills needed by the future workforce.

CONCLUSION

Industry 4.0 introduced a new manufacturing and organizational industrial model which transitions from automatic systems to connected human-machine collaborative (HMC) models. Under this model the strengths of human personnel remain distinct from those of smart technology systems while working together. Business examples support observable data which demonstrates artificial intelligence and robotics unite with human-related capabilities to drive innovative processes and operate effectively and enhance employee involvement. As workplaces transform them emerge multiple difficulties which organizations need to resolve through strategic planning together with ethical standards. The essential necessity exists to invest prompt continuous attention into solving workforce displacement issues because of automation as well as the mismatch of skills and interoperability challenges between old systems and new technologies. HMC requires organizations to sustain investments in ongoing learning programs and team-involved system development methods and ethical launch procedures focused on human innovation and choice-making and personal wellness. Partnership robots combined with intellectual support systems enhance human labor while minimizing the demands on workers' bodies and delivering decision-support tools in systems built with clarity and fairness and ethical boundaries. Organizations working in car manufacturing along with their respective sectors such as healthcare and logistics and agriculture prove that combining human thinking with robotic precision generates exceptional results through coordinated innovation. New possibilities for accountability emerge when industries combine artificial intelligence with blockchain capabilities which produces enhanced transparency and governance systems. The success of Industrial 4.0 depends on finding equilibrium through combined human-machinery operation rather than machine primacy. Industrial evolution will succeed when human priorities are combined with technical growth to develop sustainable industrial systems which create general societal advantages.

REFERENCES

Amalia, K. (2023). *Advancements in Robotic Systems and Human Robot Interaction for Industry 4.0*. .DOI: 10.53759/9852/JRS202301010

Berdal, Q., Pacaux-Lemoine, M.-P., Trentesaux, D., & Chauvin, C. (2018). *Human-Machine Cooperation in Self-organized Production Systems: A Point of View*. Springer., DOI: 10.1007/978-3-030-03003-2_9

Bonekamp, L., & Sure, M. (2015). Consequences of Industry 4.0 on Human Labour and Work Organisation. *Journal of Business and Media Psychology*, *6*(1), 33–40. https://journal-bmp.de/wp-content/uploads/04_Bonekamp-Sure_final.pdf

Charalambous, G., Fletcher, S. R., & Webb, P. (2013). *Human-automation collaboration in manufacturing: identifying key implementation factors*. https://dspace.lib.cranfield.ac.uk/handle/1826/9480

Chinthamu, N., Priya, W. D., Srinivasan, K., Vidyasagar, G., Sivarajan, S., & Srihari, T. (2025). Identification of Human-Centric Designs with Advanced Robotics in Industry 4.0 Through Deep Learning. *Social Science Research Network*. DOI: 10.2139/ssrn.5086682

Galaske, N., Arndt, A., Friedrich, H., Bettenhausen, K. D., & Anderl, R. (2017). *Workforce Management 4.0 - Assessment of Human Factors Readiness Towards Digital Manufacturing*. Springer., DOI: 10.1007/978-3-319-60474-9_10

Galin, R., & Meshcheryakov, R. (2019). *Automation and robotics in the context of Industry 4.0: the shift to collaborative robots. 537*(3), 032073. .DOI: 10.1088/1757-899X/537/3/032073

Gely, C., Trentesaux, D., Pacaux-Lemoine, M.-P., & Sénéchal, O. (2020). *Human-Machine Co-operation with Autonomous CPS in the Context of Industry 4.0: A Literature Review*. Springer., DOI: 10.1007/978-3-030-69373-2_23

Ghanghorkar, Y., & Pillai, R. (2024). *Human–Robot Coordination and Collaboration in Industry 4.0*. Springer Nature., DOI: 10.1007/978-981-99-8118-2_9

Gorecky, D., Schmitt, M., Loskyll, M., & Zühlke, D. (2014). Human-machine-interaction in the industry 4.0 era. *International Conference on Industrial Informatics*, 289–294. DOI: 10.1109/INDIN.2014.6945523

Greenhalgh, T., Rosen, R., Shaw, S. E., Byng, R., Faulkner, S., Finlay, T., Grundy, E., Husain, L., Hughes, G., Leone, C., Moore, L., Papoutsi, C., Pope, C., Rybczynska-Bunt, S., Rushforth, A., Wherton, J., Wieringa, S., & Wood, G. W. (2021). [eXtyles1.1]. Planning and Evaluating Remote Consultation Services: A New Conceptual Framework Incorporating Complexity and Practical Ethics. *Frontiers in Digital Health*, *3*(3). Advance online publication. DOI: 10.3389/fdgth.2021.726095

Inkulu, A. K., Bahubalendruni, M. V. A. R., Dara, A., & Sankaranarayanasamy, K. (2021). Challenges and opportunities in human robot collaboration context of Industry 4.0 - a state-of-the-art review. *Information Retrieval*. Advance online publication. DOI: 10.1108/IR-04-2021-0077

Krishnaveni, S., & Schwartz, E. B. (2024). *The Role of Human-Centric Solutions in Tackling Challenges and Unlocking Opportunities in Industry 4.0*. 365–375. .DOI: 10.1201/9781032694375-20

Matits, P. (2022). Industry 4.0 to your industrial campus. *Synchroinfo Journal*, *8*(5), 26–33. DOI: 10.36724/2664-066X-2022-8-5-26-33

Michalos, G., Karagiannis, P., Dimitropoulos, N., Andronas, D., & Makris, S. (2022). *Human Robot Collaboration in Industrial Environments*. Springer., DOI: 10.1007/978-3-030-78513-0_2

Morrar, R., Arman, H., & Mousa, S. (2017). [eXtyles1.1]. The Fourth Industrial Revolution (Industry 4.0): A Social Innovation Perspective. *Technology Innovation Management Review*, *7*(11), 12–20. DOI: 10.22215/timreview/1117

Nagy, B., Doka, J., Racz, S., Szabo, G., Pelle, I., Czentye, J., Toka, L., & Sonkoly, B. (2019). Towards Human-Robot Collaboration: An Industry 4.0 VR Platform with Clouds Under the Hood. *International Conference on Network Protocols*, 1–2. DOI: 10.1109/ICNP.2019.8888107

Nissoul, S., Pacaux-Lemoine, M.-P., & Chaabane, S. (n.d.). *Exploring Human-Machine Relations and Approaches for Task Management in Dynamic Environments: A Comprehensive Literature Review*. 263–273. .DOI: 10.1007/978-3-031-53445-4_22

Pullagura, L., Brahma, B., Kumari, N. V., Ravikumar, L., Katta, S. K. G., & Chiwariro, R. (2024). *Industry 4.0 Design Principles, Technologies, and Applications*. 357–388. .DOI: 10.1201/9781003581963-11

Saluja, A., & Mongia, A. (2024). Human-Machine Collaboration. *Advances in Human Resources Management and Organizational Development Book Series*, 145–170. .DOI: 10.4018/979-8-3693-9631-5.ch007

Sandini, G., Sciutti, A., & Morasso, P. (2024). Collaborative Robots with Cognitive Capabilities for Industry 4.0 and beyond. *AI*, *5*(4), 1858–1869. DOI: 10.3390/ai5040092

Sherwani, F., Asad, M. M., & Kader Ibrahim, B. S. K. (2020). *Collaborative Robots and Industrial Revolution 4.0 (IR 4.0)*. 1–5. .DOI: 10.1109/ICETST49965.2020.9080724

Singh, S., & Tretten, P. (2020). *Operator 4.0 Within the Framework of Industry 4.0*. IGI Global., DOI: 10.4018/978-1-7998-3904-0.ch007

Tan, L., Pan, Y., Wu, J., Zhou, J., Jiang, H., & Deng, Y. (2020). A New Framework for DDoS Attack Detection and Defense in SDN Environment. *IEEE Access : Practical Innovations, Open Solutions*, *8*, 161908–161919. DOI: 10.1109/access.2020.3021435

Tatarinov, K. A., Anikienko, N. N., Savchenko, I. A., & Muzyka, S. M. (n.d.). *Collaborating with robots in industry 4.0. Èkonomika i Upravlenie: Problemy*. Rešeniâ., DOI: 10.36871/ek.up.p.r.2023.10.01.001

Wuest, T., Romero, D., & Stahre, J. (2017). Introducing 'Operator 4.0,' a tech-augmented human worker. *The Conversation*. https://theconversation.com/introducing-operator-4-0-a-tech-augmented-human-worker-74117

Zarte, M., Pechmann, A., & Nunes, I. L. (2020). *Principles for Human-Centred System Design in Industry 4.0 – A Systematic Literature Review*. Springer., DOI: 10.1007/978-3-030-51369-6_19

KEY TERMS AND DEFINITIONS

Industry 4.0: Manufacturing together with other industrial operations now undergoes transformation under the fourth industrial revolution that links cyber-physical systems with automation and artificial intelligence, and smart technologies.

Human–machine collaboration: Human-machine systems use intelligent machine technologies like robots and AI systems to perform initiatives based on their individual capabilities which improves safety and innovation and productivity.

Collaborative robots: These robots are developed for operation with human workers inside common work areas ensuring safety. Industrial robots of the traditional kind exchange human abilities whereas cobots actually increase human performance.

Cyber-physical system: Integrated systems utilize computer-based algorithms that take real-time data to remotely make decisions during manufacturing and operations processes.

Human cantered design: The human-oriented design method ensures the development of technologies and systems with user-centred approaches to guarantee both safety and system acceptability.

Operator 4.0: The Operator 4.0 represents an industrial employee who enhances productivity by utilizing AI-based systems with AR/VR together with wearable devices for making strategic choices in smart factory operations.

Smart factory: The production facility depends on automation together with AI and IoT and data analytics to create a connected and digitalized facility with optimized manufacturing processes that changes dynamically according to changing conditions.

Workforce transformation: A key result of digital technology integration combined with automation features in workplaces is changes to jobs and required skills and employment patterns.

Predictive maintenance: Real-time equipment monitoring powered by AI enables businesses to detect upcoming system failures thus reducing operational disruptions and expense on maintenance.

Interoperability: Different information systems, devices and applications should have clear capabilities to access and exchange and use data properly because this is essential for AI tool integration into existing systems.

Chapter 10

Influence of Transformational Leadership on Ethiopian Smart IT Factories' Performance Through Open Innovation

Getnet Worke
https://orcid.org/0000-0003-4490-0011
Bule Hora University, Ethiopia

Shashi Kant
https://orcid.org/0000-0003-4722-5736
Lovely Professional University, India

Chalchisa Amentie Karo
Ethiopian Civil Service University, Ethiopia

ABSTRACT

Investigating the nexus among Open Innovation, Transformational Leadership, as well as Smart IT Factories performance is the goal of this paper. It tackles the burning matter of how leadership philosophies might influence overall performance as well as organizational innovation in the quickly paradigm shifting Smart IT environment. Regression analysis is used to investigate the nexus among the variables in this quantitative investigation. Data on innovation tactics, performance measures, and leadership practices were gathered from a sample of corporates in the Smart IT industry. The outcomes indicate a resilient affirmative association among Open Innovation as well as Smart IT Factories Performance and among Transformational Leadership as well as Open Innovation. Along with, the nexus among transformational leadership as well

DOI: 10.4018/979-8-3373-2985-7.ch010

as organizational performance was manifest to be somewhat mediated by open innovation. As per to the outcomes, companies may continuous augment performance by funding the synthesis of transformational leaders and nurturing a culture of innovation.

INTRODUCTION

It is impossible to overestimate the role transformative leadership has in promoting innovation in a time of swift technical paradigm shift and the need for sustainable synthesis. This idea, which places a resilient emphasis on nurturing and motivating supporters to put the needs of the organization ahead of their own, has been widely acknowledged as being essential to its sustainability (Hundie & Habtewold, 2024). The nexus of open innovation as well as transformative leadership offers a singular paradigm shift to drive technical continuous augmentation and economic prosperity in Ethiopia, a country ready to capitalize on the potential of smart IT manufacturing (Rehmani et al., 2023). James MacGregor Burns and Bernard Bass were the main contributors to the synthesis of the concept of transformative leadership in the 1970s. Burns proposed the idea of leaders who can boost their supporters' spirits and motivations, and Bass depicted this idea by defining the traits and actions of transformational leaders (Mohammed & AL-Abrrow, 2023). Empirical investigation over the years has manifested a resilient link among transformational leadership and a number of organizational outcomes, such as continuous augmented performance, employee engagement, and innovation. As per to investigations, companies with transformational leaders are more likely to use open innovation, a cooperative strategy that uses outside concepts and technology to boost competitiveness and speed up product creation (Hidayat-ur-Rehman & Alsolamy, 2023).

Practical difficulties still exist despite the substantial amount of investigation on open innovation as well as transformational leadership, especially in poor nations like Ethiopia. These difficulties include restricted technological availability, a lack of infrastructure, and cultural norms that might make it difficult to apply transformational leadership techniques sustainability (Gemeda & Lee, 2020a). As well as, there is a clear lack of investigation on how these leadership philosophies might be sustainability modified to fit the particular socioeconomic circumstances of Ethiopian smart IT firms (Cirera et al., 2023). By investigating the complex link among transformative leadership as well as Open innovation in Ethiopia's developing IT industry, this investigation seeks to close these investigation gaps. This investigation aims to offer useful insights for upcoming scholars and policymakers by pinpointing the particular difficulties faced by leaders in this setting and manifesting workable solutions.

This investigation is driven by the following main investigation question: ***"How does transformational leadership influence open innovation adoption in Ethiopia's Smart IT factories, and what tactics may be used to strengthen this association?"*** This topic highlights the need of developing an inventive culture that may propel sustainable economic progress in addition to the requirement for a better comprehension of transformative leadership in a localized context. Through tackling these obstacles and deficiencies, this investigation seeks to steer future investigations and guide policy choices that would eventually support Ethiopia's Smart IT factories' prosperity and their place in the world economy.

BACKGROUND OF INVESTIGATION

This investigation's background is the growing importance of transformational leadership as an innovation accelerator, especially in light of Ethiopia's expanding Smart IT factories. Knowing the dynamics of leadership and innovation becomes crucial as the nation seeks to establish itself as a regional center for technology (Digital Ethiopia, 2020.). Nurturing an atmosphere that encourages open innovation requires transformational leadership, which is distinguished by its emphasis on inspiring and nurturing members of staff. However, there are a number of difficulties in putting this theoretical framework into practice, which this paper aims to investigate (Hartmann et al., 2024). One noteworthy example that illustrates the difficulties in applying transformational leadership in the Ethiopian environment is the risk taking propensity of a regional tech firm called "Tech Ethiopia." Tech Ethiopia struggled with leadership techniques that were conventionally bureaucratic and hierarchical, despite its potential for innovation. By nurturing a culture of cooperation and innovation, the founder—a visionary leader—tried to embrace transformational leadership concepts (Gemeda & Lee, 2020b). However, middle management's opposition, as they were used to making decisions with authority, prevented this vision from being fully realized. The challenges of paradigm shifting corporate culture and practices in a setting where traditional leadership methods are firmly established are manifest by this scenario (Nguyen et al., 2023). The Ethiopian government's plan to build industrial parks in an effort to draw in international investment in the technology sector is another relevant example. Although the government has passed laws to promote innovation, these laws are frequently not implemented with the transformative leadership needed to motivate regional business owners (Cirera et al., 2023). Leaders at one industrial park were unable to properly engage workers, which led to low morale and little cooperation across various tech corporates. This situation highlights the discrepancy among the goals of policy and the realities on the ground, showing that attempts to promote innovation may not succeed in the absence of a transformative leadership style (Schmitz et al., 2023).

As well as, empirical investigation in a number of Ethiopian industries has manifest that many leaders do not have the support and training needed to embrace transformational leadership methods (Cirera et al., 2023). Employees acknowledged the need for creative techniques, but they frequently believed their bosses were out of touch with their problems and risk taking propensitys, as per to a investigation of IT professionals. This mismatch stifles teams' creative potential and poses a serious obstacle to putting open innovation techniques into practice by nurturing a lack of trust and involvement (Jinga et al., 2024). In conclusion, although transformational leadership has a great deal of promise to spur innovation in Ethiopia's Smart IT manufacturing. But real-world obstacles still exist. These include deeply ingrained hierarchical cultures, inadequate training for leaders, and a misalignment of the requirements of employees and leadership. To fully use transformational leadership in Ethiopia's IT industry, several obstacles must be removed (Alemu et al., 2024). In order to contribute to a more thorough knowledge of how transformational leadership may sustainability support open innovation in this developing industry, this investigation attempts to offer insights into overcoming these obstacles.

MAIN FOCUS OF THE CHAPTER

Main focus of the chapter is to thoroughly investigating the nexus among open innovation as well as transformative leadership in Ethiopia's Smart IT manufacturing, this investigation seeks to close these gaps. The investigation will help close the identified theoretical and knowledge gaps by investigating the unique difficulties faced by leaders in this setting and offering workable solutions. Along with, it will offer empirical data to guide future investigations and policy choices, ultimately directing interested parties toward the establishment of a culture of innovation that supports Ethiopia's socioeconomic objectives. By doing this, the investigation hopes to equip upcoming scholars and decision-makers with knowledge that would help the Ethiopian tech industry sustainability combine open innovation with transformative leadership.

Main focus of the chapter is framed in the imperative because Ethiopia's Smart IT manufacturers face together possibilities and problems as an outcome of the quickening pace of technological progress and the growing focus on innovation. A resilient open innovation culture led by transformational leadership that encourages invention and teamwork among members of staff would be the ideal condition for these factories. The reality of today, however, differs greatly from this ideal (Zahoor & Adomako, 2023). Investigation indicates that a large number of Ethiopian firms have difficulty sustainability implementing transformational leadership, which outcomes in less-than-ideal innovation outcomes. This disparity calls into question the elements that restrict the use of transformational leadership approaches and the ensuing influences on open innovation (Hundie & Habtewold, 2024). The comprehension of this link is complicated by contradicting facts. While some investigation shows that transformational leadership fosters creativity in a variety of settings, other investigations show that this style of leadership is sustainability in settings where hierarchical structures are prevalent (Müller et al., 2024). Since conventional leadership approaches frequently predominate in Ethiopian Smart IT factories, this discrepancy leaves gaps in the empirical investigation, especially with regard to the application of transformational leadership in these settings. Along with, there is still conflicting evidence on how well transformational leadership promotes innovation, which calls for a more thorough investigation of the contextual elements that can influence these outcomes (Buonocore et al., 2024).

There are theoretical gaps as well since Ethiopia's distinct socioeconomic and cultural dynamics could not be well captured by current frameworks. The majority of theories on transformational leadership are derived from Western settings, which might not be entirely applicable in Ethiopia (Hariharan & Anand, 2023). The capacity to incubate customized leadership models that align with regional customs and beliefs is hampered by this lack of theoretical sub structure. As well as, there are still knowledge gaps since there is a dearth of investigation that explicitly looks at how transformational leadership as well as Open innovation intersect in Ethiopia's tech sector, leaving practitioners and policymakers without useful information (Müller et al., 2024). Ethiopian Smart IT factories face additional challenges due to contextual gaps: many leaders lack the training and knowledge of transformational leadership principles, which causes a gap among leadership goals and employee risk taking propensitys; practical burning matters like poor infrastructure, limited innovative assets, and cultural resistance to paradigm shift make it difficult to implement innovative practices, making attempts to adopt transformational leadership influence (Hundie & Habtewold, 2024).

THEORETICAL FRAMEWORK

Transformational Leadership Theory

Influential leaders inspire and encourage supporters to accomplish remarkable achievements and cultivate a commitment to corporate goals, as per to the transformational leadership theory, which was first incubated by James MacGregor Burns and then expanded by Bernard Bass. Four essential elements are highlighted by this theory: intellectual stimulation, inspiring drive, idealized influence, and personalized attention (Nguyen et al., 2023). Since they foster creativity, question the status quo, and give workers a feeling of ownership, leaders who exhibit these qualities are said to foster an atmosphere where innovation thrives. Transformational leadership is essential to nurturing an open innovation culture in Ethiopia's Smart IT factories since it supports the nation's goals for economic synthesis and technical progress (Müller et al., 2024).

Open Innovation Theory

As per to Henry Chesbrough's Open Innovation Theory, companies stand to gain a great deal from incorporating outside concepts and technology into their innovation procedures. This strategy differs from conventional methods of innovation, which frequently sub structure entirely on interior innovative assets and competencies (Alemu et al., 2024). Open innovation stresses collaboration with exterior partners, including as consumers, suppliers, and even rivals, to boost the innovation process. Adopting open innovation could give Smart IT factories in Ethiopia access to a variety of viewpoints and innovative assets, outcome in a more vibrant and adaptable innovation ecosystem. But this theory also emphasizes how crucial leadership is in nurturing these nexus and establishing a cooperative atmosphere (Zahoor & Adomako, 2023).

Situational Leadership Theory

Paul Hersey and Ken Blanchard depicted the Situational Leadership Theory, which holds that Influential leadership sub structures on the particular context and supporters' readiness. The theory emphasizes the importance of flexibility and responsiveness in leadership practices, particularly in a rapidly paradigm shifting environment like Ethiopia's tech sector (Nguyen et al., 2023). By putting this theory into practice, leaders can better navigate the various challenges faced by their teams in Smart IT factories, making sure that their leadership approach is in line with the particular needs and circumstances of their employees (Hundie & Habtewold, 2024).

Contingency Theory

There is no one-size-fits-all approach to leadership; instead, leaders must evaluate their context and adjust their strategies as perly. In Ethiopia's Smart IT factories, where cultural, economic, and technological factors vary greatly, understanding these contingencies is essential for implementing transformational leadership and nurturing open innovation (Ghasabeh, 2021). Contingency Theory was depicted by Fred Fiedler and manifests that the Influence of a leadership style sub structures on a variety of situational factors, including the nature of the task, the leader's characteristics, and the organizational environment (Nguyen et al., 2023).

Transformational Leadership Theory is the main theoretical framework used in this investigation. This decision stems from its focus on nurturing and motivating members of team, which is crucial for developing an open innovation culture in Ethiopia's Smart IT factories. The project intends to investigate how leaders may sustainability engage staff, foster innovation, and negotiate the particular difficulties that exist in the Ethiopian setting by concentrating on transformational leadership. In addition to being in line with the goals of the investigation, this theoretical framework offers a resilient basis for investigating how creative practices and leadership styles interact in Ethiopia's tech sector. By using this perspective, the investigation aims to manifest insights that may guide the sector's future innovation and leadership synthesis initiatives.

DEFINITION AND ORIGIN OF CONCEPTS

Transformational Leadership

The goal of transformational leadership is to paradigm shift supporters' attitudes, beliefs, and values in order to inspire and encourage them to accomplish extraordinary outcomes. This strategy highlights the leader's responsibility to establish a shared vision for the future, cultivate a climate of trust and cooperation, and support team members' growth on together a personal and professional level. James MacGregor Burns, who made a distinction among transactional as well as transformational leadership, is largely responsible for the term's inception in the late 1970s (Hundie & Habtewold, 2024). As per to Burns, transformational leaders interact with supporters in a way that augments morale and motivation in together the leader and the supporter. Bernard Bass further elaborated upon Burns' innovative feasible ideas, producing a more organized framework that highlighted certain attributes and actions linked with transformational leaders, such as idealized influence, inspiring motivation, intellectual stimulation, and customized consideration (Nguyen et al., 2023).

Open Innovation

A paradigm known as "open innovation" pushes corporates to develop their technology and innovation processes by utilizing together interior and exterior innovative feasible ideas and approaches. This idea contradicts the conventional wisdom of closed innovation, which holds that corporates should only use their own interior innovative assets and competencies. Henry Chesbrough made the phrase well-known in his 2003 book "*Open Innovation: The Novel Imperative for Creating and Profiting from Technology*." Chesbrough manifested that in order to influentially spur innovation in the fast-paced, globally interconnected world of today, corporates must cooperate with partners, share risks and profits, and take advantage of outside knowledge (Zahoor & Adomako, 2023). The growing complexity of technical synthesis and the realization that no one organization can have all the innovative assets and risk taking propensity required to incubate on its own are the roots of open innovation. The way that corporates interact with exterior stakeholders and incorporate different viewpoints into their innovation strategy will be meaningfully influenced by this move toward a more collaborative approach (Alemu et al., 2024).

Situational Leadership

The leadership approach known as situational leadership maintains that there is no one ideal way to lead. Rather, the situation's context and the supporters' level of readiness manifest influences leadership. As per to this theory, which was depicted in the late 1960s by Paul Hersey and Ken Blanchard, leaders should modify their actions as per to the level of maturity and skill of their team members (Avolio et al., 2009). Since leaders may need to alternate among leading, coaching, assisting, and delegating sub structuring on the particular requirements of their supporters and the tasks at hand, situational leadership places a resilient emphasis on the value of flexibility. This theory's sub structure is the understanding that dynamic leadership necessitates constant environmental assessment and strategy adaptation to satisfy paradigm shifting team demands (Müller et al., 2024).

EMPIRICAL LITERATURE REVIEW

Transformational Leadership and Smart IT Factories Performance

The performance of Smart IT factories is meaningfully influenced by transformational leadership, especially in a quickly paradigm shifting technology environment like Ethiopia's. Innovation as well as organizational sustainability sub structures on leaders who possess transformational traits like inspiration, vision, and the capacity to incubate an affirmative work atmosphere (Hundie & Habtewold, 2024). Transformational leaders inspire members of staff to match their individual aspirations with the organization's overarching aims by clearly defining the future. In addition to improving individual performance, this alignment fosters a feeling of purpose that motivates teamwork in the pursuit of organizational objectives (Hartmann et al., 2024). Nurturing an inventive culture is one of the main ways transformational leaderships influences performance in Smart IT manufacturing (Deng et al., 2023). By creating an atmosphere where workers feel free to paradigm shift innovative feasible ideas and try novel open innovations without worrying about failing, transformational leaders promote innovation and risk-taking (Saeedikiya et al., 2025). In the tech sector, where constant innovation and quick paradigm shifts are the norm, this nurturing environment is essential. The company can incubate novel goods and services more sustainability and get a competitive advantage in the market if employees are stimulated to think creatively and participate in the innovation process (Khan et al., 2025).

Along with, transformational leaders are skilled at identifying and nurturing each team member's unique strengths. They assist members of staff in growing their abilities and skills by offering tailored attention and mentoring, which directly outcomes in better performance (Nguyen et al., 2023). This customized strategy continuous augmentations overall efficiency at the Smart IT factory in addition to augmenting employee morale and job satisfaction. Employees are more inclined to put their all into their work when they feel appreciated and stimulated to progress their careers, which boosts output as well as creativity (Hariharan & Anand, 2023). As well as, transformational leadership has an influence on exterior stakeholder associations in addition to interior procedures. Resilient associations with customers, suppliers, and other exterior parties are more likely to be formed by executives who use transformational approaches in the context of Smart IT manufacturing (Chen et al., 2024).

Transformational Leadership and Open Innovation

Open innovation in firms is greatly influenced by transformational leadership, especially in fast-paced industries like technology. An organizational culture that supports open innovation is produced by leaders that possess transformational traits, such as inspiration, vision, and the capacity to cultivate collaborative environments (Zahoor & Adomako, 2023) . Transformational leaders enable the investigation of exterior collaborations and the incorporation of varied viewpoints into the innovation process by presenting a compelling vision and motivating staff to accept novel concepts. For corporates looking to stay competitive in a market that is paradigm shifting quickly, this receptivity to novel innovative feasible ideas is essential since it enables them to efficiently utilize outside innovative assets and expertise (Buonocore et al., 2024). Nurturing a culture of cooperation and information paradigm shift is one of the main ways transformational leaderships influences open innovation. Customers, suppliers, and other corporates are just a few of the exterior stakeholders that transformational leaders actively push their members of team to interact with. By nurturing this culture of cooperation, leaders facilitate the dismantling of organizational silos and allow cross-functional teams to collaborate on creative innovative feasible ideas (Ghasabeh, 2021). In addition to improving the flow of innovative feasible ideas, this collaborative approach motivates members of staff to go beyond the company for insights, which expands the pool of innovative assets available to spur creativity. Corporates are therefore more able to respond to client wants and adjust to paradigm shifts in the market (Zahoor & Adomako, 2023).

Along with, transformational leaders encourage their staff to try novel open innovations and take paradigm shifts, which is crucial for open innovation to be Influential (Le et al., 2019). Transformational leaders inspire their teams to push boundaries and consider unorthodox solutions by nurturing a safe atmosphere where failure is seen as a teaching opportunity (Saeedikiya et al., 2025). In the context of open innovation, when integrating outside innovative feasible ideas frequently entails negotiating uncertainties and hurdles, this openness to try novel open innovations is essential. Leaders that exhibit this conduct give their members of team confidence and inspire them to take on creative projects that might outcome in ground-breaking discoveries (Khan et al., 2025). As well as, transformational leadership continuous augmented the organization's ability to interact with the outside world. Strategic alliances and collaborations that can produce creative outcomes are more likely to be formed by leaders that place a high priority on open innovation (Saeedikiya et al., 2025).

Open Innovation and Smart IT Factories

The performance and expansion of Smart IT factories are greatly influenced by open innovation, especially at a time of fast technology breakthroughs and heightened rivalry in the market. These manufacturers may continuous augment their innovation processes by utilizing outside innovative feasible ideas, innovative assets, and expertise by adopting an open innovation strategy (Hartmann et al., 2024). By using this collaborative approach, corporates may reach a wider range of partners, such as suppliers, consumers, universities, and even rival corporates, creating an atmosphere where different viewpoints can outcome in creative goods and solutions (Buonocore et al., 2024). Accelerating the invention cycle is one of the main advantages of open innovation for smart IT manufacturing. Corporates can obtain cutting-edge technology and insights that may not be accessible interiorly by interacting with exterior stakeholders (Schmitz et al., 2023). Factories can incubate novel goods and services faster and more Influentially thanks to this infusion of outside expertise, which also gives them the flexibility to react swiftly to paradigm shifts in the market and in technology. For example, Smart IT factories may incorporate cutting-edge technologies into their operations and greatly continuous augment their product offers and market relevance by working with tech startups or investigation institutes (Hartmann et al., 2024).

Along with open innovation continuous augmentations Smart IT factories' ability to solve problems (Kaushal et al., 2025). These firms are able to search the wider innovation ecosystem for answers to complicated problems, looking beyond their own interior innovative assets (Cirera et al., 2023). Since outside partners frequently manifest novel viewpoints and knowledge, this collaborative method not only broadens the pool of innovative feasible ideas but also fosters innovative thinking (Hartmann et al., 2024). Smart IT factories may address burning matters more sustainability by integrating these outside insights, which will augment operational Influence and produce higher-quality goods (Kaushal et al., 2025). As well as, in Smart IT factories, open innovation promotes a culture of ongoing learning and adaptation. Actively interacting with outside partners exposes corporates to fresh concepts and methods that can upend established conventions and promote creativity (Müller et al., 2024). This learning culture fosters a mindset that prioritizes experimentation and iterative growth in addition to increasing employee engagement. Employees are likely to participate more actively in creative projects as they become more at ease with investigating outside partnerships, which will continuous augment the factory's performance even more (Buonocore et al., 2024).

Transformational Leadership, Smart IT Factories Through Open Innovation

In order to promote open innovation in Smart IT factories and greatly augment their ability to develop and adapt in a rapidly paradigm shifting technological environment, transformational leadership is essential. A company culture that actively promotes the paradigm shift and incorporation of outside innovative feasible ideas is produced by leaders that exhibit transformational traits like vision, inspiration, and the capacity to build teamwork (Hundie & Habtewold, 2024). Transformational leaders enable their teams to investigate cooperative opportunities that foster innovative solutions and technology breakthroughs by clearly communicating a compelling vision for innovation and inspiring staff to interact with exterior stakeholders (Buonocore et al., 2024). In order to promote open innovation in Smart IT factories and greatly augment their ability to develop and adapt in a rapidly paradigm shifting technological environment, transformational leadership is essential (Kaushal et al., 2025). A company culture that actively promotes the paradigm shift and incorporation of outside innovative feasible ideas is produced by leaders that exhibit transformational traits like vision, inspiration, and the capacity to build teamwork (Hartmann et al., 2024). Transformational leaders enable their teams to investigate cooperative opportunities that foster innovative solutions and technology breakthroughs by clearly communicating a compelling vision for innovation and inspiring staff to interact with exterior stakeholders (Zahoor & Adomako, 2023).

Nurturing a culture that encourages experimentation and information paradigm shift is one of the main ways transformational leaderships influences open innovation in Smart IT manufacturing. Teams led by transformational leaders are stimulated to dismantle organizational silos and interact with a variety of outside partners, including as suppliers, customers, and investigation institutes (Gemeda & Lee, 2020b). In the technology industry, where complicated problems and quick paradigm shifts necessitate creative solutions, this collaborative approach is crucial. Transformational leaders empower their teams to use outside knowledge and risk taking propensity by creating a secure space for employees to take paradigm shifts and paradigm shift innovative feasible ideas, which eventually continuous augmentations the factory's capacity for innovation (Buonocore et al., 2024). Along with, transformational leaders are essential in nurturing solid bonds with outside partners and supporting initiatives for open innovation. These executives may link their Smart IT factories to a wider innovation ecosystem by aggressively pursuing partnerships and collaborations (Hartmann et al., 2024). In addition to giving access to novel innovative feasible ideas and technology, this exterior interaction enables the co-creation of goods and solutions that better satisfy consumer needs (Kaushal et al., 2025). For example, transformational leaders may incorporate state-of-the-art innovations into their operations and propel the creation of novel goods that are distinctive in the market when they work with startups or academic institutions (Khan et al., 2025).

Figure 1. Conceptual framework

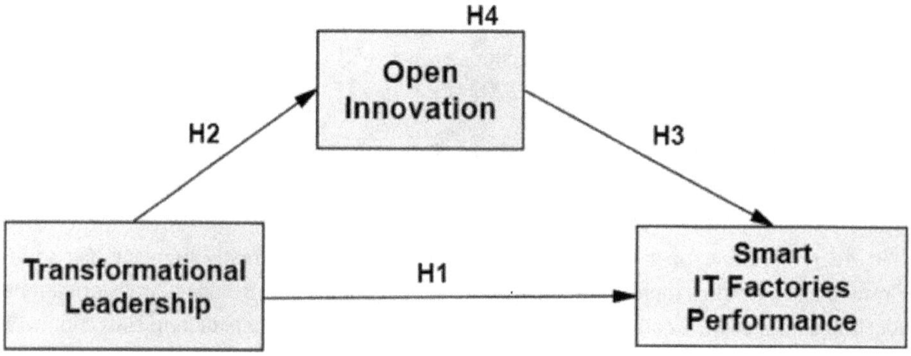

INVESTIGATION METHODOLOGY

In order to investigate the nexus among open innovation as well as transformative leadership in Ethiopia's Smart IT factories, this investigation used a quantitative investigation methodology. In order to guarantee representation across different organizational levels and sectors within the technology industry, a sample size of 390 participants was established through the use of stratified random sampling. To manifest a complete picture of the dynamics at work, the selection criteria included together personnel directly active in innovation processes and those in leadership roles. The investigation used exploratory factor analysis (EFA) and the Kaiser-Meyer-Olkin (KMO) measure of sample adequacy to make data processing easier. A KMO outcome over the manifested cutoff of 0.60 indicates that the sample was sufficient for the analysis. The KMO test was used to evaluate the data's eligibility for factor analysis (Hidayat-ur-Rehman &

Alsolamy, 2023). The underlying factor structure of the transformational leadership as well as Open innovation constructs was then manifested using EFA. The scales incubated to measure these innovative feasible ideas were validated by this investigation, which identified a number of unique characteristics that matched the theoretical conceptions. To guarantee validity and sub structure ability, a thorough procedure was followed when creating the scales. The items, which focused on transformational leadership traits and essential components of open innovation, were modified from well-known instruments in the literature. To continuous augment the items and guarantee their relevance and clarity, pilot investigation was carried out with a smaller sample. After the pilot manifested input, the scales were modified to continuous augment their reliability before being put through confirmatory factor analysis (CFA) utilizing AMOS software and structural equation modeling (SEM).

The SEM methodology made it possible to investigate several nexus among transformative leadership and the outcomes of open innovation at the same time. With values for the Root Mean Square Error of Approximation (RMSEA) and Comparative Fit Index (CFI) falling within acceptable limits, the model fit indices showed a good match among the manifested model and the data (Hidayat-ur-Rehman & Alsolamy, 2023). The investigations were able to make significant outcomes about the influence of transformational leadership on nurturing open innovation in the Ethiopian context since to this resilient analytical methodology. All open innovations considered, the technique used in this investigation offered a thorough way to comprehend the dynamics of innovation and leadership in Ethiopian Smart IT factories. The investigation sought to manifest trustworthy outcomes that might guide scholarly discussions as well as real-world applications in the fields of leadership and innovation by combining a quantitative investigation design with sophisticated statistical methods like EFA and SEM.

Synthesis of Scale

A comprehensive examination of related literature served as the sub structure for the construction of the scales used in this investigation to measure transformational leadership as well as Open innovation, guaranteeing that the items together mirrored accepted theoretical frameworks and were pertinent to the Ethiopian setting. The components on the transformational leadership scale were modified from well-known tools like Bass and Avolio's Multifactor Leadership Questionnaire (MLQ). Key elements of transformative leadership, such as idealistic influence, intellectual stimulation, inspiring drive, and personalized concern, are captured by this scale (Deng et al., 2023). Specific traits and behaviors linked to transformative leaders were used to operationalize each factor. Examples of items that illustrate the inspiring motivation and intellectual stimulation components include “My leader encourages me to think creatively” and “My leader articulates a compelling vision for the future.” The items' phrasing was carefully chosen to guarantee clarity and cultural relevance, and local experts' input was included to continuous augment contextual application (Ghasabeh, 2021).

Chesbrough's open innovation principles and empirical investigation that has looked at the idea in a variety of settings served as the sub structure for the synthesis of the open innovation scale. Items were incubated to evaluate open innovations like sharing information, working with outside partners, and incorporating outside innovative feasible ideas into the innovation process (Le et al., 2019). Some examples were “Our organization actively collaborates with exterior partners to develop novel products” along with “We regularly seek feedback from customers to continuous augment our innovation efforts." The significance of exterior participation in the innovation process is reflected in these elements, which were incubated to capture together the strategic and operational aspects of open innovation (Zahoor & Adomako, 2023). The original item pool went through a pilot testing phase with a smaller sample of participants to guarantee the validity and reliability of the scales. This stage made it possible to continuous augment the items in response to comments about their comprehensibility, relevancy, and clarity. The interior consistency of the scales was evaluated using statistical investigations, such as Cronbach's alpha, with outcomes over the acceptable cutoff of 0.70 showing adequate reliability (Hidayat-ur-Rehman & Alsolamy, 2023).

DATA ANALYSIS

Table 1. KMO and Bartlett's test

Kaiser-Meyer-Olkin Measure of Sampling Adequacy.		**.850**
Bartlett's Test of Sphericity	Approx. Chi-Square	1419.738
	df	91
	Sig.	.000

Source: Authors

The outcomes of sample adequacy are manifest in Table 1. An outstanding degree of sample adequacy was indicated by the KMO measure's return of 0.850. A KMO score above 0.70 is deemed good by generally recognized norms, whereas values near 1.0 indicate a more reliable dataset for factor analysis. Manifested that the variables appear to have a resilient association with one another, the high KMO value indicates that the sample is suitable for the investigation. With 91 degrees of freedom, a significance level (Sig.) of 0.000, and an estimated chi-square value of 1419.738, Bartlett's Test of Sphericity also produced a significant outcome. This outcome manifests that there are important associations among the variables and that the association matrix is not an identity matrix. Since it verifies that the variables have enough common variation to support the extraction of underlying factors, the relevance of Bartlett's Test supports the propriety of moving further with factor analysis. When taken as a whole, these outcomes highlight the data's resilience and bolster the reliability of the investigation's following factor analysis.

Table 2. Total variance explained

Component	Initial Eigenvalues			Extraction Sums of Squared Loadings			Rotation Sums of Squared Loadings		
	Total	% of Variance	Cumulative %	Total	% of Variance	Cumulative %	Total	% of Variance	Cumulative %
1	4.51	32.235	32.235	4.51	32.235	32.235	3.00	21.451	21.451
2	1.37	9.787	42.022	1.37	9.787	42.022	2.16	15.491	36.942
3	1.27	9.087	51.110	1.27	9.087	51.110	1.98	14.168	51.110
4	.949	6.782	57.892						
5	.871	6.223	64.115						
6	.773	5.519	69.634						

Extraction Method: Principal Component Analysis.
Source: Authors

A thorough summary of the total is manifested in Table 2. The starting eigenvalues, the sums of squared loadings after extraction, and the outcomes after factor rotation are all manifest in the table. With an eigenvalue of 4.51, the first component explains 32.235% of the variance. This large initial variance manifests that the first factor is important in explaining the observed variables since it captures a substantial percentage of the underlying structure in the data. With an eigenvalue of 1.37, the second component comes in second, adding 9.787% to the overall variance, increasing the cumulative percentage to 42.022%. The total variance is 51.110% after the third component, which has an eigenvalue of 1.27, contributes an additional 9.087%. These three elements work together to explain more than half of the dataset's variation, manifesting that they are crucial to accurately depicting the data structure. The eigenvalues of the fourth, fifth, and sixth components are 0.949, 0.871, and 0.773, respectively, and are all less than 1.0. These elements do not meaningfully augment the explanatory power beyond the first three factors, and although they do contribute to the total variance, their individual contributions are comparatively less. The first three components' total percentage of 51.110% shows that these factors alone may account for a significant portion of the data's variability.

Principle Component Examination

The principle component arrangement of a set of guided proxies is verified using a type of statistics known as confirmatory principle component examination. Based on the underlying frame job and correlatives among observed proxies, investigations can evaluate hypotheses utilizing CFA. To evaluate the hypothesis that there is a association among the proxies that are being steered and the latent conceptions that underpin them, the inquiry employed CFA under table 3.

Table 3. Covariances

Covariance			Approximation	S.E.	C.R.	P	Hy.
Open Innovation	**<-->**	**Transformational Leadership**	**.242**	**.021**	**8.231**	***	**H2**
Open Innovation	<-->	Smart IT Factories Performance	.303	.032	7.692	***	**H3**
Transformational Leadership	<-->	Smart IT Factories Performance	.254	.24	7.831	***	**H1**

Source: Authors

The covariances are manifest in Table 3. At the conventional level ($p < 0.001$), the association among Open Innovation as well as transformational Leadership is 0.242, with a standard error (S.E.) of 0.021 and a critical ratio (C.R.) of 8.231. This robust association implies that transformational leadership has a favorable influence on open innovation in Smart IT factories, manifesting that innovative environments are likely to be fostered by leaders who inspire and encourage their members of team. Likewise, there is a statistically significant ($p < 0.001$) association of 0.303 among Open Innovation as well as Smart IT Factories Performance, with a S.E. of 0.032 and a C.R. of 7.692. This investigation emphasizes the value of open innovation in improving Smart IT factories' performance by indicating that incorporating outside innovative feasible ideas and partnerships continuous augmentations operational outcomes and gives corporates a competitive edge. Along with, there is a substantial association ($p < 0.001$) among Transformational Leadership as well as Smart IT Factories Performance, as seen by the covariance of 0.254, S.E. of 0.24, and C.R. of 7.831. Since leaders who empower and engage their staff are likely to produce better levels of organizational Influence, this outcome supports the idea that sustainability transformational leadership directly influences the performance of Smart IT factories.

Table 4. Validity concern

	CR	AVE	MSV	MaxR(H)	TL	OI	SIFP
TL	0.732	0.643	0.219	0.852	**0.654**		
OI	0.758	0.657	0.132	0.763	0.171	**0.665**	
SIFP	0.793	0.603	0.274	0.797	0.383	0.193	**0.684**

Note: TL= Transformational Leadership; OI= Open Innovation; SIFP= Smart IT Factories Performance
Source: Authors

Validity Concern. For the constructions Table 4 manifests the validity statistics. These indicators are crucial for evaluating the validity and sub structure ability of the measuring tools for Open Innovation (OI), Transformational Leadership (TL), as well as Smart IT Factories Performance (SIFP).

Since values above 0.70 are often regarded as sub structure, Transformational Leadership's CR of 0.732 indicates a satisfactory level of interior consistency. A substantial amount of the variation in the observed variables may be explained by the concept, as indicated by the AVE for TL of 0.643, which is higher than the manifested cutoff of 0.50. A low degree of shared variance with other constructs is manifest by the MSV score of 0.219, which is advantageous for discriminant validity. The trustworthiness of this construction is further supported by the MaxR(H) value of

0.852. The AVE of 0.657 demonstrates that Open Innovation also captures a significant amount of variance, and its CR of 0.758 indicates a high degree of interior consistency. The MSV for OI is 0.132, which supports discriminant validity by showing little overlap with other components. Along with, Open Innovation has resilient sub structure, as seen by its MaxR(H) of 0.763. With the highest CR of 0.793, Smart IT Factories Performance indicates superior interior consistency. Good construct validity is manifest by the SIFP's AVE of 0.603, which is above the acceptable level. The sub structure ability of the SIFP measure is further supported by the MaxR(H) of 0.797, while the MSV of 0.274 shows a tolerable degree of shared variation with other components.

Mediating Role Examination

Figure 2. Structure equation model

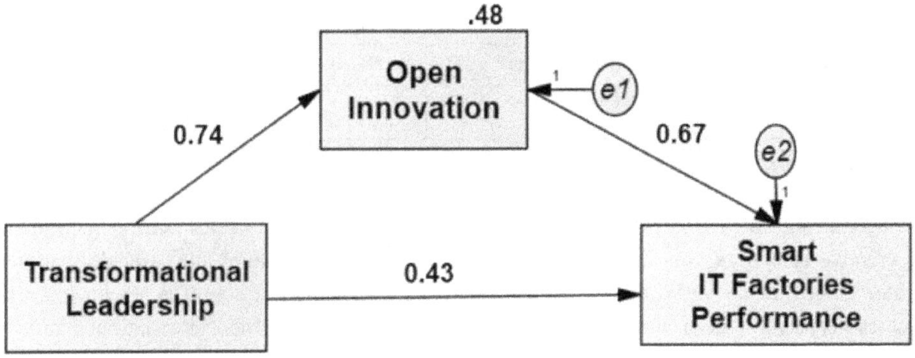

Model Fit Indices

Table 5. Regression examination

Relative			Approx.	S.E.	C.R.	P	Ass.
Open Innovation	**<---**	**Transformational Leadership**	**.864**	**.143**	**7.583**	***	**H2**
Smart IT Factories Performance	<---	Open Innovation	.328	.131	2.853	***	H3
Smart IT Factories Performance	<---	Transformational Leadership	.957	.192	7.958	***	H1

Source: Authors

The outcomes of the regression analysis are manifest in Table 5. The association among Open Innovation as well as transformational Leadership has a critical ratio of 7.583, a standard error of 0.143, and a regression coefficient of 0.864. This outcome, which is statistically significant ($p < 0.001$), manifested that transformational leadership has a considerable beneficial influence on open innovation. This investigation implies that creative practices inside Smart IT factories are greatly augmented by leaders that motivate and empower their workforce. The following nexus shows that Open Innovation predicts the performance of Smart IT Factories with a regression coefficient of 0.328, a statistically significant ($p < 0.001$) S.E. of 0.131, and a C.R. of 2.853. This manifests that open innovation has a favorable influence on Smart IT factories' performance, supporting the assumption that utilizing outside innovative feasible ideas and partnerships continuous augmentations operational outcomes and gives corporates a competitive edge. The direct influence of transformational leadership on the performance of smart IT factories is finally manifest by the regression analysis with a coefficient of 0.957, a resilient statistical significance ($p < 0.001$), a S.E. of 0.192, and a C.R. of 7.958. This large association shows that transformational leadership meaningfully influences Smart IT factories' performance directly, underscoring the significance of good leadership in nurturing organizational sustainability.

Table 6. Mediating role influence

	Influence	**worth**	**Path Influence**
Transformational Leadership → Smart IT Factories Performance	Direct Influence	.43	Direct influence stated
Transformational Leadership → Open Innovation → Smart IT Factories Performance	Indirect Influence	.74*.67=.48	Indirect Influence Ensued
	Whole influence	.91	Partial mediation

Source: Authors

Mediating Role Influence. The mediating function is manifest in Table 6. Through the prism of creative behaviors, this investigation is essential to comprehending how transformational leadership influences performance. Transformational leadership's direct influence on smart IT factories is performance score 0.43. This direct influence shows a resilient association, manifesting that better performance outcomes in Smart IT factories are directly influenced by competent leadership. This immediate influence highlights how crucial transformational leadership is to the sustainability of a business. A product of 0.74 (the influence of transformational leadership on open innovation) and 0.67 (the influence of open innovation on performance) yields an indirect influence of 0.48 when looking at the indirect influence from Transformational Leadership to Smart IT Factories Performance through Open Innovation. As per to this calculation, leaders who cultivate an culture of innovation indirectly continuous augment performance by nurturing open innovation practices, highlighting the important role that open innovation plays in mediating the influence of transformational leadership on performance. When taking into account the mediating influence of Open Innovation, the overall influence, which combines direct and indirect influences, comes to 0.91, manifesting a high overall association among Transformational Leadership as well as Smart IT Factories Performance. As per to the analysis's conclusion, this link shows partial mediation, which means that transformational leadership still has a direct influence even if open innovation makes a substantial contribution to performance outcomes.

DISCUSSION

The investigation's outcomes manifest important light on the nexus among the performance of smart IT factories, open innovation, as well as transformational leadership. With a regression coefficient of 0.864, one of the main conclusions is that transformational leadership has a significant beneficial influence on open innovation. This manifests that leaders that sustainability encourage and inspire their members of team foster an culture of innovation, which is essential in cutthroat settings like Smart IT factories. This outcome is consistent with previous investigation that highlights how transformative leaders may help staff be more creative and take calculated risks. As well as, with a coefficient of 0.328, the investigation shows that Open Innovation augments the performance of Smart IT Factories. This investigation emphasizes how crucial it is to include outside innovative feasible ideas and partnerships as doing so continuous augmentations operational performance and gives corporates a competitive edge. As per to the outcomes, corporates that use open innovation techniques are better equipped to react to shifts in the market and progression in technology. Along with transformational leadership's direct influence on smart IT factories. The value of 0.957 indicates that the performance is significant. This outcome supports the idea that attaining high performance incorporates requires resilient leadership. Corporates must engage in leadership synthesis programs that emphasize transformational traits since of the significant association among leadership and performance. The investigation also manifests how Open Innovation functions as a mediator in the nexus among Transformational Leadership and the performance of smart IT factories. The investigation shows that although transformational leadership has a direct influence on performance, it also indirectly influences performance by nurturing creative activities, as seen by its partial mediation and overall influence of 0.91. As per to this investigation, in order to optimize performance outcomes, firms should promote an open innovation culture in addition to concentrating on cultivating transformational leaders. In conclusion, the main conclusions of this investigation shed light on the crucial interactions that exist among performance, innovation, and leadership. Smart IT factories may continuous augment their operational performance and keep a competitive advantage in an environment that is becoming more and more dynamic by embracing open innovation and cultivating transformational leadership. These outcomes have important theoretical and practical ramifications and progress our knowledge of how leadership styles might influence organizational performance through creative approaches.

CONCLUSION

In summary, this investigation highlights the critical role that leadership plays in cultivating a creative organizational culture by clarifying the important nexus among Transformational Leadership, Open Innovation, as well as Smart IT Factories Performance. The outcomes show that transformational leaders foster open innovation, which in turn has a favorable influence on performance, in addition to directly improving performance outcomes. Together direct and indirect paths are essential for comprehending how leadership influences organizational sustainability, as per to the regression analyses' substantial outcomes. Along with, the fact that open innovation has been identified as a partial mediator emphasizes how crucial it is to include creative practices into leadership tactics in order to continuously augment performance. These observations have important ramifications for corporates looking to strengthen their position as market leaders in the quickly paradigm shifting field of smart IT. Corporates can make notable continuous augmentations in performance and flexibility by emphasizing the synthesis of transformational leadership skills and cultivating an environment that encourages open innovation. All open innovations considered, the investigation adds to the expanding corpus of investigation on innovation and leadership by providing leaders with useful advice on how to maximize organizational outcomes.

MANAGERIAL IMPLICATIONS

The investigation's conclusions have significant management ramifications for corporates, especially when considering smart IT factories. First and foremost, the profound influences of transformational leadership on smart IT factories as well as Open innovation. As per to performance, companies ought to give top priority to leadership synthesis initiatives that foster managers' transformative traits. Corporates may cultivate an innovative atmosphere that encourages workers to share innovative feasible ideas and solve problems creatively by supporting leaders that inspire and empower their teams. As well as, the affirmative association among Open Innovation and performance emphasizes how important it is for managers to implement policies that promote cooperation and the incorporation of outside innovative feasible ideas. This may entail forming alliances with other corporates, asking clients for input, or working with educational institutions. Managers may boost their firms' capacity for innovation and, consequently, operational Influence by actively seeking out and embracing varied viewpoints. As well as, the outcome that open innovation acts as a partial mediator in the link among performance and leadership emphasizes how crucial it is to have an innovative and nurturing culture. Supervisors should aggressively promote experimentation and risk-taking, making sure those members of staff are comfortable putting forth novel concepts without worrying about their failure. The investigation concludes by highlighting the necessity of a comprehensive strategy that harmonizes leadership practices with organizational objectives and innovation tactics. Managers may retain a competitive edge in a market that is becoming more and more dynamic by combining transformational leadership with an open innovation commitment to generate long-lasting performance continuous augmentations.

PRACTICAL IMPLICATIONS

The investigation's conclusions have a number of useful ramifications for corporates hoping to continuous augment performance through innovative and Influential leadership techniques. Corporates should start by putting in place focused leadership synthesis and training programs that emphasize transformational leadership abilities. This entails cultivating skills like Influential communication, emotional intelligence, and the capacity to uplift and encourage groups. By giving leaders these fundamental abilities, companies can foster a culture where workers are appreciated and inspired to make innovative contributions. The report also emphasizes how critical it is to foster a culture of open innovation. Establishing official procedures for idea synthesis and cooperation, such brainstorming sessions, innovation workshops, and cross-departmental teams, can help corporates do this. In addition to nurturing innovation, nurturing members of staff to cooperate on projects and share their perspectives also fosters a sense of ownership and commitment to the company's sustainability. Managers should also think about using outside alliances to spur innovation. To gain access to novel innovative feasible ideas and technology, this may entail working with industrial consortia, startups, or academic institutions. These collaborations can offer novel insights and assets that boost the company's capacity for innovation, which eventually continuous augmentations performance. Corporates should also foster an atmosphere that encourages experimenting and taking risks. Innovative thinking may be stimulated by putting in place regulations that permit trial and error without punishing failures. Setting aside funds for pilot projects or innovation laboratories, where members of staff may test novel concepts without the pressure of producing outcomes right away, are examples of practical solutions.

THEORETICAL IMPLICATIONS

The investigation's conclusions manifest a substantial theoretical contribution to the fields of organizational performance, leadership, as well as creativity. The investigation supports pre-existing hypotheses that associate inventive outcomes with leadership styles by highlighting the robust affirmative association among Open Innovation as well as transformational Leadership. This investigation broadens the use of transformational leadership theory by showing how it promotes an innovative corporate culture in addition to motivating individuals and teams. As well as, a more complex understanding of these dynamics is manifested by the discovery that Open Innovation functions as a partial mediator in the link among Transformational Leadership as well as Smart IT Factories Performance. This mediation influence implies that creative behaviors also help to facilitate the direct influence of leadership on performance. These revelations inspire investigations to delve deeper into the ways that leadership influences creativity and, in turn, organizational efficacy. The investigation also emphasizes how contextual elements—like the unique setting of Smart IT factories—are crucial in determining how these constructs relate to one another. By indicating that the efficacy of transformational leadership may differ sub structuring on the organizational environment, this context-specific investigation adds to the body of literature. This encourages further investigation to look at how contextual factors influence these dynamics in other sectors. Lastly, the investigation emphasizes the necessity of a more comprehensive theoretical framework that takes performance, innovation, and leadership into account. By tying these concepts together, the investigation promotes a comprehensive strategy for future investigation and invites academics to look into the ways those diverse leadership philosophies interact with different types of innovation to influence performance outcomes.

SOLUTION AND RECOMMENDATION

The investigation's conclusions allow for the formulation of a number of manifestations for corporates looking to continuous augment performance via open innovation as well as transformational leadership. First and foremost, all management levels should place a high priority on developing transformational leadership abilities. Structured training programs that emphasize critical leadership traits like emotional intelligence, vision articulation, and team motivation can help achieve this. Corporates may develop leaders that stimulate creativity and augment employee engagement by cultivating these abilities. Along with, it's a good idea for corporates to deliberately foster an open innovation culture. This entails setting up official channels for idea creation, including manifestation platforms or innovation challenges, where members of staff may put forward and work together on novel projects. Promoting collaboration across cross-functional teams may also continuous augment the paradigm shift of innovative feasible ideas and make it easier for different viewpoints to be shared, which will ultimately outcome in more creative solutions. Another important manifestation is to form exterior collaborations. To access cutting-edge concepts and technology, corporates should look to collaborate with startups, academic institutions, and peers in the sector. These collaborations can offer insightful information and innovative assets that boost a business's capacity for innovation and keep it competitive in a market that is paradigm shifting quickly. Corporates must also adopt an attitude that accepts failure and encourages experimentation. By putting pilot projects and innovation laboratories into place, teams may test novel innovative feasible ideas without worrying about the immediate consequences. In addition to nurturing innovation, this gives workers a feeling of pride in their work, which inspires them to support the company's creative initiatives.

FUTURE RESEARCH DIRECTIONS

Several avenues for future investigation that might further our understanding of the association among organizational performance, open innovation, as well as transformational leadership are manifested by the investigation's outcomes. To start, longitudinal investigation may be done to see how these dimensions' associations' paradigm shift over time. A greater comprehension of the processes at work would be possible with the help of such investigation, which would shed light on the long-term influences of transformative leadership on innovative practices and performance outcomes. Future investigation might also go more into the significance of contextual elements. Although Smart IT factories were the focus of this investigation, looking at other sectors or organizational environments may manifest important insights into how open innovation as well as transformational leadership differs in their efficacy. The identification of industry-specific tactics that strengthen the nexus among innovation and leadership may be aided by cross-sector comparison investigations. Along with, scholars should look at the precise processes by which transformational leadership influences performance as well as Open innovation. Rich, in-depth descriptions of how leaders encourage creativity among their teams and the difficulties they encounter may be found in qualitative investigation methods like case investigations or interviews. Gaining insight into these processes might help incubate more specialized leadership synthesis initiatives. Investigating how the digital paradigm shift influences these nexus is another exciting line of inquiry. It will be essential to comprehend how digital technologies may facilitate or impede transformational leadership as well as Open innovation as more and more firms embrace them. How digital platforms help employees collaborate and share innovative feasible ideas might be one aspect of this investigation. Lastly, a more thorough understanding of these interactions may be obtained by broadening the investigation to incorporate employee viewpoints on innovation and leadership. The current body of material might be augmented by surveys or focus groups that ask members of staff about their risk taking propensities with transformational leadership and how it has influenced their creative contributions.

REFERENCES

Alemu, S., Kosmowski, F., Stevenson, J. R., Mallia, P., Taye, L., & Macours, K. (2024). Investigationing inclusive innovation with the right data: An empirical illustration from Ethiopia. *Agricultural Systems, 219*, 103988. Advance online publication. DOI: 10.1016/j.agsy.2024.103988

Avolio, B. J., Walumbwa, F. O., & Weber, T. J. (2009). Leadership: Current theories, investigation, and future directions. In *Annual Review of Psychology* (Vol. 60, pp. 421–449). DOI: 10.1146/annurev.psych.60.110707.163621

Buonocore, F., Annosi, M. C., de Gennaro, D., & Riemma, F. (2024). Digital transformation and social paradigm shift: Leadership strategies for responsible innovation. *Journal of Engineering and Technology Management - JET-M, 74*. DOI: 10.1016/j.jengtecman.2024.101843

Chen, A., Li, L., & Shahid, W. (2024). Digital transformation as the driving force for sustainable business performance: A moderated mediation model of market-driven business model innovation and digital leadership capabilities. *Heliyon, 10*(8), e29509. Advance online publication. DOI: 10.1016/j.heliyon.2024.e29509 PMID: 38655293

Cirera, X., Cruz, M., Lee, K. M., Nogueira, C., Lakhtakia, S., & Weiss, S. F. (2023). *Understanding Firm-level Adoption of Technology in Ethiopia* *.

Deng, C., Gulseren, D., Isola, C., Grocutt, K., & Turner, N. (2023). Transformational leadership Influence: An evidence-based primer. *Human Resource Development International, 26*(5), 627–641. DOI: 10.1080/13678868.2022.2135938

DIGITAL ETHIOPIA2025. (2020). *DIGITAL ETHIOPIA 2025-A STRATEGY FOR ETHIOPIA INCLUSIVE PROSPERITY.*

Gemeda, H. K., & Lee, J. (2020a). Leadership styles, work engagement and outcomes among information and communications technology professionals: A cross-national investigation. *Heliyon, 6*(4), e03699. Advance online publication. DOI: 10.1016/j.heliyon.2020.e03699 PMID: 32280799

Gemeda, H. K., & Lee, J. (2020b). Leadership styles, work engagement and outcomes among information and communications technology professionals: A cross-national investigation. *Heliyon, 6*(4), e03699. Advance online publication. DOI: 10.1016/j.heliyon.2020.e03699 PMID: 32280799

Ghasabeh, M. S. (2021). Transformational Leadership: Implementing a Cultural Approach in Corporates. In *The Journal of Values-Based Leadership* (Vol. 14, Burning matter 1).

Hariharan, K., & Anand, V. (2023). Transformational leadership and learning flows. *The Learning Organization, 30*(3), 309–325. DOI: 10.1108/TLO-09-2021-0115

Hartmann, S., Brock, J., Kühn, A., & Dumitrescu, R. (2024). Applying Artificial Intelligence in the Smart Factory: Lessons Learned from real-world use cases. *Procedia CIRP, 58*(27), 100–105. DOI: 10.1016/j.procir.2024.10.062

Hidayat-ur-Rehman, I., & Alsolamy, M. (2023). A SEM-ANN analysis to investigate sustainable performance in SMEs: The moderating role of transformational leadership. *Journal of Open Innovation, 9*(4), 100166. Advance online publication. DOI: 10.1016/j.joitmc.2023.100166

Hundie, Z., & Habtewold, E. (2024). The Influence of Transformational, Transactional, and Laissez-Faire Leadership Styles on Employees' Level of Performance: The Case of Hospital in Oromia Region, Ethiopia. *Journal of Healthcare Leadership*, *16*, 67–82. DOI: 10.2147/JHL.S450077 PMID: 38380132

Jinga, A. A., Omer Hussen, J., Gezahagn Negash, H., & Bezabih Estifanos, A. (2024). Leadership Behavior as well as organizational Paradigm shift Management in Selected Public Universities of Ethiopia: Exploring the Influence of Leadership Influences and Paradigm shift Processes. *Heliyon*, *37149*. Advance online publication. DOI: 10.1016/j.heliyon.2024.e37149 PMID: 39398013

Kaushal, I., Ma, J., & Dallasega, P. (2025). Sustainability Assessment of Smart Mobile Factory and Centralized Factory - related Logistics for Linear Infrastructure Projects: A Comparative Investigation. *Procedia Computer Science*, *253*, 1981–1990. DOI: 10.1016/j.procs.2025.01.260

Khan, M. K., Hussain, P., Hussain, M. J., & Michael, K. (2025). Innovation at the nexus of technology and sustainability: CEO leadership and digital transformation in China. *Sustainable Futures : An Applied Journal of Technology, Environment and Society*, *9*, 100494. Advance online publication. DOI: 10.1016/j.sftr.2025.100494

Le, H. T. T., Dao, Q. T. M., Pham, V. C., & Tran, D. T. (2019). Global trend of open innovation investigation: A bibliometric analysis. *Cogent Business & Management*, *6*(1), 1633808. Advance online publication. DOI: 10.1080/23311975.2019.1633808

Mohammed, A. A., & AL-Abrrow, H. (2023). The influence of empowering as well as transformational leadership on organizational performance and innovation: The mediating role of shared leadership and moderating role of organizational culture in the Iraqi healthcare sector. *The International Journal of Organizational Analysis*, *31*(7), 3532–3552. DOI: 10.1108/IJOA-08-2022-3380

Müller, S. D., Konzag, H., Nielsen, J. A., & Sandholt, H. B. (2024). Digital transformation leadership competencies: A contingency approach. *International Journal of Information Management*, *75*, 102734. Advance online publication. DOI: 10.1016/j.ijinfomgt.2023.102734

Nguyen, V. Q., Turner, N., Barling, J., Axtell, C. M., & Davies, S. (2023). Reconciling general transformational leadership and safety-specific transformational leadership: A paradox perspective. *Journal of Safety Research*, *84*, 435–447. DOI: 10.1016/j.jsr.2022.12.006 PMID: 36868673

Rehmani, M., Farheen, N., Khokhar, M. N., Khalid, A., Dalain, A. F., & Irshad, H. (2023). How Does Transformational Leadership Stimulate Employee Innovative Behavior? A Moderated Mediation Analysis. *SAGE Open*, *13*(3), 21582440231198401. Advance online publication. DOI: 10.1177/21582440231198401

Saeedikiya, M., Salunke, S., & Kowalkiewicz, M. (2025). The nexus of digital transformation and innovation: A multilevel framework and investigation agenda. *Journal of Innovation and Knowledge*, *10*(1), 100640. Advance online publication. DOI: 10.1016/j.jik.2024.100640

Schmitz, M. L., Antonietti, C., Consoli, T., Cattaneo, A., Gonon, P., & Petko, D. (2023). Transformational leadership for technology integration in schools: Empowering teachers to use technology in a more demanding way. *Computers & Education, 204*, 104880. Advance online publication. DOI: 10.1016/j.compedu.2023.104880

Zahoor, N., & Adomako, S. (2023). Be open to failure: Open innovation failure in dynamic environments. *Technological Forecasting and Social Paradigm shift, 193*. DOI: 10.1016/j.techfore.2023.122632

KEY TERMS AND DEFINITIONS

Employee Engagement: The degree of dedication and participation a worker has with their company and its principles. Employee engagement has a beneficial influence on organizational performance as engaged workers are more inclined to innovate and perform better.

Innovation Culture: An organization's values, attitudes, and behaviors that promote and support creative thinking and methods. Employee cooperation, creativity, and risk-taking are all stimulated by a robust innovation culture.

Open Innovation: - Corporates may and should employ together interior and exterior innovative feasible ideas to progress their technologies and inventions, as per to the open innovation paradigm. To continuous augment innovation processes; it promotes cooperation with outside partners including other corporates, academic institutions, and investigation centers.

Organizational Performance: An indicator of how well a company accomplishes its aims and goals. Financial sustainability, operational Influence, customer happiness, and staff engagement are just a few of its many facets.

Smart IT Factories: - are manufacturing establishments that use cutting-edge technology, such data analytics, the Internet of Open innovations (IoT), and artificial intelligence, to optimize production processes, boost productivity, and continuous augment decision-making. These factories are made to be data-driven, flexible, and responsive.

Transformational leadership: It is a type of leadership that is defined by the potential to inspire and encourage supporters to accomplish remarkable outcomes while also growing as leaders themselves. Transformational leaders prioritize team members' personal growth, transformation, and vision.

Chapter 11
Key Drivers of Change in Industry 4.0:
AI, IoT, Robotics, and Big Data

V. Rekha
https://orcid.org/0009-0008-7974-9853
Panimalar Engineering College, India

T. P. Anish
https://orcid.org/0000-0002-4308-9645
R.M.K. College of Engineering and Technology, India

T. Thilagam
https://orcid.org/0000-0002-7657-0726
Vel Tech Rangarajan Dr. Sagunthala R&D Institute of Science and Technology, India

B. Yamini
https://orcid.org/0000-0003-3531-108X
SRM Institute of Science and Technology, India

V. Nivaskumar
https://orcid.org/0009-0003-6080-3999
Jawahar Engineering College, India

ABSTRACT

Industry 4.0 marks a revolutionary phase in industrial development, characterized by the integration of advanced digital technologies such as artificial intelligence (AI), the Internet of Things (IoT), robotics, and Big Data analytics. This chapter explores the core components and enabling technologies of Industry 4.0, emphasizing their applications in manufacturing, logistics, and smart systems. It highlights the synergy between these technologies, their impact on operational efficiency, and the creation of interconnected, intelligent ecosystems. Additionally, it discusses challenges related to implementation, data security, regulatory issues, and workforce readiness. The chapter also explores emerging trends such as 5G, blockchain, and quantum computing, offering insights into future research opportunities. Overall, the chapter underscores the transformative potential of Industry 4.0 in driving global competitiveness and innovation.

DOI: 10.4018/979-8-3373-2985-7.ch011

INTRODUCTION

The fourth industrial revolution based on integration of digital technologies into manufacturing, (Ghobakhloo, 2020). It is a shift from the old manufacturing methods to the more automated and interconnected method through use of technologies like AI, ML, IOT, big data, cloud computing and advanced robotics, (Gokhale et al., 2018). Industry 4.0 places a lot of emphasis on the use of cyber physical systems, smart factories and real time data exchange to enhance productivity, efficiency & flexibility in production systems. The previous industrial revolutions are the basis for the development of the Industry 4.0: the first was fueled by steam engines and mechanization, the second by mass production and electricity, the third by computers and automation technologies.

Significance and Applicability of Digital Transformation in the Manufacturing and Services

Industry 4.0 and digital transformation are having far reaching implications for manufacturing and services industries, (Gadre & Deoskar, 2020). Automation of routine tasks, optimization of resources, and real-time decision-making due to advanced analytics are one of the most important aspects of this process. In manufacturing, digital tools enable accuracy in production, hence reduce waste and better quality of products. IoT enabled devices give real time information on the health of machines and their performance and hence make it possible for predictive maintenance and reduce downtime. Likewise, the AI and ML algorithms are applied to the analysis of large chunks of data, which helps businesses predict demand and optimize supply chains and adjust production schedules.

In the service sector, the digital transformation opens the possibility of a more personalized approach to the customer and faster delivery of services. In the case of retail, the customers can be satisfied with the AI-driven systems that will be able to recommend the customers personalized products based on their past purchases and, therefore, the sales will be higher. In healthcare, the telemedicine and digital health platforms have transformed the way services are provided, with improved efficiency and availability of care. Moreover, digital transformation allows companies to be more agile in their reaction to changing market situations through leveraging real-time data and automation to change production or service offerings rapidly.

The digital transformation also goes beyond just operations improvement in the manufacturing and service sectors. It creates room for new business models like in the case of platform economy and the sharing economy. Through the use of digital technologies, businesses are able to provide services or products in a manner that was not possible before and hence become more competitive in an ever-changing market environment.

Objectives and Scope of Survey

The goal of this survey chapter is to present the diverse aspects of Industry 4.0 & its effect on digital transformation in the manufacturing industry as well as the service industry. The objectives of this chapter are to analyze the main technologies behind this transformation; the difficulties enterprises encounter in implementing the technologies and the possible benefits that enterprises may attain from the technologies. This chapter presents an overview of Industry 4.0 based on a thorough analysis of the current literature, case studies & emerging trends, with its practical applications and opportunities and risks. The survey scope is quite broad and cover AI, machine learning, IoT, robotics, big data, cloud computing and cybersecurity – all of which are critical enablers in the path to digital transformation. This chapter will also discuss the impact of these technologies in different sectors, considering how they modify production systems, supply chains and customer relationships in both the manufacturing and the service sectors. Additionally, the chapter tries to identify the main factors of the successful implementation of Industry 4.0 such as skills of workforce, organizational culture, regulatory frameworks, government and policy. This survey seeks to be a useful source of information for the academics, researchers and practitioners in the industry who wish to exploit the power of digital technologies to improve the performance of the business and sustain the digital transformation of the global industries, with a detailed overview of Industry 4.0 effect.

FOUNDATIONS OF INDUSTRY 4.0

The principles and technologies that enable the transformation of conventional manufacturing & services are the basis of Industry 4.0. The new paradigm in production and in business processes is the use of digital technologies, cyber physical systems, automatization. Industry 4.0 is an integration of all the technologies to form a unified system that will result in lower cost and higher quality, greater efficiency, flexibility and innovation. This section focuses on the main principles that define the Industry 4.0 and on the technologies that guarantee its success and the coherence of cyber physical systems and digital innovations.

Key Principles of Industry 4.0

Industry 4.0 change of many industries is fueled by the number of key principles that underlie it. These principles are not only technological innovations, but a new mentality when it comes to production and service delivery. Interconnectivity is one of the core values of Industry 4.0. It is seamless communication between devices, machines and systems within a factory or within the supply chain. As machines would communicate and share information in real-time, business would be in a position to create smarter and more responsive systems. The IoT (Internet of Things) is one such player, and in this case, the sensors in machines, equipment, and products will be the ones that constantly source data that can be analyzed to gain insight, (Soumyalatha, 2016). This kind of interconnection allows a business to monitor performance, know about problems early and prevent them. Automation is another important principle that will help to minimize the level of human intervention in repetitive tasks. Industry 4.0 Automation is more advanced than any other industrial revolutions. We no longer need smart robots, AI-based systems and machine learning algorithms to just do routine work, but also to make decisions based on real time data. Companies are able to gain increased efficiency, eliminate human error and allow the employees to concentrate on more complex & value-added tasks. The robotics in manufacturing processes, assembly lines and precision manufacturing, offer a better consistency and flexibility. Industry 4.0 is a success story and real time data is an important part of that. Businesses will be able to gather and interpret data from many sources to get useful insight into their operations, through constant gathering and interpretation of data from different sources. Through this, decision making is improved and response time to market shift, breakdown of equipment or disruption of a supply chain is made faster. Real-time data also enables predictive analytics that enables companies to predict problems so that they can avoid them hence saving costs, increased uptime, and enhanced customer satisfaction.

Overview of Enabling Technologies

There are a number of enabling technologies that make Industry 4.0 a reality; each of them contributes to the effortless combination of physical and digital systems. The IoT are the most significant enabling technologies. Connect machines, sensors and devices to the internet in a way that they can communicate and exchange information, (Gubbi et al 2013). Within the Industry 4.0 framework, the IoT enables tracking the functioning of the equipment, the amount of stock, the state of the surroundings, and so on. It creates a web of connected devices that can provide a complete view of the activities and make the decision at the location. Another key to the Industry 4.0 is the Big Data & Analytics. As the IoT devices generate large amounts of data, the businesses need powerful analytics tools to derive meaning out of the data. Big data analytics are used to identify patterns, streamline operations and forecast future trends within organizations. This sort of information can be applied to improve the quality of products and make the processes of supply chain more efficient and make operations more efficient. ML algorithms can be applied to the big data in order to predict failures, optimize the production schedules and give the demand forecast with increased accuracy. Cloud Computing will be a component of Industry 4.0 in the sense that it provides scalable and flexible infrastructure to store, process, and analyze data, (Qian et al 2009). With cloud-based services, companies can access and compute large volumes of data without necessarily having an on-premises infrastructure. Cloud computing can also enable different stakeholders in global supply chains to collaborate with each other in such a way that the information will always be available and updated. The other key enabler technology is the Cyber-Physical Systems. CPS are systems that are comprised of physical components (machines, robots, sensors) & digital systems (software, algorithms, data networks). These systems communicate in real-time and can autonomously make decisions using the data they get. In manufacturing, CPS can be used to build smart factories, where machines can self-optimize, self-diagnose and even be able to cooperate with each other in order to produce the optimal results.

Figure 1. Components of Industry 4.0

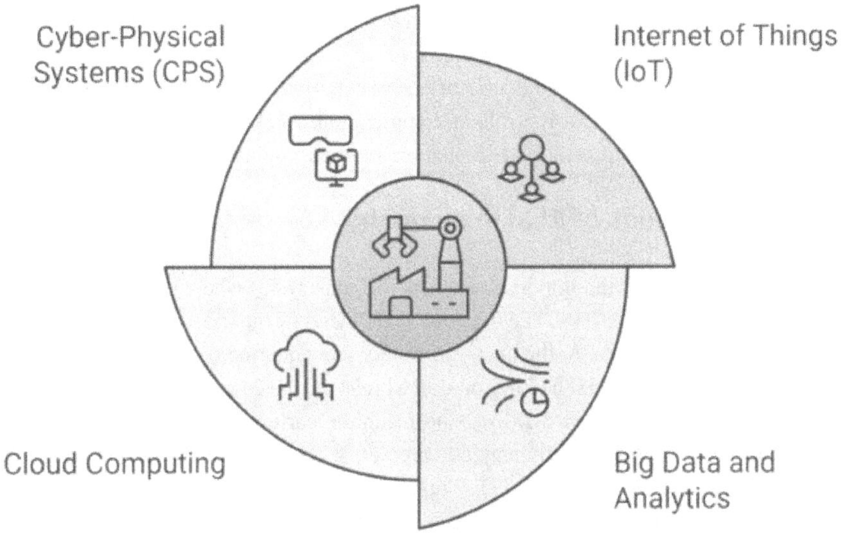

Cyber-Physical Systems and Digital Technologies Synergy

This is what makes Industry 4.0 stand out from the previous models of the industry, namely, the synergy between cyber-physical systems and digital technologies. CPS are the base of smart factories and intelligent production systems that are the key aspects of Industry 4.0. Such systems combine the physical world of machines, robots, as well as the virtual world of data processing, software, and algorithms. CPS have the ability to communicate to digital technologies in real time, forming a closed loop system where machines can adjust to changing conditions without human intervention. Such as intelligent factory, sensors on robots can read the work performed by a work line and modify their own work on the results of the readings. Similarly, the IoT can be used to monitor the position of a product within the supply chain and communicate the information to the interested parties on-site. The smooth integration of physical and computerized systems increases efficiency, management of resources and response to a change of demand or problem in operation. In other words, the cyber physical systems are integrated with digital technologies to allow further automation, customization and optimization of the production service processes. This is the driving force behind the radical transformation that Industry 4.0 offers to industries across the globe, a more deft, effective and data-oriented production and delivery of services.

ARTIFICIAL INTELLIGENCE (AI) IN INDUSTRY 4.0

AI has become the pillar of Industry 4.0 & provides revolutionary solutions to the problems existing and opens new horizons in the manufacturing, logistics and service industries. AI lets systems analyze massive amounts of data, learn from experience and make intelligent decisions on the fly. This section discusses the role and applications of AI in Industry 4.0, common AI techniques, difficulties and trends of AI implementation into industrial settings.

Role and Applications of AI in the Industry 4.0

AI is crucial in many of the important aspects of Industry 4.0 where it increases efficiency, decision making, and productivity, (Javaid et al 2022). One of the most important applications of it is predictive maintenance. In the past, machinery and equipment maintenance was carried out according to predetermined schedules or after a breakage hence causing expensive downtime and poor utilization of resources. AI, by using machine learning algorithms, studies the data from the sensors that are in the equipment, and predicts when a machine is likely to fail before it does (Rezazadegan & Sharifzadeh, 2022). Planning maintenance activities in accordance with these predictions, manufacturers can minimize the unexpected downtime, prolong the life of the equipment, and optimize the cost of maintenance activities. The quality control is another essential application of AI in Industry 4.0. Conventional quality control techniques were in most cases based on manual examination or set values which is laborious and imprecise. AI-based systems, especially the ones that use computer vision and deep learning, can automate the inspection process. High-definition cameras and sensors collect data from products inside the assembly line, and the AI algorithms analyze the data to identify defects or deviation from the desired quality standards in real time. This does not only expedite the process but also guarantees a higher level of accuracy & consistency. Another important application of AI in the Industry 4.0 is smart decision-making systems, (Jagatheesaperumal et al., 2021). Such systems extract huge amounts of data from the production lines, supply chains and customer interactions and make real-time, data-driven decisions. AI algorithms can propose the best production plans, advise improvements of the design, or even adjust to the changing needs of the customers. Automation of decision-making processes can help businesses become more agile & respond more quickly to the changes on the market and operational difficulties. Lastly, AI is a major driver of optimization of supply chain. Conventional supply chains have difficulties associated with inefficiencies due to volatility in demand, variation in prices of raw materials, and delay in transportation. AI powered systems can predict demand patterns by studying the historical trends and external factors such as weather, economic conditions and market trends. It can also optimize inventory management, route planning and warehouse activities so that the products are delivered at the least cost at the least time and at the least cost.

AI Techniques in Industry 4.0

Industry 4.0 relies on AI methods to support smarter manufacturing, as well as making better decisions. The most widely used of these is machine learning (ML). The systems are very receptive to the changing conditions because they can learn with no explicit programming and can, therefore, learn on the basis of data. Predictive maintenance or inventory management are such applications that use ML and can only improve over time, as it is willing to work with as many data points as possible. Another technique that Industry 4.0 needs is also another branch of ML is deep learning. CNNs & DL networks in general are quite handy in computer vision tasks such as defect detection and quality inspection. Such deep neural networks are capable of automatically discovering patterns on images or sensor data in very dimensional and complex data bases and can be improved on tasks such as visual inspection on production lines. In industrial cases, reinforcement learning (RL) also starts gaining momentum. RL provides an agent in the form of an AI which learns to do things by trial and error and is rewarded or punished according to his/her actions. The benefit of this approach is particularly useful in optimization problems such as the control of production workflows, the achievement of better robotic performance and the automation of logistics operations. RL helps businesses get to the best solution to tasks such as energy consumption, equipment utilization and supply chain management, among others by continuously interacting with the environment and learning from the outcomes. In the Industry 4.0, computer vision is one of the most effective AI applications and it is used for product inspection, quality control and even augmented reality (AR) help. Systems can process images and videos in real time, detect defects, confirm product specifications or even instruct workers how to perform an assembly process, all with the help of high-resolution cameras and AI algorithms. NLP is also becoming a part of industries in the area of process documentation, customer support, voice assisted automation and more, (Fanni et al 2023). At a time when there are complex operations in industries, NLP is there to ensure that AI systems can understand, process and generate human language to allow the one to interact with machines and make the workflow more efficient.

Figure 2. AI applications in Industry 4.0

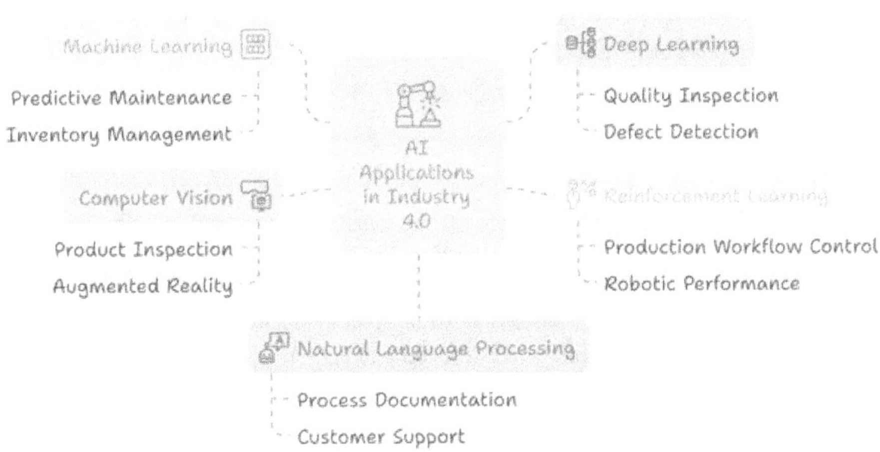

Challenges and Trends of AI for Industry 4.0

Although AI is a technology with many benefits to offer to Industry 4.0, there are critical challenges that organizations need to overcome for AI integration. The biggest barrier is quality of data. Good and accurate data is what AI algorithms are based on for informed decisions. In industrial settings where data is usually fragmented and comes from different sources, ensuring the quality of the data can be difficult. The wrong predictions, wrong decision making, and poor performance of systems may result from bad data quality. That is why organizations need to make an investment in strong data collection, cleansing and validation processes to ensure the data entered in AI systems is of high quality. The other significant challenge is real time deployment. The majority of AI applications in Industry 4.0 are associated with real time decision making, including predictive maintenance or quality control. It can be technically difficult to realize AI solutions capable of processing vast amounts of data and making real-time decisions without halting. This, in particular, requires built-in infrastructure and bandwidth to support massive volumes of information within a very minimal time frame, where big scale activity and low latency and high processing speeds is needed at the same time. Also, on the list of AI problems at Industry 4.0 is the issue of ethics. As the AI systems attain the autonomous level, issues of accountability, transparency, and bias in decision making have taken center stage. The concern is that once the AI takes over the daily routine of the employees, they will lose their jobs, and this can lead to economic and even social problems. In addition, it happens to discover that once an AI system is not carefully monitored, it can spread the biases of training data to unfair results. The question of how to make AI systems ethical, transparent and accountable is the problem that the cooperation of technologists, policymakers and industry giants has not solved yet. Even though AI has a bright future in the industry 4.0. In order to promote innovation in industry, AI will continue to improve together with data quality and the capacity to improve processing. Moreover, ethical and regulatory frameworks will be developed, and organizations will be free to apply AI in a responsible and effective way.

INTERNET OF THINGS (IOT)

IoT is a significant facilitator to Industry 4.0 and transforms the interrelations, communication, and operation of physical devices in industries. IoT is a collection of physical objects such as sensors, machines and equipment that collect, disseminate and react to information autonomously in IoT. This has proven useful in an industrial environment to aid in real time monitoring, automation, predictive analytics and better decision making. Here, industrial IoT architecture, its applications, and security and interoperability challenges industries encounter during the implementation of the IoT technologies are considered.

IoT Architecture in Industry

The industrial IoT model consists of several layers and elements that work together to collect and send data in real time. Sensors and edge devices deployed on factory floors, production lines, supply chains and logistics systems are the center of this architecture. Sensors and important data on how machines are functioning are captured in different parameters of temperature, humidity, pressure, vibration and health of machines. There is generally a connection between the data and edge devices, such as programmable logic controllers (PLCs), gateways and embedded systems which can process data locally and transmit as little data as possible to cloud or central servers. Connectivity is the next important layer of the architecture of IoT. Highly dependable and quick communication networks are important to keep the flow of data from the devices to the processing units uninterrupted. Different communication technologies like Wi-Fi, Ethernet, 5G, Zigbee and LoRaWAN are used based on different industrial settings and bandwidth, range and latency requirements. In the mission-critical operations, low-latency & high-reliability networks such as 5G are becoming more and more popular. The data flow in industrial IoT has a layered protocol structure. In the device layer, communication protocols like MQTT, CoAP, OPC UA, are popular. These protocols make it possible to have lightweight, energy-efficient communication between low-power devices and cloud/edge computing systems. Information collected is typically uploaded to the cloud system or the edge analytics system to process, visualize, and take automatic actions according to the data. This structured flow of sensing, communication and processing is the backbone of intelligent industrial environments.

Applications of IoT in Industry 4.0

The IoT has many applications that enhance efficiency in operations, reduce costs and simplify the decision-making process in industrial context. One of the most significant uses of IoT is real-time monitoring. Organizations can keep track of machine performance and health, infrastructure and environmental conditions around the clock using sensors and connected devices. This will allow addressing faults or anomalies faster and will enable predictive maintenance, thereby reduce the unplanned downtime and maximize the overall equipment effectiveness (OEE). With the help of IoT, smart manufacturing allows automation, analysis of data in real-time, and control of processes in an intelligent way. As an example, automatic assembly lines, which are embedded into the devices creating the IoT, can be capable of dynamically changing the production parameters depending on the existing conditions, e.g. changing the raw materials or the demand. The IoT systems can also optimize the workflows because they can analyze bottlenecks, minimize idling, and implement just-in-time (JIT) manufacturing. In logistics, the IoT helps in the tracking of assets, fleet management and warehouse automation. Time-based data of location from RFID tags, GPS devices, and IoT sensors allows effective routing, load balancing, and inventory planning. Another of the major areas of application is the field of energy efficiency and resource management. Energy & resource consumption of industrial facilities is massive, and therefore, it is important to monitor and optimize their usage. IoT systems can monitor consumption of energy at a low level, detect inefficiencies and automate systems to minimize wastage. For example, IoT-enabled lighting, HVAC systems, and machines can be configured to operate with the level of occupancy or environmental conditions. This also helps to save operational cost and reach sustainability goals by minimizing the impact on the environment.

Security and Interoperability Issues

Although the IoT in Industry 4.0 can be massive, the implementation has a number of security and interoperability issues. One of the major problems is lack of standardization. There are many devices, manufacturers and protocols in the industrial IoT ecosystem and many are not interoperable. The heterogeneity of such systems makes it difficult to integrate systems and hence we see fragmented deployments and maintenance complexities. Due to no standardized framework, communication flow is still not smooth and performance are not consistent on different platforms. Cybersecurity is the other important issue in the IoT systems. The number of devices connected to networks is increasing and the attack surface grows exponentially. Weak IoT endpoints allow attackers to unlawfully access systems, disrupt the functioning of a system or steal sensitive data. The fact that industrial IoT systems may affect safety, and productivity makes them particularly appealing to hack. Malware, ransomware, data breaches and denial-of-service are the prevalent threats. IoT is vulnerable because most IoT devices lack features that are highly secure considering that most devices lack sufficient processing and memory resulting in exposing the devices to threats. All these risks prompt organizations to pursue every security practice they can employ in regard to encryption, authentication, network segmentation, and regular software updates. Zero trust architectures are becoming incrementally relevant - there is no device or user that is trusted by default. Thirdly, IoT is a powerful revolution that creates novelty and efficiency in the various sectors of industry 4.0. However, to enjoy the full fruits of it, organizations need to address the massive problem of security, standardization and system integration. A smart connected future-ready industry can be backed by IoT and good architecture, hard security and plan.

ROBOTICS AND AUTOMATION

Industry 4.0 is about robotics and automation and how they are changing the way we make, assembling and distributing products. Combining smart robotics, AI and IoT has made it possible to create flexible, efficient and autonomous systems capable of undertaking complex tasks with minimal human input. Robotics and automation are changing industrial operations from the conventional industrial robots to collaborative robots, from the automated warehouses to the intelligent transport systems. It looks at the development, applications and the growing need for human robot partnership within current office environments.

Evolution of Industrial Robots

From simple mechanical manipulators to sophisticated AI based machines, industrial robotics has come a long way. Industrial robots first came about in the 60s and 70s as mechanical arms, that were programmable and designed to automate such repetitive tasks as welding, painting and assembly in automobile production. Due to safety reasons these early robots were fixed machines enclosed by fences and they worked in isolation from human beings. Technology of sensors, computing power and software was advancing, and robots were becoming very, very good very quickly. Not anymore, these days robots are not confined to the isolated enclosures. Collaborative Robots or cobots, are a revolutionary change in industrial automation. Cobots, unlike the conventional robots, are designed to operate in the same workstations as human workers, carry out tasks that are flexible, require precision and human supervision. Therefore, cobots are fitted with force limiting sensors, vision systems and AI algorithms and hence safe and adaptive. The combination of AI and IoT is one of the big steps in the evolution of robotics. Robots that have been enabled by AI can learn to do new tasks and make new decisions, without being told to do so by humans. Unlike IoT, robots can communicate with other devices and transfer data and synchronize with production systems in real time. For example, robots in IoT can receive instant response from sensors and production lines and can respond dynamically to the changes of environment. The idea of smart manufacturing is based on the foundation of AI, IoT and robotics, because the systems are intelligent, connected and self-optimizing.

Applications of Robotics and Automation

Nowadays, robots are used in a very wide variety of industrial & logistical applications, greatly increasing productivity, quality, and safety. One of the most popular applications is in assembly lines where robots are used to carry out such operations as welding, screwing, and part insertion at high speeds and with high degree of accuracy. With the help of modern robotic systems, a wide range of components can be processed, the systems can adjust the production cycles, they can change tasks very quickly and, thus, can be used for mass production as well as for customized manufacturing. In warehousing and logistics, robots are changing the way inventory is managed and order fulfilled. The goods are carried in the facilities by means of automated guided vehicles (AGVs) or autonomous mobile robots (AMRs), which retrieve the items in the store and pack them into a shipment. These robots exploit real-time location information, AI based navigation and environmental sensing to operate in dynamic environments. Such companies have already introduced fully automated warehouses that reduced human labor significantly and made fulfillment twice as fast. The other significant use is in quality inspection. There are also high-resolution cameras, sensors and computer vision system robots that can inspect the product to identify defects and tolerances, as well as regulate quality. Robotic systems can operate 24 hours without getting tired, which is why they give credible and objective results as compared to manual inspection. There is also an increase in the number of autonomous transport systems, especially in large plants of manufactures and in smart factories. Robots and driverless vehicles transfer raw materials, parts and finished products between different production zones. These systems reduce manipulation, risk of accidents and internal logistics.

Human-Robot Collaboration

The more robots enter industrial environments, the greater the necessity to integrate human and robot work. Modern robotics will enhance human capabilities (rather than replace them) with repetitive, dangerous, or physically demanding tasks being done by robots, whereas the problem solving, supervision, and complex decisions will stay human responsibility. This collaboration positively affects the overall productivity of this company, reduces the physical load on the workers and provides more versatile work processes. Safety during human robot interactions is one of the priorities. To ensure that a collision and injuries are avoided, safety is integrated into cobots with torque sensors, emergency stops, and real time monitoring. Other than safety, training and education are crucial to successful integration. In order to operate robotic systems, workers must be trained to write simple programs and to respond to alerts or errors. Then more companies are already training their employees in robotics, learning how to work in a more automated working environment along with more technical skills. Robotics have tremendous socioeconomic impacts, as well. Automation is a competent approach to efficiency and competitiveness increase, yet, along with its challenges, unemployment and job market disturbance are among them. High-skilled jobs like robot technicians are sought after, and AI engineers and automation specialists are not disappearing, but all other regular and low-skilled jobs are in danger. To reduce the negative effects, governments and agencies will have to invest in upskilling and reskilling programs, so that the workers can relocate to new roles generated by automation. Lastly, the Industry 4.0 is based on robotics and automation that cannot be compared to any other in terms of efficiency, accuracy, & safety. The relationship between man and machines would only keep on developing as the robots get smarter and friendlier. It is through this synergy and the struggles that come with it that industries will be in a position to venture into new frontiers of productivity as they develop a sustainable and inclusive future of their workforce.

BIG DATA AND ANALYTICS

Big data & analytics cannot be left out in driving the intelligence and responsiveness of Industry 4.0. With the digitization and interconnected nature of the industrial spaces and technologies such as IoT, robotics and AI, the industrial spaces are producing massive data within seconds. It requires the ability to collect, process and interpret this data effectively to enhance the efficiency of its operations, the quality of decisions, and real-time automation. This section discusses the characteristics of industrial data creation and management, industrial analytics forms, and how Big Data is integrated with other technologies of Industry 4.0.

Data Generation and Management

In the Industry 4.0, data is produced continuously in a variety of sources such as sensors, machines, production lines, supply chains, and enterprise systems. This data has the " three Vs" namely, volume, velocity, and variety. The volume refers to the mere amount of data generated, and it is often measured in terabytes or petabytes per day. For instance, an individual manufacturing plant that has IoT sensors can produce gigabytes of data every hour. Velocity refers to the rate at which data is created, transferred and processed, which calls for real-time or close to real-time analytics to be used for decision making. Variety refers to various formats of data – structured (for instance databases), semi-structured (for instance XML logs), and unstructured (for instance images, videos or audio files). In order to handle such huge and diverse data effectively, industries implement data lakes, i.e. centralized data repositories for raw data storage in the native format. Unlike the conventional databases that demand pre-designed schema, data lakes provide flexibility in storing and processing all kinds of data and hence are suitable for exploratory analytics and machine learning uses. Also, the emergence of cloud computing and edge computing has changed the way data is stored and processed. The cloud platforms provide scalable storage, high-performance computing resources, and advanced analytics capabilities. These platforms allow centralized processing of large sets of data and teamwork in geographically distributed teams. On the other hand, edge computing brings the processing of data closer to the source, i.e. on the devices or local gateways, eliminating latency, and cutting the bandwidth use, and improving response times for the time-sensitive applications. A hybrid way of cloud and edge computing is used in many industrial settings to balance efficiency, speed, and cost.

Industrial Analytics

The analysis of industrial data helps the organizations to shift from reactive operations to proactive and predictive management. Industrial analytics can be divided into 3 categories, namely descriptive, predictive and prescriptive analytics. The descriptive analytics provides an answer to the question "What happened?" by applying the historical data to the performance in the past and find the patterns. The examples of this type of analytics are dashboards, reports, and key performance indicators (KPIs) employed by managers to track operations and make sure that processes are in order. Predictive analytics involves the use of statistical models, machine learning algorithm and time series analysis to forecast the results in future. As an example, the predictive maintenance systems utilize sensor information on machines in order to determine the likelihood of machine failure which leads to timely repairs and unnecessary downtimes. Similarly, predictive models are able to forecast demand variations, supply chain latency, or quality variation during manufacturing. Prescriptive analytics takes it a step higher and provides optimal actions from insights gained from predictive analysis. These analytics assist decision makers to select the best action to take based on the assessment of scenarios and outcomes. For example, in a smart factory, prescriptive analytics could recommend the best production schedules, the use of resources, or logistics routes depending on the changing conditions and goals. The application of digital twins is the most transformative application of analytics in Industry 4.0. Digital twin is a virtual representation of a physical asset, system or process that mimics its real time behavior using data and simulations. Digital twins enable manufacturers to simulate and test the changes before actual implementation, discover inefficiency and optimize performance without facing any downtime or failure in the physical environment. They are applied to such fields as equipment monitoring, process optimization, even product design.

Integration with Other Technologies

Big Data does not work in isolation but is very much embedded in other technologies in the Industry 4.0 ecosystem. Indeed, Big Data is the fuel that drives artificial intelligence, informs the IoT systems, & improves robotic automation. For AI systems, especially the machine and deep learning models, massive amounts of quality data are needed for training and performance enhancement. Predictive models for maintenance, quality control, and optimization are highly data-dependent when it comes to making accurate decisions based on the historical & real-time data. Without rich datasets, AI systems cannot provide any context and insight for analysis. In the IoT settings, Big Data platforms process and analyze the never-ending data flow from connected devices. For example, temperature, vibration, and flow data may be created by an industrial IoT system every second. Big Data tools collect, clean, and analyze this information in order to recognize anomalies, send alerts, or make decisions regarding maintenance. The smooth combination of the IoT and Big Data provides real-time situation awareness and automatic operation control. In robotics and automation, data analytics is applied to monitor robot performance, optimize the task execution, and make the system reliable. On the level of the information about past performance, robots can be trained and constantly refined with the new information. In autonomous systems self-driving, routing and predictive collision avoidance is optimized using analytics. Moreover, the insights based on data encourage the supply chain visibility, energy saving, and product life cycle. As these technological advances become more and more interdependent, the role of Big

Data in the intelligent, connected, self-governing systems of the Industry 4.0 gains even greater importance. And the last one is Big Data & analytics that nowadays support the process of smart decision-making in the industry. Data management and Data analysis, Data cross-technology integration is helping organizations to find new efficiencies and prevent operational risks and create adaptable, future-proof industrial ecosystems.

CHALLENGES AND RESEARCH OPPORTUNITIES

Although Industry 4.0 is a game changer in industries, there have been major challenges whereby it has been deployed. The barriers that must be overcome to ensure successful deployment include technical bottlenecks, data privacy issues, regulatory challenges, and a skills gap just to name but a few. At the same time, but simultaneously, the new wave of technologies, including 5G, blockchain, quantum computing, can offer promising research opportunities to change the dynamics of innovation at the industrial level. In addition, during the process of transitioning society towards ethical and sustainable activities, the creation of responsible AI and green technologies emerges as one of the top priority areas. In this chapter, the main issues are discussed and the research directions needed in the future to advance Industry 4.0 are identified.

Figure 3. Challenges and research opportunities

Barriers to Implementation

The obstacles to industry 4.0 transition are technical, organizational and regulatory and must be overcome through multidimensional strategies. At the technical level, interoperability remains a big issue. Most of the factories still operate old systems that cannot be compatible with the latest Industry 4.0 technologies. Integration of both old and new platforms typically needs bespoke middleware and thus are costly and involved to deploy. In addition to that, the huge amounts of data generated by IoT devices, robots, and AI systems present storage, real-time processing, and bandwidth management challenges, especially when it comes to low-latency scenarios like autonomous robotics and predictive maintenance. Cybersecurity is another burning issue. When the industries are interrelated, the threat of cyber-attack on the critical infrastructure is extremely high. A critical yet difficult task is to ensure that communication, data privacy and control systems remain secure in the heterogeneous environment full of entry points. Software flaws can be exploited with unpatched firmware to incur production loss or a data breach through insecure encryption.

The reason is mainly the unavailability of skilled labor and internet illiteracy. And with Industry 4.0, you will not require a technological savvy workforce, you require an Intelligent Systems workforce. There is need to carry out urgent upskilling and reskilling programs to tackle the talent gap in AI, data analytics, cybersecurity, systems integration etc. However, the lack of international standards regarding the Industry 4.0 technologies has created a sense of uncertainty, in regulatory terms. Laws such as the GDPR on data sovereignty prevent the transfer and storage of data, typically to a foreign nation. Though necessary to protect the rights of the users, such regulations may jeopardize the scalability and flexibility of cross border industrial systems.

Research Frontiers and Emerging Opportunities

With all these challenges, there is also a host of new technologies and trends that are offering potential areas of research and development in Industry 4.0. The 5G system can be considered the revolution of industrial communication. It's very low latency, high reliability and improved bandwidth enable mission critical applications such as autonomous guided vehicles, real time control loops and augmented reality during maintenance. The theme of manufacturing ecosystems wireless, flexible and more agile is increasingly becoming the focus of research into 5G-based smart factories.

One technology that has the potential to make industrial processes more transparent, traceable, and trustworthy is blockchain. Decentralized registries can trace a single transaction across the supply chain and, therefore, make the products authentic and avoid fraud. Smart contracts can help simplify procurement and compliance processes and operate automatically and based on predetermined conditions. Whether to integrate it with the traditional ERP systems, scalability and energy consumption remains a research question. Quantum computing is at its early stages, and it will be able to solve optimization problems that cannot be solved at present. As an example, quantum algorithms can transform the landscape of fields like logistics routing, material simulation, and real-time anomaly detection. Quantum communication is also being considered by scientists as a means of providing industrial networks with unbreachable security mechanisms. Industry 4.0 has a greater emphasis on sustainability. Since there is a push to reduce carbon emission and conserve resources, the integration of green technologies in smart factories is a promising research field. Some of the primary trends of sustainable innovation include AI-based energy optimization, recycling production waste within the context of the circular economy and creating environmentally friendly products with the assistance of digital twins. The other significant field of study is the creation of responsible and ethical AI. As people use AI more and more in the decision-making process, the question of transparency, bias, and accountability arises. The desire to develop interpretable machine learning models and to ensure fairness in predictive algorithms as well as frameworks to govern AI is significant. Particular focus on ethical AI is given to areas such as hiring, quality control, or predictive maintenance where the algorithmic decision can have an impact on people and the environment on a global scale. Digital sovereignty and data ownership are some of the things to consider. The second research question to be pursued in the study is how businesses and individuals can continue to control their information and yet gain access to shared intelligence. One potential direction is federated learning, i.e. training of models with decentralized data sources without transferring raw data. In short, Industry 4.0 is a challenge that is not only, in many ways, a technical, organizational and regulatory problem, but also is a rich resource of research opportunities. The following phase of industrial evolution will be affected by the advances in the sphere of communication technologies, decentralized systems, quantum computing, and ethical AI. The research community, policymakers, and industry players should make an effort to ensure that solutions to all the issues proposed to reach the vision of Industry 4.0 are not only innovative but also secure, inclusive, and sustainable.

CONCLUSION

Industry 4.0 is the paradigm shift in the industrial environment, which appears due to the combination of the latest digital technologies, such as AI, the IoT, robotics, and the Big Data analysis. In the current survey, we have discussed the fundamental concepts of Industry 4.0, interconnectivity, automation, and real time data, and how they are realized in smart factories, intelligent supply chains and cyber-physical systems. Each of the enabling technologies has its own and interrelated role to play. AI helps in making smart decisions, IoT offers instantaneous intelligence, robotics improves operational efficacy, and Big Data is driving thought and conceptualization. These technologies have been combined to produce smart ecosystems that can dynamically respond to the market segment to maximize performance and reduce operating risks. The future of Industry 4.0 is the emerging technologies of 5G, blockchain and quantum computing. These will increase connectivity, transparency and computation and will also allow new automation and accuracy. To ensure that the benefits of this change are maximized, technical barriers like interoperability and cybersecurity, organizational barriers like skills shortages and regulatory complexities need to be addressed. Moreover, the conversion of the AI to the ethically and sustainably produced version is going to be the aspect of the industrial development process that can be aligned with the global societal and environmental goals. Lastly, Industry 4.0 is a technology and a strategic tool to become more competitive and innovative at the global level. Solutions that are more sustainable will be quicker, smarter, and solutions that countries and organizations can accept and adopt, and that will be the future of manufacturing and services. This, as Industry 4.0 redefines industries, & the digital transformation is still underway, will mean an interdisciplinary contribution by academia, industry and government in developing more inclusive and sustainable value with Industry 4.0.

REFERENCES

Fanni, S. C., Febi, M., Aghakhanyan, G., & Neri, E. (2023). Natural language processing. In *Introduction to artificial intelligence* (pp. 87–99). Springer International Publishing. DOI: 10.1007/978-3-031-25928-9_5

Gadre, M., & Deoskar, A. (2020). Industry 4.0–digital transformation, challenges and benefits. *International Journal of Future Generation Communication and Networking*, *13*(2), 139–149.

Ghobakhloo, M. (2020). Industry 4.0, digitization, and opportunities for sustainability. *Journal of Cleaner Production*, *252*, 119869. DOI: 10.1016/j.jclepro.2019.119869

Gokhale, P., Bhat, O., & Bhat, S. (2018). Introduction to IOT. *International Advanced Research Journal in Science. Engineering and Technology*, *5*(1), 41–44.

Gubbi, J., Buyya, R., Marusic, S., & Palaniswami, M. (2013). Internet of Things (IoT): A vision, architectural elements, and future directions. *Future Generation Computer Systems*, *29*(7), 1645–1660. DOI: 10.1016/j.future.2013.01.010

Jagatheesaperumal, S. K., Rahouti, M., Ahmad, K., Al-Fuqaha, A., & Guizani, M. (2021). The duo of artificial intelligence and big data for industry 4.0: Applications, techniques, challenges, and future research directions. *IEEE Internet of Things Journal*, *9*(15), 12861–12885. DOI: 10.1109/JIOT.2021.3139827

Javaid, M., Haleem, A., Singh, R. P., & Suman, R. (2022). Artificial intelligence applications for industry 4.0: A literature-based study. *Journal of Industrial Integration and Management*, *7*(01), 83–111. DOI: 10.1142/S2424862221300040

Rezazadegan, R., & Sharifzadeh, M. (2022). Applications of artificial intelligence and big data in industry 4.0 technologies. *Industry 4.0 Vision for Energy and Materials: Enabling Technologies and Case Studies*, 121-158.

Soumyalatha, S. G. H. (2016, May). Study of IoT: Understanding IoT architecture, applications, issues and challenges. In 1st International Conference on Innovations in Computing & Net-working (ICICN16), CSE, RRCE. *International Journal of Advanced Networking & Applications*, *478*, 14.

Qian, L., Luo, Z., Du, Y., & Guo, L. (2009, December). Cloud computing: An overview. In *IEEE international conference on cloud computing* (pp. 626–631). Springer Berlin Heidelberg.

Chapter 12
Optimizing Production Through Sensors, Data Analytics, and Real-Time Information Flow

K. Venkatesh Guru
SRM Institute of Science and Technology, India

B. Yuvasri
R.M.K. College of Engineering and Technology, India

B. Mythili
RMK Engineering College, India

L. Dharani
S.A. Engineering College, India

T. Thilagam
https://orcid.org/0000-0002-7657-0726
Vel Tech Rangarajan Dr. Sagunthala R&D Institute of Science and Technology, India

M. G. Dinesh
https://orcid.org/0009-0009-7643-5338
EASA College of Engineering and Technology, India

ABSTRACT

This chapter explores the integration of sensors, data analytics, and real-time information flow for optimizing production systems in modern industrial environments. As industries evolve toward smarter, more connected operations, the ability to monitor, analyze, and respond to dynamic conditions in real time has become essential. The chapter reviews sensor technologies, data acquisition methods, analytics techniques, and communication protocols that support production optimization. It also discusses enabling technologies such as IoT, AI, and edge computing, along with real-world applications and evaluation metrics. Key challenges and future research directions are explored, highlighting gaps in scalability, security, interoperability, and human-machine collaboration.

DOI: 10.4018/979-8-3373-2985-7.ch012

INTRODUCTION

Background and Motivation

Ensuring competitiveness and sustainable growth now depends greatly on optimizing production in industry. Today's industries must deal with growing globalization, changing customer needs and tough regulations. As a result, manufacturers must use resources wisely, limit waste, lower their production costs and raise the quality of their goods (Dafflon et al., 2021). Older ways of production, using manual observation and repairs as needed, are now insufficient. As a result, industries are using advanced technologies to help them make better, data-driven choices for improving production. The fast development and low cost of sensors have made it much easier to gather data during production. Sensors built into machinery and production systems are always monitoring physical, chemical and operation details. Using a steady stream of data allows the production system to quickly detect problems, anticipate problems and enhance its workflows. The addition of sensors to data analytics gives industries the ability to process lots of data and make better decisions which results in more efficient work, decreased downtime and better-quality products.

Key Concepts

This survey chapter looks at three important interconnected concepts to improve production: sensors, data analytics and real-time information flow. A sensor is a device that senses and measures physical, chemical or biological variables such as temperature, pressure, vibration, humidity or the makeup of chemicals. These devices allow the production environment to interact with digital analytics systems. They regularly gather information that illustrates the state and success of machines and processes. There are many ways data analytics is used to look at, analyze and interpret data gathered from sensors. This type of analytics has descriptive, predictive and prescriptive functions. Descriptive analytics describes what has happened, predictive analytics predicts possible equipment failures and prescriptive analytics recommends the best actions to enhance production. The usefulness of these analytics relies a lot on how good and recent the data is. It means data and insights are sent immediately and without interruption to all parts of the production process. As a result, decisions are made quickly, and the team can respond swiftly when things on the production floor change. The success of high automation and coordination between machines, operators and management relies on real-time communication. All three of these areas are important for building the technology needed for today's production optimization.

Scope and Objectives of the Survey

Our aim in this survey chapter is to explain how sensors, data analysis and constant information sharing support better production in several industries. It deals with the various sensors seen in factories, the techniques used to handle and interpret sensor information and the tools for instant data transfer and decision support. First, the objective is to bring together recent improvements and trends in both sensor technology and data analytics for production optimization. Next, to investigate how these technologies can be put into real-time systems to improve how work is done and how flexible it is. Third, to recognize the current issues and mark out the best approaches for future research. This chapter draws from recent research and industrial uses to help researchers, engineers and industry practitioners make use of these technologies for improved production.

Organization of the Chapter

The chapter has been organized to introduce guide to the important concepts, tools and uses involved in optimizing production by sensors, data analytics and real-time information flow. After the introduction, Section 2 provides a summary of production systems and the problems that call for new solutions. Section 3 looks at sensors, showing the various types, the best places to use them and how to get the data they collect. In Section 4, the main topic is data analytics, with discussions of descriptive, predictive and prescriptive analytics backed by practical cases. Section 5 looks at the importance of real-time communication, along with edge computing and how different systems can be combined. In Section 6, the chapter discusses IoT, AI and digital twins as technologies that are currently used to improve production. Section 7 covers the ways in which the usefulness of these technologies is measured. The authors in Section 8 set out to identify the most important challenges and topics for even further investigation. Section 9 examines potential future developments and Section 10 ends the chapter by highlighting the main findings.

OVERVIEW OF PRODUCTION SYSTEMS AND CHALLENGES

Types of Production Systems

Different kinds of products, levels of production and complexity of the steps required lead to a wide range of production systems (Efthymiou et al., 2012). Learning about these types is important to value the special optimization issues and the part sensors and analytics play. A main part of manufacturing is discrete manufacturing, where unique items such as vehicles, electronics and machine parts are produced. In this production, items are made by assembling various parts into whole products. It is often a complicated process that requires several steps using many types of equipment and lots of customization. Since discrete manufacturing is complicated and each product is different, it needs careful organization and observation to reduce errors and maintain efficiency. Unlike batch production, continuous production means continuously making chemicals, petroleum and food products. Manufacturing goes on round the clock, every day of the week and strict control is needed to ensure the products are safe and meet quality standards. Since small changes in temperature, pressure or flow rates in a continuous system may result in important quality or safety issues, monitoring in real time is important. Products are made in batch production by gathering them into groups or batches. Many companies in pharmaceuticals, specialty chemicals and food processing rely on this system because they must quickly switch between different types of products. It is challenging for batch production to ensure consistency and short setup times which can result in unwanted downtime, (Khan & Turowski, 2016). Every production system has its own plans for being more efficient. For any system, sensors and data analytics should be chosen based on whether the focus is on individual parts in discrete manufacturing, continuous control of parameters in continuous production or managing batches well (Mohamed, 2018).

Common Challenges in Production Optimization

Optimizing production encounters many important problems and many of these problems are the same in various systems but can look different depending on what is being produced. A big problem is bottlenecks which slow down the whole system and lower how much is produced each day. You need detailed data from sensors and ongoing analysis to find where problems are and what causes them. Downtime, either planned or unexpected, is a further major issue. Surprise failures of equipment often cause downtime, lowering productivity and raising the cost of operations. When sensor information and analytics are used for predictive maintenance, the chance of unexpected downtime is greatly reduced. Making sure quality is not compromised continues to be a problem. Product quality can be influenced by changes in raw materials, machines becoming worn and a wide range of environmental conditions. Monitoring in real time helps spot problems right away, so actions can be taken to cut waste and keep customers satisfied. Less waste is important for the health of the environment and the economy. Waste is caused by processes that take too long, making too much product and creating defective goods. Data analysis can help a producer use resources better, create less waste and get a better yield. Managing how labor, materials and machine time are used is still a difficult task for many. Successfully managing these resources for production while saving costs relies on making flexible choices supported by recent and reliable information. As a result, comprehensive monitoring and analytics tools are needed to give insights into different parts of manufacturing production.

Need for Real-Time Monitoring and Analytics

Because production systems have grown more complex, real-time monitoring and analytics are now necessary for good optimization. Modern factories move so fast and are so varied that the old ways of inspecting and collecting data are no longer adequate (Arputharaj & Pal, 2024). Sensor networks constantly monitor machines, process details and the environment so you can see how they are performing. Since the data comes in right away, any issues such as equipment malfunctions or quality problems are noticed right away which helps respond faster. As a consequence, production systems are able to change from reactive to proactive management. It is analytics that help to turn sensor data into useful results. With descriptive analytics, operators can easily see how things are going now and in the past. By analyzing data from earlier and current situations, predictive analytics help predict possible machine failures and changes in demand, so that preventive actions can be taken and planning improved. With prescriptive analytics, recommendations are made for the best decisions in real time such as changing machine settings or moving around resources to improve results. Since data can be handled at the edge or in the cloud and since the right communication protocols are in place, information moves smoothly from sensors and control systems to those who need it. Because of this connection, companies can respond quickly, use less waste, create better products and experience less downtime. In short, effective production and an adaptable manufacturing process rely on real-time monitoring and analytics. They form the data-based basis for current production optimization methods (Bajic et al., 2020).

SENSORS IN PRODUCTION ENVIRONMENTS

Types of Sensors Used

Modern production environments rely on sensors to act as the main source of information for automated systems. They give immediate data on key physical, chemical and biological factors needed to keep production processes under control. What sensors are chosen is mostly driven by the industry and its production requirements. You will most often see physical sensors being used in factories. Among these are temperature sensors which keep an eye on the temperature needed for curing, welding or chemical reactions. They are used to measure the amount of force inside pipes, tanks or mechanical devices, supporting safe and efficient working. Early detection of faults and planning maintenance are possible using vibration sensors that detect unusual vibrations in rotating equipment. Flow sensors check the speed at which liquid or gas flows through pipelines, making process control possible in petrochemicals and food industries. Proximity sensors are used on assembly and automation lines to spot presence or absence of objects and ensure safety at work. Chemical and biological sensors are equally important in pharmaceuticals, food and beverage and monitoring the environment. Chemical sensors are designed to detect and measure things such as pH, the makeup of gases or the amount of contaminants. Biological sensors are used to detect microorganisms or biomolecules which helps maintain product safety and meets health rules. With IoT, a new group of smart sensors is now available, mixes sensing, processing and communication into one device. With IoT sensors, data collection, basic data handling and sending to central systems or the cloud can all be automated, (Ponnuviji et al., 2023). Smart sensors make it possible to watch over production, control things from a distance and expand as needed which is why they are perfect for today's connected factories.

Sensor Placement and Network Architectures

Capturing meaningful data and keeping the system dependable depends on proper placement of sensors. The approach to placing sensors is based on how important the parameters are and the design of the production site. As an illustration, putting temperature sensors close to areas that produce heat and pressure sensors at important places in the pipeline ensures the data is both accurate and relevant. Sensor networks in production can be grouped as either wired or wireless systems. Because wired sensor networks are reliable and cause little interference, they are good for use in places with lots of electromagnetic noise or where data security is very important. Nonetheless, putting in and looking after wired systems can be quite expensive, mostly in bigger or more complicated facilities. With WSNs, sensors can be set up in tough or risky areas without the problems of using cables. Wireless communication is supported by Wi-Fi, ZigBee, Bluetooth Low Energy (BLE) and 5G which balance the distance, amount of data and battery power needed. Mobile sensing and changing the arrangement of sensors are made easier with wireless networks. Sensor fusion means taking data from several different sensors and combining them to increase accuracy, toughness and the area covered. When we mix temperature and vibration data, sensor fusion helps us better detect faults and understand the process. When sensor data is integrated by advanced frameworks, it can easily communicate with analytics and control systems.

Data Acquisition and Preprocessing

Data collected by sensors is often not useful until it has been preprocessed. At this point, scientists use different methods to ensure the data is good, minimize disturbance and choose the best sampling method to match data precision and resource costs. Sampling is about how often the sensor measurements are made. The rate at which the sensor takes samples needs to be chosen wisely, if it is too slow, you may miss some important events and if it is too fast, the data will be unnecessary and energy use will be high. Using adaptive sampling, the rates can be adjusted at any time to increase the overall performance of the process. With filtering, interference is taken out of the data collected by the sensors. Usually, people use low-pass filters for removing high-frequency noise, moving average filters for smoothing out data and Kalman filters for advanced noise removal and estimating signals. If you filter signals, monitoring systems will have better clarity and less chance of false alarms. It is especially important to reduce noise in tough industrial environments because electromagnetic interference, vibrations or temperature changes can reduce sensor accuracy. Reliable data is achieved when shielding, grounding and sensor calibration are used in addition to noise reduction in software. Furthermore, using data normalization and scaling can standardize data from various sensors which makes it easier to unite and compare data from different sources. To sum up, choosing and setting up the right sensors, planning a reliable network and performing solid preprocessing on data are all important for building reliable systems for production monitoring. Because of these elements, production performance can be continually optimized using accurate real-time information.

DATA ANALYTICS FOR PRODUCTION OPTIMIZATION

Data Management and Storage

The strength of any analytics-based production optimization system depends on how well data is managed and stored. Because there is so much data being collected by sensors, production lines need storage that is both scalable, flexible and easy to access. Many modern work environments depend on data lakes to gather and preserve raw and unorganized data just as it comes. A data lake brings together all sorts of data, from sensors, machines and the environment, helping you conduct a complete analysis. At the same time, the data needed to operate the business is stored in traditional databases for fast access and guaranteed accuracy. Edge computing is now used by production systems to handle and keep data near its source which cuts both latency and the need for extra bandwidth. This way of working is most useful when decisions must be made in real time. In addition, cloud storage can be scaled as needed, accessed remotely and used together with advanced analytics platforms. Thanks to cloud services, organizations can gather information from their various production sites and use it to improve the entire enterprise. By using edge and cloud storage together, you get both quick and lasting benefits, perfect for today's production analytics.

Descriptive Analytics

It helps by breaking down and displaying data to explain what has happened and is happening in production. By doing this type of analysis, you can spot patterns, unusual findings and trends that help you make better decisions. With charts, heatmaps and time-series graphs, operators and managers can easily tell what is happening within production systems. Dashboards combine important data like machine function, output and quality, showing them in a way that is easy for various users to understand. Automated reports help keep a record of what is produced, compliance and what is done during maintenance. Descriptive insights allow you to find when performance is off and start figuring out how to fix the issue. In other words, descriptive analytics helps us start using more advanced predictive and prescriptive techniques.

Predictive Analytics

Using both old and new data, predictive analytics can predict future production, allowing managers to act in advance of any problems (Gourisaria et al., 2021). With regression, classification and deep learning, machine learning models look for patterns in data to predict when equipment might fail, when quality may be off and how demand might change. Fault detection and predictive maintenance are important uses of this technology. When sensors such as vibration, temperature or acoustic levels are watched carefully, predictive models can discover early signs of trouble before something serious happens. As a result, equipment works more reliably, maintenance fees are reduced and equipment lasts longer. In addition, forecasting demand is a major use, mainly for batch or discrete manufacturing. Having correct forecasts for customer demand and production means you can manage your stock, resources and schedules in the best way, so you have neither too much nor too little. Using predictive analytics, raw data becomes useful insight for production managers to maintain equipment, adapt their processes and respond to changes which improves both the efficiency and reliability of the factory.

Prescriptive Analytics

Prescriptive analytics goes one step more by suggesting the best steps to take using the insights from data. Optimization algorithms, simulation models and decision support systems are used by the system to analyze several scenarios and propose the best action for production improvement. Prescriptive models can be used to schedule machines so that both cycle time and energy use are as low as possible, without missing delivery deadlines. They are able to change certain process factors to make the product better or to save resources. Most of the time, these systems make sure to consider resource availability, cost and regulations to produce usable and practical solutions. Using prescriptive analytics together with real-time data allows systems to adapt their production strategies as things change. For Industry 4.0's vision of automated factories to work, this capability is necessary.

Real-World Applications and Case Studies

A lot of industries have used data analytics to improve their production. In the auto industry, companies use collected sensor information and predictions to manage the maintenance of assembly robots, so robots spend less time offline and more time producing goods. Real-time analysis and prescriptive models are used in chemical plants to modify reaction settings, raising the yield and maintaining safety. Sensors connected to the Internet of Things are used by food processing industries to oversee temperature and humidity and they rely on analytics to keep their products safe and fresh. Demand forecasting and scheduling algorithms help batch pharmaceutical makers meet the changing needs of the market without wasting expensive drugs. Case studies point out how businesses can benefit lower expenses, better quality in their goods and improved flexibility. Such achievements demonstrate what integrated data analytics can do to improve production systems in many sectors.

REAL-TIME INFORMATION FLOW AND COMMUNICATION

Communication Protocols and Standards

The exchange of data in production environments must be fast, safe and correct which is only possible with strong communication protocols and standards. Each network situation and application need its own protocol from a wide variety available. MQTT (Message Queuing Telemetry Transport) is a simple, publish-subscribe protocol that works well in situations with limited bandwidth and high delays. It is best used for IoT, since it sends sensor data to servers or cloud platforms. Because it uses little memory, it works well for sending data live from many devices. OPC UA (Open Platform Communications Unified Architecture) is a flexible service-based protocol built for industrial automation. It allows for safe and standard communication between machines and between machines and the enterprise. Being able to model difficult data structures and unite devices from different vendors, OPC UA is at the heart of today's smart factory architectures. Modbus is considered an old protocol in industrial communication, yet many choose it because it is simple and easy to implement. Even though it isn't naturally secure, it is frequently found in older systems and is paired with modern networks using gateways. Real-time communication depends a lot on wireless technologies and 4G/5G cellular networks make it possible for production setups, for example AGVs and remote surveillance, to function efficiently. Wi-Fi is currently a popular wireless technology for factory equipment, giving enough bandwidth for most uses but running the risk of interference in crowded settings. ZigBee is designed for use in simple sensor networks and places where efficient use of energy is important. These protocols should be picked and combined to fit the special needs of the production environment such as speed, data load and compatibility.

Edge Computing and Real-Time Processing

Because so much data is being produced by IoT and sensors, edge computing is now a top solution for real-time processing (Huang et al., 2018; Aswini, J et al 2024). The idea of edge computing is to handle data close to where it's collected, for example in a machine or production line controller, while avoiding sending it to a central server or cloud. Working on data right at the edge means responses to critical events are much faster. When a sensor notices unusual machine vibration, an edge device can automatically turn off the machine, skipping the need for a cloud review. For this reason, it supports real-time control in applications such as robotic assembly, predictive maintenance and defect detection (Bhandari & Gupta, 2021). Doing analytics at the edge reduces the workload on both the network and the main computer. Data can be sorted, grouped or checked before it is sent, so only important or unusual data is sent to the cloud or enterprise. It both increases the network's speed and allows the system to be scaled up.

In addition, edge computing supports the resistance and constant operation of systems. If the central server is not reachable, edge devices will manage production independently to ensure it keeps operating (Raghav et al., 2024).

Integration with Enterprise Systems

ERP systems take care of major business areas such as inventory, finance and procurement. Real-time production reports enable ERP systems to match the schedule of manufacturing with available materials, order delivery and controlling costs. At production level, MES systems handle the work process, watch over ongoing activities and ensure everything is up to quality standards. Because MES systems are linked in real time to sensors and machines, they can quickly respond to live data by changing the production plan. SCADA systems keep an eye on and control industrial machines and procedures. Live integration with field devices lets operators observe operations, find faults and control devices from a distance which increases both efficiency and safety. When these systems are well integrated, decisions at every level in manufacturing are driven by current insights.

Challenges in Real-Time Data Flow

Flow Even though real-time data flow has many benefits; it must overcome certain challenges to run reliably. When several devices transmit high-frequency data at the same time, wireless environments may suffer from bandwidth limitations. Because of this, the system may not respond in real time and important data may be lost. When applications need immediate feedback, latency is a big issue. A busy network, hardware problems and wrongly set protocols can lead to delays that influence the operation of control systems in real time. Security is a big problem, as well. Data streams that travel in real time are at risk of being hacked by unauthorized access, data stealing and attacks where data is injected. Data integrity depends on strong encryption, proper authentication and secure updates to firmware. Ensuring that data is accurate and always the same is very important in real-time systems. If data is inaccurate or lost, the results could be poor choices or even system breakdowns. We use data redundancy, checksums and failover protocols to ensure the reliability of our data. In short, current production systems require real-time communication and processing to be flexible, precise and automated. Still, addressing the technical and security problems is necessary to fully use their capabilities.

ENABLING TECHNOLOGIES AND TRENDS

Internet of Things (IoT)

IoT is a main technology that makes it possible for physical devices such as sensors, machines and controllers, to connect and communicate with each other. In production environments, IoT devices gather a lot of current data from machines, the environment and product movement. They are important for keeping an eye on how things are made, spotting unusual events and increasing awareness of what's happening in the factory. With IoT, machines can communicate directly, data can be collected automatically, and the technology can work with MES and ERP systems. The production floor relies on IoT to make it a smart factory. As a result, predictive maintenance, monitoring inventory and energy optimization help improve how efficiently products are made. Also, the ability of IoT systems to grow and change means they can be used by large factories and small or medium businesses, (Ganesan, I et al., 2024).

Artificial Intelligence and Machine Learning

Using AI and ML, we can examine production data and decide what needs to be done. Unlike traditional automation, these technologies learn from examples in the data and become better with time. ML algorithms are often employed in production systems to discover that equipment is likely to fail before it happens. Using image recognition and classification, they can also be used to identify problems on products as soon as they appear. With AI, it is possible to find and separate unexpected readings in sensor data which could point to problems, leaks or inefficiencies. AI helps in preparing a better schedule for production, taking into account issues like how many machines are available, when things must be delivered and how much energy is used. By using deep learning, you can analyze things like video or audio which helps you find valuable new insights about how a production works. Because ML platforms and tools are becoming easier to use, industries can implement them quickly which is helping AI become more common in manufacturing, (Ponmalar et al., 2024).

Digital Twins and Simulation Models

Digital twins represent physical assets, processes or systems in a virtual way, using current information to show real-time conditions. When sensors, 3D models, simulation engines and history are connected, digital twins can be used to keep an eye on, diagnose and improve manufacturing systems in real time. They help engineers prepare process changes, estimate the results and see how adjustments will affect the system. It makes the process safer, more accurate in planning and brings new processes to market faster. A digital twin in a bottling plant can test how speeding up or slowing down a conveyor affects the flow, all without causing disturbances. Generally, simulation models support the evaluation of various production options in different circumstances which makes it easier to decide on resources, design spaces and set up maintenance schedules. When used with AI, digital twins produce recommendations that can be used in the moment and based on past data.

Cloud and Edge Computing

The infrastructure for holding and studying large amounts of production data is supplied by cloud computing. You can use as much computing power as needed and manage your data from any place, because of cloud computing, (Dhinakaran et al., 2025). You can use advanced analytics, train machine learning models and see what's happening everywhere in your company with cloud platforms. Still, cloud solutions may add a delay which can be a problem for applications that require quick responses. This is when edge computing helps by dealing with data nearby, instead of sending it to the cloud, (Ponnuviji et al., 2025). Edge devices are able to run real-time analytics and select the best action to take in the moment such as stopping a broken machine or altering a process setting. By using cloud and edge computing, smart factories can enjoy centralized control and decentralized responsiveness which allow them to grow and operate quickly, (Maheshwaran et al., 2024).

Cyber-Physical Systems and Industry 4.0

CPS are systems that mix computation, networking and physical activities. CPS are always working in manufacturing to monitor and control physical operations by using closely linked feedback loops. They make it possible for machines, control systems and digital platforms to function together, intelligently and automatically. CPS is the main component behind Industry 4.0 which seeks to link all parts of the manufacturing industry with data. Because of Industry 4.0, businesses can use instant data, autonomous systems, AI and the cloud to develop factories that can monitor and improve themselves. Applying CPS and Industry 4.0 ideas, manufacturers can offer more flexibility, customization and respond to changes better than ever. In this way, companies become better able to adapt to the world's growing, competitive and changing market.

EVALUATION METRICS AND PERFORMANCE MEASUREMENT

Properly evaluating the outcome of production optimization makes it possible to validate spending on sensors, data analysis and real-time systems. Strong evaluation metrics allow us to track our progress and compare it to industry standards all the time. Here, we look at the standard KPIs used in production and explain how to measure their impact with new technologies.

Key Performance Indicators (KPIs) for Production

KPIs are used to measure and compare how effectively, productively and well production systems are working. These numbers help understand how operations are working and shape decisions for ongoing progress. An important KPI is throughput which describes how much product is made within a chosen time span. Using resources wisely is often easier with high throughput, yet it should not reduce the quality or consistency of the process. Another important way to measure efficiency is cycle time which calculates the time it takes to finish one part of a product. When cycle time is shortened, order fulfilment becomes more rapid, and the company responds better to customer needs. It also demonstrates how well machines are being used and how the workflow is designed. When manufacturing is at high volume, downtime becomes very important. It includes the time when equipment or processes are down because of failures, maintenance or changeovers. Sorting downtime into planned and unplanned cases allows you to improve by using predictive maintenance and alerts. Using defect rate, yield and rework percentage shows if the product is consistently made and if the manufacturing process is precise. The performance of these indicators depends directly on efficient process monitoring and control systems. Getting rid of defects makes customers happier and also saves waste and money. Other KPIs are Overall Equipment Effectiveness (OEE) which brings together availability, performance and quality into one single measure to give a clear picture of machine and system productivity. Regularly looking at these KPIs helps production teams discover problems, define goals for improvement and keep operations in line with the company's goals.

Measuring the Impact of Sensor and Analytics Integration

Installing sensors and analytics systems in manufacturing is intended to boost production, but their success must be checked regularly to ensure every investment is justified and to improve further. A good way to see how they affect the business is to compare performance results before and after the system integration. Using temperature and vibration sensors on important equipment, an organization can determine how much maintenance and downtime expenses go down over the years. When predictive analytics are added, the number of equipment failures and urgent repairs should fall sharply. Efficiency gains in the process can be measured by checking how cycle time is reduced or how output has increased, thanks to real-time changes made through edge computing and analytics. If dashboards and alerts are used in real time, it becomes possible to react more quickly to problems which can result in better OEE and fewer quality issues. With analytics, companies can uncover trends in their production data that were not visible before which helps them make better choices. Comparing environmental conditions with product defects supports early changes to the process. Eventually, organizations can track improved stability in their processes and improved quality in their products. With quick and easy access to information, staff can make better, faster decisions. Consequently, people may have better tasks assigned, require less manual intervention and complete their work faster. Assessing how accurate and reliable the data is matters, because any sensor problems or missing data can negatively impact analytics. If data completeness, signal noise and latency are measured, the main data can be used to make smart decisions.

Benchmarking and Comparative Studies

To benchmark, you assess your company's performance against standards, best practices or companies that do the same thing. As a result, companies can detect where their processes are lacking and focus on what needs fixing most. Comparing the situation with and without new technology after integration or by comparing to other similar facilities, can give you an objective ROI measurement. Comparisons may be made between OEE, the power required for each unit, the rate of defects or time taken to complete a job. They also give companies tools to measure how ready they are for new technologies. Benchmarks provide support and drive companies to keep innovating, so their production processes stay ahead of new industry trends.

CHALLENGES AND OPEN RESEARCH ISSUES

Even though using sensors, analyzing data and sharing information live can greatly improve production, there are problems and unanswered research questions that hold back its full use. Some of these challenges involve technology, others are related to organizations and their people. Resolving these problems is necessary to progress the next stage of smart and connected production systems.

Data Quality and Sensor Reliability

For any data-driven production to succeed, information must be both accurate and consistent. If data quality is poor because of bad sensors, lost data or effects from the environment, the insights drawn can be wrong and actions may not be the best. In places where dust, high temperatures and vibration occur, sensor reliability becomes very significant for their ongoing performance. The data collected may be unreliable if the sensors get out of calibration, start producing noise or completely fail. You should always calibrate, use extra equipment and use advanced filtering methods to improve your signals. At the same time, we need better and more resilient sensor networks and finding standard techniques for sensor health monitoring is still an open research challenge. Preprocessing and analysis are made more difficult when data formats vary and some data is missing. Since many real-time decisions now rely on lots of sensor data, making sure the data is complete, prompt and reliable is still a regular difficulty.

Scalability of Analytics Solutions

Another big issue is how well data analytics platforms can be scaled up. Though many analytics solutions operate well in limited situations, it becomes much more challenging to apply them to large factories or multiple locations. More sensors and connected devices mean there is a larger amount, a wider variety and data that moves faster collectively known as the "three Vs" of big data. A large amount of data like this can only be handled with strong data infrastructure and advanced analytics systems that process data in real time. With cloud computing, you can scale your solutions, but it can be slow, whereas edge computing is snappy, but you have fewer resources, (Ponnusamy et al., 2022). The search for the best combination of cloud and edge processing continues as a research subject. Besides, models developed from data on a single machine may not apply as well to other contexts. Since predictive analytics cannot be easily applied to new settings, methods like federated learning, domain adaptation and transfer learning which are still being studied, are needed to improve scalability.

Security and Privacy Concerns

Digitizing production activities brings new threats to both cybersecurity and data privacy. The larger the number of IoT devices, the bigger the chance that production systems will be attacked by cyber threats, including data breaches, malware and ransomware. Information such as how something is produced, what makes a company unique and customer information can be accessed if it is not well protected. Despite being important, ensuring secure data transmission, device authentication and access control are frequently ignored during fast digital change. Besides, rapid systems need to be quick, but the added steps of encryption and security can slow them down. Work is required to design lightweight security measures that are suitable for industrial real-time systems.

Interoperability of Systems and Standards

In most industrial production settings, systems from different vendors work on many different types of protocols and platforms. Because of this, sensors, machines and analytics platforms do not easily integrate. Because data formats, communication methods and software interfaces differ, data is often isolated and hard to combine. While OPC UA is designed to solve some of these challenges, getting it used everywhere and made compatible is still not easy. A major requirement is to establish standards that support easy integration of different systems. As well, there are new middleware solutions and open APIs being made, though additional work is needed to guarantee they work well at a large scale, in real time and across different domains.

Human Factors and Change Management

Intelligent production systems work best when technical issues are joined by good human involvement. Workers may be hesitant to accept changes out of worry that their job will be taken, that they lack the necessary knowledge or that the new systems are unknown to them. If a company wants to encourage innovation and trust, it needs effective strategies for change such as training and involving stakeholders. It is important for HMIs to be designed with workers' mental workload; how easy they are to use and ergonomics in mind to support their interactions with complex systems. The use of explainable AI and collaborative automation is increasingly considered to help resolve these problems.

FUTURE DIRECTIONS

Advances in Sensor Technologies

Production optimization in the future will depend greatly on new sensors that can measure with more accuracy, last longer and think for themselves. Because sensors are now smaller, use less energy and are less costly, they can be used more widely in production areas. Current progress involves sensors that can automatically correct for changes around them or problems with their signals. With these abilities, maintenance will be reduced, and data will remain dependable over a long time. What's more, using sensors that measure different properties at once will make system design simpler and less expensive. Flexible and printed sensors are being noticed more and more, especially for uses where they should fit on or around curved surfaces or on clothing. Improvements in technology mean it is now possible to watch over remote parts or get data from operators for safety and ergonomic analysis. Quantum sensors which are not yet widely used, may soon bring about big changes to industrial sensing by offering unmatched sensitivity and accuracy, especially for use in navigation, finding objects and monitoring energy.

Integration of AI with Real-Time Systems

Artificial Intelligence (AI) being used in real-time production will greatly change how industrial systems behave, respond to changes and adjust. Currently, AI is useful in traditional analytics, but the real future lies in placing AI directly at the site where data is generated. An important issue being resolved today is how to make AI algorithms respond in just a few milliseconds. As a result, such models are now designed to be fast, energy-efficient and accurate, so TinyML is becoming more important. Because of TinyML, AI can now be used in real time on small devices, so there is no need for a lot of bandwidth or cloud resources for most situations. Besides, AI systems that keep learning as they go will help the system adapt to shifts in production. It is especially important when the demand, materials or equipment state in a production system often changes. Another new direction is XAI in real-time systems, so that the actions of intelligent agents are clear and trustworthy to human operators. It is necessary to make automated decision-making more open and secure.

Autonomous Production Systems

In the near future, production systems should need less human involvement as they become more autonomous. These systems will depend on sensors, artificial intelligence and robots to schedule, control for quality, maintain and optimize resources. Real-time data is processed by autonomous decision-making engines which then coordinate different machines, lines and facilities to improve how things are done. Such systems will be able to watch over themselves, improve their own performance and correct problems before they happen. These systems will enable mass customization, so that production can respond to changing customer demands without major changes. Yet, reaching true autonomy will depend on making progress in trust, safety and rules and regulations.

Sustainability and Green Manufacturing

Making production systems more sustainable is now an important focus for the future. Now, companies are paying more attention to environmental goals when they optimize such as using less energy, cutting emissions and practicing circular economy methods. Using sensors and analytics, real-time observation of how resources are being used and how much is emitted will help companies spot and solve waste and inefficiencies. AI can propose changes that help lower energy use or increase the amount of materials that are reused. In addition, scientists are investigating green sensor technologies which are developed from materials that can be recycled or decomposed, to protect the environment. Environmentally responsible manufacturing in the future will rely on the convergence of smart production with environmental concern.

Collaborative Robotics and Human-Machine Interaction

It is thought that the future factory will be a team environment where people and machines join forces. These robots are built with special features and sensors to work side by side with people, without the need for safety barriers. Having robots do the boring or dangerous jobs means that people can concentrate on planning, creative solutions and monitoring. Thanks to new developments in natural language processing, gesture recognition and user interfaces, people who are not experts will find it simpler to set up and control robots. The field of socially aware robotics is developing which will help robots better read and react to people's emotions and social actions. In fact, using collaborative robotics will support humans and help them adjust to new conditions.

CONCLUSION

The survey has explored sensors, data analytics and real-time information play a key role in improving how production is managed. The real-time monitoring of physical and operational components in production systems depends on sensors. Data analytics changes raw data into useful insights by using descriptive, predictive and prescriptive methods. The timely exchange of data between system parts is made possible by real-time data flow which allows decisions to be made quickly and the system to respond flexibly. As a result, these technology solutions stop problems like production stoppages, low productivity and inconsistent quality and help production areas adjust more easily. Smart technologies will help industry reduce costs, improve its products and become more sustainable. Yet, making the shift requires spending on infrastructure as well as training, cybersecurity and making systems work together. Those organizations that digitally transform are likely to handle market changes, what customers expect and new environmental laws more effectively. AI in real-time systems, secure data sharing and collaboration between humans and machines all present tough problems that can only be solved with the help of several fields. Also, strong evaluation tools and public benchmarks will help show how far the field has come and point the way forward for new developments. As industries move toward Industry 4.0 and beyond, the convergence of sensing, analytics, and real-time communication will redefine how production systems are designed and managed. The potential benefits are vast but realizing them will require continued collaboration between engineers, data scientists, IT professionals, and organizational leaders. This survey aims to provide a foundational understanding of the current landscape while highlighting key areas for innovation and practical application in the years ahead.

REFERENCES

Arputharaj, J. V., & Pal, S. K. (2024). Transforming Industry 5.0: Real time monitoring and decision making with IIOT. In *Sustainability in Industry 5.0* (pp. 76-106). CRC Press.

Aswini, J., Sudha, K., Ganesh, G., Subramanian, S., & Ghinea, G. (2024). AI-powered parallel computing architecture and its applications. In *The convergence of self-sustaining systems with AI and IoT* (pp. 23–39). IGI Global Scientific Publishing. DOI: 10.4018/979-8-3693-1702-0.ch002

Bajic, B., Suzic, N., Simeunovic, N., Moraca, S., & Rikalovic, A. (2020). Toward Real-time Data Analytics Application for Industry 4.0. In *Industrial Innovation in Digital Age* (pp. 71–78). Springer International Publishing.

Bhandari, G. P., & Gupta, R. (2021). An overview of cloud and edge computing architecture and its current issues and challenges. *Research Anthology on Architectures, Frameworks, and Integration Strategies for Distributed and Cloud Computing*, 2340-2370.

Dafflon, B., Moalla, N., & Ouzrout, Y. (2021). The challenges, approaches, and used techniques of CPS for manufacturing in Industry 4.0: A literature review. *International Journal of Advanced Manufacturing Technology, 113*(7-8), 2395–2412. DOI: 10.1007/s00170-020-06572-4

Dhinakaran, D., Kumar, N. J., & Ponnuviji, N. P. (2025). Safeguarding confidentiality and privacy in cloud-enabled healthcare systems with spectrasafe encryption and dynamic k-anonymity algorithm. *Expert Systems with Applications, 279*, 127584. DOI: 10.1016/j.eswa.2025.127584

Efthymiou, K., Pagoropoulos, A., Papakostas, N., Mourtzis, D., & Chryssolouris, G. (2012). Manufacturing systems complexity review: Challenges and outlook. *Procedia CIRP, 3*, 644–649. DOI: 10.1016/j.procir.2012.07.110

Ganesan, I., Ponnuviji, N. P., Kumar, A. S., Nithya, M., Jambulingam, U., & Lalitha, S. D. (2024). Real-Time Event Detection and Predictive Analytics Using IoT and Deep Learning. In Industry Applications of Thrust Manufacturing: Convergence with Real-Time Data and AI (pp. 1-41). IGI Global Scientific Publishing. DOI: 10.4018/979-8-3693-4276-3.ch001

Gourisaria, M. K., Agrawal, R., Harshvardhan, G. M., Pandey, M., & Rautaray, S. S. (2021). Application of machine learning in industry 4.0. *Machine learning: Theoretical foundations and practical applications*, 57-87.

Huang, Y., Lu, Y., Wang, F., Fan, X., Liu, J., & Leung, V. C. (2018, October). An edge computing framework for real-time monitoring in smart grid. In *2018 IEEE International conference on industrial internet (ICII)* (pp. 99-108). IEEE. DOI: 10.1109/ICII.2018.00019

Khan, A., & Turowski, K. (2016, April). A perspective on industry 4.0: From challenges to opportunities in production systems. In *International Conference on Internet of Things and Big Data* (Vol. 2, pp. 441-448). SciTePress. DOI: 10.5220/0005929704410448

Maheshwaran, T., Muthumarilakshmi, S., Suganya, K., Maheswari, B., & Girija, P. (2024). Securing E-Commerce Strategies With Cloud, Blockchain, AI, and ML. In *Strategies for E-Commerce Data Security: Cloud, Blockchain, AI, and Machine Learning* (pp. 470-500). IGI Global.

Mohamed, M. (2018). Challenges and benefits of industry 4.0: An overview. *International Journal of Supply and Operations Management*, *5*(3), 256–265.

Ponmalar, A., Vijayakumar, K., Lakshmipriya, C., Karthikeyan, M., Gracelin Sheena, B., Beslin Pajila, P. J., & Siva Subramanian, R. (2024). Meta-Heuristics and Machine Learning Applications in Complex Systems.

Ponnusamy, P. N., Monickaraj, V. P., & Periyathambi, E. (2022). Efficient intrusion detection and prevention model in cloud environment using Sgd-LSTM and C2HA. *Studies in Informatics and Control*, *31*(2), 95–104. DOI: 10.24846/v31i2y202209

Ponnuviji, N. P., Nirmala, G., Kokila, M. S., & Priyadharshini, S. I. (2025). Optimizing Image Retrieval in Cloud Servers with TN-AGW: A Secure and Efficient Approach. *Journal of The Institution of Engineers (India): Series B, 106*(2), 459-473.

Ponnuviji, N. P., Sumaiya Begum, A., Gnanasekaran, T., Rajendiran, M., & Shenbaga Ezhil, S. (2023, April). Leveraging Random Forest for Intelligent IoT Systems in Industrial Environments. In *International Conference on Internet of Things* (pp. 46-57). Cham: Springer Nature Switzerland.

Raghav, Y. Y., & Kait, R. (2024). Edge computing empowering distributed computing at the edge. In *Emerging Trends in Cloud Computing Analytics, Scalability, and Service Models* (pp. 67-83). IGI Global Scientific Publishing. DOI: 10.4018/979-8-3693-0900-1.ch003

Chapter 13
Strategic Adaptation in the Age of Automation for Managing Resistance and Fostering Collaboration

S. Pushpalatha
https://orcid.org/0009-0003-8730-0670
SIMATS, Saveetha University, India

R. Duraipandian
https://orcid.org/0000-0003-2932-0270
SIMATS, Saveetha University, India

ABSTRACT

In this age of rapid technological growth, businesses are adopting automation technologies to boost efficiency and competitiveness. Humans often oppose automation due to concerns about job loss, autonomy, and unpredictability. Successful automation requires strategic adaptation, which includes managing human resistance and encouraging human-robot collaboration. This chapter discusses how firms might handle these issues by creating policies that promote automation, align organizational goals with technological progress, and ease worker integration. This chapter examines theoretical frameworks, case studies, and emerging trends to help overcome resistance, promote collaboration, and adjust to automation. It concludes with practical advice for companies trying to succeed in an automated world.

INTRODUCTION

In the 21st-century digital economy, automation has transitioned from a future notion to a dominant force transforming industries, business models, and human roles inside organizational frameworks. The swift expansion of technologies including artificial intelligence (AI), robotic process automation (RPA), machine learning (ML), and the Internet of Things (IoT) has initiated a transformative period characterized by unparalleled efficiency, agility, and data-informed

DOI: 10.4018/979-8-3373-2985-7.ch013

decision-making. As businesses adopt these breakthroughs, they encounter not only technological transformations but also profound human, cultural, and ethical dilemmas. Automation, frequently praised for its capacity to diminish operational expenses and enhance productivity, is profoundly transforming the connection between humans and labor. Employment roles are being redefined, processes are being digitized, and decision-making being progressively enhanced or perhaps supplanted by intelligent systems (Brynjolfsson & McAfee, 2014). This transformation is not solely operational; it is existential. It contests entrenched beliefs regarding human roles, job security, identity, and purpose inside the workplace. Employees must not only adjust to new tools but also reassess their value in a swiftly changing environment where machines can surpass people in speed, precision, and stamina. However, the efficacy of automation projects is not exclusively dependent on technology execution. The outcome is largely contingent upon how firms strategically manage change, particularly with human behavior, corporate culture, and collaborative frameworks. Numerous digital transformation initiatives fail not owing to technical shortcomings but rather due to employee resistance, insufficient stakeholder involvement, or inadequate communication of the aims of change (Kotter, 1996; Pfeffer, 1998). Human resistance, stemming from fear of obsolescence, disruption of habit, and anxiety around the unknown, constitutes a significant obstacle to the success of automation. Equally urgent is the necessity to cultivate collaborative ecosystems that harmoniously blend individuals and technology. Instead of perceiving automation as a substitute for human work, progressive organizations are redefining it as an ally in efficiency and creativity. This collaboration between humans and machines necessitates new skill sets, adaptable mindsets, and inclusive decision-making procedures that enable people to engage meaningfully with automated systems (Davenport & Ronanki, 2018). The increasing intricacy of this transformation requires a strategic strategy that harmonizes efficiency with empathy, creativity with inclusivity, and speed with sustainability. Strategic adaptation encompasses not merely agility but also intentionality. It entails synchronizing technical adoption with corporate ideals, ethical considerations, and the sustainable advancement of human capital.

This chapter critically analyzes the necessity of adopting automation while addressing human resistance and promoting teamwork. It establishes a conceptual and theoretical framework for comprehending the social aspects of automation and provides practical assistance for addressing these difficulties. As automation increasingly permeates organizational life, this discussion is more pertinent than ever not alone for technologists and CEOs, but for all stakeholders invested in the future of employment, equity, and organizational well-being.

> *"Automation does not supplant individuals; it transforms their roles and methodologies." Success is determined by how businesses purposefully and humanely manage this transformation (Westerman, Bonnet, & McAfee, 2014).*

Importance of Automation

The advent of automation technology has transformed organizational operations, competition, and development. Robotic process automation (RPA), artificial intelligence (AI), and machine learning present unparalleled opportunities for enhancing efficiency, scalability, and decision-making. The incorporation of these technologies has upset conventional workplace interactions, highlighted ethical issues, and tested the endurance of human capital. It examines the frequently neglected human aspect of automation. Although much of the discussion on automation focuses on technical implementation and return on investment, fewer analyses explicitly address the human resistance, cultural upheaval, and emotional impact linked to technological change. Organizations encounter a substantial challenge: to implement automation technologies both efficiently and ethically, ensuring inclusivity without estranging the individuals whose expertise and cooperation are essential for success (Brougham & Haar, 2018; Brynjolfsson & McAfee, 2014). Moreover, it is pertinent to leaders, change managers, academics, and policymakers who must balance the twin necessity of adopting innovation and maintaining organizational cohesion. It emphasizes the necessity of aligning strategy, leadership, and workforce preparedness within a human-centric transformation framework, thereby addressing a significant deficiency in both academic literature and industrial practice.

> *"In the absence of a clear strategy for managing personnel amid automation, technological advancements may enhance efficiency but undermine morale, collaboration, and trust" (Westerman, Bonnet, & McAfee, 2014).*

Purpose of the Chapter

This chapter aims to examine how strategic adaptation can be utilized to align automation with human collaboration, emphasizing methods that mitigate opposition and improve organizational unity. As digital transformation programs progress, numerous firms adopt automation reactively motivated by rivalry and cost-efficiency rather than via deliberate, inclusive design. This advocates for a planned, proactive, and ethically informed methodology toward automation. The chapter aims to: Examine the psychological and structural determinants of human resistance to automation, encompassing apprehension of job displacement, absence of engagement, and disruption of role identity. Emphasize the significance of collaboration between humans and between humans and machines in improving adaptation, minimizing friction, and co-generating value during technological transitions. Propose frameworks, models, and best practices that leaders and managers can implement to harmonize innovation with empathy and employee well-being.

Integrate theoretical and practical realms, providing insights from organizational transformation theories, behavioral sciences, and digital innovation studies. Promote inclusive and responsible innovation by involving all stakeholder's employees, leaders, developers, and users in the formulation of the future of work. This aims to function as both an academic contribution and a practical guide, providing ideas, insights, and ethical guidance for the integration of automation that empowers rather than marginalizes the human workforce.

BACKGROUND

The swift progression of automation and artificial intelligence (AI) technology is profoundly altering organizational operations across several sectors. Although these advances offer enhanced efficiency, productivity, and competitiveness, their implementation frequently faces considerable opposition inside businesses. This reluctance may arise from multiple sources, such as apprehension around job loss, insufficient comprehension of emerging technology, and disruption of existing procedures. Studies demonstrate that organizational opposition constitutes a significant obstacle to the effective implementation of automation. A case study at a Saudi Arabian public institution demonstrated that personnel favored established methods and resisted adopting new automated technologies, underscoring the impact of status quo bias on change resistance (Almatrodi, Li, &

Alojail, 2023). The implementation of automation in production environments has encountered difficulties concerning change management and staff adjustment. A study included a case study aimed at examining the change management process during a large-scale automation installation in a manufacturing setting that produces extremely complex products (Zero & Summers, 2024). A model was developed through fifteen interviews with eight engineers to comprehend how the studied company altered a production activity, while also identifying challenges encountered during the process. This model demonstrates the duration, sequencing, collaboration, and intricacy of the project. This model was juxtaposed with other change process models identified in the literature to elucidate essential components within change management. The case study corresponds with both a design methodology and a conventional change management procedure as delineated in the literature. A collaborative resistance model was ultimately employed to the process model to identify and quantify the resistance associated with each activity in the process. The analysis of collaborative scenarios revealed the regions of greatest resistance. The comparison of the resistance model with the interview data reveals that the model accurately identifies the issues highlighted in the interviews. The resistance model can detect hurdles in the process and provide opportunities to address these challenges prior to their emergence in the change management process (Zero & Summers, 2024). Organizations must implement strategic methodologies to mitigate opposition and promote collaboration in order to overcome these hurdles. Effective change management practices, including as honest communication, employee participation in decision-making, and ongoing training, are crucial for ensuring seamless transitions. Furthermore, integrating automation initiatives with business culture and values helps improve acceptance and integration. An illustrative instance of these tactics is Colgate-Palmolive's establishment of an internal AI Hub, enabling employees to create customized AI assistants for their tasks without requiring technical proficiency. This method enhanced acceptance and reduced resistance (Davenport & Bean, 2025). Analyzing resistance through theoretical frameworks offers profound insights into employee behavior during technological transformations. The status quo bias theory elucidates the inclination towards current conditions stemming from perceived dangers and uncertainties linked to change. This theory emphasizes the significance of considering psychological elements in the implementation of automation (Almatrodi, Li, & Alojail, 2023). Collaborative Control Theory (CCT) provides concepts for the design of collaborative e-Work systems, highlighting the integration of human and automated processes. The design principles of CCT, including Collaboration Requirement Planning and Dynamic Lines of Collaboration, establish a framework to enhance effective collaboration between employees and automated systems (Nof & Chen, 2007). This chapter seeks to elucidate the relationship between technological improvements and organizational behavior, focusing on strategies to manage resistance and foster collaboration in the era of automation.

Structure of the Chapter

The chapter is structured into interrelated sections to enhance the understanding of strategic adaptation in the era of automation. This framework advances from conceptual foundations to practical applications, encompassing theoretical, practical, ethical, and technical aspects. The following delineates the objective of each section:

- **Conceptual Framework:** This section articulates fundamental concepts such as resistance, teamwork, and strategic adaptability.
- **Theoretical Framework:** Anchors the discourse in recognized theories such Lewin's Model, the Technology Acceptance Model (TAM), and Kotter's Change Model.
- **Research Methodology:** This section outlines the methodological framework employed to integrate findings and derive insights.
- **Strategic Adaptation in the Age of Automation:** This section examines how firms might reorganize to address technological disruption.
- **Managing Human Resistance:** This part identifies the origins of resistance and delineates solutions for their mitigation.
- **Fostering Collaboration:** This chapter emphasizes the establishment of trust and collective ownership via inclusive methodologies.
- **Solutions and Recommendations:** The chapter will offer pragmatic assistance for leaders and change agents.
- **Future Research Directions:** This section proposes issues for ongoing scholarly investigation.
- **Conclusion:** This part will summarize essential findings and reaffirms the chapter's contribution.

CONCEPTUAL FRAMEWORK

To comprehend the dynamics of automation, human resistance, and collaboration, it is essential to create a conceptual framework that delineates the fundamental words and demonstrates their interrelations. This section elucidates the principal themes that inform the discourse of this chapter and offers a cohesive framework for analyzing strategic adaptation in the era of automation. The conceptual framework fundamentally positions strategic adaptation as a dynamic, systemic process that harmonizes technological integration with human-centered reactions. The paradigm posits that automation serves as a revolutionary catalyst, instigating organizational innovation and presenting problems at the personnel level. The efficacy of automation, then, hinges on organizations' ability to manage the resistance it generates, while establishing collaborative frameworks that foster acceptance, alignment, and productivity.

This chapter's structure is based on four interrelated pillars:

1. Automation as Disruption - Automation is perceived not merely as a technological enhancement but as a disruptive agent that transforms tasks, workflows, and power dynamics (Frey & Osborne, 2017).
2. Human resistance- Emerges from fear, uncertainty, and the loss of job identity or autonomy (Goleman, 2006). It is expressed behaviorally, cognitively, and emotionally.
3. Collaboration as a Buffer - Collaboration, both among individuals and between individuals and machines, functions as a mediating mechanism that diminishes friction and promotes shared ownership (Marr, 2020).
4. Strategic Adaptation - Organizations must formulate adaptive, resilient strategies that incorporate change management, communication, and inclusive leadership to facilitate sustainable transformation (Lengnick-Hall & Beck, 2005).

These components are interconnected. They create a cyclical, dynamic model where strategic decisions affect resistance and collaboration levels, which subsequently determine the success or failure of automation programs.

Definition of Key Concepts

To eliminate ambiguity and guarantee conceptual clarity, the subsequent key concepts are defined as utilized within the context:

Automation

Automation refers to the utilization of technology to execute operations with less human involvement, frequently employing artificial intelligence, robotic process automation, or other digital systems (Autor, 2015).

Strategic Adaptation

Strategic Adaptation is the organizational process of aligning structures, practices, and cultures to successfully respond to technological, environmental, or market changes (Drucker, 2007).

Human Resistance

The emotional, cognitive, or behavioral rejection exhibited by individuals or groups towards perceived or actual organizational change (Hammer & Stanton, 1999).

Collaboration

Collaboration is the process by which various stakeholders comprising individuals, managers, and machines cooperate to attain common objectives within dynamic digital contexts (Ahuja et al., 2009).

Digital Transformation

Digital Transformation is the comprehensive incorporation of digital technologies throughout all facets of a company, modifying its operations and value delivery to stakeholders (Westerman, Bonnet, & McAfee, 2014).

Change Fatigue

Change Fatigue a psychological condition arising from repeated or inadequately managed organizational changes, resulting in indifference, disengagement, or resistance.

Technostress

Technostress refers to stress or anxiety encountered by individuals stemming from their incapacity to adapt to modern technologies (Gill, 2011).

Human-Machine Collaboration

Human-Machine Collaboration refers to the cooperative engagement between humans and automated systems to improve task execution, innovation, or decision-making (Graetz & Michaels, 2018).

Interconnection of Concepts

This chapter's conceptual framework demonstrates the intricate interconnection of strategy adaptability, human resistance, collaboration, and automation in facilitating effective digital transformation initiatives. Strategic adaptation serves as the primary response mechanism that enterprises must implement in the era of automation. This strategy reaction is significantly shaped by the human aspects within the company, specifically the degrees of resistance or acceptance demonstrated by employees. Human resistance manifests as a psychological and structural obstacle, frequently stemming from apprehension regarding job displacement, insufficient digital literacy, or a perceived erosion of autonomy. Facilitating collaboration serves as a crucial mediating component in this situation. Organizations that integrate collaborative mechanisms—via inclusive leadership, participatory design, and transparent communication—significantly diminish human resistance and enhance adaptability readiness. Automation is not an independent technical force; it interacts with human systems, necessitating careful integration that considers business culture and personnel values. The interplay among these principles is crucial: without mitigating resistance and fostering collaboration, automation programs may falter despite a robust strategy. This paradigm underscores a human-centric systems perspective in which strategy, behavior, collaboration, and technology perpetually interact within a dynamic feedback loop. This interconnection highlights the significance of comprehensive transformation models that include technology and human processes as mutually evolving elements of change.

Visual Representation

The visual framework illustrates a dynamic, interconnected paradigm with Strategic Adaptation at its center, serving as the key organizing principle. It is encircled by three fundamental, interrelated elements: Human Resistance, Fostering Collaboration, and Automation. All these aspects are interconnected with the fundamental notion and with each other, creating a circular structure of reciprocal effect.

1. Strategic Adaptation, positioned at the core, signifies the organization's response system to external and internal challenges stemming from automation. It is influenced by and influences the surrounding factors.
2. The Human Resistance is linked to Strategic Adaptation through a bidirectional arrow, indicating that resistance presents a challenge to adaptation while simultaneously evolving

in reaction to adaptive tactics. Significant resistance may impede or distort adaptation, but effective strategies can mitigate fear, uncertainty, and opposition.

3. Fostering Collaboration is strategically positioned between Human Resistance and Automation, representing its intermediary function. It serves as a buffer and facilitator, bridging the chasm between human and technology realms. Collaborative initiatives such as participative decision-making, transparent communication, and cross-functional teams mitigate opposition and facilitate seamless automation integration.
4. Automation is linked to both Strategic Adaptation and Human Resistance. Automation necessitates adaptation via technical enhancements, business model evolution, and process reconfiguration. Conversely, it frequently incites human opposition because to perceived dangers to employment stability and personal identity. Nonetheless, when implemented jointly, automation can improve employee experience rather than diminish it.
5. The entire system is encapsulated in a feedback loop, underscoring that this is neither a linear nor a singular process. Alterations in one domain (e.g., heightened automation) influence others (e.g., necessitating renewed collaborative endeavors and strategic realignment).

This demonstrates that successful strategic adaptation in the era of automation is not merely a technical or managerial task, but a human-centered, systemic undertaking necessitating alignment among innovation, emotion, culture, and collaboration.

Case Study

Siemens AG – Human-Centric Automation and Strategic Adaptation

Siemens AG, a prominent entity in industrial manufacturing and digital solutions, exemplifies the interrelation of strategic adaptability, human resistance, collaboration, and automation in practice.

Siemens implemented advanced automation and AI-driven technology in its manufacturing facilities as part of its transition to Industry 4.0. This transition incited initial human opposition, particularly among factory-floor workers, who feared job displacement and lacked confidence in managing intricate technologies. Siemens adopted a strategic adaption methodology based on inclusive innovation rather than enforcing automation as a top-down directive. The corporation initiated a comprehensive workforce development program, encompassing reskilling efforts, digital academies, and ongoing feedback mechanisms that engage employees in the automation process. These initiatives aimed to promote collaboration and alleviate opposition. Employees were not only instructed but also invited to collaboratively create new workflows, enhancing their faith in the process and elevating engagement. Simultaneously, Siemens employed automation not to supplant labor but to enhance human capabilities specifically by implementing collaborative robots (cobots) to assist technicians with physically strenuous jobs. This fostered a vision of technology as an enabler rather than a menace. Siemens has shown that a unified strategy, which acknowledges and integrates human considerations while fostering technical adoption, results in enduring transformation. The company's sustained success in maintaining this equilibrium underscores the necessity of perceiving automation as an integral component of a socio-technical system, rather than merely a technological enhancement.

THEORETICAL FRAMEWORK

A comprehensive theoretical framework is crucial to elucidate the mechanisms of strategic adaptability within the intricate dynamics of automation, human resistance, and collaboration. This chapter utilizes various existing organizational and behavioral theories to underpin its research and provide a multifaceted perspective. These theories offer the conceptual framework for comprehending how individuals and institutions react to technological change and how leadership can effectively navigate these transitions.

Lewin's Change Management Theory

Kurt Lewin's Change Management Theory is a fundamental model utilized in this framework. This model delineates three phases Unfreeze, Change, and Refreeze that facilitate comprehension of how organizations transition from an existing condition to a desired future state (Parasuraman & Riley, 1997). Within the realm of automation: Unfreeze entails equipping the organization to acknowledge the necessity of change, including the recognition of the inevitability of automation. Change involves the introduction of new systems, procedures, and technology. Refreeze aims to stabilize the organization post-change by institutionalizing new behaviors and norms. Strategic adaptation is strongly aligned with this approach, especially in addressing the resistance frequently encountered throughout the unfreezing and transformation phases.

Technology Acceptance Model (TAM)

The Technology Acceptance Model, asserts that an individual's intention to utilize a new technology is determined by two elements: Perceived Usefulness (PU) and Perceived Ease of Use (PEOU) (Orlikowski, 2000). This approach elucidates why certain employees oppose automation; they may view the technology as cumbersome or inconsequential to their responsibilities. In strategic adaptation, enhancing perceptions of utility and usability is crucial through training, participation in decision-making, and iterative feedback mechanisms.

Theory of Socio-Technical Systems

The Socio-Technical Systems (STS) Theory underscores the interrelation of social and technical components within an organization. STS posits that effective change initiatives must account for both technical elements (e.g., software, hardware) and social systems (e.g., culture, communication, power dynamics) (Moore, 1995). Automation cannot not be considered in isolation. It must be assimilated with human roles, team configurations, and organizational principles. STS theory posits that collaboration is crucial for mitigating the negative impacts of automation and fostering systemic coherence.

Theory of Psychological Contracts

The Psychological Contract Theory addresses the implicit expectations between employers and employees (Nadler & Tushman, 1997). Automation frequently undermines these psychological contracts by altering job requirements or abolishing positions, resulting in mistrust or disengagement. Strategic adaptation must consequently entail clear communication and the renegotiation of employee expectations. Leaders ought to emphasize the significance of human contributions despite the automation of work.

Adaptive Structuration Theory

Adaptive Structuration Theory (AST), examines the appropriation and adaptation of technologies within organizational contexts (Porter & Heppelmann, 2014). The theory asserts that technologies are not only adopted as designed but rather reconfigured by users' actions, intents, and interpretations. This approach elucidates how human resistance or collaboration can either bolster or weaken the strategic implementation of automation. It underscores the significance of engaging users in the development of technology utilization, so diminishing resistance and cultivating a sense of ownership. These ideas jointly facilitate a multifaceted comprehension of automation's effects and offer a framework for measures that reconcile technological advancement with human involvement. The theoretical framework supports the forthcoming talks on practical approaches, strategic interventions, and collaborative models.

> *"The intricacy of automation necessitates theories that elucidate not only the utilization of technology but also encompass human significance, conduct, and opposition." (Orlikowski, 2000)*

Applications for Managing Human Resistance and Fostering Collaboration through Automation

To adeptly manage organizational change in the era of automation, firms are utilizing a range of intelligent technologies that fulfill two functions: mitigating human resistance and improving collaborative participation. These applications frequently utilize adaptive technologies such as machine learning, robotic process automation (RPA), collaborative robots (cobots), and natural language processing (NLP) that are specifically engineered to automate routine tasks while also engaging meaningfully with human workers. The objective has shifted from complete automation to augmented intelligence, wherein automation enhances human functions instead of supplanting them. This alignment reduces employee fear, promotes inclusivity, and facilitates simpler transitions to digitally connected workplaces.

Overview of Algorithms Utilized in Automation

Numerous algorithmic models underpin these applications. Among these, the most significant are:

Supervised Machine Learning Algorithms. These are employed in predictive systems that forecast human behavior, including resistance levels, learning curves, or productivity declines. Algorithms such as decision trees, random forests, and support vector machines facilitate the personalization of change interventions.

Natural Language Processing (NLP). NLP algorithms examine employee feedback, chat bot interactions, and internal communications to identify resistance indicators, emotional tone, or obstacles to collaboration. Sentiment analysis assists managers in pinpointing the origins of resistance.

Reinforcement Learning. Reinforcement learning, utilized in autonomous systems that learn via interaction, enables collaborative robots to modify their behavior according to human feedback or performance results.

Robotic Process Automation (RPA). RPA employs rule-based logic to automate monotonous business operations. When implemented judiciously, it alleviates burden and stress frequently a primary cause of opposition enabling individuals to concentrate on strategic or creative endeavors.

Human-in-the-Loop Artificial Intelligence Models. These algorithms facilitate the collaborative decision-making process between humans and AI systems. This enhances transparency and trust essential elements in diminishing resistance and encouraging participation.

Framework for Implementing Automation to Address Resistance and Facilitate Collaboration

An effective implementation framework must adopt a staged, human-centric strategy, guaranteeing that the deployment of automation is both technically robust and socially inclusive. This method is delineated in a five-step framework:

- **Evaluate Organizational Preparedness:** Perform workforce sentiment analysis utilizing NLP techniques. Assign roles and responsibilities to pinpoint opportunities for automation and areas necessitating human discernment. Evaluate digital maturity and pinpoint skill deficiencies.
- **Formulate a Human-Centric Automation Strategy:** Synchronize automation objectives with employee principles. Establish the limits of automation while maintaining human decision-making opportunities. Engage employees in co-design workshops to comprehend their issues and suggestions.
- **Develop Training and Change Management Protocols:** Implement tailored training modules with AI to individualize educational trajectories. Utilize chat bots and collaborative platforms for immediate assistance and peer mentorship. Facilitate clear communication, demonstrating how automation augments rather than supplants roles.
- **Implement Automation Gradually:** Initiate pilot programs in locations with minimal resistance. Employ RPA bots or collaborative robots to manage mundane tasks while employees monitor or engage in collaboration. Continuously gather feedback and modify algorithms according on user experience.
- **Observe, Assess, and Expand:** Utilize dashboards and artificial intelligence monitoring tools to evaluate performance, collaboration metrics, and resistance patterns. Consistently enhance automation algorithms and workflow architecture informed by data and employee input. Only proceed with scaling the effort if social and technological alignment has been verified.

This human-aligned application paradigm transcends mere task automation; it fosters digital trust, enhances collaboration, and mitigates psychological resistance by empowering employees with agency, clarity, and collective ownership in the change process.

RESEARCH METHODOLOGY

This chapter utilizes a qualitative, exploratory research methodology to examine the intricate relationship between automation, human resistance, and collaborative methods in the realm of strategic adaptation. A qualitative methodology was selected for its efficacy in elucidating the intricacies and contexts of human experiences, organizational behaviors, and cultural reactions to technology transformation. In contrast to quantitative research, which aims to generalize results through numerical data, this method facilitates a comprehensive and nuanced comprehension of how individuals and organizations respond to the challenges and opportunities presented by automation. The research utilizes secondary data sources, including academic articles, industry reports, white papers, and practical case studies from industries such as manufacturing, healthcare, information technology, and finance. These materials were chosen from reputable academic databases and organizational repositories, emphasizing documents released within the last decade to address current concerns and developments. The chosen literature offers insights into global perspectives and local practices, enhancing the analysis with varied viewpoints on automation and its effects. The analysis employed theme analysis, a prevalent qualitative method that entails detecting, organizing, and interpreting patterns within the data. Principal topics, including automation as a disruptive force, categories and origins of human resistance, and strategies for promoting collaboration, were identified and aligned with the previously stated conceptual and theoretical frameworks. This facilitated the development of a cohesive and interrelated comprehension of how strategic adaptation occurs across many organizational settings.

The selected methodology facilitates thorough investigation; however, it possesses inherent limits. The utilization of secondary data may exhibit the biases or constraints of the original writers, and the qualitative aspect of the research limits its generalizability across all industries or geographic areas. Nonetheless, the depth and contextual richness it provides are crucial for examining emerging and complex issues such as automation and human dynamics. Ethical issues were upheld throughout the research process. Only publicly accessible or suitably licensed materials were utilized, and meticulous attention was given to ensuring the correctness and integrity of the contained information. This analytical framework underpins the formulation of actionable insights and strategies for enterprises maneuvering through the swiftly evolving technological environment.

STRATEGIC ADJUSTMENT IN THE AGE OF AUTOMATION

Strategic adaptation denotes the realignment of organizational strategies and structures in reaction to evolving internal and external contexts, especially technology disruptions like automation (Porter & Heppelmann, 2014). The implementation of automation technologies, including AI, machine learning, and robotics, necessitates that firms modify their management strategies, processes, and employee engagement practices. In the absence of strategic adaptation, automation may lead to operational inefficiencies, employee discontent, and organizational misalignment (Brynjolfsson & McAfee, 2014).

Principal Factors Influencing Strategic Adaptation

Multiple essential factors affect strategic adaptation in the era of automation:

Technological Advancements: Innovations in artificial intelligence, robots, and the Internet of Things necessitate that enterprises modify their plans to include automation technologies. Businesses must remain competitive by utilizing these advances to improve operational efficiency and consumer satisfaction (Gill, 2011).

- **Market Competition:** Automation frequently emerges as a strategic imperative as organizations endeavor to secure a competitive advantage. Companies that neglect automation risk lagging behind rivals that use new technology to optimize processes and reduce expenses (Zero & Summers, 2024).
- **Customer Expectations:** Clients increasingly want expedited, tailored services, which automation can deliver. Organizations must align their strategies to leverage automation for enhancing customer experiences as they adjust to these demands (Porter & Heppelmann, 2014).

Elements of Strategic Adaptation in Automation

- **Leadership and Vision:** Leaders must possess a distinct vision for automation and articulate it effectively throughout the organization. The integration of automation technology may encounter resistance or misunderstanding in the absence of robust leadership and a well-articulated strategic goal (Kotter, 1996). The efficacy of strategic adaptation relies on the alignment of the organization's culture with automation objectives. Organizations that cultivate an environment of innovation and perpetual learning are more inclined to excel in the adoption of new technology (Kranz et al., 2016). Modifications in organizational structure may involve reconfiguring workflows and establishing new roles to integrate automated technology (Harari, 2018).
- **Training and Development:** The staff must possess the requisite abilities to collaborate effectively with automated technologies. This necessitates continuous training initiatives and upskilling opportunities to equip individuals for new jobs arising from automation (Davenport & Ronanki, 2018).

Case Studies on Strategic Adaptation

Numerous companies have effectively managed the shift to automation by strategic adaptation. General Electric (GE) implemented new manufacturing technologies, such as 3D printing and the Internet of Things (IoT), to improve its manufacturing capabilities. By synchronizing their organizational strategies with these technological innovations, GE successfully optimized production, enhanced quality, and minimized costs (Chui, Manyika, & Miremadi, 2016). Likewise, Amazon has persistently modified its methods to integrate robotics and artificial intelligence into its delivery facilities. By means of strategic adaptation, Amazon has preserved its competitive advantage in the e-commerce sector by enhancing delivery speed and operating efficiency (Cappelli & Tavis, 2018).

MANAGING HUMAN RESISTANCE

Comprehending Human Resistance to Automation

Human resistance to automation is a multifaceted issue stemming from several psychological, social, and organizational elements. Employees may have apprehension regarding job displacement, diminished autonomy, or a decline in their significance within the firm (Ford, 2016). If not adequately addressed, these anxieties may result in disengagement, diminished morale, and heightened turnover (Susskind & Susskind, 2015). Resistance is likewise founded on a lack of comprehension regarding automation and its prospective advantages, together doubt concerning its execution (Ghosh, 2020).

Psychological Motivations for Resistance

A primary psychological motivator of resistance is employment instability. As automated technologies execute duties previously managed by humans, employees frequently express concerns around potential job loss or diminished work hours (Acemoglu, & Restrepo, 2019). The apprehension about being supplanted by machines can diminish employee motivation and engagement. A further psychological factor is the loss of control. Employees accustomed to overseeing their work processes may experience alienation when automation systems assume essential functions (Susskind & Susskind, 2015). The perception of losing control might result in resistance, since employees may view the change as a threat to their identity and autonomy.

Approaches to Mitigating Resistance

Organizations can implement various measures to alleviate human opposition to automation:

- **Effective Communication:** Articulate and transparent discourse regarding the rationale for automation and its prospective advantages can alleviate apprehension and foster trust. Employees are more inclined to embrace change when they comprehend how automation would augment their employment instead of supplanting it (Kotter, 1996).
- **Engagement in Decision-Making:** Including employees in the decision-making process fosters a sense of ownership and control during the shift. By seeking input and addressing concerns promptly, organizations can mitigate feelings of exclusion (Tapscott & Tapscott, 2016).
- **Retraining and Upskilling:** Providing individuals with opportunity to develop new competencies through training programs can facilitate their transfer into emerging positions resulting from automation (Buchanan & Huczynski, 2019). This ensures employees that they are not being abandoned but are instead being equipped for new challenges.

FOSTERING COLLABORATION IN AUTOMATION

The Notion of Human-Robot Collaboration

Human-robot cooperation (HRC) denotes the cooperative interaction between humans and robots, wherein both entities enhance each other's capabilities to accomplish shared objectives (Kaplan & Haenlein, 2020). In the realm of automation, cooperation manifests in several forms, including co-working with robots in production settings and joint decision-making procedures that incorporate AI (Davenport & Ronanki, 2018).

Advantages of Human-Robot Collaboration

- Numerous benefits exist in promoting collaboration between people and robots:

Enhanced Efficiency: The effective collaboration between people and robots enables the completion of tasks more swiftly and correctly than each could achieve independently (Brynjolfsson & McAfee, 2014).

- **Improved Safety:** Robots can assume perilous duties, such as handling hazardous substances or operating in harsh environments, enabling people to concentrate on activities that necessitate cognitive abilities or inventiveness (Landers & Marin, 2021).
- **Innovation and Creativity:** Automation technology can manage monotonous chores, allowing human workers to concentrate on advanced creative and strategic endeavors that necessitate critical thinking and problem-solving abilities (Levy & Murnane, 2013).

SOLUTIONS AND RECOMMENDATIONS

As enterprises increasingly incorporate automation technology, it is essential to establish effective ways to mitigate human resistance and cultivate a collaborative work environment. This section provides practical answers and recommendations to assist firms in surmounting obstacles to automation adoption.

Incorporating Change Management Principles

Integrating change management principles is one of the most successful techniques for overseeing the shift to automation. Kotter's (1996) eight-step model of change management offers a systematic framework to assist firms in navigating technological transformation. The following steps encompass:

1. **Establishing a sense of urgency:** Leaders must convey the necessity of automation to staff, highlighting the advantages it will provide to the firm and their individual positions.
2. **Building a guiding coalition:** Effective implementation of automation necessitates endorsement from essential stakeholders, including executives, managers, and prominent personnel.

3. **Formulating a vision and strategy:** A definitive vision for automation must be developed, describing the objectives and advantages of the transition.
4. **Articulating the vision:** It is imperative to consistently convey the vision for automation to all employees and stakeholders to guarantee alignment and commitment.
5. **Facilitating comprehensive action:** Employees must be provided with the resources and authority to make decisions concerning automation.
6. **Producing immediate victories:** Showcasing first achievements of automation may sustain momentum and garner staff support.
7. **Consolidating achievements and fostering further transformation:** Automation must be embedded within the organization's fundamental operations and subject to ongoing enhancement.
8. **Institutionalizing new methodologies inside the culture:** Ultimately, automation must be ingrained as a lasting component of the company culture, sustained by ongoing leadership endorsement.

Executing these measures will enable firms to address resistance more efficiently and facilitate the seamless incorporation of automated technology (Kotter, 1996).

Fostering Employee Engagement and Inclusion

A crucial advice is to enhance employee engagement during the automation process. Employees that actively participate in the change process are more inclined to accept automation, perceiving it as an opportunity rather than a threat (Zuboff, 2019). This can be accomplished by:

1. **Facilitate open discussions and feedback sessions:** Encouraging employees to express their problems and suggestions foster trust and transparency.
2. **Incorporating employees in decision-making:** Automation fosters a sense of ownership, hence increasing their likelihood of endorsing the changes.
3. **Incentive and recognition programs:** Offering awards to employees who effectively adjust to automation can encourage others to emulate their behavior.

Utilizing Training and Development Initiatives

Investing in training and development is essential for the success of automation projects. By providing ongoing learning opportunities, organizations can assist individuals in acquiring the skills required to collaborate with automated systems. Programs ought to concentrate on:

1. **Enhancing the skills of current employees:** This involves instructing personnel on new tools, technologies, and systems that will be incorporated into their responsibilities.
2. **Reskilling for new roles:** As automation alters employment, workers may require a move to different positions. Offering reskilling opportunities guarantees that employees maintain their employability and contribute value to the firm (OECD, 2019).
3. **Promoting a culture of continuous learning:** Motivating employees to perceive learning as a perpetual process rather than a singular occurrence can enhance their competitiveness in a constantly evolving technology environment (Vial, 2019).

Highlighting the Human Component

Notwithstanding the proliferation of automation, firms must prioritize the human component in their operations. Although automated systems can do numerous jobs, humans retain distinct abilities, like creativity, critical thinking, and emotional intelligence, which are not easily emulated by robots (Mckinsey & Company, 2021). Leaders should motivate employees to utilize these abilities to enhance the existing automation systems. Collaboration between human workers and robots can yield superior innovation and outcomes compared to exclusive reliance on technology.

BEST PRACTICES FOR EFFECTIVELY INCORPORATING AUTOMATION

The incorporation of automation technology into business processes necessitates meticulous design and implementation. Numerous best practices can assist firms in effectively implementing automation, reducing opposition, and promoting collaboration.

Initiate Modestly and Expand Incrementally

An excellent best practice is to commence modestly and expand incrementally. Organizations should initiate pilot projects or restricted automation initiatives instead of aiming to automate all business activities simultaneously. This enables employees to engage with automation directly without being daunted by the transition. Incremental scaling allows firms to enhance their automation plans, informed by initial achievements and obstacles (Deloitte, 2020).

Facilitate Unambiguous Communication

Lucid and coherent communication is essential for the effective implementation of automation. Organizations must convey the objectives, advantages, and effects of automation to all personnel. This transparency helps mitigate apprehensions and foster trust in the process (Kotter, 1996). Furthermore, consistent updates regarding the advancement of automation programs and the results attained may ensure staff remain informed and interested.

Promote a Cooperative Environment

Promoting a collaborative culture is crucial for the integration of automation within the workplace. This entails establishing a setting in which personnel collaborate with automated systems and assist one another in attaining shared objectives. Promoting collaboration and knowledge exchange between humans and robots can lead to enhanced automation and increased job satisfaction for individuals (Davenport & Bean, 2025).

Ensure Sufficient Support

Organizations must furnish employees with essential support during the automation process. This encompasses providing training, guidance, and tools to assist staff in acclimating to new technologies. Offering explicit guidance and support can facilitate a smoother transition and diminish resistance (Susskind & Susskind, 2015).

ADDRESSING ETHICAL CHALLENGES IN AUTOMATION

As automation technologies progressively transform industries, it is imperative to address the ethical consequences. This section examines the ethical dilemmas associated with automation, encompassing issues of employment displacement, privacy, and bias, and suggests ways for mitigating these obstacles.

Employment Displacement and Economic Disparity

A primary ethical concern about automation is job displacement. As machines and robots assume duties historically executed by people, several workers apprehend job displacement, perhaps resulting in heightened economic inequality (Autor, 2015). The ethical quandary pertains to how enterprises can adopt automation without worsening social inequality or disregarding the welfare of displaced employees. Organizations can use retraining and reskilling programs to facilitate employee transitions into new roles. Governments and educational institutions are essential in equipping workers with the skills required for future employment (Brynjolfsson & McAfee, 2014). Moreover, enterprises might consider universal basic income (UBI) as a viable remedy to mitigate the economic disturbances induced by automation (Alvesson & Syeningsson, 2015).

Concerns Regarding Privacy and Data Security

The extensive implementation of AI, IoT, and machine learning in automated systems generates considerable privacy and data security issues. These technologies frequently gather extensive personal data, prompting inquiries regarding its utilization, storage, and safeguarding (Zuboff, 2019). Organizations must comply with stringent data privacy requirements and establish comprehensive security processes to safeguard sensitive information. Transparency is essential in managing data privacy. Organizations must communicate to employees and consumers the intended use of their data and grant them authority over their information. Ethical data practices must be integrated into the organizational culture, and data collecting should be restricted to what is essential for the effective operation of the automation system (Westerman, Bonnet, & McAfee, 2014).

Algorithmic Bias

A significant ethical issue is the possibility of prejudice in automated decision-making systems. Artificial intelligence and machine learning algorithms may inadvertently reinforce existing prejudices if the training data is faulty or prejudiced (Brougham & Haar, 2018). This may result in discriminatory practices in recruitment, advancement, and customer service, among other domains. To alleviate algorithmic bias, enterprises must utilize varied datasets in the training of their AI systems. Systematic evaluations of AI systems must be performed to identify and rectify any biases that may arise. Organizations must prioritize ethical AI development by implementing frameworks that highlight fairness, responsibility, and openness (Kranz et al., 2016).

Ethical Governance and Accountability

Ethical governance is essential for the responsible deployment of automation technology. Organizations must formulate ethical principles for automation and develop governance frameworks to supervise the use of these technologies. Ethical committees or advisory boards can offer supervision, ensuring that automation is implemented in accordance with the organization's values and social duties (Vial, 2019).

ENHANCING ACCESSIBILITY IN THE AGE OF AUTOMATION

Automation can provide substantial advantages to enterprises; but, these advantages must be attainable for all individuals, including those with disabilities or restricted access to new technologies. This section examines how firms can guarantee that automation technologies are inclusive and accessible to a varied array of employees and customers.

Creating Accessible Automation Systems

As automation increasingly permeates company processes, it is crucial to ensure that these systems are developed with accessibility as a priority. This entails ensuring that automated systems, including AI interfaces and robotic technologies, are accessible and user-friendly for individuals with diverse abilities. Voice-activated systems, screen readers, and adaptive devices enhance automation accessibility for those with visual or motor disabilities (Tapscott & Tapscott, 2016).

Organizations want to perform accessibility audits to assess if their automation systems comply with accessibility guidelines. This may involve evaluating automation systems with various user demographics to detect possible obstacles and guarantee inclusivity (Marr, 2020).

Comprehensive Workforce Strategies

Alongside the development of accessible technologies, enterprises must implement inclusive workforce practices that allow all employees to engage in the automation process. This encompasses providing training programs specifically designed for individuals with disabilities and guaranteeing that automation does not adversely impact specific workforce demographics (Levy & Murnane, 2013). Employees with impairments ought to receive equitable opportunity for reskilling and professional advancement in automated settings.

Client Accessibility

Automation technologies must be developed with client accessibility as a priority. Customer service automation, including chat bots and AI-driven support systems, must be accessible to those with disabilities. Offering several communication channels and guaranteeing that automation does not marginalize specific customer segments is essential for maintaining fair access to services (Landers & Marin, 2021).

FUTURE RESEARCH DIRECTIONS

Future study in strategic adaptation and automation should extend beyond traditional concerns of efficiency and productivity to investigate the profound, systemic effects of automation on organizational ecosystems. Researchers and professionals must examine the effects of ongoing automation on workplace identity, human dignity, and the sustainability of the workforce throughout time. As AI and robotics become more integrated into decision-making processes, academics must examine their impact on psychological safety, employee autonomy, and job happiness. Specifically, research should investigate the impact of automation on organizational trust pertaining to both leadership and the technology itself. Moreover, forthcoming study ought to employ a cross-sectoral and cross-cultural methodology, examining the disparities in automation adoption across established and emerging countries, public and private sectors, as well as small and large firms. Analyzing adaptive responses across various institutional, regulatory, and socio-economic contexts will yield more inclusive and internationally pertinent frameworks. Research might be broadened to evaluate the influence of automation on organizational resilience in crises such as pandemics or global market disruptions where automation may either enhance stability or intensify vulnerabilities. Examining the relationship between automation and emerging workforce patterns, such as gig employment and remote work, may enhance comprehension of strategic adaptability in this contemporary era.

EMERGING TRENDS IN AUTOMATION

The automation sector is advancing swiftly, with growing trends that are influencing the future of commerce and society. This section analyzes key trends in automation.

Artificial Intelligence and Machine Learning

Artificial intelligence and machine learning are leading the automation revolution. These technologies allow machines to learn from data, adjust to novel circumstances, and make judgments autonomously (Cappelli & Tavis, 2018). Artificial intelligence is employed in diverse applications, ranging from customer service chat bots to predictive analytics in supply chain management.

Robotic Process Automation (RPA)

Robotic Process Automation (RPA) represents a notable trend in automation. Robotic Process Automation (RPA) employs software robots to automate monotonous, rule-governed processes. This technology is progressively being embraced in sectors including banking, healthcare, and manufacturing to optimize operations and minimize expenses (Almatrodi, Li, & Alojail, 2023).

Autonomous Systems

Autonomous technologies, such as self-driving automobiles and drones, are anticipated to be pivotal in automation in the forthcoming years. These systems have the capacity to transform sectors including transportation, logistics, and agriculture (Chui, Manyika, & Miremadi, 2016). As autonomous systems advance, they will necessitate additional legal frameworks and ethical concerns to guarantee their safe and responsible utilization.

Proposed Domains for Subsequent Investigation Numerous deficiencies remain in the existing literature, indicating potential avenues for further research. The intersection of leadership and emotional intelligence within the realm of automation is yet inadequately examined. Although technical competencies are crucial, it is the emotional and interpersonal qualities of leaders that frequently dictate the efficacy of change communication and adoption. Research should investigate the impact of various leadership styles on employee resilience, adaptability, and performance during transitions to automation. Secondly, the psychological ramifications of automation such as worry, identity erosion, fatigue, and diminished motivation require more thorough investigation, particularly among underrepresented populations and individuals in low-skilled positions. The digital gap and disparities resulting from automation require more focus to guarantee inclusive growth and fair access to reskilling possibilities. Research should investigate hybrid organizational models that integrate automation with human supervision in intricate decision-making processes. In crucial industries like healthcare and law enforcement, it is vital to comprehend how trust is formed in human-machine collaboration and which governance systems are most effective in maintaining accountability. An additional area of focus is the cultivation of adaptive organizational cultures. Cultural agility the capacity of businesses to modify their values, beliefs, and practices in reaction to technological disruption may be a crucial factor in achieving success. Case studies may investigate how corporate culture facilitates or obstructs strategic adaptation and whether culture can be deliberately cultivated to enhance creativity and psychological safety during periods of transition. Ultimately, comparative analyses of policy frameworks, regulatory support mechanisms, and national policies for automation adaption can illuminate macro-level actions that facilitate micro-level organizational transformation.

Incorporation of Novel Technologies Incorporating new technologies into business processes entails more than mere deployment; it necessitates the reengineering of workflows, the alignment of digital tools with strategic goals, and the cultivation of a culture that promotes innovation. As automation technologies advance, firms must implement a comprehensive integration strategy that encompasses human resource development, organizational learning, and change management.

A crucial element of integration pertains to interoperability and scalability. Technologies including robotic process automation (RPA), machine learning algorithms, cloud computing, and enterprise resource planning (ERP) systems must integrate with current infrastructure to facilitate seamless operations. Scalability enables firms to adjust these technologies as business requirements change, so avoiding premature obsolescence or underutilization. Employee empowerment using digital tools is equally significant. Organizations that involve employees in the design and implementation of automation systems frequently achieve elevated adoption rates and enhanced performance outcomes. Collaborative design approaches and the incorporation of user feedback mechanisms improve system usability and foster employee engagement. Furthermore, cybersecurity and data governance are essential factors. Given that automation frequently entails the management of sensitive data, safeguarding data privacy and security is essential for preserving trust and guaranteeing compliance. In industries such as healthcare and finance, characterized by stringent regulatory oversight, ethical data practices can function as both a compliance need and a competitive edge. Moreover, effective integration necessitates concurrent strategic talent development. This encompasses establishing internal avenues for skill enhancement, providing digital literacy initiatives, and forming interdisciplinary teams proficient in collaborating with machines. The integration of virtual reality (VR) and augmented reality (AR) in training programs is an emerging approach that improves experiential learning and optimizes retention. Ultimately, integration is an ongoing process necessitating feedback, iteration, and a readiness to accept both the potential and ambiguities of technological advancement.

CONCLUSION

In conclusion, strategic adaptation in the era of automation is no longer optional but essential. Organizations that do not adapt face obsolescence, but those that adopt a deliberate, human-centered strategy are more likely to succeed in the digital economy. The adaptation process is intricate and multifaceted, encompassing the modification of technology, individuals, procedures, and objectives. This chapter has illustrated that addressing human resistance is crucial to effective automation. Resistance is not indicative of weakness but rather a natural reaction to uncertainty and the perception of loss. It must be approached with compassion, clarity, and involvement. Equally significant is the development of collaborative methods that dismantle silos and promote co-creation between humans and robots, as well as within other organizational units.

Strategic adaptation necessitates ongoing investment in education, leadership, and ethical governance. As automation technologies advance from predictive analytics to generative AI and further our frameworks for guiding their integration must also develop. By forecasting future trends, doing proactive research, and emphasizing inclusive innovation, firms can establish themselves at the vanguard of this shift. The genuine potential of automation resides not in supplanting humans but in enhancing human creativity. Organizations that prioritize their adaptation strategies on individuals equipping them with resources, confidence, and a sense of purpose will not only adeptly manage change but also cultivate a more resilient, equitable, and motivated future of work.

REFERENCES

Acemoglu, D., & Restrepo, P. (2019). Artificial intelligence, automation, and work. NBER Working Paper No. 24196.

Ahuja, V., Yang, J., & Shankar, R. (2009). Study of ICT adoption for building project management in the Indian construction industry. *Automation in Construction, 18*(4), 415–423. DOI: 10.1016/j.autcon.2008.10.009

Almatrodi, I., Li, F., & Alojail, M. (2023). Organizational resistance to automation success: How status quo bias influences organizational resistance to an automated workflow system in a public organization. *Systems, 11*(4), 191. DOI: 10.3390/systems11040191

Alvesson, M., & Sveningsson, S. (2015). *Changing organizational culture: Cultural change work in progress.* Routledge. DOI: 10.4324/9781315688404

Autor, D. H. (2015). Why are there still so many jobs? The history and future of workplace automation. *The Journal of Economic Perspectives, 29*(3), 3–30. DOI: 10.1257/jep.29.3.3

Brougham, D., & Haar, J. (2018). Smart technology, artificial intelligence, robotics, and algorithms (STARA): Employees' perceptions of our future workplace. *Journal of Management & Organization, 24*(2), 239–257. DOI: 10.1017/jmo.2016.55

Brynjolfsson, E., & McAfee, A. (2014). *The second machine age: Work, progress, and prosperity in a time of brilliant technologies.* W.W. Norton.

Buchanan, D. A., & Huczynski, A. A. (2019). *Organizational behaviour* (10th ed.). Pearson Education.

Cappelli, P., & Tavis, A. (2018). HR goes agile. *Harvard Business Review, 96*(2), 46–52.

Chui, M., Manyika, J., & Miremadi, M. (2016). Where machines could replace humans—And where they can't (yet). *The McKinsey Quarterly.*

Davenport, T. H., & Bean, R. (2025). The GenAI focus shifts to innovation at Colgate-Palmolive. *MIT Sloan Management Review.*

Davenport, T. H., & Ronanki, R. (2018). Artificial intelligence for the real world. *Harvard Business Review, 96*(1), 108–116.

Deloitte. (2020). *Global Human Capital Trends: The social enterprise at work—Paradox as a path forward.* Deloitte Insights.

Drucker, P. F. (2007). *Management: Tasks, responsibilities, practices.* Transaction Publishers.

Ford, M. (2016). *Rise of the robots: Technology and the threat of a jobless future.* Basic Books.

Frey, C. B., & Osborne, M. A. (2017). The future of employment: How susceptible are jobs to computerisation? *Technological Forecasting and Social Change, 114*, 254–280. DOI: 10.1016/j.techfore.2016.08.019

Ghosh, R. (2020). Human capital management in the age of automation. *Journal of Human Resource and Sustainability Studies*, *8*(2), 191–204.

Gill, R. (2011). *Theory and practice of leadership*. SAGE Publications.

Goleman, D. (2006). *Emotional intelligence: Why it can matter more than IQ*. Bantam Books.

Graetz, G., & Michaels, G. (2018). Robots at work. *The Review of Economics and Statistics*, *100*(5), 753–768. DOI: 10.1162/rest_a_00754

Hammer, M., & Stanton, S. (1999). How process enterprises really work. *Harvard Business Review*, *77*(6), 108–118. PMID: 10662000

Harari, Y. N. (2018). *21 Lessons for the 21st Century*. Jonathan Cape.

Kaplan, A., & Haenlein, M. (2020). Rulers of the world, unite! The challenges and opportunities of artificial intelligence. *Business Horizons*, *63*(1), 37–50. DOI: 10.1016/j.bushor.2019.09.003

Kotter, J. P. (1996). *Leading change*. Harvard Business Press.

Kranz, J. J., Hanelt, A., & Kolbe, L. M. (2016). Understanding the influence of absorptive capacity and ambidexterity on the process of business model change. *Technological Forecasting and Social Change*, *103*, 104–113.

Landers, R. N., & Marin, S. (2021). Theory and design of digital nudges for behavior change: Implications for AI in HRM. *Applied Psychology*, *70*(3), 889–921.

Lengnick-Hall, C. A., & Beck, T. E. (2005). Adaptive fit versus robust transformation: How organizations respond to environmental change. *Journal of Management*, *31*(5), 738–757. DOI: 10.1177/0149206305279367

Levy, F., & Murnane, R. J. (2013). *Dancing with robots: Human skills for computerized work*. Third Way Foundation.

Marr, B. (2020). *Artificial intelligence in practice: How 50 successful companies used AI and machine learning to solve problems*. Wiley.

McKinsey & Company. (2021). The future of work after [McKinsey Global Institute.]. *COVID*, *•••*, 19.

Moore, M. H. (1995). *Creating public value: Strategic management in government*. Harvard University Press.

Nadler, D. A., & Tushman, M. L. (1997). *Competing by design: The power of organizational architecture*. Oxford University Press. DOI: 10.1093/acprof:oso/9780195099171.001.0001

Nof, S. Y., & Chen, F. F. (2007). Collaborative control theory for e-Work, e-Production, and e-Service. *Computers & Industrial Engineering*, *53*(2), 252–268.

OECD. (2019). *Preparing for the future of work*. OECD Publishing.

Orlikowski, W. J. (2000). Using technology and constituting structures: A practice lens for studying technology in organizations. *Organization Science, 11*(4), 404–428. DOI: 10.1287/orsc.11.4.404.14600

Parasuraman, R., & Riley, V. (1997). Humans and automation: Use, misuse, disuse, abuse. *Human Factors, 39*(2), 230–253. DOI: 10.1518/001872097778543886

Pfeffer, J. (1998). *The human equation: Building profits by putting people first.* Harvard Business School Press.

Porter, M. E., & Heppelmann, J. E. (2014). How smart, connected products are transforming competition. *Harvard Business Review, 92*(11), 64–88.

Susskind, R., & Susskind, D. (2015). *The future of the professions: How technology will transform the work of human experts.* Oxford University Press. DOI: 10.1093/oso/9780198713395.001.0001

Tapscott, D., & Tapscott, A. (2016). *Blockchain revolution: How the technology behind bitcoin is changing money, business, and the world.* Portfolio.

Vial, G. (2019). Understanding digital transformation: A review and a research agenda. *The Journal of Strategic Information Systems, 28*(2), 118–144. DOI: 10.1016/j.jsis.2019.01.003

Westerman, G., Bonnet, D., & McAfee, A. (2014). *Leading digital: Turning technology into business transformation.* Harvard Business Review Press.

Zero, N., & Summers, J. D. (2024). Aligning a collaborative resistance model with a change management process: A case study in automation in industry. *Journal of Engineering Design, 35*(2), 1–25. DOI: 10.1080/09544828.2024.2387218

Zuboff, S. (2019). *The age of surveillance capitalism: The fight for a human future at the new frontier of power.* PublicAffairs.

KEY TERMS AND DEFINITIONS

Strategic Adaptation: The intentional organizational process of modifying objectives, procedures, and structures in reaction to environmental and technological transformations, particularly automation.

Automation: The application of technologies, including artificial intelligence, robotics, and software, to execute jobs typically performed by humans.

Human Resistance: The psychological and behavioral opposition exhibited by employees in response to change, particularly in relation to technological transition.

Digital Transformation: The incorporation of digital technology across all facets of an organization, significantly altering its operations and value delivery.

Human-Centric Design: A framework for developing systems and processes that prioritize the needs, competencies, and experiences of individuals in their interface with technology.

Generative AI: A branch of artificial intelligence capable of producing text, graphics, and various media in response to prompts, frequently employed in creative and strategic endeavors.

Organizational Culture: The collective values, beliefs, and norms that shape the interactions and operations of an organization's members.

Agile Transformation: The transition in organizational structure and mentality towards agile principles—adaptability, collaboration, and iterative development—to improve responsiveness to change.

Upskilling: Educating individuals in new competencies or enhancing existing skills to meet evolving job demands.

Workplace Collaboration: Cooperative endeavors among employees or between people and machines to attain shared objectives effectively and creatively.

Ethical AI: AI systems designed and utilized with a pronounced focus on fairness, transparency, accountability, and the reduction of bias.

Digital Literacy: The proficiency in utilizing digital tools efficiently and securely within a professional or educational context.

Emotional Intelligence: The capacity to identify, comprehend, and regulate one's own emotions as well as those of others, particularly vital in leadership during periods of transition.

Human-Machine Collaboration: Synergistic interaction between humans and intelligent machines designed to optimize productivity and decision-making.

Change Management: The practice of directing and assisting individuals, teams, and organizations during organizational transformations.

Reskilling: Instructing employees in new competencies to undertake alternative roles, frequently as a reaction to automation or shifts in industry requirements.

Adaptive Leadership: A leadership paradigm that prioritizes responsiveness, learning, and the capacity to manage intricate transformations.

Remote Work: A work arrangement in which people execute their job responsibilities outside conventional office settings, frequently facilitated by digital means.

Process Innovation: The adoption of novel or substantially enhanced production or delivery techniques that influence efficiency and value generation.

Organizational Resilience: The ability of an organization to foresee, prepare for, react to, and recuperate from disturbances while preserving essential functions.

Chapter 14
The Augmented Operator: Enhancing Human Capabilities

Aditya Saha
https://orcid.org/0009-0004-1783-8587
Brainware University, India

Ranjan Banerjee
https://orcid.org/0009-0003-1950-7530
Brainware University, India

Debesh Das
https://orcid.org/0009-0009-0287-4995
Brainware University, India

Misha Samanta
https://orcid.org/0009-0000-6037-2760
Brainware University, India

Malay Maity
https://orcid.org/0009-0001-9173-6763
Brainware University, India

Milan Sahoo
https://orcid.org/0009-0007-0619-3682
Brainware University, India

Salankara Sarkar
https://orcid.org/0009-0005-0123-3724
Brainware University, India

ABSTRACT

The concept of the "augmented operator" represents a paradigm shift in how humans interact with technology, aiming to enhance rather than replace human capabilities. This involves the integration of advanced technologies, such as augmented reality (AR), artificial intelligence (AI), and wearable sensors, into the operator's workflow. By overlaying digital information onto the real-world environment, AR provides operators with real-time guidance, instructions, and data, enabling them to perform complex tasks with increased accuracy and efficiency. AI algorithms can analyze vast amounts of data to provide predictive insights, optimize decision-making, and automate repetitive tasks, freeing up operators to focus on more critical and creative aspects of their work. Wearable sensors monitor the operator's physiological state, providing valuable data on their well-being and performance, and enabling proactive interventions to prevent fatigue or injury.

DOI: 10.4018/979-8-3373-2985-7.ch014

INTRODUCTION

The evolution of technology has consistently reshaped the landscape of human work. From the industrial revolution to the digital age, each technological advancement has presented both challenges and opportunities. Today, we stand at the cusp of a new era defined by the "augmented operator," a concept that reimagines the human-technology relationship. Instead of simply automating tasks and potentially displacing human workers, the augmented operator framework focuses on enhancing human capabilities through the seamless integration of advanced technologies. This paper explores the multifaceted nature of this concept, examining the key technologies involved, its applications across various industries, and its implications for the future of work.

KEY TECHNOLOGIES ENABLING THE AUGMENTED OPERATOR

Several key technologies underpin the development and implementation of the augmented operator paradigm.

Augmented Reality (AR)

Augmented Reality (AR) is no longer a futuristic concept confined to science fiction. It has permeated various aspects of our lives, transforming how we interact with information and the world around us. At its core, AR is a technology that overlays digital content onto the real-world environment, creating an enhanced or augmented experience. This is achieved through devices like smartphones, tablets, and specialized AR headsets, which utilize cameras, sensors, and software to detect and interpret the user's surroundings. The result is a seamless integration of virtual elements, such as images, videos, 3D models, and text, into the user's perception of reality.

The fundamental principle behind AR lies in its ability to bridge the gap between the physical and digital domains. Unlike Virtual Reality (VR), which immerses users in a completely simulated environment, AR enhances the existing reality by adding layers of digital information. This augmentation can range from simple overlays, like displaying directions on a smartphone screen, to complex interactive experiences, such as placing virtual furniture in a room or simulating surgical procedures. The key is that the user remains aware of their physical surroundings while simultaneously engaging with digital augmentation.

One of the defining characteristics of AR is its real-time interactivity. The digital content adapts dynamically to the user's movements and the changes in their environment. This responsiveness is crucial for creating a truly immersive and engaging experience. For instance, when using an AR app to identify constellations in the night sky, the virtual labels and outlines of the stars adjust in real-time as the user moves their device. This interactivity is powered by sophisticated algorithms that track the user's position and orientation, ensuring that the digital content remains accurately aligned with the physical world (Acikgoz, 2019).

Key Aspects of Augmented Reality

- **Overlay of Digital Information:** AR superimposes virtual elements onto the real world, enriching the user's perception. This can include text, images, videos, 3D models, and interactive elements.
- **Real-Time Interactivity:** The digital content responds dynamically to the user's movements and changes in the environment, creating a seamless and engaging experience.
- **Context-Awareness:** AR leverages location, surroundings, and user actions to provide relevant and personalized information.
- **Device Independence:** AR can be accessed through various devices, including smartphones, tablets, and specialized AR headsets.
- **Enhancement of Reality:** AR enhances the existing reality rather than replacing it, allowing users to remain aware of their physical surroundings.

Applications of Augmented Reality

The applications of AR are diverse and continue to expand as technology evolves. Here are some prominent examples:

- **Gaming and Entertainment:** AR has revolutionized gaming by bringing virtual characters and environments into the real world. Games like Pokémon GO popularized the concept of location-based AR gaming, while other AR apps offer immersive storytelling experiences.
- **Retail and E-commerce:** AR enables customers to virtually try on clothing, visualize furniture in their homes, and explore product details in an interactive way. This enhances the shopping experience and reduces the likelihood of returns.
- **Education and Training:** AR can create interactive and engaging learning experiences by overlaying 3D models and simulations onto textbooks and real-world objects. This is particularly valuable for subjects like science, engineering, and medicine.
- **Healthcare:** AR is used in surgical procedures, medical training, and patient rehabilitation. It can overlay vital information onto the surgeon's view, guide medical students through complex procedures, and assist patients with physical therapy.
- **Navigation and Transportation:** AR-powered navigation systems provide turn-by-turn directions and real-time traffic information directly onto the user's view of the road. AR is also being explored for applications in autonomous vehicles and public transportation.
- **Manufacturing and Maintenance:** AR can provide workers with real-time instructions and guidance for complex tasks, improving efficiency and reducing errors. It can also be used for remote maintenance and troubleshooting.

Challenges and Future Directions

Despite its immense potential, AR faces several challenges. These include:

- **Technological Limitations:** Current AR devices can be bulky and uncomfortable to wear for extended periods. The accuracy and reliability of AR tracking and rendering also need to be improved.
- **Content Creation:** Creating high-quality AR content can be time-consuming and expensive. Tools and platforms that simplify content creation are needed.
- **Privacy Concerns:** AR devices can collect vast amounts of data about the user's surroundings and behavior, raising privacy concerns. Secure data management and user consent are crucial.
- **Social Acceptance:** AR's potential to blur the lines between reality and virtuality raises ethical and social concerns. Addressing these concerns through education and open dialogue is essential.

Looking ahead, AR is poised to become an even more integral part of our lives. Advancements in hardware, software, and artificial intelligence will drive further innovation and expand the range of applications. The development of lightweight and comfortable AR glasses will enable seamless integration of AR into everyday activities. The integration of AR with other emerging technologies, such as 5G, cloud computing, and AI, will unlock new possibilities, (Ambrose & Kulik, 1993). As AR technology continues to evolve, it will undoubtedly transform how we interact with information, communicate with each other, and experience the world around us.

Artificial Intelligence (AI)

Artificial Intelligence (AI) has transitioned from a science fiction concept to a pervasive force shaping our present and future. It encompasses a broad spectrum of technologies aimed at replicating human intelligence in machines, enabling them to perform tasks that typically require human cognition. From the algorithms powering search engines to the sophisticated systems driving autonomous vehicles, AI's influence is undeniable. Understanding its core principles, applications, and ethical considerations is crucial in navigating this rapidly evolving landscape.

At its heart, AI seeks to create machines capable of learning, reasoning, problem-solving, perception, and even creativity. This is achieved through various techniques, most notably machine learning (ML), which allows systems to learn from data without explicit programming. Deep learning (DL), a subset of ML, employs artificial neural networks with multiple layers to analyze complex patterns, enabling breakthroughs in areas like image recognition and natural language processing.

Key Components and Concepts

- **Machine Learning (ML):**
    - o Focuses on enabling computers to learn from data without being explicitly programmed.
    - o Algorithms identify patterns and make predictions based on data.
    - o Types include supervised learning, unsupervised learning, and reinforcement learning.
- **Deep Learning (DL):**

- A subset of ML using artificial neural networks with multiple layers (deep neural networks).
  - Excels at processing complex data like images, audio, and text.
  - Drives advances in areas like computer vision and natural language processing.
- **Natural Language Processing (NLP):**
  - Enables computers to understand, interpret, and generate human language.
  - Powers applications like chatbots, virtual assistants, and language translation.
  - Crucial for human-computer interaction.
- **Computer Vision:**
  - Enables computers to "see" and interpret images and videos.
  - Used in applications like facial recognition, object detection, and autonomous driving.
  - Relies on deep learning for accurate image analysis.

Applications Across Industries

AI's versatility has led to its adoption across numerous sectors, revolutionizing how businesses operate and how individuals interact with technology.

- **Healthcare:**
  - AI-powered diagnostic tools can analyze medical images and patient data to detect diseases earlier and more accurately.
  - Personalized medicine uses AI to tailor treatments to individual patients based on their genetic makeup and medical history.
  - Robotic surgery enhances precision and minimizes invasiveness.
  - AI assists in drug discovery and development.
- **Finance:**
  - AI algorithms detect fraudulent transactions and prevent financial crimes.
  - Algorithmic trading uses AI to execute trades based on market trends and data analysis.
  - Chatbots provide customer support and financial advice.
  - Risk assessment is improved through AI analysis.
- **Transportation:**
  - Autonomous vehicles use AI to navigate roads and avoid obstacles.
  - Traffic management systems use AI to optimize traffic flow and reduce congestion.
  - AI-powered navigation systems provide real-time traffic updates and route planning.
- **Education:**
  - Personalized learning platforms tailor educational content to individual student needs.
  - AI-powered grading systems automate the assessment of assignments.
  - Virtual tutors provide personalized instruction and support.

Ethical Considerations and Challenges

As AI becomes more powerful, it raises significant ethical considerations that must be addressed.

- **Bias and Fairness:** AI algorithms can inherit biases from the data they are trained on, leading to discriminatory outcomes. Ensuring fairness and transparency in AI systems is crucial.
- **Job Displacement:** Automation driven by AI has the potential to displace workers in certain industries. Addressing the impact of AI on employment requires proactive measures like retraining and upskilling.
- **Privacy and Security:** AI systems collect and process vast amounts of data, raising concerns about privacy and security. Robust data protection measures are essential.
- **Autonomous Weapons:** The development of autonomous weapons systems raises ethical concerns about the potential for unintended consequences and the loss of human control.
- **Explainability and Transparency:** Understanding how AI algorithms arrive at their decisions is crucial for building trust and accountability. Black-box AI systems can be difficult to interpret, raising concerns about transparency.
- **The risk of AI being used for malicious purposes:** Deepfakes, and other AI powered tools can be used to spread misinformation.

The Future of AI

The field of AI is constantly evolving, with ongoing research pushing the boundaries of what is possible. Future advancements are likely to include:

- **Artificial General Intelligence (AGI):** The development of AI systems with human-level intelligence and the ability to perform a wide range of tasks.
- **Explainable AI (XAI):** Focus on making AI algorithms more transparent and understandable.
- **Edge AI:** Processing AI algorithms directly on devices, reducing reliance on cloud computing.
- **Quantum AI:** Combining quantum computing with AI to solve complex problems that are beyond the capabilities of classical computers.
- **Neuro-inspired AI:** Developing AI systems that mimic the structure and function of the human brain.

AI is a transformative technology with the potential to revolutionize numerous aspects of our lives. Understanding its principles, applications, and ethical implications is crucial for harnessing its benefits while mitigating its risks. As AI continues to evolve, it is essential to engage in ongoing dialogue and collaboration to ensure that it is developed and used responsibly for the benefit of humanity, (Braun & Clarke, 2006).

Wearable Sensors

Wearable sensors have emerged as transformative technology, revolutionizing how we monitor and interact with our health, environment, and daily activities. These devices, seamlessly integrated into clothing, accessories, or directly onto the skin, provide continuous and real-time data acquisition, offering unprecedented insights into human physiology and behavior. The proliferation of wearable sensors stems from advancements in microelectronics, material science, and wireless communication, enabling the creation of compact, energy-efficient, and user-friendly devices.

The fundamental principle behind wearable sensors lies in their ability to detect and translate physical or chemical signals into measurable data. These signals can range from simple motion and acceleration to complex physiological parameters like heart rate, blood glucose, and even brain activity. The data collected is then processed and analyzed, often using sophisticated algorithms, to extract meaningful information that can be used for various applications.

Key Applications of Wearable Sensors

- **Healthcare and Wellness:**
  - **Continuous Health Monitoring:** Wearable sensors enable continuous monitoring of vital signs, such as heart rate, blood pressure, oxygen saturation, and body temperature, facilitating early detection of health anomalies and personalized healthcare interventions.
  - **Chronic Disease Management:** Devices like continuous glucose monitors (CGMs) and wearable ECGs empower individuals with chronic conditions to manage their health proactively.
  - **Rehabilitation and Physical Therapy:** Wearable sensors track movement patterns, gait analysis, and muscle activity, aiding in rehabilitation programs and monitoring patient progress.
  - **Sleep Monitoring:** Sensors embedded in wristbands or mattresses track sleep stages, movement, and heart rate variability, providing insights into sleep quality and identifying potential sleep disorders.
  - **Mental Health Monitoring:** Wearable sensors are being explored for monitoring stress levels, anxiety, and mood fluctuations by analyzing physiological markers like heart rate variability and skin conductance.
- **Environmental Monitoring:**
  - **Air Quality Monitoring:** Wearable sensors detect pollutants like particulate matter, volatile organic compounds, and gases, providing real-time air quality data.
  - **UV Exposure Monitoring:** Sensors integrated into wearables track UV radiation exposure, alerting users to potential sunburn risks.
  - **Noise Monitoring:** Wearable devices can measure noise levels, providing insights into environmental noise pollution and its impact on health.

Types of Wearable Sensors

- **Accelerometers and Gyroscopes:** These sensors measure acceleration and angular velocity, enabling the tracking of movement, orientation, and activity.
- **Heart Rate Sensors:** Optical sensors, such as photoplethysmography (PPG) sensors, measure heart rate by detecting changes in blood volume.
- **Electromyography (EMG) Sensors:** EMG sensors measure electrical activity produced by muscles, providing insights into muscle function and fatigue.
- **Electrocardiography (ECG) Sensors:** ECG sensors measure the electrical activity of the heart, enabling the detection of cardiac arrhythmias and other heart conditions.

- **Temperature Sensors:** Thermistors and infrared sensors measure body temperature, providing insights into health status.
- **Chemical Sensors:** These sensors detect specific chemicals, such as glucose, sweat electrolytes, and gases, enabling monitoring of physiological parameters and environmental pollutants.

Challenges and Future Directions

Despite the immense potential of wearable sensors, several challenges remain:

- **Data Privacy and Security:** The vast amount of personal data collected by wearable sensors raises concerns about privacy and security. Robust data encryption and access control mechanisms are crucial.
- **Power Consumption:** Wearable sensors need to be energy-efficient to ensure long battery life. Advancements in low-power electronics and energy harvesting technologies are essential.
- **Sensor Accuracy and Reliability:** Ensuring the accuracy and reliability of sensor data is critical for accurate monitoring and analysis. Calibration and validation procedures are necessary.
- **User Comfort and Wearability:** Wearable sensors need to be comfortable and unobtrusive to ensure user compliance. Flexible and stretchable materials are being developed to improve wearability.
- **Data Integration and Analysis:** Integrating data from multiple sensors and analyzing complex datasets requires sophisticated algorithms and data analytics platforms.

The future of wearable sensors is bright, with ongoing research and development focusing on:

- **Miniaturization and Integration:** Developing smaller and more integrated sensors that can be embedded into everyday objects.
- **Advanced Materials:** Utilizing biocompatible, flexible, and stretchable materials to improve wearability and sensor performance.
- **Artificial Intelligence (AI) and Machine Learning (ML):** Leveraging AI and ML algorithms to analyze complex sensor data and provide personalized insights and recommendations.
- **Energy Harvesting:** Developing energy harvesting technologies, such as solar, thermal, and kinetic energy harvesting, to power wearable sensors.
- **Closed-Loop Systems:** Creating closed-loop systems that integrate wearable sensors with actuators to provide real-time feedback and control, such as insulin delivery systems for diabetes management.
- **Personalized Medicine:** Utilizing wearable sensor data to develop personalized medicine approaches that tailor treatments to individual patient needs.

Wearable sensors are poised to revolutionize various aspects of our lives, from healthcare and fitness to environmental monitoring and human-computer interaction. As technology continues to advance, we can expect to see even more innovative and impactful applications of wearable sensors in the years to come, further blurring the lines between humans and the digital world, (Bryman & Bell, 2011).

APPLICATIONS ACROSS INDUSTRIES

The augmented operator concept has broad applicability across various industries.

Manufacturing

Manufacturing, the cornerstone of industrial societies, is undergoing a profound transformation. From its historical roots in artisanal workshops to the modern era of automated production lines, the sector has consistently adapted to technological advancements and evolving market demands. This evolution is not merely a linear progression but a complex interplay of innovation, globalization, and sustainability concerns, shaping the future of how goods are produced and consumed.

Historical Foundations and the Rise of Mass Production

The earliest forms of manufacturing were characterized by handcrafted goods, produced by skilled artisans. The Industrial Revolution, however, ushered in a paradigm shift, driven by the advent of steam power, mechanization, and the factory system. This period witnessed the birth of mass production, spearheaded by pioneers like Henry Ford, who introduced the assembly line, revolutionizing automotive manufacturing. The principles of standardization, specialization, and efficiency became paramount, leading to unprecedented increases in output and a decline in production costs.

The 20th century further refined manufacturing processes, with the introduction of electricity, automation, and advanced materials. World War II accelerated technological advancements in manufacturing, particularly in aerospace and defense industries. The post-war era saw the rise of consumerism, fueling demand for mass-produced goods and solidifying manufacturing's role as a key driver of economic growth, (Carson et. Al., 2001)

The Digital Revolution and Industry 4.0

The late 20th and early 21st centuries have witnessed the rise of the digital revolution, profoundly impacting manufacturing. The integration of information technology, automation, and data analytics has led to the emergence of Industry 4.0, a concept that envisions a fully connected and intelligent manufacturing ecosystem.

Key technologies driving Industry 4.0 include:

- **Internet of Things (IoT):** Connecting machines, sensors, and devices, enabling real-time data collection and analysis.
- **Cloud Computing:** Providing scalable and accessible computing resources for data storage, processing, and analysis.
- **Artificial Intelligence (AI) and Machine Learning (ML):** Enabling predictive maintenance, process optimization, and intelligent decision-making.

- **Robotics and Automation:** Enhancing efficiency, precision, and flexibility in production processes.
- **Additive Manufacturing (3D Printing):** Enabling on-demand production of customized parts and prototypes.
- **Big Data Analytics:** Extracting valuable insights from vast datasets to improve operational efficiency and product quality.
- **Cyber-Physical Systems (CPS):** Integrating physical and digital systems to create intelligent and autonomous production environments.

Industry 4.0 promises to revolutionize manufacturing by enabling:

- **Increased Efficiency and Productivity:** Optimizing processes, reducing waste, and minimizing downtime.
- **Enhanced Flexibility and Agility:** Adapting to changing market demands and producing customized products.
- **Improved Product Quality:** Implementing real-time monitoring and control systems.
- **Reduced Costs:** Optimizing resource utilization and minimizing operational expenses.
- **Enhanced Sustainability:** Optimizing energy consumption and reducing environmental impact.

Globalization and Supply Chain Dynamics

Globalization has significantly reshaped the manufacturing landscape, leading to the fragmentation of supply chains and the relocation of production facilities to countries with lower labor costs. This has resulted in a complex network of suppliers, manufacturers, and distributors spanning across continents.

However, recent geopolitical events, such as trade disputes and the COVID-19 pandemic, have highlighted the vulnerabilities of global supply chains. This has led to a growing trend towards reshoring and nearshoring, with companies seeking to bring production closer to their end markets to enhance resilience and reduce risks.

Sustainability and the Circular Economy

Environmental concerns have become increasingly prominent in the manufacturing sector. The traditional linear model of "take, make, dispose" is being challenged by the principles of the circular economy, which aims to minimize waste and maximize resource utilization, (Kaplan, 2016)

Sustainable manufacturing practices include:

- **Reducing Energy Consumption:** Implementing energy-efficient technologies and processes.
- **Minimizing Waste Generation:** Implementing waste reduction and recycling programs.
- **Using Sustainable Materials:** Sourcing materials from renewable and recycled sources.
- **Designing for Durability and Recyclability:** Creating products that are long-lasting and easy to disassemble and recycle.

- **Implementing Closed-Loop Systems:** Recovering and reusing materials and components.

The adoption of sustainable manufacturing practices is not only environmentally responsible but also economically advantageous, as it can lead to cost savings and enhanced brand reputation.

The Future of Manufacturing: Trends and Challenges

The future of manufacturing will be shaped by several key trends and challenges:

- **Increased Automation and Robotics:** Advancements in AI and robotics will lead to further automation of manufacturing processes, potentially displacing some jobs but also creating new opportunities in areas like robotics programming and maintenance.
- **Personalized Manufacturing:** Additive manufacturing and other technologies will enable the production of customized products tailored to individual needs and preferences.
- **Data-Driven Decision-Making:** The increasing availability of data will enable manufacturers to make more informed decisions regarding production planning, quality control, and supply chain management.
- **Focus on Sustainability:** Environmental concerns will continue to drive the adoption of sustainable manufacturing practices.
- **Skills Gap:** The rapid pace of technological change will require a workforce with new skills in areas like data analytics, robotics, and AI.
- **Cybersecurity:** The increasing connectivity of manufacturing systems will require robust cybersecurity measures to protect against cyberattacks.
- **Geopolitical Instability:** Trade disputes and geopolitical tensions can disrupt supply chains and create uncertainty for manufacturers.

Human-Machine Collaboration

While automation and robotics are transforming manufacturing, the role of human workers remains crucial. The future of manufacturing will likely involve a collaborative approach, where humans and machines work together to optimize production processes. Humans possess unique skills in areas like creativity, problem-solving, and critical thinking, which are essential for innovation and complex decision-making.

Manufacturing is a dynamic and evolving sector that plays a vital role in the global economy. The transition to Industry 4.0, coupled with the growing emphasis on sustainability and the challenges of globalization, is transforming the way goods are produced and consumed. Manufacturers must embrace innovation, adapt to changing market demands, and prioritize sustainability to remain competitive in the future. The successful integration of humans and machines, along with a focus on developing a skilled workforce, will be critical for navigating the evolving landscape of manufacturing and ensuring its continued contribution to economic growth and societal well-being.

Healthcare

Healthcare, a fundamental pillar of any civilized society, is undergoing a profound transformation in the 21st century. Driven by technological advancements, shifting demographics, and evolving societal needs, the healthcare landscape is becoming increasingly complex and dynamic. This essay will explore key aspects of this evolution, examining the challenges and opportunities that lie ahead.

Technological Advancements: Revolutionizing Healthcare Delivery

Technological innovation is arguably the most significant driver of change in modern healthcare.

- **Artificial Intelligence (AI) and Machine Learning (ML):** AI and ML are revolutionizing diagnostics, treatment planning, and drug discovery. AI-powered algorithms can analyze vast datasets of patient information, identify patterns, and predict disease risks with unprecedented accuracy. In radiology, AI algorithms can assist radiologists in detecting anomalies in medical images, improving diagnostic accuracy and efficiency. In drug discovery, AI can accelerate the process of identifying and developing new medications by analyzing molecular structures and predicting their potential efficacy and safety.
- **Telemedicine and Remote Patient Monitoring:** Telemedicine platforms enable remote consultations, allowing patients to access healthcare services from the comfort of their homes. This is particularly beneficial for patients in rural areas or those with limited mobility. Remote patient monitoring devices, such as wearable sensors and smart home devices, allow healthcare providers to continuously monitor patients' vital signs and other health metrics, enabling early detection of potential health issues and proactive interventions.
- **Robotics and Surgery:** Robotic surgery systems offer enhanced precision and flexibility, enabling minimally invasive procedures with shorter recovery times and reduced risks of complications. These systems are particularly beneficial in complex surgeries such as neurosurgery and cardiac surgery, (Mueller & Baum, 2011).

Addressing Global Health Challenges

The 21st century presents a unique set of global health challenges, including:

- **Aging Populations:** As populations age, the prevalence of chronic diseases such as diabetes, heart disease, and Alzheimer's is increasing. This necessitates a shift towards long-term care and preventive care strategies.
- **Emerging Infectious Diseases:** The emergence and re-emergence of infectious diseases, such as COVID-19, pose significant threats to global health. Effective surveillance systems, rapid diagnostic tests, and the development of vaccines and antiviral medications are crucial for mitigating the impact of these outbreaks.
- **Non-communicable Diseases (NCDs):** NCDs, such as heart disease, cancer, and chronic respiratory diseases, are the leading causes of death worldwide. Addressing these chal-

lenges requires a multifaceted approach, including lifestyle interventions, early detection, and improved access to effective treatments.

Reimagining Healthcare Systems

To effectively address these challenges, healthcare systems must undergo significant transformation:

- **Shifting from a Fee-for-Service Model to Value-Based Care:** The traditional fee-for-service model incentivizes providers to perform more procedures, rather than focusing on patient outcomes. Value-based care models, on the other hand, incentivize providers to deliver high-quality, cost-effective care that improves patient outcomes.
- **Improving Access to Healthcare:** Ensuring equitable access to quality healthcare for all individuals, regardless of their socioeconomic status, is a critical challenge. This requires addressing social determinants of health, such as poverty, discrimination, and lack of access to healthy food and safe living environments.
- **Empowering Patients:** Empowering patients to take an active role in their own health is essential for improving health outcomes. This includes providing patients with access to their own health data, supporting patient education and self-management, and encouraging patient-centered care.
- **Strengthening Primary Care:** Primary care plays a crucial role in preventive care, early disease detection, and chronic disease management. Strengthening primary care systems is essential for improving the overall health of the population.

Ethical Considerations in Healthcare

The rapid advancements in healthcare technology raise important ethical considerations:

- **Data Privacy and Security:** The collection and use of patient data raise concerns about privacy and security. Robust data protection measures are essential to ensure the confidentiality and security of patient information.
- **Algorithmic Bias:** AI algorithms are trained on data that may reflect existing biases, potentially leading to discriminatory outcomes. It is crucial to ensure that AI systems are developed and deployed in a fair and equitable manner.
- **Access and Equity:** The benefits of new technologies may not be equally accessible to all individuals, exacerbating existing health disparities. Efforts must be made to ensure equitable access to these technologies for all populations.
- **Autonomy and Informed Consent:** As technology become increasingly integrated into healthcare, it is crucial to respect patient autonomy and obtain informed consent for the use of new technologies.

The Future of Healthcare

The future of healthcare is likely to be characterized by:

- **Increased personalization:** Personalized medicine, tailored to an individual's unique genetic makeup and lifestyle, will become increasingly prevalent.
- **Greater integration of technology:** Technology will continue to play an increasingly important role in all aspects of healthcare, from diagnosis and treatment to patient monitoring and care coordination.
- **A stronger focus on prevention and wellness:** There will be greater emphasis on preventive care and promoting healthy lifestyles to prevent chronic diseases.
- **Increased collaboration and interdisciplinary care:** Collaboration among healthcare providers, researchers, and patients will be essential for improving health outcomes.

The healthcare landscape is undergoing a period of unprecedented change. Technological advancements, evolving societal needs, and global health challenges are driving the transformation of healthcare delivery and creating new opportunities for improving health outcomes. By embracing innovation, addressing global health challenges, and reimagining healthcare systems, we can build a more equitable, efficient, and effective healthcare system for all, (Mantere & Ketokivi, 2013)

This essay addresses the prompt by:

- **Providing a comprehensive overview of the topic:** It covers key aspects of the evolving healthcare landscape, including technological advancements, global health challenges, reimagining healthcare systems, ethical considerations, and the future of healthcare.
- **Analyzing the impact of key factors:** It analyzes the impact of technological advancements, such as AI, telemedicine, and genomics, on healthcare delivery. It also examines the impact of aging populations, emerging infectious diseases, and non-communicable diseases on global health.
- **Offering insights into the future of healthcare:** It provides insights into the future of healthcare, including the increasing importance of personalization, the continued integration of technology, and the growing emphasis on prevention and wellness.

This essay aims to provide a comprehensive and insightful analysis of the evolving landscape of healthcare, while also considering the ethical and social implications of these changes.

Logistics

Logistics, the intricate web of processes that ensure the efficient and timely movement of goods and services from origin to destination, plays a pivotal role in the modern globalized economy. It encompasses a diverse range of activities, including transportation, warehousing, inventory management, packaging, and customs clearance. In an increasingly interconnected world, the effectiveness of logistics operations can significantly impact a company's competitiveness, profitability, and customer satisfaction.

The Foundations of Logistics

At its core, logistics is about optimizing the flow of goods and information. It involves making strategic decisions regarding:

- **Transportation Modes:** Selecting the most suitable mode of transport (road, rail, air, sea, or a combination) based on factors like cost, speed, and environmental impact.
- **Warehousing and Inventory Management:** Strategically locating warehouses, optimizing inventory levels, and implementing efficient inventory tracking systems to minimize holding costs and ensure product availability.
- **Packaging and Material Handling:** Designing and selecting appropriate packaging to protect goods during transit and facilitate efficient handling throughout the supply chain.
- **Customs Clearance and Documentation:** Navigating complex customs regulations and ensuring accurate and timely documentation to minimize delays and costs.
- **Information Flow:** Leveraging technology to track shipments, monitor inventory levels, and communicate effectively with all stakeholders in the supply chain.

Key Trends Shaping the Logistics Industry

The logistics landscape is constantly evolving, driven by several key trends:

- **Globalization and Supply Chain Complexity:** The increasing globalization of trade has led to longer and more complex supply chains, presenting challenges in terms of coordination, visibility, and risk management.
- **Technological Advancements:** Technological innovations such as artificial intelligence (AI), the Internet of Things (IoT), blockchain, and robotics are transforming logistics operations, enabling greater efficiency, automation, and real-time visibility.
- **E-commerce Growth:** The rapid growth of e-commerce has placed significant pressure on logistics providers to deliver goods quickly and efficiently to consumers' doorsteps.
- **Sustainability and Environmental Concerns:** Growing environmental awareness has driven the need for more sustainable logistics practices, including the use of alternative fuels, reduced emissions, and improved packaging efficiency.
- **Supply Chain Disruptions:** Global events such as pandemics, natural disasters, and geopolitical instability can disrupt supply chains, highlighting the need for greater resilience and flexibility.

The Role of Technology in Modern Logistics

Technology is playing an increasingly crucial role in modern logistics, enabling:

- **Improved Visibility and Tracking:** Real-time tracking of shipments using GPS, RFID, and other technologies provides valuable insights into the location and status of goods, enabling proactive issue resolution and improved customer service.
- **Enhanced Decision-Making:** AI-powered analytics can analyze vast amounts of data to identify patterns, predict demand, and optimize routes and schedules, leading to more efficient and cost-effective operations.
- **Increased Automation:** The use of robots, drones, and automated guided vehicles (AGVs) can automate various tasks in warehouses and distribution centers, improving efficiency and reducing labor costs.

- **The Internet of Things (IoT):** IoT sensors can be embedded in various assets, such as shipping containers and vehicles, to monitor conditions, track movements, and improve operational efficiency.

Logistics and Supply Chain Management

Logistics is an integral part of supply chain management, which encompasses the entire process of planning, implementing, and controlling the flow of goods, services, and information from the point of origin to the point of consumption. Effective supply chain management requires a holistic approach that considers all aspects of the logistics process, including:

- **Demand Forecasting:** Accurate demand forecasting is essential for effective inventory planning and resource allocation.
- **Procurement and Sourcing:** Selecting and managing suppliers to ensure the timely and cost-effective procurement of raw materials and components.
- **Production Planning:** Coordinating production schedules with demand forecasts and ensuring the timely availability of necessary materials.
- **Order Fulfillment:** Processing orders accurately and efficiently and ensuring timely delivery to customers.
- **Customer Service:** Providing excellent customer service by addressing inquiries, resolving issues, and managing returns effectively.

Challenges and Opportunities in the Logistics Industry

The logistics industry faces a number of challenges, including:

- **Rising Costs:** Increasing fuel prices, labor costs, and regulatory compliance costs can significantly impact profitability.
- **Skills Shortages:** A shortage of skilled workers in areas such as logistics and transportation can hinder operational efficiency.
- **Competition:** The logistics industry is highly competitive, with companies constantly seeking to gain a competitive advantage through innovation, cost optimization, and improved customer service.
- **Environmental Concerns:** The environmental impact of logistics operations is a growing concern, requiring companies to adopt more sustainable practices.

Despite these challenges, the logistics industry also presents numerous opportunities:

- **Growth of E-commerce:** The continued growth of e-commerce will create significant demand for efficient and reliable delivery services.
- **Technological Advancements:** The adoption of new technologies such as AI, IoT, and blockchain can revolutionize logistics operations and create new opportunities for innovation.
- **Focus on Sustainability:** Companies that prioritize sustainability can differentiate themselves in the market and attract environmentally conscious customers.
- **Globalization:** Continued globalization will create new opportunities for logistics providers to expand their operations into new markets.

The Future of Logistics

The future of logistics will be characterized by:

- **Increased Automation and Robotics:** Automation will play an increasingly important role in all aspects of logistics, from warehouse operations to long-haul transportation.
- **Greater Sustainability:** The focus on sustainability will continue to grow, with companies adopting more eco-friendly practices and reducing their carbon footprint.
- **Data-Driven Decision Making:** Data analytics will play a crucial role in optimizing logistics operations, improving efficiency, and reducing costs.
- **Enhanced Customer Experience:** Customers will expect faster, more reliable, and more personalized delivery experiences, putting pressure on logistics providers to innovate and improve their services.
- **The Rise of the "Gig Economy":** The rise of the gig economy may lead to a shift towards more flexible and on-demand logistics solutions, such as crowd-sourced delivery and peer-to-peer transportation.

Logistics is a critical function in today's globalized economy, playing a vital role in connecting businesses with their customers and ensuring the smooth flow of goods and services. By embracing technological advancements, adopting sustainable practices, and focusing on customers, (Salin & Winston, 1992)

Emergency Response

Emergency response encompasses a complex and dynamic system designed to mitigate the impact of unforeseen events that threaten human life, property, and the environment. These events, ranging from natural disasters to technological failures and acts of terrorism, necessitate a coordinated and rapid response to minimize damage and facilitate recovery. Effective emergency response relies on a robust framework of preparedness, planning, communication, and execution, involving a diverse array of stakeholders from individuals to international organizations.

The Spectrum of Emergencies

Emergencies can be broadly categorized into several types, each demanding specific response strategies:

- **Natural Disasters:** These include earthquakes, hurricanes, floods, wildfires, tsunamis, volcanic eruptions, and extreme weather events. The unpredictability and scale of these disasters often overwhelm local resources, requiring regional, national, or even international assistance.
- **Human-Caused Emergencies:** This category includes acts of terrorism, armed conflicts, civil unrest, and large-scale public health emergencies (pandemics, epidemics).
- **Public Health Emergencies:** These are events that pose a serious and imminent threat to the health of a population, requiring rapid and coordinated public health measures.

The Pillars of Emergency Response

A comprehensive emergency response system rests on several key pillars:

- **Preparedness:** This involves proactive measures taken to anticipate and prepare for potential emergencies. It includes:
  - o Risk assessment and hazard mapping: Identifying potential threats and vulnerabilities.
  - o Developing and implementing emergency plans: Outlining procedures for evacuation, communication, resource allocation, and recovery.
  - o Conducting drills and exercises: Testing the effectiveness of emergency plans and training personnel.
  - o Stockpiling essential supplies: Ensuring availability of food, water, medical supplies, and other critical resources.
  - o Public education and awareness campaigns: Informing the public about potential hazards and emergency procedures.
- **Communication:** Effective communication is crucial for disseminating timely and accurate information to the public, emergency responders, and other stakeholders. This includes:
  - o Establishing reliable communication channels: Utilizing multiple communication platforms, such as radio, television, internet, and mobile phones.
  - o Developing clear and concise messaging: Providing accurate and understandable information to the public.
  - o Establishing a central command center: Coordinating communication and information flow among different agencies.
  - o Use of early warning systems.
- **Recovery:** This involves the long-term process of restoring affected communities to their pre-disaster state or better. This includes:
  - o Providing financial assistance: Supporting individuals and businesses in rebuilding their lives and livelihoods.
  - o Rebuilding infrastructure: Repairing or replacing damaged roads, bridges, buildings, and utilities.
  - o Providing psychological support: Addressing the emotional and mental health needs of affected individuals.
  - o Long term community rebuilding.
  - o Review of the emergency response, to improve future responses.

The Role of Technology in Emergency Response

Technology plays an increasingly vital role in enhancing emergency response capabilities. Some key technological advancements include:

- **Geographic Information Systems (GIS):** GIS enables the mapping and analysis of spatial data, providing critical information for hazard assessment, evacuation planning, and resource allocation.

- **Remote Sensing:** Satellite imagery and aerial photography provide real-time information about the extent and impact of disasters.
- **Communication Technologies:** Mobile phones, satellite phones, and internet-based communication platforms facilitate rapid communication and information sharing.
- **Early Warning Systems:** Sensor networks and monitoring systems provide early warnings of impending disasters, such as earthquakes, tsunamis, and floods.

The Importance of Interagency Coordination

Effective emergency response requires seamless coordination among various agencies, including:

- **Government Agencies:** Local, state, and national government agencies play a crucial role in planning, coordinating, and executing emergency response operations.
- **Emergency Responders:** Firefighters, police officers, paramedics, and other emergency responders are on the front lines of emergency response.
- **Non-Governmental Organizations (NGOs):** NGOs provide critical support in areas such as search and rescue, medical assistance, and humanitarian aid.
- **Private Sector:** Businesses and industries play a vital role in providing resources, expertise, and logistical support.

Challenges and Future Directions

Despite advancements in emergency response, several challenges remain:

- **Climate Change:** The increasing frequency and intensity of extreme weather events pose significant challenges for emergency response systems.
- **Urbanization:** Rapid urbanization and population growth increase the vulnerability of communities to disasters.
- **Cybersecurity Threats:** Critical infrastructure is increasingly vulnerable to cyberattacks, which can disrupt emergency response operations.
- **Misinformation and Disinformation:** The spread of misinformation and disinformation during emergencies can hinder effective response efforts.
- **Funding and Resource Constraints:** Many communities and countries face funding and resource constraints, limiting their ability to prepare for and respond to emergencies.
- **Pandemic Preparedness:** The covid 19 pandemic showed the weaknesses of many emergency response systems concerning global pandemics.

Future directions for emergency response include:

- **Investing in Resilience:** Building resilient infrastructure and communities that can withstand the impacts of disasters.
- **Enhancing Early Warning Systems:** Improving the accuracy and timeliness of early warning systems.

- **Strengthening Interagency Coordination:** Enhancing communication and collaboration among different agencies.
- **Leveraging Technology:** Utilizing advanced technologies, such as AI and data analytics, to improve emergency response capabilities.
- **Building Community Capacity:** Empowering communities to take ownership of their own safety and preparedness.

Emergency response is a multifaceted and evolving field that requires continuous improvement and adaptation. By investing in preparedness, strengthening coordination, leveraging technology, and building community resilience, we can enhance our ability to protect lives, property, and the environment in the face of unforeseen events.

Human-Centric Systems and the Future of Work

A critical aspect of the augmented operator is the focus on creating human-centric systems. This means designing technologies that adapt to humans, rather than forcing the human to adapt to technology. This approach is essential for ensuring user acceptance and maximizing the benefits of augmentation.

The rise of digital native generations entering the workforce, who are accustomed to seamless technology integration, further emphasizes the importance of human-centric design. These generations expect intuitive interfaces, personalized experiences, and technologies that enhance their capabilities, (Searle, 2006).

Challenges and Considerations

While the potential benefits of the augmented operator are significant, several challenges and considerations must be addressed. These include:

Data Privacy and Security. In an era defined by ubiquitous connectivity and the relentless generation of data, the twin pillars of data privacy and security have ascended to paramount importance. They are not merely technical considerations, but fundamental rights, shaping the contours of individual autonomy and societal trust. The intricate dance between these two concepts, while often intertwined, demands a nuanced understanding to effectively safeguard sensitive information in a landscape rife with evolving threats.

Data Privacy: The Right to Control. Data privacy, at its core, embodies the individual's right to control their personal information. It encompasses the ability to determine what data is collected, how it is used, and with whom it is shared. This right extends beyond mere confidentiality, encompassing concepts of transparency, accountability, and individual agency.

The modern data ecosystem, however, presents formidable challenges to achieving meaningful privacy. The sheer volume of data generated, coupled with the sophisticated techniques used for collection and analysis, often leaves individuals with limited visibility into the fate of their personal information. From browsing histories and social media interactions to biometric data and financial transactions, the digital footprint we leave behind is vast and multifaceted.

Key aspects of data privacy include:

- **Informed Consent:** Individuals must be fully informed about the data being collected, the purposes for which it will be used, and their rights regarding its management. This necessitates clear and concise privacy policies, avoiding opaque legalese that obfuscates rather than clarifies.
- **Data Minimization:** Organizations should only collect the data that is strictly necessary for the stated purpose. Excessive data collection not only increases the risk of privacy breaches but also undermines the principle of proportionality.
- **Purpose Limitation:** Data should only be used for the specific purposes for which it was collected, unless explicit consent is obtained for alternative uses. This prevents function creep, where data collected for one purpose is repurposed for unrelated activities.
- **Data Subject Rights:** Individuals must have the right to access, rectify, erase, and restrict the processing of their personal data. This empowers individuals to exercise control over their information and hold organizations accountable.

Data Security: Protecting the Fortress. Data security, on the other hand, focuses on the technical and organizational measures employed to protect data from unauthorized access, use, disclosure, disruption, modification, or destruction. It is the fortress that safeguards the sensitive information entrusted to organizations.

In an environment characterized by persistent cyber threats, including malware, ransomware, phishing attacks, and data breaches, robust data security measures are indispensable. The consequences of security failures can be devastating, ranging from financial losses and reputational damage to identity theft and national security risks.

Essential components of data security include:

- **Encryption:** Converting data into an unreadable format using cryptographic algorithms, ensuring that even if intercepted, it remains inaccessible to unauthorized parties. Encryption should be applied both in transit and at rest.
- **Access Control:** Implementing strict access controls, such as strong passwords, multi-factor authentication, and role-based access, to limit access to sensitive data to authorized personnel.
- **Network Security:** Deploying firewalls, intrusion detection systems, and other network security measures to protect against unauthorized access and malicious traffic.
- **Vulnerability Management:** Regularly scanning for and patching vulnerabilities in software and systems to prevent exploitation by attackers.
- **Incident Response:** Developing and implementing a comprehensive incident response plan to effectively handle security breaches and minimize their impact.
- **Physical Security:** Securing physical infrastructure, such as data centers and servers, to prevent unauthorized access and physical tampering.
- **Data Backup and Recovery:** Regularly backing up critical data and implementing robust recovery procedures to ensure business continuity in the event of data loss or system failure.
- **Security Awareness Training:** Educating employees about security best practices, such as recognizing phishing emails and avoiding unsafe websites, to mitigate human error, which is a major point of security weakness.

The Interplay: Privacy and Security in Harmony. While distinct, data privacy and security are inextricably linked. Security measures are essential for protecting confidentiality, integrity, and availability of personal data, thereby upholding privacy principles. Conversely, privacy considerations should inform the design and implementation of security measures, ensuring that they are proportionate and respect individual rights.

For example, implementing strong encryption can protect the confidentiality of personal data, but it is also essential to ensure that access to decryption keys is restricted to authorized personnel. Similarly, implementing access controls can prevent unauthorized access to personal data, but it is also crucial to ensure that these controls are not overly intrusive and do not unreasonably restrict legitimate access, (Thomas, 2006).

Furthermore, the legal and regulatory landscape is increasingly recognizing the interdependence of privacy and security. Regulations such as the General Data Protection Regulation (GDPR) and the California Consumer Privacy Act (CCPA) mandate that organizations implement appropriate technical and organizational measures to protect personal data, reflecting the understanding that security is a prerequisite for privacy.

The Future: Navigating Emerging Challenges

The rapid pace of technological innovation presents ongoing challenges to data privacy and security. The rise of artificial intelligence, the Internet of Things, and blockchain technology, among other advancements, raises new questions about data collection, processing, and protection.

Furthermore, the increasing globalization of data flows necessitates international cooperation and harmonization of privacy and security standards. Addressing these challenges requires a multi-faceted approach, involving technological innovation, regulatory reform, and public awareness.

Data privacy and security are not merely technical or legal issues, but fundamental rights that are essential for building a trustworthy and equitable digital society. By understanding the intricacies of these concepts and implementing robust measures to safeguard personal information, we can navigate the labyrinth of the digital age and ensure that technology empowers individuals rather than undermines their autonomy.

Ethical Considerations

Ethics, the philosophical study of morality, permeates every facet of human existence. From individual choices to complex societal structures, ethical considerations serve as a compass, guiding actions and shaping values. In an increasingly interconnected and rapidly evolving world, the imperative to understand and apply ethical principles has never been more critical. This exploration delves into the multifaceted nature of ethical considerations, examining their significance, challenges, and evolving applications.

At its core, ethical consideration involves evaluating actions, decisions, and policies based on principles of right and wrong, good and bad. It necessitates a critical examination of values, duties, and consequences, often requiring the balancing of competing interests and the navigation of complex moral dilemmas. The significance of ethical considerations stems from their capacity to foster fairness, justice, and well-being, both for individuals and society as a whole.

Foundational Ethical Theories. Several foundational ethical theories provide frameworks for understanding and applying ethical principles. These theories offer diverse perspectives on the nature of morality and the criteria for ethical decision-making.

- **Deontology:** This theory, championed by Immanuel Kant, emphasizes the importance of duty and adherence to moral rules. Deontologists believe that actions are inherently right or wrong, regardless of their consequences. The categorical imperative, a central concept in Kantian ethics, posits that individuals should act only according to principles that they would want to become universal laws.
- **Consequentialism:** In contrast to deontology, consequentialism focuses on the outcomes of actions. Utilitarianism, a prominent form of consequentialism, advocates for actions that maximize overall happiness and minimize suffering. The greatest good for the greatest number is the guiding principle.
- **Virtue Ethics:** This approach, rooted in the philosophy of Aristotle, emphasizes the development of virtuous character. Rather than focusing on rules or consequences, virtue ethics emphasizes the cultivation of traits such as honesty, courage, and compassion. It posits that virtuous individuals will naturally make ethical choices.

Ethical Considerations in Diverse Domains. Ethical considerations are not confined to philosophical discourse; they are integral to various domains of human activity.

- **Business Ethics:** In the corporate world, ethical considerations are paramount for maintaining trust, fostering responsible practices, and ensuring long-term sustainability. Ethical dilemmas arise in areas such as corporate social responsibility, fair labor practices, environmental protection, and marketing ethics. Transparency, accountability, and stakeholder engagement are essential for promoting ethical business conduct.
- **Medical Ethics:** The field of medicine is rife with ethical challenges, from end-of-life decisions and informed consent to the allocation of scarce resources and the use of emerging technologies. Principles such as autonomy, beneficence, non-maleficence, and justice guide ethical decision-making in healthcare.
- **Technological Ethics:** The rapid advancement of technology has raised a host of new ethical questions. Artificial intelligence, data privacy, cybersecurity, and genetic engineering are just a few areas where ethical considerations are crucial. Ensuring that technological advancements are used for the benefit of humanity and that potential harms are mitigated requires careful deliberation and ongoing dialogue.
- **Environmental Ethics:** With increasing awareness of the environmental crisis, ethical considerations regarding human interactions with the natural world have become increasingly important. Environmental ethics explores issues such as climate change, biodiversity loss, and sustainable resource management, advocating for a shift towards a more responsible and harmonious relationship with the environment.

Challenges and Evolving Perspectives. Ethical considerations are not static; they evolve in response to changing societal values, technological advancements, and emerging challenges. Several key challenges and evolving perspectives warrant attention.

- **Cultural Relativism vs. Ethical Universalism:** The debate between cultural relativism, which asserts that ethical values are culturally determined, and ethical universalism, which posits the existence of universal moral principles, remains a central issue in ethical philosophy. Finding a balance between respecting cultural diversity and upholding fundamental human rights is a critical challenge.
- **The Impact of Technology:** The rapid pace of technological innovation presents unprecedented ethical challenges. Issues such as algorithmic bias, data privacy, and the potential for autonomous weapons systems require careful consideration and the development of ethical frameworks that can keep pace with technological advancements.
- **Global Ethics:** In an increasingly interconnected world, ethical considerations must extend beyond national borders. Global issues such as climate change, poverty, and human rights violations require a global ethical framework that promotes cooperation and shared responsibility.

Ethical considerations are fundamental to individual and societal well-being. By embracing ethical principles, fostering critical thinking, and promoting dialogue, we can navigate the labyrinth of moral dilemmas and create a more just and compassionate world. The ongoing evolution of ethical thought and practice is essential for addressing the challenges of the 21st century and beyond, (Tybout & Artz, 1994).

Training and Skill Development

In an era defined by rapid technological advancements, evolving market demands, and a globally interconnected workforce, the significance of robust training and skill development initiatives cannot be overstated. It is no longer sufficient to rely on traditional educational models; the contemporary landscape necessitates a dynamic and adaptive approach to fostering competence, ensuring individuals and organizations remain relevant and competitive. Training and skill development, therefore, emerge as the cornerstones of individual empowerment and organizational efficacy, driving innovation, productivity, and sustainable growth.

The Evolving Paradigm of Skill Acquisition. The traditional notion of a static skillset acquired through formal education is rapidly becoming obsolete. The modern workforce demands a continuous learning paradigm, characterized by adaptability and a commitment to lifelong learning. This shift is driven by several factors:

- **Technological Disruption:** Automation, artificial intelligence, and machine learning are transforming industries, requiring workers to acquire new technical skills and adapt to evolving job roles.
- **Globalization:** The interconnected global economy necessitates a workforce with cross-cultural competence, language proficiency, and an understanding of international business practices.
- **Changing Demographics:** An aging workforce, coupled with a growing emphasis on diversity and inclusion, requires organizations to develop training programs that cater to diverse learning styles and needs.

- **The Rise of the Gig Economy:** The increasing prevalence of freelance and contract work demands individuals to be highly adaptable and possess a diverse range of skills to remain employable.

Key Components of Effective Training and Skill Development. To navigate this dynamic landscape, organizations and individuals must adopt a strategic approach to training and skill development, encompassing the following key components:

- **Needs Assessment:** The foundation of any successful training program lies in a thorough needs assessment. This involves identifying the specific skills gaps that exist within an organization or among individuals, aligning training objectives with strategic goals.
- **Curriculum Design:** Based on needs assessment, a comprehensive curriculum must be developed that incorporates relevant content, engaging delivery methods, and effective assessment strategies. This may involve a blend of traditional classroom instruction, online learning modules, hands-on workshops, and experiential learning opportunities.
- **Delivery Methods:** The choice of delivery methods should be tailored to the target audience and the specific learning objectives. Online learning platforms, virtual reality simulations, and mobile learning applications offer flexible and accessible options for delivering training content.
- **Assessment and Evaluation:** Regular assessments and evaluations are crucial for measuring the effectiveness of training programs and identifying areas for improvement. This may involve pre- and post-training assessments, performance evaluations, and feedback surveys.
- **Continuous Learning:** Skill development is not a one-time event but an ongoing process. Organizations should foster a culture of continuous learning, providing employees with access to resources and opportunities for professional development.
- **Mentorship and Coaching:** Mentorship and coaching programs can provide individuals with personalized guidance and support, helping them to develop their skills and advance their careers.

The Role of Technology in Skill Development. Technology plays a pivotal role in revolutionizing training and skill development. Online learning platforms, learning management systems (LMS), and virtual reality (VR) simulations offer a wealth of opportunities for delivering engaging and effective training experiences.

- **Online Learning Platforms:** Platforms like Coursera, edX, and Udemy provide access to a vast library of online courses and resources, enabling individuals to learn at their own pace and on their own schedule.
- **Learning Management Systems (LMS):** LMS platforms streamline the delivery and management of training programs, providing tools for tracking progress, assessing performance, and delivering personalized learning experiences.
- **Virtual Reality (VR) Simulations:** VR simulations offer immersive and interactive learning experiences, enabling individuals to practice skills in a safe and realistic environment. This is particularly valuable for training in high-risk industries, such as healthcare and aviation.

- **Artificial Intelligence (AI):** AI-powered learning platforms can personalize learning experiences, adapt to individual learning styles, and provide real-time feedback.

The Impact of Skill Development on Individuals and Organizations. Investing in training and skill development yields significant benefits for both individuals and organizations.

- **Individual Benefits:**
  - Enhanced employability and career advancement opportunities.
  - Increased earning potential.
  - Improved job satisfaction and morale.
  - Greater personal and professional growth.
  - Increased confidence and self-efficacy.
- **Organizational Benefits:**
  - Increased productivity and efficiency.
  - Improved quality of products and services.
  - Enhanced innovation and creativity.
  - Reduced employee turnover.
  - Improved customer satisfaction.
  - Enhanced competitiveness.

The Future of Training and Skill Development. The future of training and skill development will be characterized by a continued emphasis on personalized learning, continuous learning, and the integration of emerging technologies. Organizations and individuals that embrace these trends will be well-positioned to thrive in the dynamic and competitive landscape of the 21st century.

The commitment to training and skill development is not merely an investment in the present, but a strategic imperative for the future. By cultivating a culture of continuous learning and embracing innovative training methodologies, we can empower individuals, strengthen organizations, and build a more prosperous and equitable society.

User Acceptance and Ergonomics

User acceptance and ergonomics are intertwined concepts crucial for the successful implementation of any technology or system. Ergonomics, the science of designing products and systems to optimize human well-being and overall performance, directly influences user acceptance. When a system aligns with human capabilities and limitations, users are more likely to find it comfortable, efficient, and satisfactory.

Effective ergonomic design considers factors like physical comfort, cognitive load, and usability. A well-designed interface, for instance, minimizes physical strain and cognitive effort, leading to a positive user experience. Conversely, poorly designed systems can result in frustration, errors, and ultimately, rejection, (Zang & Ye, 2015).

User acceptance, on the other hand, reflects the degree to which individuals are willing to use and adopt a particular technology. Factors influencing acceptance include perceived usefulness, ease of use, and social influence. Ergonomic principles directly bolster these factors. A system that is easy to use, physically comfortable, and reduces cognitive load is perceived as more useful and easier to adopt, thereby enhancing user acceptance.

CONCLUSION

The augmented operator concept represents a transformative approach to human-technology interaction. By integrating AR, AI, and wearable sensors, we can enhance human capabilities and unlock new levels of potential. The applications of this concept are vast, spanning across various industries and promising to improve productivity, safety, and efficiency. As we move forward, it is crucial to prioritize human-centric design, address ethical considerations, and invest in training and skill development. By doing so, we can create a future where humans and technology work together in a truly synergistic partnership.

REFERENCES

Acikgoz, Y. (2019). Employee recruitment and job search: Towards a multi-level Integration. *Human Resource Management Review*, *29*(1), 1–13. DOI: 10.1016/j.hrmr.2018.02.009

Ambrose, M., & Kulik, C. (1993). Old friends, new faces: Motivation research in the 1990's. *Journal of Management*, *25*(3), 231–292. DOI: 10.1177/014920639902500302

Braun, V., & Clarke, V. (2006). Using thematic analysis in psychology. *Qualitative Research in Psychology*, *3*(2), 77–101. DOI: 10.1191/1478088706qp063oa

Bryman, A., & Bell, E. (2011). *Business research methods* (3rd ed.). Oxford University Press.

Carson, D., Gilmore, A., Perry, C., & Gronhaug, K. (2001). *Qualitative marketing research*. Sage. DOI: 10.4135/9781849209625

Kaplan, J. (2016). *Artificial Intelligence: What everyone needs to know*. Oxford University press. DOI: 10.1093/wentk/9780190602383.001.0001

Mantere, S., & Ketokivi, M. (2013). Reasoning in Organization Science. *Academy of Management Review*, *38*(1), 70–89. DOI: 10.5465/amr.2011.0188

Mueller, J. R., & Baum, B. (2011). The definitive guide to hiring right, Journal of applied. *Business Economics (Cleveland, Ohio)*, *12*(3), 140–153.

Salin, E. D., & Winston, P. H. (1992). Machine Learning and Artificial Intelligence. *Analytical Chemistry*, *64*(1).

Searle, R. H. (2006). New technology: The potential impact of surveillance techniques in Recruitment practices. *Personnel Review*, *35*(3), 336–351. DOI: 10.1108/00483480610656720

Thomas, D. R. (2006). A General Inductive Approach for Analyzing Qualitative Evaluation Data. *The American Journal of Evaluation*, *27*(2), 237–246. DOI: 10.1177/1098214005283748

Tybout, A. M., & Artz, N. (1994). Consumer psychology. *Annual Review of Psychology*, *45*(1), 131–169. DOI: 10.1146/annurev.ps.45.020194.001023

Zang, S., & Ye, M. (2015). Human Resource Management in the Era of Big Data. *Journal of Human Resource and Sustainability Studies*, *3*(01), 41–45. DOI: 10.4236/jhrss.2015.31006

Chapter 15
The Evolution of Manufacturing and the Role of Human Resources in the Technological Era

Vannie Naidoo
https://orcid.org/0000-0001-8435-4348
University of KwaZulu-Natal, South Africa

Baldev Singh
https://orcid.org/0009-0006-8041-1812
University of Delhi, India

Kaushal Mehta
Delhi Skill and Entrepreneurship University, India

ABSTRACT

The manufacturing industry has undergone radical transformation over the past few decades, driven primarily by technological advancement. From the rise of automation and robotics to the integration of artificial intelligence (AI), the sector has evolved from labour - intensive processes to highly digitized and efficient systems. This transformation has significantly impacted the workforce and redefined the role of Human Resources (HR). As manufacturing becomes increasingly automated and intelligent, the human element remains crucial. Human Resources (HR) professionals are at the forefront of ensuring that people are equipped, engaged, and empowered to thrive in this evolving landscape. The synergy between technology and human capital will define the next chapter of manufacturing excellence.

INTRODUCTION

The manufacturing sector has undergone major transformation in recent years. The manufacturing organisation had to become more technologically viable in their processes and business

DOI: 10.4018/979-8-3373-2985-7.ch015

operations. Technological advancements has been on the rise in the manufacturing sector, to grow and sustain their businesses, many manufactures have moved from traditional manufacturing to smart manufacturing. The use of Information and Communication Technology (ICTs), Artificial Intelligence (AI), Robotics, Algorithms, Automation, Internet of Things (IoTs) and Block chain. The following key themes will be unpacked in this chapter:

- Manufacturing an historic overview
- The Macro - environment and technological variable that impacts the organisation
- Technological advancements and how it has impacted manufacturing
- Human Resources (HR) in the 4th industrial revolution
- Training and developing and motivation staff in the new era of technology
- Challenges in using new technologies in manufacturing
- Areas of future interest

The Macro - Environment and Technological Variable That Impacts the Organisation

The macro environment refers to the external factors that are beyond the direct control of an organisation but can significantly impact its performance and strategic decisions (Bulgurcu and Mashatan, 2024). According to Thakur (2020) the external factors are those that stem outside of organisations and are usually beyond the organisations control. These variables can be very volatile and directly affect the business, even those in the manufacturing sector. According to PESTEL analysis, the main factors that affect a business in the macro environment is the political, economic, social, technological, environmental and legal environment variables. Technology elements form a fundamental part of the macro environment and shape how organisations do business, compete, and create value for customer (Dana et al., 2022). The macro environment is more aligned with the overall business cycle than with the performance of individual sectors and plays a crucial role in shaping how businesses operate, make decisions, and develop strategies. Since it is dynamic and ever - changing, organisations must continuously monitor these external conditions despite their inability to control them, as they still exert considerable influence (Gini & Agala, 2023).

The digital era has forced companies to adopt digital transformation for companies to stay pertinent (Wyszomirski, 2023). The role of technology is to enable operational efficiencies, allowing organisations to be able to standardize processes, lower costs, and ensure that each project is more productive (Shen, Yeh & Lin, 2022). Innovation driven by technology is what allows businesses to create new products and improve older ones (Wyszomirski, 2023). New discoveries and new processes of competing with existing technology, which can transform entire industries, change competitive dynamics, and impact consumer behaviour. Vial (2019) argues that Industry 4.0 technologies offer significant benefits to manufacturing by enabling more efficient and flexible production, allowing mass customization without sacrificing competitiveness or increasing costs. Acknowledging technological elements is crucial for management as technology can wield a powerful influence over strategy, processes, and business performance (Zafar & Kantola, 2020). For example, the development of Artificial Intelligence (AI) and machine learning has led to new product development opportunities in many sectors, including manufacturing (Shen, Yeh & Lin, 2022). Houti, Abbadi, and Abouabdellah (2019) emphasize the importance of effectively implementing lean manufacturing to maximize its benefits. Xin, Lang, and Mishra (2022) highlight that Industry 4.0 technologies — like Artificial Intelligence (AI), Internet of Things (IoT), and blockchain — are particularly advantageous for the manufacturing and logistics sectors. Naidoo (2020) notes that many manufacturing Small and Medium Enterprises (SMEs) fail due to slow adoption of new technologies, which are crucial for meeting evolving customer demands. Embracing these technologies is also key for sustainable entrepreneurship, enabling the development of innovative, eco - friendly products. Hill, Schilling, and Jones (2020) posit that the technological environment is a constantly changing macro environmental element that perpetually reshapes industries via innovations like Artificial Intelligence (AI), automation, and digital platforms. Technological change is a powerful driver of industry evolution. These changes may lead to increased productivity, cost reductions, and new product offerings, while also rendering existing products or services obsolete. Technological changes can also upheave entire sectors, necessitating companies to consistently invest in research and development, improve infrastructure, and boost digital capabilities.

Table 1. PESTEL analysis of factors influencing manufacturing in the technological era

Factor	Key Issues / Variables	Impact on Manufacturing Sector
Political	Government stability, trade policies, taxation, industrial regulations	Affects investment decisions, global supply chains, and import / export strategies. Political uncertainty can disrupt manufacturing planning.
Economic	Inflation, interest rates, GDP growth, currency fluctuations	Determines production costs, pricing strategies, and overall demand for manufactured goods. Economic downturns reduce capital investment.
Social	Workforce demographics, education levels, changing consumer preferences	Drives demand for sustainable, customized products and influences labour skill requirements. Social awareness increases Corporate Social Responsibility (CSR) expectations.
Technological	Automation, Artificial Intelligence (AI), Internet of Things (IoT), 3D printing	Enhances efficiency, precision, and scalability; creates smart factories but requires digital skills and cyber security.
Environmental	Climate change, energy use, waste management, emissions policies	Encourages green manufacturing, renewable energy use, and circular economy adoption; non - compliance can harm reputation.
Legal	Labour laws, intellectual property rights, data protection, safety regulations	Ensures compliance and fair labour practices; affects HR policies and technology deployment; non - compliance risks fines or legal action.

Source: Adapted from Thakur (2020), Dana et al. (2022), and Gini & Agala (2023).

BACKGROUND

Manufacturing: An Historic Overview

Before the Industrial Revolution, manufacturing largely depended on manual craftsmanship, with artisans and small workshops producing goods using rudimentary tools and traditional techniques. Production was limited in scale and scope, relying heavily on human and animal labour to meet local demand (Mokyr, 1990). This pre - industrial economy was characterized by low productivity and minimal technological innovation, laying the groundwork for the transformative changes that would follow during the Industrial Era.

The First Industrial Revolution, beginning in Great Britain around 1760, signalled a profound transition from agrarian economies to industrialized societies. Among the most transformative innovations was mechanized textile production, exemplified by inventions like the spinning jenny and the power loom, which dramatically increased productivity in the textile sector. Steam power also played a pivotal role; James Watt's improvements to the steam engine enabled more efficient and widespread use of mechanical energy in factories and transportation. In parallel, advancements in iron smelting — particularly the use of coke in blast furnaces — facilitated the mass production of iron, a key material for machinery, railways, and infrastructure (Stearns, 2013).

The Second Industrial Revolution, or Technological Revolution, occurred from the late 19th to early 20th century and was marked by major breakthroughs in industrial technology and production methods. One of the hallmark developments was mass production, exemplified by Henry Ford's implementation of the moving assembly line in 1913, which significantly lowered the cost of automobile manufacturing and made cars accessible to a broader population. Electrification also transformed industry, as factories increasingly utilized electric power to operate machinery, enhancing productivity and operational flexibility. In addition, the Bessemer process revolutionized steel production by enabling the efficient, large - scale manufacture of steel, which became a cornerstone of modern infrastructure and heavy industry (Mokyr, 1998).

The Third Industrial Revolution, often referred to as the Digital Revolution, emerged in the late 20th century and fundamentally reshaped manufacturing through the integration of digital technologies. One key development was the adoption of Computer - Aided Design (CAD), which enabled engineers to create highly precise and modifiable digital models, thereby improving design efficiency and reducing production errors. Computer Numerical Control (CNC) technology further advanced manufacturing by allowing for the automated operation of machine tools with exceptional speed and accuracy. Additionally, increased automation through robotics and computerized systems significantly decreased the reliance on manual labour for repetitive tasks, leading to greater consistency and productivity in manufacturing operations (Rifkin, 2011).

The Fourth Industrial Revolution, emerging in the 21st century, integrates digital, physical, and biological systems in manufacturing. It involves smart technologies like Internet of Things (IoT), sensors, and data analytics to optimize production in real time. Collaborative robots (cobots) enhance productivity and safety, while 3D printing (additive manufacturing) enables efficient, flexible production with reduced waste (Schwab, 2017).

Manufacturing has evolved through technological advancements that have improved efficiency and capabilities. In the Fourth Industrial Revolution, the integration of digital technologies is set to further transform the industry, making it smarter, more adaptable, and more sustainable.

MAIN FOCUS OF THE CHAPTER

Technological Advancements and How It Has Impacted Manufacturing

Technological advancements have profoundly transformed the manufacturing industry, ushering in an era characterized by increased efficiency, precision, and adaptability. The integration of digital technologies into manufacturing processes has not only optimized production but also redefined operational paradigms.

Automation and Robotics

The introduction of automation and robotics has revolutionized manufacturing by performing repetitive and labour - intensive tasks with high precision and speed. This shift has led to significant reductions in production times and labour costs, while enhancing product consistency. For instance, companies like Priestley's Gourmet Delights in Australia have implemented Artificial Intelligence (AI) - powered smart factories, utilizing autonomous vehicles and collaborative robots to double production capacity and improve operational efficiency (World Economic Forum, 2021).

Internet of Things (IoT) and Smart Manufacturing

The Internet of Things (IoT) has facilitated the creation of interconnected systems within manufacturing environments, enabling real - time data collection and analysis. This connectivity allows for predictive maintenance, where sensors monitor equipment conditions and alert for maintenance before failures occur, thereby minimizing downtime and reducing costs (McKinsey & Company, 2020). Additionally, Internet of Things (IoT) enhances supply chain management by providing real - time data on inventory levels and production schedules, leading to more efficient operations.

Artificial Intelligence (AI) and Machine Learning

Artificial Intelligence (AI) and machine learning have become integral in optimizing manufacturing processes. These technologies analyse vast amounts of data to improve decision - making, enhance quality control, and streamline production schedules. For example, Artificial Intelligence (AI) algorithms can detect defects early in the production process, reducing waste and ensuring consistent product quality (Accenture, 2020). Furthermore, Artificial Intelligence (AI) aids in demand forecasting and inventory management, allowing manufacturers to respond more effectively to market changes.

Additive Manufacturing (3D Printing)

Additive manufacturing, commonly known as 3D printing, enables the creation of complex parts and components by adding material layer by layer. This method allows for rapid prototyping, customization, and reduced material waste. Industries such as aerospace and healthcare have leveraged 3D printing to produce lightweight components and personalized medical devices, respectively (Deloitte, 2019).

Digital Twins

Digital twins are virtual representations of physical assets or processes that allow manufacturers to simulate, analyse, and optimize operations in a virtual environment. This technology enables proactive maintenance by predicting equipment failures and optimizing production processes before physical implementation, leading to improved efficiency and reduced costs (Investopedia, 2023).

Augmented Reality (AR) and Virtual Reality (VR)

Augmented Reality (AR) and Virtual Reality (VR) technologies have enhanced training and maintenance procedures in manufacturing. Augmented Reality (AR) overlays digital information onto the physical world, guiding workers through complex tasks, while Virtual Reality (VR) provides immersive simulations for training purposes. These technologies improve skill acquisition and reduce errors, leading to more efficient operations (PwC, 2020).

Big Data and Analytics

The utilization of big data analytics has empowered manufacturers to make informed decisions based on comprehensive data analysis. By examining production data, manufacturers can identify inefficiencies, optimize processes, and enhance product quality. This data - driven approach has led to more agile and responsive manufacturing systems (Forbes Insights, 2022).

Cloud and Edge Computing

Cloud computing allows manufacturers to store and process data remotely, reducing the need for on - site infrastructure and enabling scalability. Edge computing, on the other hand, processes data closer to where it's generated, reducing latency and improving real - time decision - making. Both technologies contribute to more flexible and efficient manufacturing operations (Gartner, 2021).

The integration of technological advancements in manufacturing has led to smarter, more efficient, and adaptable production systems. As industries continue to embrace these technologies, the future of manufacturing promises further innovations that will enhance productivity, sustainability, and competitiveness in the global market.

Human Resources (HR) in the 4th Industrial Revolution

The Fourth Industrial Revolution (4IR) is characterized by the fusion of physical, digital, and biological technologies, leading to profound changes in industries and societies. Human Resources (HR) functions are at the forefront of this transformation, adapting to new technologies and methodologies to manage the evolving workforce effectively.

Automation and Artificial Intelligence (AI) are transforming Human Resources (HR) by optimizing operational efficiency and enabling data - driven strategies. Routine administrative tasks such as payroll processing, benefits administration, and employee data management are increasingly managed through Artificial Intelligence (AI) - powered systems, significantly reducing manual errors and allowing Human Resources (HR) professionals to concentrate on strategic planning. In the realm of recruitment, Artificial Intelligence (AI) facilitates resume screening, initial assessments, and candidate suitability predictions, which contributes to faster, more objective hiring decisions. Furthermore, big data analytics empower Human Resources

(HR) departments to analyse employee performance, forecast retention risks, and support organizational development initiatives with greater precision (Bersin, 2018).

The accelerating pace of technological change has made reskilling and up skilling essential components of workforce development strategies. To remain competitive and future - ready, organizations are increasingly investing in training programs that emphasize digital literacy, data analysis, and other high - demand competencies. E - Learning platforms have become vital tools in this effort, offering flexible, on - demand access to educational resources that support continuous learning and adaptability among employees. Moreover, collaboration between businesses and educational institutions plays a crucial role in aligning academic curricula with evolving industry requirements, thereby helping to bridge the growing skills gap (World Economic Forum, 2020).

In the context of the Fourth Industrial Revolution (4IR), enhancing employee experience and engagement has become a strategic priority for organizations aiming to attract and retain top talent. Personalized on boarding programs, which are tailored to individual roles and needs, play a key role in helping new hires integrate effectively into the company culture and fostering early engagement. Flexible work arrangements, including remote work and adaptable schedules, have emerged as vital tools for supporting employee well - being and accommodating diverse personal and professional needs. Furthermore, continuous feedback mechanisms — such as pulse surveys and real - time feedback tools — allow Human Resources (HR) departments to monitor employee satisfaction, identify issues proactively, and foster a culture of open communication and responsiveness (Gallup, 2021).

Diversity, Equity, and Inclusion (DEI) have become central to organizational strategies in the era of the Fourth Industrial Revolution (4IR), where innovation thrives on varied perspectives and equitable practices. Advanced technologies such as Artificial Intelligence (AI) and data analytics are increasingly employed to detect and reduce biases in recruitment and performance evaluations, thereby fostering more objective and equitable outcomes. Organizations are also implementing inclusive policies aimed at supporting diverse talent pools and creating environments in which all employees feel respected, empowered, and represented. Additionally, the establishment of Employee Resource Groups (ERGs) provides underrepresented groups with platforms for community building, mentorship, and advocacy, reinforcing a culture of inclusion across the workforce (McKinsey & Company, 2020).

While the Fourth Industrial Revolution (4IR) has introduced significant advancements in Human Resources (HR) practices, it also presents a range of challenges that organizations must navigate. One major obstacle is resistance to change, as employees may be hesitant to embrace new technologies, requiring effective change management strategies to ensure smooth and inclusive transitions. The growing dependence on digital platforms also heightens cyber security and data privacy concerns, compelling Human Resources (HR) departments to implement stringent security protocols to protect sensitive employee information. Additionally, the rapid pace of technological advancement often results in skill shortages, underscoring the urgent need for ongoing learning and development initiatives to maintain workforce readiness (Cleopatra Enterprise, 2023).

The Fourth Industrial Revolution presents both opportunities and challenges for Human Resources (HR). By embracing technological advancements, investing in employee development, and fostering inclusive environments, Human Resources (HR) can play a pivotal role in shaping the future workforce. Continuous adaptation and strategic planning are essential for organizations to thrive in the evolving landscape of 4IR.

Training, Development and Motivation of Staff in the New Era of Technology

Training focuses on improving employees' current job performance through structured learning of specific skills and knowledge (Noe, 2017). In contrast, development is a broader, long - term approach aimed at preparing individuals for future roles and responsibilities, emphasizing personal and professional growth through methods like mentorship, coaching, job rotation, and formal education (Dessler, 2019).

The authors think that integrating key aspects of training and development is imperative within a manufacturing organisation, as this will enable the employees to upgrade their technological skills so that they can easily migrate to the new manufacturing processes and smart factory technologies. This will further ensure employee engagement within the manufacturing business and foster staff motivation, which fuels staff commitment and loyalty towards the organisation. Baroudi et al. (2018) highlight the effectiveness of the 70 - 20 - 10 training model in preparing employees to independently manage technological changes. Jeni et al. (2021) emphasize that employee training is vital for human resource development, as tailored programs improve skills, attitudes, and performance.

Employee development is a strategic investment in human capital, essential for driving organizational growth and competitiveness (Khan et al., 2011). To grow, management should invest capital into training and development programs for staff within manufacturing organisations. Through continuous training and development, organizations can enhance employee skills, increase productivity, and address performance gaps. In manufacturing, this means equipping staff with the knowledge and abilities needed to perform tasks efficiently. Identifying and addressing skill weaknesses among employees ensures improved performance and helps organizations achieve their goals (Koval, 2015; Ganesh & Indradevi, 2015).

Human Capital Theory views training as an investment because employees' skills and knowledge add value to an organization, especially in manufacturing where technical skills are vital. Training improves worker competency, reduces errors, and boosts efficiency (Tan, 2014; Hatch & Dyer, 2004). Additionally, Maslow's Hierarchy of Needs links training to employee motivation, showing that skills - based training enhances self-esteem and that advanced training supports self - actualization, helping staff reach their full potential (Benson & Dundis, 2003; Boon et al., 2011).

Memon et al. (2021) found a positive link between training and development and employee performance, highlighting that skill acquisition keeps work engaging and employees involved. Similarly, Kishore and Fonceca (2023) emphasized the significant impact of training and development on performance in a paper manufacturing company with 150 employees. The findings highlight a strong positive correlation between training and development and improved employee performance. Research by Chaudhry et al. (2017) and Siddiqui & Sahar (2019) supports this, showing that effective training not only enhances performance but also boosts employee engagement. The studies recommend a systematic training approach to ensure consistent and successful outcomes.

According to Madhumitha, S. & Asrafi, S. (2024) in his study the following were highlighted as key ways in which technology is integrated into training and development:

- E - Learning Platforms;
- Mobile Learning (M - Learning);

- Virtual Reality (VR) and Augmented Reality (AR);
- Gamification;
- Learning Management Systems (LMS):
- Micro learning; and
- Personalised Learning Paths

According to Madhumitha & Asrafi (2024), technology plays a significant role in modern training and development through several key methods. E - Learning platforms offer digital courses and training materials that employees can access anytime and anywhere, using a variety of content formats such as videos, interactive modules, quizzes, and assessments. This approach supports self - paced and personalised learning experiences. Mobile learning, or M - learning, leverages smartphones and tablets to deliver content on the go, making it especially useful for remote workers or those with tight schedules. Technologies like Virtual Reality (VR) and Augmented Reality (AR) create immersive and interactive training environments, particularly valuable for high - risk or complex scenarios such as medical procedures or emergency response (Madhumitha & Asrafi, 2024). Gamification introduces elements like points, badges, leader boards, and rewards to boost engagement and motivation, transforming training into a fun and interactive experience. Learning Management Systems (LMS) serve as centralised platforms for organising and tracking training content, progress, and assessments, supporting scalability and regulatory compliance. Micro learning delivers short, focused modules that target specific skills or knowledge areas, ideal for quick updates or reinforcement. Lastly, personalised learning paths use data analytics and Artificial Intelligence (AI) to tailor training based on individual skill levels, learning preferences, and career goals, ensuring that employees receive relevant and impactful development opportunities aligned with both personal and organisational objectives (Madhumitha & Asrafi, 2024).

Another area to look at is employee engagement. Employee engagement as a key factor influencing motivation and performance. Braune and Roodt (2011) define engagement as an employee's long - term motivation and willingness to go beyond basic duties to help the organization thrive. Naidoo (2025) highlights that engagement is a current and critical workplace issue that directly affects performance. In South Africa, relevant laws and policies support strong employee engagement practices. The text stresses that employees are valuable assets, and to ensure their productivity and efficiency, organizations must provide a supportive work environment, recognize their efforts, build trust, and invest in skill development. Engaged and skilled employees ultimately contribute to competitive advantage.

Tarafdar et al. (2007) argue that employees often lack the necessary skills to use new technologies, requiring them to invest extra time and effort to understand technical aspects and complete tasks. It should be advised that not all workers understand and comprehend advanced machines, algorithms and processes within smart factories. So training is very important, but the mind - set of employees and their ability to adapt and understand no matter their qualifications, is not a given for smooth adaption to new technologies.

Hewitt (2015) examined employee engagement as the state in which personnel are passionately and logically devoted to a group affiliation otherwise the firm, as corresponding to three primary deeds namely: Say (personnel speak respectfully about the firm to people within and outside the organisation), Stay (staff boast a fierce desire to be a part of the firm) and Strive (staff employ additional energy and seize opportunities to contribute to business success). Chiumento (2007) referred to his study as Happiness at Work Research, which depicts engagement as a positive, bilateral relationship between worker and organisation. Engaged staff, have organisational commitment and do something extra for each other, as they view investing in their affiliation as mutually beneficial. According to Naidoo (2025) Employee engagement is essential to employees within an organisation. An employee's sense of self and feelings toward their job are key indicators of job performance. Motivation plays a crucial role in strengthening their commitment to the organization. This emotional connection and commitment lead employees to care deeply about their work and the organization, motivating them to go beyond their basic responsibilities. Their performance is driven not merely by financial rewards, but by a genuine sense of purpose, loyalty, and personal investment in the success of the organization.

In the new era of technology, training, development, and motivation of staff must evolve to keep pace with digital transformation. Here's a structured approach to tackle each of these elements effectively:

Training in the Digital Age

Training in the digital age has evolved to meet the dynamic needs of a tech - driven workforce, with a focus on personalization, flexibility, and engagement. Artificial Intelligence (AI) - driven learning platforms, such as Coursera for Business, LinkedIn Learning, and Udemy for Business, enable the creation of personalized learning paths tailored to individual roles, performance levels, and learning preferences. On - demand micro learning — delivered through short videos, interactive quizzes, and simulations — supports continuous development by allowing employees to learn in the flow of work. Immersive technologies like Virtual Reality (VR) and Augmented Reality (AR) offer highly effective training environments for complex or high - risk scenarios, including technical skills and on boarding experiences. Additionally, the use of gamification strategies, including point systems, levels, and achievement badges, increases learner engagement and motivation, resulting in improved knowledge retention and training outcomes (Pillai, 2020).

Developing Staff in a Tech - Driven Workplace

In a technology - driven work environment, staff development strategies must focus on cultivating digital fluency, cross - disciplinary capabilities, and transparent career progression. Establishing digital competency frameworks — outlining essential skills such as data literacy, remote collaboration, and cyber security awareness — allows organizations to benchmark employee capabilities and guide targeted development efforts. Cross - functional development opportunities, such as assigning employees to projects beyond their core roles, promote adaptability, foster innovation, and strengthen both technical and business acumen. Mentorship and coaching programs, particularly those that incorporate peer learning and reverse mentoring, are effective for accelerating digital learning and bridging generational knowledge gaps. Furthermore, digital career pathing platforms empower employees by mapping out advancement opportunities and identifying the skills required to achieve their career goals, thereby enhancing engagement and retention (Bersin, 2020).

Motivating Staff in the Technology Era

Motivating employees in the technology era requires organizations to leverage digital tools in ways that support purpose, autonomy, recognition, well - being, and inclusive communication. Emphasizing how technology initiatives contribute to broader organizational goals — such as sustainability, innovation, or social responsibility — helps employees connect their work to a larger sense of purpose. Empowerment is further enhanced through flexible digital work environments, enabling individuals to choose how, when, and where they work. Artificial Intelligence (AI) - powered recognition platforms like Bonusly and Kudos facilitate real - time acknowledgment of contributions, fostering a culture of appreciation and innovation. To support employee well - being, companies are integrating digital wellness tools, virtual therapy options, and stress management resources, while promoting responsible tech use through initiatives such as digital detox policies. Finally, communication platforms like Slack, Microsoft Teams, and Human Resources (HR) chatbots ensure that internal communication is timely, transparent, and accessible, strengthening engagement across all levels of the organization (Gartner, 2022).

Table 2. Training, development, and motivation of staff in the new era of technology

Area	Strategy	Tool / Method
Training	Personalized & on - demand learning	Artificial Intelligence (AI) - driven LMS, Micro learning
Development	Cross - skilling, digital competency, mentorship	Virtual Reality (VR) / Augmented Reality (AR), Career maps, Mentorship
Motivation	Autonomy, recognition, well - being	Slack, Recognition platforms, Apps

Source: Own

Sustainable Manufacturing

Sustainable manufacturing involves producing goods through economically viable processes that reduce environmental impact, conserve energy and resources, and ensure safety for workers, communities, and consumers.

Sustainable manufacturing is guided by a set of core principles aimed at minimizing environmental impact while maintaining economic and social responsibility. Resource efficiency is central, emphasizing the reduction of energy, water, and raw material consumption, alongside the adoption of renewable inputs where feasible. Pollution prevention is achieved through cleaner production methods that limit emissions, effluents, and waste. A product life cycle perspective ensures that environmental considerations span from design and material sourcing to usage and end - of - life disposal. Process optimization techniques, such as Lean manufacturing and Six Sigma, contribute to waste reduction and operational efficiency. Integrating circular economy principles further promotes sustainability by prioritizing reuse, remanufacturing, and recycling over traditional linear production models. Lastly, ensuring worker safety and accounting for the broader societal impacts of manufacturing processes underscore the importance of ethical and community - conscious production practices (U. S. EPA, 2020).

Implementing effective strategies is essential for advancing sustainable manufacturing practices across industries. Green product design emphasizes eco - design principles aimed at reducing material consumption, improving durability, and utilizing recyclable and non - toxic materials. Energy management initiatives involve the adoption of high - efficiency technologies and a transition toward renewable energy sources such as solar and wind power. Waste minimization is achieved through zero - waste targets, reuse of production scrap, and the adoption of industrial symbiosis, where one company's waste serves as another's input. Water conservation efforts include the reuse of process water via filtration and the harvesting of rainwater for non - potable applications. Digital technologies — especially those associated with Industry 4. 0, including Internet of Things (IoT), Artificial Intelligence (AI), and big data analytics — enable real - time monitoring of resource use and environmental performance. For example, smart sensors can detect machine inefficiencies and prevent energy or material overuse. Additionally, sustainable supply chain management practices involve sourcing from certified green suppliers, conducting sustainability audits, and promoting low - emission logistics to minimize environmental footprints across the value chain (Bocken, Short, Rana, & Evans, 2014).

Table 3. Benefits of sustainable manufacturing

Environmental	Economic	Social
Reduced emissions & waste	Lower operational costs	Improved brand image
Conservation of resources	Increased efficiency & innovation	Employee satisfaction
Biodiversity protection	Access to green markets & subsidies	Community support

Source: Own

Toyota serves as a prominent example of sustainable manufacturing through its integration of environmental stewardship into core business operations. The company's adoption of lean manufacturing principles has significantly reduced waste and improved overall resource efficiency across its production lines. Toyota has also led innovation in environmentally conscious product development, particularly through its hybrid vehicle portfolio, which emphasizes fuel economy and reduced emissions. In its facilities, Toyota implements green building strategies, including the use of solar panels for renewable energy generation and water recycling systems to conserve and reuse process water. These initiatives reflect Toyota's broader commitment to sustainability and serve as a benchmark for other manufacturers seeking to align operational excellence with environmental responsibility (Toyota Motor Corporation, 2022).

Here are two compelling case studies that illustrate sustainable manufacturing practices in the textile and automotive industries:

Textile Industry: Tirupur, India - A Model of Sustainable Transformation

Tirupur, located in Tamil Nadu, India, has emerged as a global exemplar of sustainable transformation within the textile industry. Historically plagued by severe environmental degradation due to unregulated industrial effluents, the region has implemented pioneering sustainability practices that have reversed its ecological footprint. A cornerstone of this transformation is the widespread adoption of Zero Liquid Discharge (ZLD) systems, now operational in approximately 300 dyeing units. This effort is supported by 18 Common Effluent Treatment Plants (CETPs) and 60 Individual Effluent Treatment Plants (IETPs), which together recycle over 120 million litres of water each day. In addition to advanced wastewater management, the region has embraced renewable energy, notably solar and wind power, to minimize dependence on fossil fuels. Rainwater harvesting systems have also been integrated into industrial operations to conserve local water resources. These collective measures have drastically reduced water consumption and greenhouse gas emissions, positioning Tirupur as a leading model of sustainable fashion manufacturing on a global scale (Centre for Science and Environment [CSE], 2021).

Automotive Industry: Ford Motor Company's Use of Renewable Materials

Ford Motor Company has emerged as a leader in sustainable innovation within the automotive industry through its pioneering use of renewable materials in vehicle manufacturing. One of its most notable achievements is the replacement of petroleum - based polyol with soybean - based foam in car seat cushions, a move that significantly reduces the carbon footprint of its vehicles. In addition to soy - based components, Ford has integrated a total of nine sustainable materials into various parts of its automobiles, including recycled plastics, wheat straw, and coconut fibres. These initiatives underscore the company's commitment to reducing reliance on fossil fuels and promoting the circular economy. Ford's success demonstrates that renewable materials can be effectively scaled in mass automotive production, paving the way for broader industry adoption (Ford Motor Company, 2022).

Both case studies highlight the importance of innovation and commitment to sustainability in manufacturing. By adopting eco - friendly practices and materials, industries can significantly reduce their environmental impact while maintaining economic viability.

SUGGESTIONS & RECOMMENDATIONS

Strategies to Enhance Human Resources (HR) in Aligning To the New Technological Advancements

Enhancing Human Resources (HR) to align with new technological advancements requires a multi - faceted strategy that addresses skills development, cultural adaptation, organizational structure, and the integration of digital tools. Here are several key strategies:

Up Skilling and Reskilling Programs

Up skilling and reskilling programs have become critical in preparing the workforce for the demands of the Fourth Industrial Revolution. Organizations are increasingly implementing continuous learning initiatives that focus on emerging technologies such as Artificial Intelligence (AI), machine learning, data analytics, cyber security, and cloud computing to enhance employee competencies and adaptability. Micro learning and e - learning platforms provide flexible, bite - sized content that enables employees to engage in self - paced learning, making skill development more accessible and efficient. Furthermore, partnerships with technology providers and online education platforms like Coursera and LinkedIn Learning offer organizations access to curated, industry - relevant training content tailored to evolving workforce needs (Deloitte, 2021).

Digital Mindset and Culture Shift

Fostering a digital mind - set and initiating a cultural shift are essential components of successful digital transformation in the workplace. Structured change management programs, such as the ADKAR model or Kotter's 8 - Step Change Model, provide frameworks to guide employees through transitions and reduce resistance to new technologies. Leadership plays a critical role in this process, as senior executives who actively model and endorse digital initiatives help reinforce their importance and build organizational trust. Cultivating a culture of innovation — where experimentation is encouraged, failure is treated as a learning opportunity, and technological advancement is celebrated — further supports the development of a resilient and forward - thinking workforce (Westerman, Bonnet, & McAfee, 2014).

Strategic Workforce Planning

Strategic workforce planning is vital for aligning human capital with the evolving demands of the digital age. Organizations are increasingly conducting tech - readiness audits to assess existing workforce competencies and identify skill gaps in relation to emerging technologies. Talent forecasting, often powered by predictive analytics, enables companies to anticipate future roles and skill requirements, thereby guiding recruitment and employee development strategies. Additionally, agile talent models are gaining prominence, characterized by the integration of flexible staffing approaches such as gig workers, external consultants, and Artificial Intelligence (AI) - driven workflows to enhance organizational adaptability and responsiveness (Manyika et al., 2017).

Artificial Intelligence (AI) and Automation Integration

The integration of Artificial Intelligence (AI) and automation in Human Resources (HR) is transforming how organizations attract, manage, and retain talent. Modern Human Resources (HR) technologies now incorporate Artificial Intelligence (AI) - driven tools for various functions, such as applicant tracking systems for recruitment, sentiment analysis for gauging employee engagement, and intelligent platforms for performance management and feedback. Importantly, the emphasis is shifting toward augmentation rather than replacement — using Artificial Intelligence (AI) to support and enhance human decision - making, reduce administrative burden, and enable Human Resources (HR) professionals to focus on more strategic, value - adding activities. This approach reinforces the collaborative potential of humans and machines in the future of work (Davenport & Ronanki, 2018).

Employee Engagement and Experience

Employee engagement and experience have become key priorities in digitally transforming workplaces, with technology playing a pivotal role in enhancing these dimensions. Tech - enhanced on boarding processes, such as virtual reality (VR) simulations and interactive platforms, offer immersive and engaging introductions to organizational culture and workflows. Digital feedback mechanisms, including pulse surveys and real - time sentiment analysis tools, enable Human Resources (HR) teams to continuously monitor employee satisfaction and adapt strategies proactively. Furthermore, the shift toward hybrid work models is supported by collaboration platforms like Slack, Microsoft Teams, and Asana, which facilitate communication, productivity, and flexibility in increasingly decentralized work environments (Gartner, 2021).

Data - Driven Human Resources (HR) Decision Making

Data - driven decision - making has become a cornerstone of modern Human Resources (HR) strategy, enabling more informed and strategic choices across all facets of workforce management. Human Resources (HR) analytics tools allow organizations to extract actionable insights from employee data, supporting initiatives such as turnover prediction, performance evaluation, and workforce planning. The use of real - time dashboards and reporting tools further enhances managerial decision - making by providing immediate visibility into key metrics, facilitating proactive interventions and continuous improvement. This evidence - based approach empowers Human Resources (HR) to align closely with business objectives and demonstrate measurable impact (Bassi, Carpenter, & McMurrer, 2012).

Cyber Security and Compliance Training

As organizations increasingly rely on digital infrastructure, cyber security and compliance training have become essential components of workforce development. Regular training on data protection policies, acceptable technology use, and regulatory compliance ensures that employees understand their responsibilities in safeguarding sensitive information. To reinforce this knowledge, many organizations deploy simulated cyber - attacks — such as phishing exercises — and conduct awareness campaigns to test and improve employee readiness against real - world threats. These proactive measures not only reduce organizational risk but also foster a security - conscious culture across all levels of the workforce (SANS Institute, 2021).

Key Components of the Human Resources (HR) Digital Transformation Framework

A successful Human Resources (HR) digital transformation framework is built on several interrelated components that enable organizations to adapt effectively to the digital era. The adoption of digital innovation involves transitioning from traditional human resource management (HRM) systems to electronic HRM (e - HRM) by integrating advanced technologies such as Artificial Intelligence (AI), social media platforms, and cloud - based solutions. Supporting this shift requires a robust technology infrastructure capable of handling secure data storage and high - speed connectivity to ensure seamless access to digital tools. Furthermore, organizational structures must be redesigned to support cross - functional collaboration and agile digital workflows. Enhancing employee engagement through digital platforms enables continuous feedback, real - time performance monitoring, and individualized development opportunities. Finally, data - driven decision - making powered by Human Resources (HR) analytics allows for evidence - based strategies in talent acquisition, employee retention, and workforce planning, aligning Human Resources (HR) practices with broader organizational goals (Stone, Deadrick, Lukaszewski, & Johnson, 2015).

Implementation Stages

The successful implementation of Human Resources (HR) digital transformation requires a structured, multi - stage approach to ensure alignment with organizational objectives and long - term sustainability. The process begins with assessment and strategy development, wherein organizations evaluate existing Human Resources (HR) processes to identify opportunities for digital enhancement and strategic alignment. Following this, technology selection and implementation involve choosing digital tools and platforms that best support the company's goals and workforce needs. In the next phase, process redesign and automation help streamline workflows, reduce manual errors, and improve overall efficiency. Centralizing Human Resources (HR) data and leveraging analytics enables data - driven insights into workforce trends, facilitating informed decision - making. Change management and training are critical to ensuring employee buy - in, requiring structured plans and comprehensive training programs to ease the transition. Lastly, continuous improvement and innovation involve regularly evaluating the effectiveness of digital initiatives and adapting to emerging technologies to maintain competitiveness (Ulrich, Younger, Brockbank, & Ulrich, 2019).

This framework serves as a comprehensive guide for Human Resources (HR) departments aiming to align with technological advancements, ensuring a structured and effective transformation process.

FUTURE RESEARCH DIRECTIONS

Studies on Africa and how the continent is dealing with keeping abreast of technological changes in the manufacturing industry should be reviewed, as there is not much research conducted in this area, and it can add to research in the African continent. Another complicated research area that should be highlighted in the future is cybercrime and how it can disrupt a smart factory that is highly reliant on technology. Challenges faced by Asian, African and European manufacturers should be investigated in the future. This research can bring forth important research in the manufacturing sector, as each continent has their unique circumstances that can impact the sector and pose challenges for the industry. Further future areas of research should also be in the area of Human Resources (HR) and training for the employees within the manufacturing sector.

CONCLUSION

The evolution of manufacturing has been profoundly shaped by technological advancements — from mechanization and mass production to automation, Artificial Intelligence (AI), and Industry 4.0. These transformations have not only redefined production processes but have also significantly altered the workforce landscape. In this rapidly changing environment, Human Resources (HR) plays a pivotal role in bridging the gap between technological progress and human adaptability. Human Resources (HR) has moved beyond just handling basic administrative tasks like payroll, attendance, and hiring paperwork. Today, Human Resources (HR) plays a much bigger and more important role in helping a company succeed. It works closely with leadership to plan for the company's future workforce needs (workforce planning), develop employees' skills and careers (talent development), and help the organization adapt to changes such as new technologies or market shifts (change management). In other words, Human Resources (HR) is now a key player in shaping the company's strategy and growth, not just managing routine office tasks. By fostering a culture of continuous learning, supporting digital skill acquisition, and promoting organizational agility, Human Resources (HR) ensures that human capital remains aligned with the evolving demands of the manufacturing sector. Ultimately, the future of manufacturing will depend not just on technology, but on how effectively organizations leverage the human element to drive innovation and sustainable growth.

REFERENCES

Accenture. (2020). AI: Built to scale - Rewire your business for AI. Accenture. https://www. accenture. com/us - en/insights/artificial - intelligence/ai - investments - built - to - scale

Baroudi, S., Abdulaal, D., & Rampur, M. (2018). Training effectiveness during technological changes in Dubai based companies: Perspectives from managers and employees. *Development and Learning in Organizations*, *32*(3), 1–3. DOI: 10.1108/DLO-10-2017-0085

Bassi, L., Carpenter, R., & McMurrer, D. (2012). *HR analytics handbook*. McBassi & Company.

Benson, S. G., & Dundis, S. P. (2003). Understanding and motivating health care employees: Integrating Maslow's hierarchy of needs, training and technology. *Journal of Nursing Management*, *11*(5), 315–320. DOI: 10.1046/j.1365-2834.2003.00409.x PMID: 12930537

Bersin, J. (2018). HR technology disruptions for 2018: Productivity, design, and intelligence reign. Deloitte Insights. https://www2. deloitte. com/insights/us/en/focus/human - capital - trends/2018/ hr - technology - disruptions. html

Bersin, J. (2020). Building digital talent: The new workforce strategy. Josh Bersin Academy. https://joshbersin. com/2020/01/building - digital - talent - the - new - workforce - strategy/

Bocken, N. M. P., Short, S. W., Rana, P., & Evans, S. (2014). A literature and practice review to develop sustainable business model archetypes. Journal of Cleaner Production, 65, 42 - 56. . jclepro. 2013. 11. 039DOI: <ALIGNMENT.qj></ALIGNMENT>10. 1016/j

Boon, C. T., Low, D. F. L., Lim, S. Y., Ng, X. Q., & Wong, P. P. (2011). *The influence of core self - evaluation (CSE) on employees' motivation in continuing education*. UTAR.

BULGURCU, B. & MASHATAN, A. A. (2024). Environmental Factors that Hinder an O0rganization's Ability to Learn from Cyber Incidents: A Case Study on SolarWinds.

Centre for Science and Environment (CSE). (2021). Tirupur's water - wise success: A zero liquid discharge model for the textile industry. https://www. cseindia. org/tirupur - zld - textile - model

Chaudhry, N. I., Jariko, M. A., Mushtaque, T., Mahesar, H. A., & Ghani, Z. (2017). Impact of Working Environment and Training & Development on Organization. *European Journal of Training and Development Studies*, *4*(2), 33–48.

Chiumento. (2007). Happiness at Work. Research Report 2007. Retrieved from https://www. scribd. com/document/72220871/Happiness - at - Work - Index - 2007

Cleopatra Enterprise. (2023). Top HR challenges in the fourth industrial revolution. https://www. cleopatraenterprise. com/hr - challenges - in - the - 4ir/

Dana, L. P., Salamzadeh, A., Mortazavi, S., & Hadizadeh, M. (2022). Investigating the impact of international markets and new digital technologies on business innovation in emerging markets. *Sustainability (Basel)*, *14*(2), 983. DOI: 10.3390/su14020983

Davenport, T. H., & Ronanki, R. (2018). Artificial intelligence for the real world. *Harvard Business Review*, *96*(1), 108–116.

Deloitte. (2019). 3D opportunity for production: Additive manufacturing makes its (business) case. Deloitte Insights. https://www2. deloitte. com/insights/us/en/focus/3d - opportunity/overview. html

Deloitte. (2021). 2021 Global Human Capital Trends: The social enterprise in a world disrupted. https://www2. deloitte. com/global/en/pages/human - capital/articles/gx - human - capital - trends. html

Dessler, G. (2019). *Human Resource Management*. Pearson.

Forbes Insights. (2022). 2022 state of manufacturing report. Forbes & Oracle NetSuite. https:// www. forbes. com/sites/forbestechcouncil/2022/08/16/the - state - of - manufacturing - 2022/? Sh=6b9823666b3b

Ford Motor Company. (2022). Sustainability report 2022: Driving the future of sustainability. https://corporate. ford. com/microsites/sustainability - report - 2022

Gallup. (2021). State of the Global Workplace: 2021 Report. https://www. gallup. com/workplace/349484/state - of - the - global - workplace - 2021 - report. aspx

Ganesh, M., & Indradevi, R. (2015). Importance and effectiveness of training and development. *Mediterranean Journal of Social Sciences*, *6*(1), 334–338. DOI: 10.5901/mjss.2015.v6n1p334

Gartner. (2021). Hype cycle for edge computing. Gartner Research. https://www. gartner. com/ en/documents/4009496

Gartner. (2021). Top 5 priorities for HR leaders in 2021. https://www. gartner. com/en/human - resources/trends/top - priorities - for - hr - leaders

Gartner. (2022). redesigning work for well - being: The future of employee experience. https:// www. gartner. com/en/articles/redesigning - work - for - well - being

Gini, K. B., & Agala, H. O. (2023). The Impact of Macro - Environmental Factors on Business Performance. *International Journal of Research and Innovation in Social Science*, *7*(1), 1837–1843. DOI: 10.47772/IJRISS.2023.7011145

Hatch, N. W., & Dyer, J. H. (2004). Human capital and learning as a source of sustainable competitive advantage. *Strategic Management Journal*, *25*(12), 1155–1178. DOI: 10.1002/smj.421

Hewitt, A. (2015). Trends in Global Employee Engagement - Making engagement happen. Aon Hewitt cooperation. https://www. aon. com/attachments/human - capital - consulting/2015 - trends - in - global - employee - engagement - report. pdf, Hill, C. W., Schilling, M. A. & Jones, G. R. (2020). Strategic management: an integrated approach: theory and cases, Cengage Learning: Boston.

Houti, M., Abbadi, L. E., & Abouabdellah, A. (2019). CSFs for the Implementation of the Hybrid Lean ERP System. Stakeholders Interactions. *Advances in Science, Technology and Engineering Systems Journal*, *4*(6), 443–447. DOI: 10.25046/aj040655

Investopedia. (2023). How generative AI is revolutionizing manufacturing. https://www. investopedia. com/how - generative - ai - is - revolutionizing - manufacturing - 7480630

Jeni, F., & AlAmin, M. (2021). The impact of training and development on employee performance and productivity:an empirical study on private bank of Noakhali region in Bangladesh. *South Asian Journal of Social Studies and Economics, 1*(1), 1–18. DOI: 10.9734/sajsse/2021/v9i230234

Khan, R. A. G., Khan, F. A., & Khan, M. A. (2011). Impact of training and development on organisational performance. *Global Journal of Management and Business Research, 11*(7), 63–68.

Kishore, S., & Fonceca, C. M. (2023). Impact of Training and Development on Employee Performance and Productivity. *Journal of Academia and Industrial Research, 11*(3), 43–47.

Koval, S. (2015). Employee training and development [Undergraduate thesis, University College Prague]. IS UCP. https://is. ucp. cz/th/u9zm0/

Madhumitha, S., & Asrafi, S. (2024). The Role of Technology in Revolutionizing Training & Development. *International Journal of Business and Management Invention, 13*(4), 134–137.

Manyika, J., Lund, S., Chui, M., Bughin, J., Woetzel, J., Batra, P., & Sanghvi, S. (2017). Jobs lost, jobs gained: Workforce transitions in a time of automation. McKinsey Global Institute. https:// www. mckinsey. com/featured - insights/future - of - work/jobs - lost - jobs - gained - what - the - future - of - work - will - mean - for - jobs - skills - and - wages

McKinsey & Company. (2020). Diversity wins: How inclusion matters. https://www. mckinsey. com/featured - insights/diversity - and - inclusion/diversity - wins - how - inclusion - matters

McKinsey & Company. (2020). the internet of things: Catching up to an accelerating opportunity. https://www. mckinsey. com/business - functions/mckinsey - digital/our - insights/the - internet - of - things - catching - up - to - an - accelerating - opportunity

Memon, M. A., Salleh, R., Mirza, M. Z., Cheah, J. H., Ting, H., Ahmad, M. S., & Tariq, A. (2021). Satisfaction matters: The relationships between HRM practices, work engagement and turnover intention. *International Journal of Manpower, 42*(1), 21–50. DOI: 10.1108/IJM-04-2018-0127

Mokyr, J. (1990). *The Lever of Riches: Technological Creativity and Economic Progress*. Oxford University Press.

Mokyr, J. (1998). The Second Industrial Revolution, 1870 - 1914. In Floud, R., & Johnson, P. (Eds.), *the Cambridge Economic History of Modern Britain* (Vol. 2, pp. 1–29). Cambridge University Press.

Naidoo, V. (2020). *Creativity and Innovation for Entrepreneurs in the Circular Economy, IGI Global, Handbook of Research on Entrepreneurship Development and Opportunities in Circular Economy* (Baporikar, N., Ed.).

Naidoo, V. (2025). Employee Engagement: A Key Catalyst Within an Organization During COVID - 19 and Beyond the Pandemic - The South African Perspective. In T. Kapoor, A. Singh, K. Sahay, H. Kaur, & R. Verma (Eds.), Sustainable Management Practices for Employee Retention and Recruitment (pp. 127 - 150). IGI Global Scientific Publishing. . ch007DOI: 10.4018/979-8-3693-8679-8

Noe, R. A. (2017). *Employee Training and Development.* McGraw - Hill Education.

Pillai, R. (2020). Digital learning trends in the workplace: Personalization, gamification, and immersive technologies. *Journal of Workplace Learning, 32*(3), 181–198. DOI: 10. 1108/JWL - 01 - 2020 - 0010

PwC. (2020). Seeing is believing: How virtual reality and augmented reality are transforming business and the economy. PricewaterhouseCoopers. https://www. pwc. com/gx/en/technology/ seeing - is - believing. html

Rifkin, J. (2011). *The Third Industrial Revolution: How Lateral Power Is Transforming Energy, the Economy, and the World.* Palgrave Macmillan.

SANS Institute. (2021). Security awareness report: Managing human cyber security risk. https:// www. sans. org/white - papers/security - awareness - report/

Schwab, K. (2017). *The Fourth Industrial Revolution.* Crown Business.

Shen, C. C., Yeh, C. C., & Lin, C. N. (2022). Using the perspective of business information technology technicians to explore how information technology affects business competitive advantage. *Technological Forecasting and Social Change, 184*(1), 121973. DOI: 10.1016/j. techfore.2022.121973

Siddiqui, D. A., & Sahar, N. (2019). The Impact of Training & Development and Communication on Employee Engagement. A Study of Banking Sector. *Business Management and Strategy, 10*(1), 23. DOI: 10.5296/bms.v10i1.14592

Stearns, P. N. (2013). *The Industrial Revolution in World History* (4th ed.). Westview Press.

Stone, D. L., Deadrick, D. L., Lukaszewski, K. M., & Johnson, R. (2015). The influence of technology on the future of human resource management. Human Resource Management Review, 25 (2), 216 - 231. . hrmr. 2015. 01. 002DOI: <ALIGNMENT.qj></ALIGNMENT>10. 1016/j

Tan, E. (2014). Human capital theory: A holistic criticism. *Review of Educational Research, 84*(3), 411–445. DOI: 10.3102/0034654314532696

Tarafdar, M., Tu, Q., Ragu-Nathan, B. S., & Ragu-Nathan, T. S. (2007). The impact of technostress on role stress and productivity. *Journal of Management Information Systems, 24*(1), 301–328. DOI: 10.2753/MIS0742-1222240109

Thakur, V. (2020). Framework for PESTEL dimensions of sustainable healthcare waste management: Learnings from COVID - 19 outbreak. *Journal of Cleaner Production, 287*(1), 125562. PMID: 33349739

Toyota Motor Corporation. (2022). Sustainability Data Book 2022. https://global. toyota/en/sustainability/report/sdb

U. S. Environmental Protection Agency. (2020). Sustainable manufacturing: Principles and practices. https://www. epa. gov/sustainability/sustainable - manufacturing

Ulrich, D., Younger, J., Brockbank, W., & Ulrich, M. (2019). *HR Transformation: Building Human Resources from the Outside In*. McGraw - Hill Education.

Vial, G. (2019). Understanding digital transformation: A review and a research agenda. *The Journal of Strategic Information Systems*, *28*(2), 118–144. DOI: 10.1016/j.jsis.2019.01.003

Westerman, G., Bonnet, D., & McAfee, A. (2014). *Leading Digital: Turning Technology into Business Transformation*. Harvard Business Review Press.

World Economic Forum. (2020). the Future of Jobs Report 2020. https://www. weforum. org/reports/the - future - of - jobs - report - 2020

World Economic Forum. (2021). the future of jobs report 2020. World Economic Forum. https://www. weforum. org/reports/the - future - of - jobs - report - 2020

Wyszomirski, A. (2023). THE ECONOMIC MACRO - ENVIRONMENT AS A FACTOR SHAPING THE USE OF NEW TECHNOLOGIES. SELECTED ASPECTS. Management & Quality/Zarządzanie i Jakość, 5 (4).

Xin, L., Lang, S., & Mishra, A. R. (2022). Evaluate the challenges of sustainable supply chain 4. 0 implementation under the circular economy concept using new decision making approach. *Operations Management Research : Advancing Practice Through Research*, *15*(1), 773–792. DOI: 10.1007/s12063-021-00243-7

Zafar & Kantola. (2020). The effect of macro environmental and firm - level factors on open innovations in product development and their impact on firms' performance. In Advances in Human Factors, Business Management and Leadership: Proceedings of the AHFE 2019 International Conference on Human Factors, Business Management and Society, and the AHFE International Conference on Human Factors in Management and Leadership, July 24 - 28, 2019, Washington DC, USA 10 (pp. 61 - 71). Springer International Publishing.

KEY TERMS AND DEFINITIONS

Artificial Intelligence (AI): Artificial Intelligence (AI) is a type of technology that allows machines or computer systems to mimic human intelligence. It enables them to learn from experience, adapt to new information, and carry out tasks like decision - making, problem - solving, or even understanding language.

Automation: Automation refers to the use of machines, software, or technology to perform tasks with little or no human involvement. It helps streamline work processes, reduce manual effort, and improve efficiency in both simple and complex operations.

Employee Engagement: Employee engagement is the emotional and professional connection that employees feel toward their work and workplace. When employees are engaged, they are more motivated, committed, and likely to contribute positively to the organization's success.

Human Resources (HR): Human Resources (HR) is the department or team within an organization responsible for managing people. This includes hiring, training, performance management, employee relations, and ensuring that workplace policies support both employees and business goals.

Chapter 16

The Future of Collaborative Manufacturing

Milan Sahoo
https://orcid.org/0009-0007-0619-3682
Brainware University, India

Aditya Saha
https://orcid.org/0009-0004-1783-8587
Brainware University, India

Ranjan Banerjee
https://orcid.org/0009-0003-1950-7530
Brainware University, India

Misha Samanta
https://orcid.org/0009-0000-6037-2760
Brainware University, India

Malay Maity
https://orcid.org/0009-0001-9173-6763
Brainware University, India

Debesh Das
https://orcid.org/0009-0009-0287-4995
Brainware University, India

Salankara Sarkar
https://orcid.org/0009-0005-0123-3724
Brainware University, India

ABSTRACT

The world of manufacturing is undergoing a dramatic transformation, away from static, linear modes of production toward dynamic, interdependent systems. Central to this transformation is the concept of collaborative manufacturing, wherein human imagination and sophisticated automation collaborate to create an elastic and extremely responsive production space. "The Future of Collaborative Manufacturing" looks at this revolution, focusing on the main technologies, social implications, and ethical factors that will drive the next wave of intelligent factories. An important part of this vision is the seamless embedding of cognitive artificial intelligence (AI). AI's ability to process vast amounts of data, identify patterns, and make intelligent decisions will enable machines to anticipate needs, automate processes, and adapt to changing conditions in real time. Adaptive robotics, with advanced sensors and AI-based control systems, will further enhance this collaborative environment.

DOI: 10.4018/979-8-3373-2985-7.ch016

INTRODUCTION

The realm of manufacturing is experiencing a seismic shift, moving away from traditional, static production lines towards a more dynamic and interconnected ecosystem. At the epicenter of this metamorphosis is the burgeoning concept of collaborative manufacturing, a revolutionary approach that seamlessly blends the creative prowess of human ingenuity with the sophisticated precision of advanced automation. This synergy cultivates an exceptionally agile and responsive production environment, capable of adapting swiftly to the ever-fluctuating demands of the contemporary marketplace.

This paper embarks on an exploration of "The Future of Collaborative Manufacturing," delving into the pivotal technologies, societal implications, and ethical considerations that are poised to shape the next wave of intelligent factories. A cornerstone of this visionary landscape is the seamless integration of cognitive artificial intelligence (AI). AI's remarkable ability to process vast troves of data, discern intricate patterns, and execute intelligent decisions in real time empowers machines to anticipate production needs, automate complex processes, and adapt to rapidly changing conditions.

Adaptive robotics, equipped with advanced sensor technology and AI-driven control systems, further amplifies this collaborative environment. These robots transcend the limitations of repetitive tasks, operating in flexible and dynamic configurations alongside human workers, adapting to evolving production requirements while maintaining stringent safety protocols. Furthermore, the emergence of immersive digital twin technology is poised to revolutionize the way we design, simulate, and optimize manufacturing processes. Digital twins, virtual replicas of physical systems, facilitate real-time monitoring, predictive maintenance, and virtual testing, thereby minimizing downtime and maximizing overall efficiency. This technology also enhances human-machine interaction, providing workers with intuitive visualizations and data-driven insights.

The efficacy of human-machine interfaces is paramount in fostering seamless collaboration. The cumbersome and often obtuse interfaces of the past are being superseded by personalized and adaptive interfaces. Voice commands, hand and finger gesture recognition, and augmented reality (AR) are becoming increasingly prevalent, enabling more natural and effortless interactions. This paper also intends to explore the social and ethical ramifications of these technological advancements, considering the impact on the workforce, the potential for job displacement, and the imperative of ensuring equitable access to these transformative technologies. We will address the importance of responsible innovation and the need for ethical guidelines to govern the development and deployment of collaborative manufacturing systems. Finally, we will consider the environmental impact of this shift, exploring how these technologies can contribute to more sustainable and resource-efficient manufacturing practices. By examining these multifaceted dimensions, we aim to provide a comprehensive understanding of the transformative potential of collaborative manufacturing and its implications for the future of industry.

BACKGROUND

The future of collaborative manufacturing paints a vivid picture of interconnected ecosystems where human ingenuity and technological prowess merge seamlessly. This evolution transcends mere automation, delving into a realm of synergistic partnerships between humans and machines, each contributing their unique strengths. We're witnessing a departure from traditional, siloed manufacturing processes, moving towards a dynamic environment where real-time data flow, adaptive technologies, and human-centric design converge. Industry 4.0, with its intricate web of IoT sensors, cloud computing, and AI algorithms, serves as the backbone of this transformation. Digital twins, virtual replicas of physical assets, allow for meticulous simulations and optimizations, enabling manufacturers to anticipate and mitigate potential issues before they materialize. Advanced robotics, particularly cobots, are no longer confined to isolated cages; they work alongside human operators, handling repetitive tasks with precision, freeing up human workers to focus on complex problem-solving and creative endeavors. This human-robot collaboration fosters a safer and more efficient work environment, where the physical burden is shared, and human cognitive abilities are maximized. The deluge of data generated within these connected environments is not merely collected but analyzed, interpreted, and utilized to drive informed decision-making. AI and machine learning algorithms sift through vast datasets, identifying patterns, predicting maintenance needs, and optimizing production flows. This data-driven approach empowers manufacturers to anticipate challenges, adapt to market fluctuations, and continuously improve their processes. Flexibility and agility become paramount in this dynamic landscape. Manufacturers are no longer bound by rigid production lines; they can rapidly reconfigure their operations to meet evolving customer demands, producing customized products with unprecedented speed and efficiency. The concept of distributed and decentralized production gains traction, enabling localized manufacturing, reducing transportation costs, and enhancing supply chain resilience. This shift towards localized production fosters closer relationships with customers and suppliers, creating a more responsive and sustainable manufacturing ecosystem. However, the human element remains indispensable. The future of collaborative manufacturing necessitates a skilled workforce capable of navigating the complexities of advanced technologies and adapting to new ways of working. This requires significant investment in training and education, bridging the skills gap and equipping workers with the knowledge and abilities to thrive in this evolving landscape. Ethical considerations surrounding AI and automation, as well as the need for robust cybersecurity measures, also demand careful attention. The future of collaborative manufacturing is not just about technology; it's about creating a harmonious balance between human potential and technological capabilities, fostering innovation, and driving sustainable economic growth.

MAIN FOCUS OF THE CHAPTER

The central focus of "The Future of Collaborative Manufacturing" revolves around the paradigm shift from traditional, isolated manufacturing practices to an interconnected, dynamic ecosystem where human and technological capabilities are seamlessly integrated. This chapter emphasizes the transition beyond mere automation, highlighting the critical importance of synergistic collaboration between humans and advanced technologies, particularly robotics, artificial intelligence, and the Internet of Things. It delves into how this collaboration fosters a more agile, responsive, and efficient manufacturing environment, driven by real-time data analysis and adaptable production systems. A key aspect is the exploration of how these technologies enable flexible manufacturing, customized production, and distributed networks, allowing for greater responsiveness to market demands and enhanced supply chain resilience. Moreover, the chapter underscores the crucial role of the human element in this evolving landscape, emphasizing the need for a skilled workforce capable of navigating and leveraging these advanced technologies. It addresses the implications for workforce development, highlighting the necessity for training and education to bridge the skills gap and prepare workers for the future of manufacturing. Finally, the chapter considers the broader impacts of this transformation, examining the potential for increased productivity, enhanced sustainability, and the creation of more resilient and innovative manufacturing ecosystems, while also acknowledging the ethical considerations and challenges associated with the integration of AI and automation. Essentially, the core focus is on how the convergence of human ingenuity and technological advancements is reshaping manufacturing, creating a future characterized by collaboration, adaptability, and sustainable growth.

SOLUTIONS AND RECOMMENDATIONS

Navigating the intricate landscape of the future of collaborative manufacturing demands a multifaceted approach, blending technological advancement with human-centric strategies. A primary focus must be placed on cultivating a workforce adept at operating within this dynamic environment. This necessitates substantial investment in comprehensive training programs, equipping individuals with proficiencies in robotics, artificial intelligence, data analytics, and digital twin technology. Embracing lifelong learning becomes crucial, fostering continuous professional development to enable adaptation to the ever-evolving technological terrain of manufacturing. Collaboration between industry and academia is paramount, fostering partnerships to develop curricula that directly address the sector's evolving needs. Apprenticeships and vocational training, providing practical, hands-on experience, must be prioritized. Simultaneously, building a robust digital infrastructure is essential. Secure and scalable IoT networks must be established to ensure seamless data flow across the manufacturing ecosystem, while robust cybersecurity measures are implemented to safeguard sensitive data and critical infrastructure from cyber threats. Investment in cloud and edge computing is vital to facilitate real-time data processing and analysis, enabling proactive decision-making and optimization. Standardizing data protocols and ensuring interoperability are crucial for the seamless integration of diverse systems and technologies. Fostering a collaborative ecosystem is equally vital. Open innovation and knowledge sharing must be encouraged, promoting partnerships between manufacturers, technology providers, and research institutions. Industry consortia and partnerships should be established to develop shared standards and best practices, while platforms for supply chain collaboration enhance transparency and communication across the value chain. Supporting the development of regional manufacturing hubs fosters innovation and economic growth through collaborative efforts. Agile and flexible manufacturing practices should be embraced, implementing modular and reconfigurable production systems to enable rapid adaptation to market demands. Digital twins should be utilized for simulation and optimization, minimizing downtime and improving efficiency. Additive manufacturing facilitates rapid prototyping and customization, enhancing product innovation and responsiveness. Real-time monitoring and predictive maintenance optimize asset utilization and reduce downtime. Ethical considerations and sustainability must be prioritized. Developing ethical guidelines for AI and automation ensures responsible and equitable deployment. Sustainable manufacturing practices reduce waste, minimize environmental impact, and optimize resource utilization. Investing in circular economy initiatives promotes product lifecycle management and resource recovery. Worker well-being and safety are paramount, necessitating human-centric design and ergonomics in collaborative environments. Finally, supporting small and medium-sized enterprises (SMEs) is crucial. Providing access to affordable technology and training, offering financial incentives and grants, developing platforms for knowledge sharing and collaboration, and simplifying regulatory requirements are essential for their successful integration into the future of collaborative manufacturing.

THE ROLE OF EDGE COMPUTING IN REAL-TIME COLLABORATIVE MANUFACTURING

The advent of real-time collaborative manufacturing hinges on the ability to process and act upon data with minimal latency. In this context, edge computing emerges as a pivotal technology, shifting computational power closer to the source of data generation. Traditional cloud-centric models, while powerful, often introduce delays that hinder the responsiveness demanded by dynamic production environments, (Wang et al., 2018). This is where edge computing's decentralized architecture proves invaluable.

Within a collaborative manufacturing setting, a multitude of sensors, robots, and human-machine interfaces generate a constant stream of data. This data, encompassing everything from machine performance to environmental conditions, requires immediate analysis to enable swift decision-making. Edge computing facilitates this by deploying processing capabilities directly onto local devices or servers situated near the factory floor. By reducing the distance data travels to be processed, latency is significantly minimized, allowing for near-instantaneous feedback loops.

This low-latency processing is crucial for applications like adaptive robotics, where robots must react in real-time to changes in their environment or human interactions. For instance, if a robot detects an unexpected obstacle or a deviation from its programmed path, edge computing enables it to make an immediate adjustment, ensuring safety and maintaining production efficiency. Similarly, in predictive maintenance scenarios, edge devices can analyze sensor data to detect anomalies and trigger alerts before equipment failures occur, minimizing downtime and optimizing resource utilization.

Furthermore, edge computing enhances the reliability of collaborative manufacturing systems. By distributing processing power, the reliance on a single, centralized cloud server is reduced. This decentralized approach mitigates the risk of system-wide failures and ensures that critical operations can continue even in the event of network disruptions. This is particularly important in remote or geographically dispersed manufacturing facilities where network connectivity may be intermittent.

In essence, edge computing empowers real-time collaborative manufacturing by enabling localized data processing, reducing latency, enhancing reliability, and facilitating immediate responses to dynamic production demands. As manufacturing continues to evolve towards increasingly intelligent and interconnected systems, the role of edge computing will only become more critical in driving efficiency, adaptability, and innovation.

The burgeoning realm of real-time collaborative manufacturing hinges significantly on the transformative power of edge computing. In this dynamic landscape, where split-second decisions and immediate responses are paramount, the traditional reliance on centralized cloud infrastructure proves inadequate. Edge computing, by strategically placing computational resources closer to the data source, directly addresses the latency and bandwidth limitations inherent in cloud-centric models. This localized processing capability is not merely a technological enhancement; it's a fundamental shift that enables a new paradigm of responsiveness and efficiency within the manufacturing ecosystem. Within the context of collaborative manufacturing, where human operators, robotic systems, and intelligent machines interact in real-time, the ability to process and analyze data at the edge becomes crucial. Imagine a scenario where a collaborative robot, working alongside a human, needs to react instantaneously to a change in the environment or a human gesture. The latency associated with sending data to a distant cloud server and awaiting a response would render such real-time interaction impractical. Edge computing, however, facilitates on-the-spot processing, enabling the robot to react with the required immediacy. This capability extends beyond robotic collaboration, influencing every aspect of the manufacturing process. Real-time monitoring of machine performance, predictive maintenance based on immediate data analysis, and dynamic adjustments to production lines based on sensor inputs all benefit from the low-latency processing that edge computing provides. In collaborative scenarios, where multiple devices and systems interact simultaneously, the sheer volume of data generated can overwhelm traditional networks. Edge computing mitigates this by processing data locally, filtering out irrelevant information, and transmitting only essential data to the cloud for further analysis or long-term storage. This reduction in data transmission not only alleviates network congestion but also enhances security by minimizing the exposure of sensitive data to external networks. Furthermore, edge computing enables the creation of localized control loops, allowing for autonomous decision-making and immediate corrective actions within the manufacturing environment. This capability is particularly valuable in dynamic and unpredictable scenarios, where centralized control systems may struggle to adapt quickly enough. For instance, in a flexible manufacturing system that produces customized products, edge computing can enable real-time adjustments to production parameters based on immediate feedback from sensors and human operators. This level of responsiveness is essential for meeting the demands of a rapidly changing market. Moreover, the integration of edge computing with AI and machine learning algorithms opens up new possibilities for intelligent automation. By deploying AI models at the edge, manufacturers can enable real-time analysis of sensor data and immediate decision-making, without relying on constant communication with a centralized cloud server. This localized AI processing enhances the autonomy and adaptability of manufacturing systems, enabling them to respond to changing conditions with greater speed and precision. In essence, the role of edge computing in real-time collaborative manufacturing is not just about faster data processing; it's about enabling a new level of intelligence, responsiveness, and efficiency within the manufacturing environment. It's about empowering machines to make immediate decisions, fostering seamless human-machine collaboration, and enabling the creation of dynamic and adaptable production systems that can meet the demands of the future.

DEVELOPING OPEN-SOURCE PLATFORMS FOR COLLABORATIVE ROBOTICS IN SMALL TO MEDIUM ENTERPRISES (SMES)

The transformative potential of collaborative robotics is undeniable, yet its widespread adoption within Small to Medium Enterprises (SMEs) remains hindered by significant cost barriers and the complexities of proprietary systems. To address these challenges, the development of open-source platforms for collaborative robotics presents a compelling solution. By fostering a collaborative ecosystem, these platforms can democratize access to advanced automation, empowering SMEs to enhance their productivity and competitiveness, (Brettel et al., 2014).

Open-source platforms for collaborative robotics offer several key advantages. Firstly, they significantly reduce the initial investment required for robotic integration. By eliminating licensing fees and providing access to readily available software and hardware components, SMEs can overcome the financial hurdles associated with traditional robotic solutions. Secondly, open-source platforms promote customization and adaptability. SMEs can tailor robotic systems to their specific needs and production processes, leveraging the collective knowledge and expertise of a vibrant community of developers and users.

Furthermore, open-source platforms foster innovation and collaboration. By providing access to modular software libraries, hardware designs, and simulation tools, SMEs can experiment with new robotic applications and contribute to the ongoing development of the platform. This collaborative approach accelerates the pace of innovation and ensures that the platform remains relevant to the evolving needs of the manufacturing industry.

The development of such platforms necessitates a focus on user-friendliness and accessibility. Intuitive interfaces, comprehensive documentation, and robust community support are crucial(Lee et. Al., 2015) for enabling SMEs with limited robotics expertise to effectively integrate and utilize these systems. Additionally, the platform should prioritize interoperability, allowing SMEs to seamlessly integrate existing machinery and software systems. Security and reliability are also paramount considerations. Open-source platforms must incorporate robust security measures to protect sensitive data and prevent unauthorized access. Rigorous testing and validation procedures are essential for ensuring the reliability and stability of robotic systems deployed in production environments.

THE INTEGRATION OF BLOCKCHAIN TECHNOLOGY FOR SECURE AND TRANSPARENT SUPPLY CHAINS IN COLLABORATIVE MANUFACTURING

In the era of interconnected manufacturing, the integrity and transparency of supply chains are paramount. Collaborative manufacturing, with its complex network of partners, necessitates robust mechanisms to ensure data security and provenance. Blockchain technology offers a compelling solution, providing a decentralized and immutable ledger for tracking materials, processes, and transactions throughout the supply chain. The inherent characteristics of blockchain, such as its distributed nature and cryptographic security, address the critical challenges of data tampering and single points of failure. By recording every transaction and movement of goods on a shared, tamper-proof ledger, blockchain establishes an auditable trail that enhances trust and accountability among supply chain participants.

In collaborative manufacturing, where multiple stakeholders contribute to the production process, blockchain can facilitate seamless information sharing and real-time visibility, (Tao et al., 2018). For instance, manufacturers can use blockchain to track the origin and quality of raw materials, ensuring ethical sourcing and compliance with industry standards. Similarly, logistics providers can leverage blockchain to monitor the movement of goods, reducing delays and minimizing the risk of counterfeiting. Furthermore, blockchain can streamline complex financial transactions within the supply chain. Smart contracts, self-executing agreements embedded within the blockchain, can automate payments and enforce contractual obligations, reducing administrative overhead and minimizing disputes. This automation also contributes to faster settlement times and improved cash flow management. The integration of blockchain also enhances the security of intellectual property and proprietary data. By encrypting sensitive information and controlling access through permissioned blockchains, manufacturers can protect their designs and processes from unauthorized disclosure, (Romero et al., 2016). This is particularly crucial in collaborative environments where multiple partners share confidential information. Moreover, blockchain can empower consumers by providing them with access to verified information about the products they purchase. By scanning QR codes or accessing blockchain-based platforms, consumers can trace the journey of a product from its origin to its final destination, ensuring authenticity and sustainability. However, the successful implementation of blockchain in collaborative manufacturing requires addressing challenges such as scalability, interoperability, and regulatory compliance. Industry-wide collaboration and standardization are essential for establishing a robust and interoperable blockchain ecosystem.

ADVANCEMENTS IN HAPTIC FEEDBACK SYSTEMS FOR ENHANCED HUMAN-ROBOT COLLABORATION IN COMPLEX ASSEMBLY TASKS

Complex assembly tasks, often requiring intricate manipulations and precise force control, pose significant challenges for human-robot collaboration. To bridge the gap between human dexterity and robotic precision, advancements in haptic feedback systems are crucial. These systems provide tactile and force-related information to human operators, enabling them to perceive and interact with the robotic environment in a more intuitive and effective manner.

Traditional robotic systems often rely solely on visual feedback, which can be insufficient for tasks requiring fine motor skills and nuanced force control. Haptic feedback systems, on the other hand, provide a sense of touch, allowing operators to feel the forces exerted by the robot and the textures of objects being manipulated. This enhanced sensory input significantly improves the operator's ability to control the robot's movements and adapt to unforeseen situations.

Recent advancements in haptic feedback technology have focused on improving the fidelity and responsiveness of tactile and force feedback, (El Zaatari et al., 2019). High-resolution tactile sensors, capable of detecting minute changes in pressure and texture, are being integrated into robotic end-effectors. These sensors provide detailed information about the contact between the robot and the workpiece, enabling operators to perceive subtle variations in surface properties and adjust their grip accordingly. Furthermore, advancements in force feedback algorithms and actuators have led to more realistic and immersive haptic experiences. By accurately replicating the forces exerted by the robot, these systems allow operators to feel the weight and resistance of objects being manipulated. This is particularly important for tasks involving delicate or deformable materials, where precise force control is essential to prevent damage.

The integration of haptic feedback systems into collaborative robotic platforms also necessitates the development of intuitive interfaces. These interfaces should seamlessly translate tactile and force information into meaningful feedback that can be easily understood and acted upon by human operators. This may involve the use of visual cues, auditory signals, or even direct stimulation of the operator's skin. The application of haptic feedback systems extends beyond simple assembly tasks. They can be used to enhance human-robot collaboration in a wide range of complex scenarios, including surgical procedures, remote maintenance, and hazardous environment operations. By providing operators with a more immersive and intuitive experience, haptic feedback systems can significantly improve the efficiency, safety, and precision of human-robot collaboration.

THE EVOLUTION OF AI-DRIVEN PREDICTIVE MAINTENANCE IN DIGITAL TWIN-ENABLED COLLABORATIVE FACTORIES

The pursuit of operational excellence within collaborative factories has driven a paradigm shift towards proactive maintenance strategies. At the forefront of this evolution lies the convergence of artificial intelligence (AI) and digital twin technology, enabling a new era of AI-driven predictive maintenance. This integration transforms reactive and even preventative maintenance into a dynamic, data-informed process, significantly enhancing efficiency and minimizing disruptions.

Digital twins, virtual replicas of physical assets and processes, serve as the foundation for this transformation. By continuously mirroring real-time sensor data from the factory floor, these digital counterparts provide a comprehensive and dynamic view of equipment health and performance. AI algorithms then analyze this wealth of data, identifying subtle patterns and anomalies that may indicate impending failures. The evolution of AI-driven predictive maintenance can be characterized by several key advancements. Initially, basic statistical models were employed to detect deviations from normal operating parameters. However, with the proliferation of sensor data and the development of advanced machine learning techniques, AI systems have become increasingly sophisticated. Deep learning algorithms, for instance, can analyze complex time-series data to identify intricate relationships and predict failures with greater accuracy.

Furthermore, the integration of AI with digital twins enables the simulation of various operating scenarios. This allows manufacturers to assess the impact of different maintenance strategies and optimize their schedules accordingly. By simulating potential failures and evaluating the effectiveness of various interventions, manufacturers can minimize downtime and extend equipment lifespan.

The collaborative nature of modern factories further amplifies the benefits of AI-driven predictive maintenance. By sharing data and insights across the production network, manufacturers can gain a holistic view of equipment health and optimize maintenance schedules for the entire facility. This collaborative approach also facilitates the development of shared AI models and best practices, accelerating the adoption of predictive maintenance across the industry.

However, the successful implementation of AI-driven predictive maintenance requires addressing several challenges. These include ensuring data quality and security, developing robust AI models, and integrating these systems seamlessly into existing workflows. As AI and digital twin technology continue to evolve, we can expect to see even more sophisticated and effective predictive maintenance solutions emerge, further enhancing the efficiency and resilience of collaborative factories.

5G NETWORKS AND THE REVOLUTION OF LATENCY-SENSITIVE APPLICATIONS IN COLLABORATIVE MANUFACTURING

The advent of 5G networks is poised to dramatically reshape the landscape of collaborative manufacturing, particularly in its ability to address the critical issue of latency. Latency, the delay in data transmission, has long been a bottleneck for real-time applications within industrial settings. In collaborative manufacturing, where seamless communication and instantaneous responses are paramount, the low-latency capabilities of 5G present a transformative opportunity.

Traditional wired or even 4G networks often struggle to deliver the near-zero latency required for applications like remote robotics control, augmented reality (AR) guided assembly, and real-time sensor data analysis. These applications demand immediate feedback loops to ensure precision, safety, and efficiency. 5G, with its ultra-reliable low-latency communication (URLLC) capabilities, promises to bridge this gap, enabling a new level of responsiveness and agility in collaborative manufacturing.

Impact on Latency-Sensitive Applications

Remote Robotics Control

5G's low latency allows for precise and real-time control of robots from remote locations. This is crucial for tasks performed in hazardous environments or for enabling expert operators to assist with complex procedures from afar. The reduced latency minimizes the risk of errors and ensures smooth, coordinated movements.

Augmented Reality (AR) Guided Assembly

AR overlays real-time information onto the physical world, guiding workers through complex assembly processes. 5G's low latency ensures that AR overlays are synchronized with the worker's movements, providing seamless and accurate guidance. This improves efficiency, reduces errors, and enhances worker safety.

Real-Time Sensor Data Analysis

Collaborative manufacturing environments generate vast amounts of sensor data, which needs to be analyzed in real time to monitor equipment health, optimize processes, and detect anomalies. 5G's low latency enables rapid data processing and analysis, allowing for immediate responses to changing conditions.

Digital Twin Interaction

Interacting with a digital twin in real time requires very low latency. 5G helps to make this interaction feel more like interacting with the real world, and therefore more useful.

Autonomous Guided Vehicles (AGV)

AGV's require very low latency networks to operate safely and efficiently. 5G networks enable the vehicles to react to changes in the environment in real time.

Benefits of 5G in Collaborative Manufacturing

I. Enhanced Responsiveness: 5G's low latency enables near-instantaneous feedback loops, improving the responsiveness of collaborative manufacturing systems.
II. Increased Efficiency:** Real-time data analysis and remote-control capabilities optimize processes and minimize downtime.
III. Improved Safety:** Low-latency communication ensures timely responses to potential hazards, enhancing worker safety.
IV. Greater Flexibility:** 5G's wireless connectivity allows for flexible deployment of robots and other equipment, adapting to changing production needs.

Challenges and Considerations

- **Infrastructure Deployment**: The widespread adoption of 5G requires significant infrastructure investment.
- **Security:** Ensuring the security of 5G networks is crucial for protecting sensitive data.
- **Interoperability:** Seamless integration of 5G with existing manufacturing systems is essential.

5G networks are poised to revolutionize collaborative manufacturing by enabling a new era of latency-sensitive applications. By addressing the critical issue of latency, 5G empowers manufacturers to create more responsive, efficient, and safe production environments.

CRAFTING EQUITABLE INTERACTIONS

Designing Inclusive Human-Machine Interfaces (HMIs) for Diverse Workforces in Collaborative Manufacturing The advent of collaborative manufacturing heralds a new era of human-machine symbiosis, where seamless interaction is paramount for optimizing productivity and ensuring worker well-being. However, the promise of this collaborative future hinges on the design of Human-Machine Interfaces (HMIs) that are not only efficient but also inherently inclusive, catering to the diverse needs of a modern workforce. This necessitates a departure from traditional, standardized interfaces towards a more nuanced and adaptable approach, recognizing the inherent heterogeneity in human capabilities, backgrounds, and experiences.

In the context of collaborative manufacturing, HMIs serve as the crucial bridge between human operators and complex automated systems. They encompass a wide spectrum of technologies, from traditional control panels and displays to advanced modalities like voice commands, gesture recognition, and augmented reality (AR). The challenge lies in ensuring that these interfaces are accessible and usable by all members of the workforce, regardless of their age, physical abilities, cognitive styles, language proficiency, or cultural backgrounds.

Understanding Diversity in the Workforce

Designing inclusive HMIs begins with a deep understanding of the diverse characteristics that shape user experiences. This includes:

1. **Physical Abilities:** HMIs must accommodate individuals with varying degrees of physical dexterity, strength, and sensory perception. This may involve incorporating adjustable controls, tactile feedback, and alternative input methods.
2. **Cognitive Styles:** Individuals process information and learn in different ways. HMIs should cater to diverse cognitive styles by offering multiple modalities of information presentation, customizable learning pathways, and adaptive feedback mechanisms.
3. **Language Proficiency:** In a globalized workforce, language barriers can pose significant challenges. HMIs should support multilingual interfaces, incorporate visual aids, and utilize clear and concise language.
4. **Cultural Backgrounds:** Cultural differences can influence user expectations and preferences. HMIs should be designed with cultural sensitivity, avoiding potentially offensive symbols or gestures and adapting to diverse interaction styles.
5. **Age:** An age diverse workforce brings a range of experience, but also potential differences in comfort levels with technology. HMIs must be flexible to suit the needs of both younger and older workers.
6. **Education and Experience:** HMIs must be adaptable to the skill level of the user. This means that a new employee should be able to understand the HMI, and that an experienced employee should be able to use advanced features.

Principles of Inclusive HMI Design

To create truly inclusive HMIs, designers should adhere to the following principles:

- **Universal Design:** Adopt a universal design approach, aiming to create interfaces that are usable by all people, to the greatest extent possible, without the need for adaptation or specialized design.
- **Modularity and Customization:** Offer modular interfaces that can be customized to individual user preferences and needs. This may involve allowing users to adjust font sizes, color schemes, input methods, and feedback modalities.
- **Multimodal Interaction:** Incorporate multiple modalities of interaction, such as voice commands, gesture recognition, and tactile feedback, to accommodate diverse user preferences and abilities.
- **Adaptive Learning and Feedback:** Implement adaptive learning systems that adjust to the user's skill level and provide personalized feedback.
- **Clear and Concise Communication:** Utilize clear and concise language, avoiding jargon and technical terms. Incorporate visual aids, such as icons and diagrams, to enhance understanding.

- **Error Prevention and Recovery:** Design interfaces that minimize the risk of errors and provide clear and intuitive error recovery mechanisms.
- **Contextual Awareness:** Design systems that are aware of the user's context, such as their location, task, and skill level, and adapt accordingly.
- **Accessibility Standards:** Adhere to established accessibility standards, such as WCAG (Web Content Accessibility Guidelines), to ensure that HMIs are usable by individuals with disabilities.
- **User-Centered Design:** Prioritize user-centered design methodologies, involving diverse user groups in the design and testing process. This ensures that HMIs are designed with the needs and preferences of real users in mind.

Technologies for Inclusive HMIs

Several technologies can contribute to the development of inclusive HMIs:

- **Voice Recognition and Natural Language Processing (NLP):** Enables users to interact with machines using natural language, overcoming language barriers and providing hands-free control.
- **Gesture Recognition:** Allows users to interact with machines using hand and body gestures, providing an intuitive and natural interaction modality.
- **Augmented Reality (AR):** Overlays digital information onto the physical world, providing contextual guidance and enhancing situational awareness.
- **Haptic Feedback**: Provides tactile and force-related information, enhancing the user's perception of the robotic environment.
- **Eye Tracking:** Enables users to interact with machines using eye movements, providing an alternative input method for individuals with limited mobility.
- **AI-Powered Personalization:** AI algorithms can be used to personalize HMIs based on individual user preferences and needs.

Benefits of Inclusive HMIs

- **Increased Productivity:** Inclusive HMIs enhance user efficiency and reduce errors, leading to increased productivity.
- **Improved Safety:** Accessible interfaces minimize the risk of accidents and ensure worker safety.
- **Enhanced Job Satisfaction:** Inclusive HMIs empower workers and create a more positive and engaging work environment.
- **Expanded Workforce Participation:** Inclusive HMIs enable individuals with diverse abilities and backgrounds to participate in collaborative manufacturing.
- **Reduced Training Costs:** Intuitive and easy-to-use interfaces minimize the need for extensive training.

Designing inclusive HMIs is not merely a matter of technological innovation, it is a fundamental ethical imperative. By embracing diversity and prioritizing user-centered design, we can create collaborative manufacturing environments that are not only efficient and productive but also equitable and empowering for all members of the workforce.

THE ETHICAL IMPLICATIONS OF AI-DRIVEN DECISION-MAKING IN COLLABORATIVE MANUFACTURING: ENSURING FAIRNESS AND TRANSPARENCY

The integration of artificial intelligence (AI) into collaborative manufacturing is ushering in an era of unprecedented efficiency and automation. However, the increasing reliance on AI-driven decision-making raises profound ethical concerns, particularly regarding fairness and transparency. As AI algorithms assume more critical roles in production processes, it becomes imperative to scrutinize their potential biases and establish robust ethical guidelines to ensure responsible deployment.

Collaborative manufacturing environments, characterized by human-robot interaction and data-driven optimization, are particularly susceptible to the ethical challenges posed by AI. AI algorithms are often trained on vast datasets that may reflect existing societal biases, leading to discriminatory outcomes in areas such as resource allocation, task assignment, and performance evaluation. This can perpetuate inequalities and undermine the principles of fairness and inclusivity that are essential for a harmonious and equitable workplace.

Potential Biases in AI Algorithms

AI algorithms, particularly those based on machine learning, learn from the data they are trained on. If this data is biased, the resulting algorithms will also be biased, leading to discriminatory outcomes. Common sources of bias include:

- **Historical Bias:** Datasets may reflect historical patterns of discrimination, such as gender or racial bias in hiring or promotion practices.
- **Sampling Bias:** Datasets may not be representative of the population they are intended to model, leading to skewed results.
- **Measurement Bias:** Data collection methods may introduce systematic errors or inaccuracies, leading to biased results.
- **Algorithmic Bias:** The design of AI algorithms themselves may introduce biases, such as favoring certain types of data or decision-making strategies.
- **Interpretative Bias:** How humans interpret the data output by AI systems can also introduce bias.

In collaborative manufacturing, these biases can manifest in various ways. For example, an AI algorithm used to assign tasks to workers may unfairly favor certain demographics based on historical performance data. Similarly, an AI-powered quality control system may be biased against products manufactured by certain teams or individuals.

The Need for Ethical Guidelines

To mitigate the risks associated with AI-driven decision-making, it is essential to establish clear ethical guidelines that address the following key areas:

- **Fairness and Non-Discrimination:** AI algorithms should be designed and trained to ensure fairness and non-discrimination, avoiding biases that could lead to discriminatory outcomes.
- **Accountability and Responsibility:** Clear lines of accountability and responsibility should be established for AI-driven decisions. This includes identifying who is responsible for the design, training, and deployment of AI algorithms, as well as who is responsible for the consequences of their decisions.
- **Human Oversight and Control:** AI algorithms should be subject to human oversight and control, ensuring that humans retain the ultimate authority over critical decisions.
- **Data Privacy and Security:** AI algorithms should be designed and deployed in a manner that protects data privacy and security, adhering to relevant regulations and best practices.
- **Social Impact Assessment:** Before deploying AI algorithms, manufacturers should conduct thorough social impact assessments to identify potential risks and benefits.
- **Stakeholder Engagement:** Manufacturers should engage with stakeholders, including workers, unions, and community groups, to ensure that AI-driven decision-making is aligned with their values and concerns.
- **Continuous Monitoring and Evaluation:** AI algorithms should be continuously monitored and evaluated to identify and address potential biases and unintended consequences.
- **Ethical Training and Education:** Workers and managers should receive ethical training and education to ensure that they understand the ethical implications of AI-driven decision-making.
- **Regulatory Frameworks:** Governments and industry associations should develop regulatory frameworks to govern the development and deployment of AI in collaborative manufacturing.

Ensuring Fairness and Transparency in Practice

Implementing these ethical guidelines requires a multi-faceted approach, involving technical solutions, organizational policies, and cultural changes. Key strategies include:

- **Data Auditing and Preprocessing:** Regularly audit training data to identify and mitigate biases. Implement data preprocessing techniques to balance datasets and reduce the impact of outliers.
- **Algorithmic Fairness Techniques:** Employ algorithmic fairness techniques, such as fairness-aware machine learning, to mitigate biases in AI algorithms.
- **Explainable AI (XAI):** Utilize XAI techniques to provide insights into how AI algorithms make decisions, enabling users to understand and trust their outputs.
- **Human-In-the-Loop Systems:** Design systems that allow humans to review and override AI-driven decisions, ensuring human oversight and control.

- **Diversity and Inclusion in AI Development Teams:** Foster diversity and inclusion in AI development teams to ensure that diverse perspectives are represented in the design and training of AI algorithms.
- **Whistleblowing Mechanisms:** Establish whistleblowing mechanisms to encourage workers to report potential ethical violations.
- **Ethical Review Boards:** Establish ethical review boards to oversee the development and deployment of AI in collaborative manufacturing.
- **Industry Standards and Best Practices:** Develop industry standards and best practices for ethical AI in collaborative manufacturing.

The ethical implications of AI-driven decision-making in collaborative manufacturing are complex and multifaceted. By proactively addressing these challenges, manufacturers can ensure that AI is used in a responsible and ethical manner, promoting fairness, transparency, and inclusivity in the workplace. This will not only enhance the well-being of workers but also contribute to the long-term sustainability and success of collaborative manufacturing.

THE IMPACT OF COLLABORATIVE MANUFACTURING ON THE FUTURE OF WORK

Collaborative manufacturing, a paradigm where human ingenuity and advanced automation converge, is poised to fundamentally transform the future of work. This shift transcends mere technological integration, ushering in an era of redefined job roles, evolving skill requirements, and a reimagined relationship between humans and machines. The impact of this transformation extends beyond the factory floor, influencing workforce dynamics, educational paradigms, and societal structures.

At the core of collaborative manufacturing lies the seamless integration of technologies like robotics, artificial intelligence (AI), digital twins, and the Internet of Things (IoT). These technologies are not merely replacing human labor but rather augmenting it, creating a symbiotic relationship where humans and machines leverage their respective strengths. Robots handle repetitive and physically demanding tasks, while humans provide critical thinking, creativity, and complex problem-solving. This collaboration necessitates a workforce that is not only technically proficient but also adept at working alongside intelligent machines.

Transformation of Job Roles

The traditional assembly line worker, performing repetitive tasks, is gradually being replaced by a more versatile and skilled workforce. New job roles are emerging, focusing on:

- **Robotics Technicians and Programmers:** These professionals are responsible for designing, programming, and maintaining robotic systems.
- **Data Analysts and Interpreters:** As manufacturing becomes increasingly data-driven, these roles are crucial for extracting insights from sensor data and optimizing processes.
- **Digital Twin Engineers:** These experts create and manage virtual replicas of physical systems, enabling simulation and optimization.

- **Human-Robot Collaboration Specialists:** These professionals focus on designing and implementing safe and efficient human-robot interaction strategies.
- **Cybersecurity Specialists:** With increased connectivity, cybersecurity skills are vital to protect manufacturing systems from cyberattacks.
- **Remote Maintenance Technicians:** Low latency networks enable expert technicians to assist with equipment maintenance from remote locations.

Evolving Skill Requirements

The changing nature of work demands a shift in skill requirements. While technical proficiency remains essential, the emphasis is moving towards a blend of technical and soft skills.

- **Technical Skills:** Proficiency in robotics, automation, data analytics, digital twin technology, and cybersecurity is crucial.
- **Soft Skills:** Critical thinking, problem-solving, adaptability, collaboration, communication, and creativity are increasingly valued.
- **Lifelong Learning:** Continuous learning and upskilling are essential to keep pace with technological advancements.

Impact on Workforce Dynamics

Collaborative manufacturing is influencing workforce dynamics in several ways:

- **Job Displacement and Creation:** While some traditional jobs may be displaced, new, higher-skilled jobs are being created.
- **Skills Gap:** The rapid pace of technological change is creating a skills gap, requiring proactive upskilling and reskilling initiatives.
- **Increased Productivity and Efficiency:** Human-robot collaboration can significantly enhance productivity and efficiency.
- **Improved Workplace Safety:** Robots can handle hazardous tasks, reducing the risk of injuries to human workers.
- **Flexible Work Arrangements:** Remote monitoring and control capabilities enable more flexible work arrangements.

Impact on Educational Paradigms

Educational institutions must adapt to the changing needs of the workforce by:

- **Integrating Technology into Curricula:** Incorporating robotics, AI, and data analytics into educational programs.
- **Focusing on STEM Education:** Emphasizing science, technology, engineering, and mathematics education.
- **Promoting Lifelong Learning:** Encouraging continuous learning and professional development.

- **Developing Partnerships with Industry:** Collaborating with industry to develop relevant training programs.

Societal Implications

The impact of collaborative manufacturing extends beyond the workplace, influencing societal structures:

- **Economic Inequality:** Addressing the potential for increased economic inequality due to job displacement is crucial.
- **Ethical Considerations:** Establishing ethical guidelines for AI and automation is essential to ensure responsible deployment.
- **Social Responsibility:** Manufacturers have a social responsibility to invest in upskilling and reskilling initiatives.
- **Accessibility:** Ensuring equitable access to technology and training is vital to prevent further divisions in society.

The Human Element

Despite the increasing automation, the human element remains crucial in collaborative manufacturing. Human creativity, problem-solving skills, and ethical judgment are irreplaceable. The focus should be on creating a harmonious partnership between humans and machines, where each leverages their unique strengths.

THE PSYCHOLOGICAL IMPACT OF HUMAN-ROBOT COLLABORATION: BUILDING TRUST AND FOSTERING POSITIVE WORK ENVIRONMENTS

The integration of robots into collaborative manufacturing environments is not merely a technological shift; it's a profound social and psychological transformation. Human-robot collaboration, while promising increased efficiency and productivity, necessitates a nuanced understanding of its psychological impact on workers. Building trust and fostering positive work environments are paramount for ensuring the successful integration of robots and the well-being of the workforce.

The introduction of robots into the workplace can evoke a range of emotional responses in humans, from excitement and curiosity to anxiety and fear. These responses are influenced by various factors, including the perceived capabilities of the robots, the nature of the tasks being performed, and the level of human control. Understanding these psychological dynamics is crucial for creating a harmonious and productive human-robot collaborative environment.

Building Trust in Human-Robot Collaboration

Trust is the cornerstone of effective human-robot collaboration. Without trust, workers may feel apprehensive, hesitant, or even resistant to working alongside robots. Building trust requires a multi-faceted approach, focusing on:

- **Transparency and Explainability:** Robots should be transparent in their actions, providing clear explanations for their decisions and movements. This helps workers understand the robot's behavior and build confidence in its capabilities.
- **Reliability and Consistency:** Robots should perform consistently and reliably, minimizing errors and unpredictable behavior. This reinforces the perception that robots are dependable and trustworthy.
- **Predictability and Controllability:** Robots should behave in a predictable manner, allowing workers to anticipate their actions and maintain a sense of control. This reduces anxiety and fosters a sense of agency.
- **Communication and Feedback:** Robots should be able to communicate effectively with humans, providing feedback on their actions and responding to human input. This facilitates clear communication and enhances coordination.
- **Safety and Security:** Ensuring the physical safety of workers is paramount. Robots should be equipped with safety features and protocols to prevent accidents and injuries.
- **Empathy and Social Cues:** While robots may not possess genuine empathy, they can be designed to exhibit social cues that foster a sense of connection and understanding. This may involve using non-verbal communication, such as gestures and facial expressions.
- **Demonstrated Competence:** Robots which demonstrate competence in the tasks they perform will gain trust from the human workers.
- **Ethical Behavior:** Robots should be programmed to follow ethical guidelines, and act in a manner that is perceived to be fair.

Fostering Positive Work Environments

Beyond building trust, it's essential to create positive work environments that promote collaboration and well-being. This involves:

- **Human-Centered Design:** Designing robotic systems and workspaces that prioritize human needs and preferences. This includes considering ergonomics, accessibility, and user experience.
- **Collaborative Task Design:** Designing tasks that leverage the strengths of both humans and robots, fostering a sense of shared purpose and collaboration.
- **Open Communication and Feedback Channels:** Establishing open communication channels for workers to express their concerns, provide feedback, and share ideas.
- **Training and Education:** Providing comprehensive training and education to workers on how to work effectively with robots. This includes addressing any anxieties or misconceptions about robots.
- **Social Integration:** Integrating robots into the social fabric of the workplace, fostering a sense of inclusion and belonging. This may involve creating opportunities for workers to interact with robots in informal settings.
- **Addressing Job Security Concerns:** Addressing workers' concerns about job displacement by providing opportunities for upskilling and reskilling.
- **Promoting a Culture of Learning:** Fostering a culture of continuous learning and adaptation, encouraging workers to embrace new technologies and challenges.

- **Addressing Emotional Needs:** Recognizing that workers may experience a range of emotions, and providing support systems to address those needs.
- **Creating a Sense of Shared Purpose:** Emphasizing the shared goals and benefits of human-robot collaboration, fostering a sense of collective achievement.
- **Allowing Human Agency:** Ensure that humans retain a sense of agency, and are not simply treated as extensions of the robotic system.

Psychological Considerations

The psychological impact of human-robot collaboration is influenced by several factors:

- **Perceived Threat:** Workers may perceive robots as a threat to their jobs or physical safety.
- **Anxiety and Fear:** The introduction of robots can evoke anxiety and fear, particularly among workers who are unfamiliar with the technology.
- **Sense of Control:** Workers need to feel a sense of control over their interactions with robots.
- **Social Acceptance:** The social acceptance of robots in the workplace is crucial for fostering positive work environments.
- **Cognitive Load:** The cognitive load associated with interacting with robots can impact worker performance and well-being.
- **Emotional Attachment:** Over time, workers may develop emotional attachments to robots, which can influence their interactions.

The Future of Human-Robot Collaboration

As robots become increasingly sophisticated, the psychological impact of human-robot collaboration will become even more pronounced. Researchers and practitioners must continue to explore these dynamics, developing strategies to build trust, foster positive work environments, and ensure the well-being of the workforce.

By prioritizing human-centered design, promoting open communication, and addressing the psychological needs of workers, we can create a future where humans and robots work together in harmony, enhancing both productivity and quality of life.

DATA SECURITY, PRIVACY, AND WORKPLACE SAFETY IN COLLABORATIVE MANUFACTURING

Collaborative manufacturing, with its interconnected networks and data-driven processes, presents unique challenges and opportunities regarding data security, privacy, and workplace safety. Establishing a robust framework for responsible data sharing and ensuring a safe working environment are paramount for the successful and ethical implementation of this manufacturing paradigm.

Responsible Data Sharing and Privacy Protection

In collaborative manufacturing ecosystems, data flows freely between various stakeholders, including suppliers, manufacturers, and customers. This interconnectedness, while fostering efficiency and innovation, also raises concerns about data security and confidentiality. Sensitive information, such as intellectual property, customer data, and proprietary processes, must be protected from unauthorized access and misuse.

Challenges

- **Interconnected Networks:** The distributed nature of collaborative manufacturing makes it challenging to control data access and prevent breaches.
- **Data Volume and Variety:** The vast amounts of data generated by sensors, robots, and other devices require sophisticated security measures.
- **Data Ownership and Access:** Clarifying data ownership and access rights among stakeholders is essential for responsible data sharing.
- **Regulatory Compliance:** Manufacturers must comply with various data privacy regulations, such as GDPR and CCPA.
- **Cybersecurity Threats:** The increasing reliance on digital technologies makes manufacturing systems vulnerable to cyberattacks.

Framework for Responsible Data Sharing

- **Data Governance Policies:** Develop clear data governance policies that define data ownership, access rights, and security protocols.
- **Data Encryption and Access Control:** Implement robust encryption and access control mechanisms to protect sensitive data.
- **Blockchain Technology:** Leverage blockchain technology to create immutable and transparent data ledgers.
- **Data Minimization:** Collect and store only the data that is necessary for specific purposes.
- **Privacy by Design:** Incorporate privacy considerations into the design of manufacturing systems and processes.
- **Incident Response Planning:** Develop comprehensive incident response plans to address data breaches and security incidents.
- **Auditing and Monitoring:** Regularly audit and monitor data security practices to identify vulnerabilities.
- **Data Sharing Agreements:** Establish clear data sharing agreements with all stakeholders, outlining data usage and protection protocols.

The Effects of Collaborative Manufacturing on Workplace Safety

Collaborative manufacturing has the potential to enhance workplace safety by automating hazardous tasks and improving real-time monitoring. However, it also introduces new safety considerations.

Positive Effects

- **Automation of Hazardous Tasks:** Robots can perform tasks that are dangerous or physically demanding for humans.
- **Real-Time Monitoring:** Sensors and AI-powered systems can monitor equipment and environmental conditions, detecting potential hazards.
- **Predictive Maintenance:** Predictive maintenance can prevent equipment failures that could lead to accidents.
- **Remote Operation:** Remote operation capabilities enable workers to perform tasks from a safe distance.

Potential Risks

- **Human-Robot Interaction:** Close collaboration between humans and robots requires careful planning and safety protocols.
- **Cybersecurity Risks:** Cyberattacks can disrupt manufacturing systems and create safety hazards.
- **Data Privacy Concerns:** The collection and sharing of worker data may raise privacy concerns.
- **Psychological Impact:** The introduction of robots can evoke anxiety and fear among workers.

Ensuring Workplace Safety

- **Safety Standards and Regulations:** Adhere to established safety standards and regulations for collaborative manufacturing environments.
- **Risk Assessments:** Conduct thorough risk assessments to identify potential hazards.
- **Safety Training:** Provide comprehensive safety training to workers on how to work safely with robots and automated systems.
- **Emergency Shutdown Systems:** Implement emergency shutdown systems to quickly stop robotic operations in case of accidents.
- **Safety Sensors and Protocols:** Equip robots with safety sensors and protocols to prevent collisions and injuries.
- **Ergonomic Design:** Design workstations and interfaces that are ergonomically sound.
- **Clear Communication:** Establish clear communication protocols between humans and robots.
- **Continuous Improvement:** Regularly evaluate and improve safety practices.
- **Worker Feedback:** Actively seek and respond to worker feedback regarding safety concerns.

By addressing the challenges of data security, privacy, and workplace safety, manufacturers can create a responsible and sustainable collaborative manufacturing ecosystem that benefits both businesses and workers.

FUTURE RESEARCH DIRECTIONS

The dynamic nature of collaborative manufacturing necessitates continuous exploration and research to fully realize its potential. Future research directions should focus on addressing emerging challenges and capitalizing on evolving technological advancements, pushing the boundaries of what's possible.

Advanced Human-Robot Interaction and Cognitive Collaboration

- Investigate the development of more intuitive and adaptive human-robot interfaces, enabling seamless collaboration in complex tasks.
- Explore cognitive AI systems that can understand and anticipate human intentions, fostering more natural and efficient teamwork.
- Research the impact of emotional intelligence in collaborative robotics, enhancing human-robot trust and cooperation.
- Develop methodologies for measuring and optimizing the cognitive load of human operators in collaborative manufacturing environments.

AI-Driven Autonomous and Adaptive Manufacturing Systems

- Explore the development of self-learning and self-optimizing manufacturing systems that can adapt to dynamic production environments.
- Research the application of reinforcement learning and generative AI for autonomous design and process optimization.
- Investigate the integration of digital twins and AI for real-time predictive maintenance and fault detection.
- Develop robust AI-powered decision-making frameworks for distributed and decentralized manufacturing networks.

Sustainable and Circular Collaborative Manufacturing

- Research the development of closed-loop manufacturing systems that minimize waste and maximize resource utilization.
- Explore the application of AI and IoT for real-time monitoring and optimization of energy consumption and environmental impact.
- Investigate the integration of bio-inspired materials and processes in collaborative manufacturing.
- Develop methodologies for assessing and improving the social and environmental sustainability of collaborative manufacturing supply chains.

Enhanced Cybersecurity and Data Privacy

- Research the development of robust cybersecurity frameworks for interconnected manufacturing systems, addressing vulnerabilities associated with IoT and cloud computing.

- Explore the application of blockchain technology for secure data sharing and supply chain traceability.
- Investigate the development of privacy-preserving AI algorithms for data analysis in collaborative manufacturing.
- Develop methodologies for assessing and mitigating the risks associated with data breaches and cyberattacks in interconnected manufacturing environments.

Social and Ethical Implications of Collaborative Manufacturing

- Research the impact of automation and AI on the future of work, focusing on job displacement and skills development.
- Explore the ethical considerations surrounding the use of AI and robotics in manufacturing, including issues of bias, transparency, and accountability.
- Investigate the development of inclusive and equitable collaborative manufacturing ecosystems, addressing the needs of diverse populations.
- Develop frameworks for assessing and mitigating the potential social and economic inequalities associated with the adoption of advanced manufacturing technologies.

Next-Generation Digital Twin and Simulation Technologies

- Research the creation of highly accurate and real-time digital twins that incorporate multi-physics simulations and AI-driven predictive capabilities.
- Explore the development of immersive and interactive virtual reality (VR) and augmented reality (AR) environments for collaborative design and training.
- Investigate the application of quantum computing for complex simulations and optimization in manufacturing processes.
- Develop methodologies for integrating digital twins with real-time sensor data and machine learning algorithms for adaptive control and decision-making.

By focusing on these research directions, we can unlock the full potential of collaborative manufacturing, creating a more efficient, sustainable, and human-centric industrial future.

Navigating the intricate landscape of the future of collaborative manufacturing demands a multifaceted approach, blending technological advancement with human-centric strategies. A prim the evolution of collaborative manufacturing, a confluence of human ingenuity and technological advancement, necessitates a robust research agenda to navigate its uncharted territories. As we stride towards a future defined by seamless human-machine interaction, adaptable production systems, and sustainable practices, it becomes imperative to delineate research pathways that will unlock the full potential of this transformative paradigm.

The Symphony of Human and Robot: Advanced Interaction and Cognition

The core of collaborative manufacturing lies in the harmonious interplay between human operators and robotic systems. Future research must transcend the limitations of basic task execution, delving into the realm of true cognitive collaboration. This entails developing intuitive and adaptive human-robot interfaces, enabling fluid communication and shared understanding. AI-driven cognitive systems, capable of interpreting human intentions and anticipating needs, are crucial for fostering natural and efficient teamwork. The role of emotional intelligence in robotics warrants exploration, focusing on how robots can perceive and respond to human emotions, thereby building trust and enhancing cooperation. Furthermore, methodologies for quantifying and optimizing the cognitive load experienced by human operators in collaborative environments are essential to ensure both efficiency and well-being. This research will pave the way for human-robot teams that operate as a unified entity, maximizing the strengths of both.

The Intelligent Factory: AI-Driven Autonomy and Adaptability

Artificial intelligence holds the key to unlocking the full potential of autonomous and adaptive manufacturing systems. Future research should focus on developing self-learning and self-optimizing systems that can dynamically adjust to fluctuating production demands and unforeseen disruptions. Reinforcement learning and generative AI can be leveraged to autonomously design and optimize manufacturing processes, leading to unprecedented levels of efficiency and flexibility. The integration of digital twins with AI algorithms for real-time predictive maintenance and fault detection will minimize downtime and maximize asset utilization. Moreover, the development of robust AI-powered decision-making frameworks for distributed and decentralized manufacturing networks is crucial for enabling efficient coordination and resource allocation. This research will usher in an era of intelligent factories capable of autonomously adapting and optimizing their operations.

The Sustainable Imperative: Circularity and Environmental Responsibility

As environmental consciousness intensifies, the imperative for sustainable manufacturing practices becomes paramount. Future research should focus on developing closed-loop manufacturing systems that minimize waste generation and maximize resource utilization. AI and IoT technologies can be leveraged for real-time monitoring and optimization of energy consumption and environmental impact, leading to more sustainable production processes. The integration of bio-inspired materials and processes, mimicking nature's efficiency and resilience, offers promising avenues for sustainable manufacturing. Furthermore, methodologies for assessing and improving the social and environmental sustainability of collaborative manufacturing supply chains are essential for ensuring responsible and ethical practices. This research will pave the way for a circular economy, where resources are utilized sustainably and waste is minimized.

The Fortress of Trust: Cybersecurity and Data Privacy

The interconnected nature of collaborative manufacturing systems presents significant cybersecurity challenges. Future research must focus on developing robust cybersecurity frameworks that address the vulnerabilities associated with IoT and cloud computing. Blockchain technology can be explored for secure data sharing and supply chain traceability, ensuring transparency and accountability. Privacy-preserving AI algorithms are crucial for data analysis in collaborative manufacturing, safeguarding sensitive information while enabling valuable insights. Methodologies for assessing and mitigating the risks associated with data breaches and cyberattacks are essential for building trust and ensuring the security of interconnected manufacturing environments. This research will ensure the integrity and security of collaborative manufacturing systems, fostering trust and enabling secure data exchange.

The Human Factor: Social and Ethical Implications

The rapid adoption of automation and AI in manufacturing raises profound social and ethical questions. Future research should explore the impact of these technologies on the future of work, focusing on job displacement, skills development, and the creation of new employment opportunities. Ethical considerations surrounding the use of AI and robotics, including issues of bias, transparency, and accountability, must be addressed. Research should also focus on developing inclusive and equitable collaborative manufacturing ecosystems that address the needs of diverse populations and mitigate potential social and economic inequalities. Frameworks for assessing and mitigating the potential social and economic inequalities associated with the adoption of advanced manufacturing technologies must be developed. This research will ensure that the benefits of collaborative manufacturing are distributed equitably and that the transition to this new paradigm is socially responsible.

The Virtual Frontier: Next-Generation Digital Twins and Simulation

Digital twins, virtual replicas of physical assets and processes, hold immense potential for optimizing manufacturing operations. Future research should focus on creating highly accurate and real-time digital twins that incorporate multi-physics simulations and AI-driven predictive capabilities. Immersive and interactive virtual reality (VR) and augmented reality (AR) environments can be developed for collaborative design and training, enhancing human understanding and collaboration. The application of quantum computing for complex simulations and optimization in manufacturing processes offers promising avenues for innovation. Methodologies for integrating digital twins with real-time sensor data and machine learning algorithms are crucial for enabling adaptive control and decision-making. This research will unlock the full potential of digital twins, transforming them into powerful tools for optimizing manufacturing operations.

By embarking on these research journeys, we can shape the future of collaborative manufacturing, creating a more efficient, sustainable, and human-centric industrial landscape. The pursuit of knowledge and innovation will not only redefine manufacturing but also pave the way for a more prosperous and equitable future for all.ary focus must be placed on cultivating a workforce adept at operating within this dynamic environment. This necessitates substantial investment in comprehensive training programs, equipping individuals with proficiencies in robotics, artificial intelligence, data analytics, and digital twin technology. Embracing lifelong learning becomes crucial, fostering continuous professional development to enable adaptation to the ever-evolving technological terrain of manufacturing. Collaboration between industry and academia is paramount, fostering partnerships to develop curricula that directly address the sector's evolving needs. Apprenticeships and vocational training, providing practical, hands-on experience, must be prioritized. Simultaneously, building a robust digital infrastructure is essential. Secure and scalable IoT networks must be established to ensure seamless data flow across the manufacturing ecosystem, while robust cybersecurity measures are implemented to safeguard sensitive data and critical infrastructure from cyber threats. Investment in cloud and edge computing is vital to facilitate real-time data processing and analysis, enabling proactive decision-making and optimization. Standardizing data protocols and ensuring interoperability are crucial for the seamless integration of diverse systems and technologies. Fostering a collaborative ecosystem is equally vital. Open innovation and knowledge sharing must be encouraged, promoting partnerships between manufacturers, technology providers, and research institutions. Industry consortia and partnerships should be established to develop shared standards and best practices, while platforms for supply chain collaboration enhance transparency and communication across the value chain. Supporting the development of regional manufacturing hubs fosters innovation and economic growth through collaborative efforts. Agile and flexible manufacturing practices should be embraced, implementing modular and reconfigurable production systems to enable rapid adaptation to market demands. Digital twins should be utilized for simulation and optimization, minimizing downtime and improving efficiency. Additive manufacturing facilitates rapid prototyping and customization, enhancing product innovation and responsiveness. Real-time monitoring and predictive maintenance optimize asset utilization and reduce downtime. Ethical considerations and sustainability must be prioritized. Developing ethical guidelines for AI and automation ensures responsible and equitable deployment. Sustainable manufacturing practices reduce waste, minimize environmental impact, and optimize resource utilization. Investing in circular economy initiatives promotes product lifecycle management and resource recovery. Worker well-being and safety are paramount, necessitating human-centric design and ergonomics in collaborative environments. Finally, supporting small and medium-sized enterprises (SMEs) is crucial. Providing access to affordable technology and training, offering financial incentives and grants, developing platforms for knowledge sharing and collaboration, and simplifying regulatory requirements are essential for their successful integration into the future of collaborative manufacturing.

CONCLUSION

In conclusion, "The Future of Collaborative Manufacturing" heralds a transformative era, where the synergy between human intellect and advanced technologies redefines industrial paradigms. This evolution transcends mere automation, emphasizing a collaborative ecosystem driven by real-time data, adaptable systems, and human-centric design. The integration of edge computing, AI, and robotics enables responsive production, while digital twins and flexible systems cater to dynamic market demands. Crucially, human capital remains paramount, requiring continuous reskilling and ethical considerations to navigate this evolving landscape. By embracing collaboration, fostering innovation, and prioritizing sustainability, manufacturers can unlock unprecedented efficiency, resilience, and growth, shaping a future where technology empowers human potential and drives sustainable industrial progress.

REFERENCES

Brettel, M., Friederichsen, N., Keller, M., & Rosenberg, M. (2014). How virtualization, decentralization and network building change the manufacturing landscape: An industry 4.0 perspective. *International Journal of Mechanical, Aerospace, Industrial, Mechatronic and Manufacturing Engineering, 8*(1), 37-44. (Focuses on the broader Industry 4.0 context.)

El Zaatari, S., Marei, M., Li, W., & Usman, Z. (2019). The impact of Industry 4.0 technologies on supply chains. [Connects collaborative manufacturing to supply chain implications.]. *Journal of Manufacturing Technology Management*, *30*(5), 851–879.

Lee, J., Bagheri, B., & Kao, H. A. (2015). A cyber-physical systems architecture for industry 4.0-based manufacturing systems. [Explores the architecture underpinning collaborative systems.]. *Manufacturing Letters*, *3*, 15–23. DOI: 10.1016/j.mfglet.2014.12.001

Romero, D., Bernus, P., Noran, O., Stahre, J., & Fast-Berglund, Å. (2016). The Operator 4.0: Human Cyber-Physical Systems & Adaptive Automation Towards Human-Automation Symbiosis Work Systems. *IFIP Advances in Information and Communication Technology*, *488*, 677–686. DOI: 10.1007/978-3-319-51133-7_80

Tao, F., Qi, Q., Zhang, D., & Zhang, H. (2018). Digital twin and Industry 4.0: What are they, and how to apply them? [Essential for understanding digital twin technology.]. *Applied Sciences (Basel, Switzerland)*, *8*(10), 1819.

Wang, L., Gao, R. X., & Sivardière, F. (2018). Collaborative manufacturing: State-of-the-art and future trends. *Journal of Manufacturing Systems, 48*(Part B), 283-294. (Provides a comprehensive overview and future directions.)

Chapter 17
Smart Factories, Skills Gap, Well-Being, and Overcoming Barriers to Human-Machine Collaboration:
Significance of Smart Factories

Bhagawan Chandra Sinha
https://orcid.org/0000-0003-2466-1316
Sharda University, Greater Noida, India

Nitu Singhi
Kirloskar Institute of Management, Pune, India

Pankaj C. Sharma
https://orcid.org/0009-0000-0968-1816
School of Open Learning, University of Delhi, India

Bhardwaj Shukla
https://orcid.org/0009-0007-6270-5654
School of Open Learning, University of Delhi, India

Upasana Sharma
Sharda University, Greater Noida, India

Sushil Laddhu
https://orcid.org/0009-0006-9856-8033
Medicaps University, India

Arunpreet Singh
ESC School of Business, France

Shweta Chandra Gupta
Sharda University, Greater Noida, India

ABSTRACT

Through the lens of Design Thinking—a problem-solving framework that prioritizes empathy, ideation, and prototyping—we present a structured, yet flexible approach to navigating the complexities of this transition. When leveraged effectively, Design Thinking can help organizations co-create solutions with employees, rather than for them, resulting in workplaces that are both technologically advanced and emotionally intelligent.The evolving dynamics of human-machine collaboration in smart factories and seeks to propose actionable, human-centric solutions to

DOI: 10.4018/979-8-3373-2985-7.ch017

ensure that technology serves as a partner—not a replacement. The transformation is not merely technological; it is deeply human. As machines take on repetitive, precision-heavy tasks, human roles are shifting toward creative problem-solving, system oversight, ethical decision-making, and innovation.

INTRODUCTION

When Ananya stepped into the manufacturing plant where she had worked for over a decade, something felt different. The familiar hum of machinery was still there, but alongside it, sleek robotic arms moved with precision and grace, sorting, welding, and assembling. A screen near her workstation displayed live data analytics, and a collaborative robot—or cobot—waited patiently for her next input. Ananya was not just operating machines anymore; she was working *with* them. And yet, despite the futuristic environment, a nagging thought lingered in her mind—*Was she still needed in this factory of the future?*

Ananya's story is emblematic of a global shift in the industrial landscape. The advent of smart factories—where automation, artificial intelligence (AI), and robotics merge to redefine manufacturing—promises remarkable gains in efficiency, cost-effectiveness, and innovation. These technologies are not just transforming machines but reshaping the very essence of work. While organizations envision a future where humans and machines collaborate in seamless harmony, the reality for many employees is filled with uncertainty, adaptation pressure, and fear of becoming obsolete (Lasi, Fettke, Kemper, Feld, & Hoffmann, 2014).

FOCUS OF THE CHAPTER

This chapter explores the evolving dynamics of human-machine collaboration in smart factories and seeks to propose actionable, human-centric solutions to ensure that technology serves as a partner—not a replacement. The transformation is not merely technological; it is deeply human. As machines take on repetitive, precision-heavy tasks, human roles are shifting toward creative problem-solving, system oversight, ethical decision-making, and innovation.

This chapter delves into several critical dimensions for fostering this collaboration:

- The importance of workforce training and reskilling to bridge skill gaps.
- Ensuring employee safety, mental well-being, and ergonomic adaptability.
- Psychological and economic resistance to automation, including the fear of job displacement.
- The barriers of trust, communication, and adaptability that hinder human-machine synergy.
- And most importantly, the role of Design Thinking in crafting inclusive, empathetic solutions that place people at the heart of this industrial evolution.

Through the lens of Design Thinking—a problem-solving framework that prioritizes empathy, ideation, and prototyping—we present a structured, yet flexible approach to navigating the complexities of this transition. When leveraged effectively, Design Thinking can help organizations co-create solutions with employees, rather than for them, resulting in workplaces that are both technologically advanced and emotionally intelligent.

BACKGROUND

As we move deeper into the smart factory era, the goal is clear: *to build a diverse, adaptable, and empowered workforce that thrives not despite automation, but because of it.*

BRIDGING THE SKILLS GAP: TRAINING AND RESKILLING EMPLOYEES

Employees must upskill and learn continuously because of the change to Industry 4.0 to stay relevant in the workforce. Rapid technological progress and a shortage of workers with high technical and cognitive abilities are the main causes of the growing skills gap (Ejsmont, Gladysz, & Kluczek, 2020).

Key Aspects of Bridging the Skills Gap

Why the Skills Gap Exists in the Era of Industry 4.0?

The skills gap refers to the mismatch between the skills that employers need and the skills that workers currently possess. According to the McKinsey Report (2018), To measure skill shifts from automation and AI, they modelled skill shifts going forward to 2030—and found that they accelerated. While the demand for technological skills has been growing since 2002, it will gather pace in the 2016 to 2030 period. The increase in the need for social and emotional skills will similarly accelerate. According to the McKinsey Global Institute Workforce Skill Model (2018). By contrast, the need for both basic cognitive skills and physical and manual skills will decline.

As Industry 4.0 transforms how, we work—with automation, AI, IoT, and advanced data systems—this gap is widening (Gorecky, Schmitt, Loskyll, & Zühlke, 2014). Here is why:

- Rapid Technological Advancement Outpacing Education & Training: Technologies are evolving faster than training systems can adapt. Schools, colleges, and corporate training programs are often slow to update curricula to reflect the emerging skills needed for AI, cloud computing, robotics, and cybersecurity. According to the World Economic Forum's Future of Jobs Report (2023), "Technological adoption is expected to remain a key driver of business transformation, yet the skills needed for these roles are rapidly changing—creating both emerging jobs and skills gaps." For example, manufacturers implementing smart factories need workers skilled in robotics and predictive maintenance, but traditional mechanical engineering courses rarely cover these.
- Mismatch Between Academic Output and Industry Needs: Universities and training institutes often focus on theoretical knowledge rather than practical, job-ready skills. Graduates may hold degrees but lack the ability to apply digital tools or think critically in agile business environments. As per a McKinsey report, *'Bridging the Skills Gap,'* (McKinsey & Company, 2020) highlights:43% of employers say they cannot find the

talent they need, even for entry-level positions. An example may be an engineering graduate may not know how to work with real-time data platforms or tools like Power BI or Python—essential for many tech-enabled roles.

- Increased Demand for Cognitive and Social Skills: While machines are taking over routine and manual tasks, there's a greater demand for cognitive skills like complex problem solving, adaptability, and collaboration with AI systems—skills that are not taught through traditional models. As noted by OECD Skills Outlook (2021) that "the demand for workers who can manage information, communicate effectively, and solve non-routine problems is growing across all sectors."
- Changing Job Roles & Emergence of Hybrid Roles: Industry 4.0 creates new hybrid job roles—such as a marketing analyst who also needs data science skills or an HR manager working with AI recruitment tools. These roles require cross-disciplinary knowledge not typically taught in silos.
- Uneven Access to Lifelong Learning and Reskilling Opportunities: Many workers—especially in developing countries or smaller organizations—lack access to high-quality training, internet infrastructure, or financial resources to upskill. UNESCO's Global Education Monitoring Report (2022), emphasizes that there is a digital divide not only in infrastructure but also in digital skills training opportunities.

Table 1. Summary of the reports

Reason	Description	Reports
Tech advancing rapidly	Training can't keep up with evolving tools	WEF (2023)
Academic-industry mismatch	Theoretical focus over real-world skill needs	McKinsey (2021)
Cognitive/social skills in demand	Critical thinking, adaptability > manual skills	OECD (2021)
Unequal access	Digital divide limits reskilling	UNESCO (2022)

DETERMINING SKILLS THAT ARE FUTURE-READY

Industry 4.0, the fourth industrial revolution, is marked by automation, AI, robotics, IoT (Internet of Things), and big data. While it enhances productivity, it also creates a skills gap, as many workers are not trained in these new technologies. Determining future-ready skills is essential to equip the workforce for evolving industry demands, especially within smart factories. Smart factories—powered by IoT, AI, robotics, and big data—require employees to adapt to new roles involving digital systems, automation, and data-driven decision-making. These advanced manufacturing environments demand a combination of technical, soft, and digital skills for seamless human-machine collaboration. The table 2 below summarizes the essential skills required to thrive in smart factories operating within the industry 4.0 environment.

Table 2. Summary of essential skills required to thrive in smart factories

Skill Category	Description	Example in Smart Factory Context
Technical & Digital Skills	Proficiency in automation systems, robotics, IoT devices, data analytics, programming, cloud computing, and cybersecurity.	A technician uses Python to analyze sensor data from IoT-enabled machines and adjusts production parameters in real time
Data Analysis & Interpretation	Ability to collect, interpret, and act on large data sets from interconnected devices and production lines.	A quality control analyst uses AI-powered dashboards to spot defects and predict maintenance needs
Adaptability & Continuous Learning	Willingness and ability to learn new technologies and adapt to changing roles and systems.	An operator participates in microlearning modules on new collaborative robot (cobot) interfaces as they are introduced on the factory floor
Critical Thinking & Problem Solving	Skill in identifying, analyzing, and solving complex, non-routine problems, often in collaboration with AI systems.	An engineer troubleshoots an unexpected production halt by interpreting machine learning alerts and collaborating with IT to resolve the issue
Collaboration & Communication	Working effectively in cross-disciplinary teams, often including both humans and machines; clear communication of data-driven insights.	A production supervisor leads a team that includes both human workers and cobots, ensuring smooth workflow and resolving misunderstandings between departments
Creativity & Innovation	Developing new solutions, optimizing processes, and leveraging technology for continuous improvement.	A process engineer designs a new workflow using 3D printing to rapidly prototype and test custom machine parts
Emotional Intelligence & Empathy	Understanding and addressing human concerns in tech-driven environments; supporting well-being and change management.	A team leader uses empathy mapping to understand workers' anxieties about automation and co-designs training programs to build confidence
System Oversight & Ethical Decision-Making	Supervising autonomous systems, making ethical choices regarding AI deployment, and ensuring safety and compliance.	An operations manager oversees the deployment of AI-driven quality checks, ensuring they meet safety and fairness standards

Why These Skills Matter

- **Bridging the Skills Gap:** As automation takes over repetitive tasks, human roles shift toward oversight, problem-solving, and innovation, requiring higher-order cognitive and social skills.
- **Hybrid Job Roles:** New positions-such as Smart Quality Analyst or Data-Driven Production Supervisor-require cross-disciplinary expertise, blending traditional manufacturing knowledge with digital fluency.
- **Continuous Change:** The rapid pace of technological advancement means that adaptability, resilience, and a commitment to lifelong learning are as important as technical know-how.

HUMAN-CENTERED TRANSFORMATION WITH DESIGN THINKING

Despite these challenges, the emergence of Design Thinking—a human-centered innovation framework—offers a promising approach to workforce transformation. Unlike traditional top-down models of technological change, Design Thinking involves employees directly in the ideation, development, and testing of new solutions. By placing empathy, collaboration, and experimentation at the core of innovation, it helps organizations to align technology adoption with real human needs (Brown, 2009; Norman, 2013). In smart factory environments, this approach allows for the co-creation of inclusive, intuitive systems that reduce resistance, build trust, and enhance worker engagement. Technology alone does not future-proof a workforce—*mindset and empathy* do. Design Thinking helps companies align tech adoption with human needs, ensuring employees feel included, supported, and empowered. Design Thinking, a human-centric problem-solving approach, plays a crucial role in reshaping workforce training, fostering collaboration, and addressing resistance to automation in smart factories. By prioritizing empathy and iterative innovation, this methodology ensures that technological transitions align with employee needs and organizational goals.

Notable Industry Examples

Design thinking is increasingly used in smart factories to drive innovation, solve practical problems, and ensure that technology solutions are user-centric. Here are several industry examples illustrating how design thinking is applied in manufacturing and smart factory contexts:

Adient-Infosys: Collaborative Problem Solving for Industry 4.0

- Process: Adient, a global automotive seating supplier, partnered with Infosys to conduct design thinking workshops at the Infosys innovation center. These workshops brought together executives and plant supervisors to identify key operational challenges and opportunities, focusing on real-world use cases rather than abstract savings targets.
- Outcome: The team prioritized practical, high-impact projects, such as a mobile app for floor performance, a track-and-trace solution for inventory accuracy, and analytics-driven business insights. Each solution was prototyped and tested with direct input from plant staff, ensuring relevance and usability.

Sonoco Metal Packaging: Predictive Maintenance Through Cross-Functional Teams

- Process: Sonoco used a smart manufacturing platform to implement a condition-based maintenance program. Cross-functional teams representing IT, operations, reliability, and quality collaborated to analyze machine and operator data, identify pain points, and design predictive maintenance workflows.
- Outcome: The design thinking approach led to the creation of a proactive repair system, resulting in significant cost avoidance and improved equipment uptime.

Rittal Smart Factory: User-Centric IoT Integration

- Process: At Rittal's smart factory in Haiger, Germany, design thinking principles guided the integration of IoT technologies. Interdisciplinary teams worked closely with operators to understand workflow bottlenecks and usability challenges.
- Outcome: The factory implemented digital tools and dashboards that were iteratively refined based on operator feedback, leading to higher adoption rates and smoother transitions to automated processes.

Procter & Gamble (P&G): Consumer-Centric Product Innovation

- Process: P&G embedded design thinking into its product development by sending teams to observe consumers in their homes. Insights from these observations led to the identification of cumbersome cleaning routines.
- Outcome: The Swiffer mop was developed through ideation, prototyping, and direct user testing, resulting in a breakthrough product and a new market category.

Cross-Functional Data Teams for Problem Solving

- Process: In many smart factories, cross-functional teams (IT, operations, engineering, quality) use design thinking to turn manufacturing data into actionable insights. These teams hold regular workshops to interpret big data, define problems, ideate solutions, and test new processes.
- Outcome: This approach leads to continuous improvement in areas such as predictive maintenance, supply chain resilience, and big data management, ensuring solutions are both innovative and practical.

Smart factories at the forefront of Industry 4.0 are increasingly adopting Design Thinking to ensure that technology implementations are not just efficient, but also empathetic and user-focused. This approach places people-operators, engineers, and managers-at the center of digital transformation, resulting in solutions that are both technologically advanced and emotionally intelligent.

5 Phases of Design Thinking

A. **Empathize:** In the empathize stage, a smart factory might conduct in-depth interviews and floor-walks with machine operators, maintenance engineers, and supervisors to understand their daily pain points and aspirations. For instance, Makino Asia used AR-assisted reality devices to let engineers in Singapore observe and guide shop-floor workers overseas, uncovering challenges in remote troubleshooting and communication. Similarly, OXO's Good Grips product line was born from observing people with arthritis struggling with kitchen tools, leading to universally ergonomic designs. In smart factories, empathy mapping and direct observation help reveal fears (like job loss due to automation) and hopes (such as safer, less repetitive tasks).

B. **Define:** After gathering insights, the factory team synthesizes findings to define the core problem. For example, GE's Brilliant Factory initiative identified that inconsistent adoption of predictive maintenance tools was due to unclear roles and lack of confidence among operators. The team set measurable goals, such as increasing the use of AI-driven maintenance systems by 40% and boosting operator confidence through targeted support. Defining the problem with clear metrics ensures that the next steps address the real barriers to transformation.

C. **Ideate:** With a clear problem statement, the team brainstorms innovative solutions. At Mitsubishi Electric's smart factory, cross-functional teams proposed using AR and VR for immersive training, enabling workers to practice complex tasks in a risk-free environment. Other ideas might include deploying AI-powered virtual mentors that offer real-time guidance, or developing new human-AI collaboration workflows where operators oversee and fine-tune automated processes. This stage encourages creativity and draws on diverse perspectives to generate a wide range of potential solutions.

D. **Prototype:** Next, the factory tests small-scale versions of the most promising ideas. For instance, Makino Asia piloted AR-based remote assistance, allowing overseas workers to receive step-by-step visual instructions from experts. Another example is the use of interactive digital twins-virtual models of the production line-where employees can simulate changes and observe outcomes without disrupting actual operations. Prototyping in this way allows teams to quickly validate concepts and gather early feedback.

E. **Test:** Finally, the prototype is put into real-world use with a select group of employees, and feedback is collected to refine the solution. At Zuellig Pharma, AR headsets and drone inventory solutions were tested for picking and inventory management, resulting in a 30% improvement in picking productivity and 100% pick accuracy. Data and user feedback from these pilots informed further adjustments, such as interface tweaks or additional training modules, ensuring the final deployment was both effective and embraced by the workforce.

Table 3. Summary table: Design Thinking in smart factory workforce transformation

Stage	Action/Example	Outcome
Empathize	Interviews, empathy maps, floor-walks with workers	Uncover fears, needs, and training gaps
Define	Analyze feedback, set clear problem statements and goals	Focused on confidence and adoption
Ideate	Co-design MR simulations, role-based learning, peer mentoring, improved dashboards	Innovative, user-driven solutions
Prototype	Pilot MR training, new dashboards, wearables	Test ideas in real settings
Test	Collect feedback, measure confidence/adoption, iterate designs	Refined solutions, higher engagement

In summary, the background of this chapter situates smart factories not just as technological marvels, but as human systems—where the future of manufacturing depends on the successful integration of machine intelligence with human capability, emotion, and purpose. Organizations that acknowledge and address the socio-technical dynamics of this transformation will be better positioned to harness the full potential of Industry 4.0 while safeguarding workforce dignity and relevance.

Solutions and Recommendations

To successfully transition to smart factories, companies need to balance technological upgrades with a deep focus on human needs. This shift is not just about robots or AI-it's about reimagining workflows, roles, and workplace culture to keep people at the heart of innovation. Below are key strategies that blend technical solutions with human-centered support:

CORPORATE TRAINING PROGRAMS AND LIFELONG LEARNING

To build a future-ready workforce, corporations must invest in lifelong learning models that support continuous skill acquisition. Modern corporate training emphasizes personalization, engagement, and practical learning.

- **On-the-job Training**: Employees learn directly while operating cobots (collaborative robots), smart sensors, and AI interfaces. Learning becomes contextual and immediately applicable. Example: Bosch utilizes cobot-based training where employees simulate production tasks using AI-assisted interfaces.
- **Microlearning**: This involves delivering learning content in small, focused segments. It supports faster retention and adaptability. Research shows microlearning boosts engagement and completion rates by up to 90% (Bersin, 2021).
- **Gamification**: AI-powered simulations and gamified learning modules enhance engagement, especially in technical upskilling. PwC (2022) found that gamified training led to a 70% increase in learner participation.
- **Public-Private Partnerships (PPPs)**: Collaborations between government, academia, and industry foster specialized programs. Germany's dual system and India's Skill India initiative are prime examples.

RESKILLING AND UPSKILLING STRATEGIES

Smart factories, driven by Industry 4.0 technologies, demand a workforce that is both adaptable and continuously learning. Organizations must deploy targeted strategies to address the widening skills gap, distinguishing between reskilling (preparing employees for entirely new roles) and *upskilling* (enhancing current capabilities to keep pace with technological change).

To navigate the fast-changing landscape of Industry 4.0, especially in smart factories, organizations are turning to agile and tech-enabled workforce development methods. According to a report European companies expect to resolve skill mismatch by retraining workers.

According to LinkedIn's 2023 Workplace Learning Report, 94% of employees say they would stay longer at a company that invests in their development.

Reskilling: Preparing for New Roles

What it Involves

Reskilling is about equipping employees with entirely new skill sets so they can transition into roles that are in demand but outside their current expertise. In the context of smart factories, this often means moving from traditional, manual roles to tech-enabled positions.

Example

An assembly line worker is retrained as an IoT Integration Technician. This transition requires foundational knowledge in automation, cloud computing, and data systems. The worker might start with basic courses in digital literacy, followed by hands-on training in sensor calibration, PLC programming, and IoT device management. AI-driven learning platforms can personalize this journey, recommending modules based on performance and identified gaps.

Key Strategies

- **Personalized Learning Paths:** AI tools analyze employee performance and recommend specific learning modules (e.g., sensor calibration, PLC programming).
- **Certifications and Micro-Credentials:** Workers pursue recognized credentials in new domains (e.g., Industrial IoT certification).
- **Mentorship Programs:** Senior technicians guide reskilled employees, helping them integrate human skills with smart systems (e.g., interpreting predictive maintenance alerts).
- **On-the-Job Training:** Employees learn by operating cobots, smart sensors, and AI interfaces in real production environments.

Upskilling: Enhancing Current Capabilities

What it Involves

Upskilling focuses on deepening or updating the skills of employees in their existing roles, ensuring they remain effective as technologies evolve. This is crucial for roles that are not disappearing but are transforming due to automation and digitalization.

Example

A maintenance engineer is upskilled to become a Predictive Analytics Specialist. This path involves advanced training in AI, diagnostics, and robotics, enabling the engineer to leverage data analytics for predictive maintenance and system optimization. Microlearning modules, gamified training, and role-based learning paths support this continuous development.

Key Strategies

- **Microlearning:** Short, focused modules on new technologies (e.g., predictive maintenance, AI diagnostics) to boost retention and adaptability.
- **Gamification:** AI-powered simulations and gamified content to enhance engagement and technical proficiency.
- **Advanced Certifications:** Earning micro-credentials in areas like machine learning or robotics to validate expertise.
- **Continuous Learning Culture:** Encouraging lifelong learning as a core organizational value, supported by accessible and affordable training opportunities.

The below Table 4 illustrates the strategic distinction between reskilling for new roles and upskilling for growth within existing roles, particularly in the context of smart factory transitions.

Table 4. Reskilling vs upskilling in smart factories

Aspect	Reskilling	Upskilling
Definition	Learning new skills for a different role	Enhancing existing skills to match evolving technologies
Objective	Career shift into future-ready roles	Stay competitive and efficient in current roles
Smart Factory Example	Assembly Line Worker → IoT Integration Technician	Maintenance Engineer → Predictive Analytics Specialist
Learning Path	Foundational training in automation, cloud, data systems	Advanced training in AI, diagnostics, and robotics

DESIGN THINKING FOR WORKFORCE TRANSFORMATION

To effectively apply a human-centered, design thinking approach to workforce adaptation in smart factories, the LXD Framework by Boller and Fletcher (2020) offers a comprehensive model that integrates learner needs, business objectives, and technology-enabled experiences. By leveraging this framework, organizations can systematically design and deliver engaging, measurable, and adaptive learning solutions tailored to the evolving demands of Industry 4.0 environments. It begins with understanding the problem and progresses through iterative cycles of ideation, prototyping, and refinement

- **Get Perspective** (Empathize): In smart factories, the empathize stage means going beyond surface-level surveys to deeply engage with frontline workers, engineers, and managers. For instance, a factory implementing autonomous guided vehicles (AGVs) might conduct floor-walk interviews and empathy mapping sessions to uncover real concerns-such as the anxiety operators feel when interpreting live sensor data or interfacing with new autonomous systems. In one smart manufacturing setup, supervisors used co-observation, shadowing maintenance staff as they diagnosed IoT-enabled machines, revealing hesitancy in trusting automated diagnostics and a desire for more transparent system feedback. These insights shape the direction of workforce adaptation by ensuring solutions are grounded in authentic user experience.
- **Refine Problem** (Define): After gathering these insights, the team synthesizes feedback to pinpoint the core workforce challenges. For example, in a smart factory deploying predictive maintenance, it might become clear that employees lack confidence in handling machine learning outputs or struggle to interpret complex IoT dashboards. Others may be hesitant to trust AI-enabled systems for critical decision-making. By clearly defining these pain points-such as "increase operator confidence in using predictive analytics by 30% within six months"-the factory sets measurable goals that guide the next phases of solution development.
- **Ideate**: After gathering these insights, the team synthesizes feedback to pinpoint the core workforce challenges. For example, in a smart factory deploying predictive maintenance, it might become clear that employees lack confidence in handling machine learning outputs or struggle to interpret complex IoT dashboards. Others may be hesitant to trust AI-enabled systems for critical decision-making. By clearly defining these pain points-such

as "increase operator confidence in using predictive analytics by 30% within six months"- the factory sets measurable goals that guide the next phases of solution development.

- **Prototype**: The most promising ideas are then piloted on a small scale. A factory might introduce voice-guided machine operations, where wearable headsets provide AI-assisted alerts and learning nudges as workers interact with new equipment. For example, a pilot group of operators could use smart wearables that vibrate to signal maintenance needs, while embedded surveys and haptic response tracking collect feedback on usability and effectiveness. In another scenario, interactive digital twins-virtual replicas of the production line-allow employees to simulate adjustments and troubleshoot issues in a safe, controlled environment before changes are made in real life.
- **Iterate**: During iteration, usability tests and performance analytics are used to refine the solutions. If, for instance, data shows that employees are not engaging effectively with digital twins, the factory might redesign the user interface for greater clarity or adjust the feedback mechanisms to be more intuitive. Continuous feedback loops-such as short post-task surveys or real-time analytics on error rates-ensure that learning experiences and digital tools are continually improved based on actual user behavior and needs.
- **Implement**: Once refined, the solution is scaled across the smart factory workforce. For example, the MR simulation training or AI-guided wearables are rolled out to all relevant teams. The factory monitors key metrics such as machine uptime, operator error rates, and employee digital engagement scores to assess impact. In one documented case, the adoption of MR-based training led to faster onboarding, a measurable reduction in assembly errors, and higher confidence among workers handling complex robotics. Ongoing monitoring and support ensure that the workforce remains adaptable and engaged as the factory continues to evolve.

This iterative, human-centered approach ensures that workforce adaptation in smart factories is not only technologically advanced but also deeply responsive to the real needs and experiences of employees. This loop enables continuous co-creation, ensuring workforce interventions are relevant, tech-integrated, and people-centric in the context of smart factories. By embedding Design Thinking into workforce development and collaboration strategies, organizations can cultivate a future-ready workforce that sees automation as an enabler rather than a disruptor.

To ensure seamless integration of automation and build a resilient workforce, organizations must go beyond traditional training. By integrating AI, microlearning, and gamified content within a Design Thinking framework, companies can foster a growth mindset among employees. This approach empowers them to view automation as an opportunity, not a threat—creating a future-ready ecosystem where humans and machines collaborate productively.

EMPLOYEE WELL-BEING AND SAFETY IN HUMAN-MACHINE COLLABORATIVE ENVIRONMENTS

As smart factories evolve with the integration of artificial intelligence, IoT, and collaborative robotics, human workers increasingly find themselves interfacing directly with machines. This hybrid working environment, while enabling operational excellence, introduces a unique set of challenges concerning physical safety and psychological well-being. Ensuring both is paramount to achieving sustainable workforce adaptation and productivity.

Ensuring Physical Safety in Human-Machine Collaboration

- **Ergonomic Design and Workplace Adaptation**: Smart factories must prioritize human-centered design to prevent injuries and optimize worker efficiency:
- **Wearable Devices:** Smart wearables, such as biometric wristbands and posture trackers, monitor fatigue levels, heart rate variability, and environmental conditions to alert workers and supervisors.
- **AI-Based Hazard Detection:** Machine learning algorithms analyze real-time sensor data to identify unsafe practices and potential hazards, enabling predictive risk mitigation.
- **Cobotic Safety Features:** Collaborative robots (cobots) are equipped with emergency stop buttons, force sensors, and vision-based object detection to minimize the risk of collision or injury.

Figure 1. Process of involving collaborative robots to ensure safety

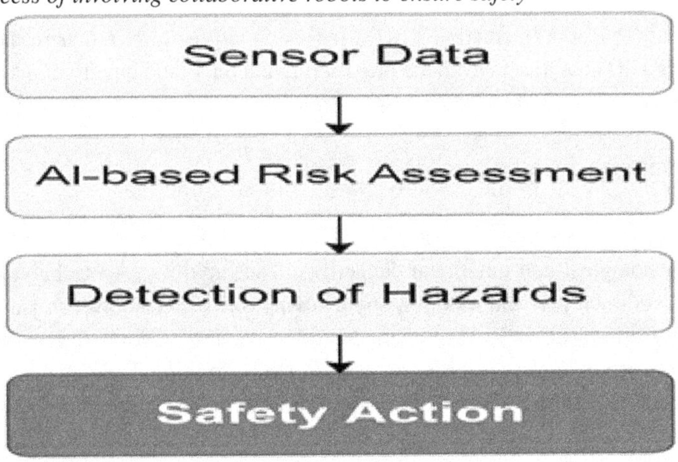

- **Regulatory and Compliance Standards:** To safeguard employees working alongside machines in smart factories, compliance with globally recognized health and safety frameworks is critical. These standards offer systematic guidelines for minimizing risk, enhancing operational safety, and promoting a culture of continuous safety improvement.
- **Adherence to ISO 45001 standards for occupational health and safety:** ISO 45001 is an international standard developed by the International Organization for Standardization to help organizations proactively improve employee safety, reduce workplace risks, and create better, safer working conditions. Key principles applicable to smart factories include:

- Hazard Identification through Data Analytics: Integration of IoT sensors and AI enables real-time hazard detection aligned with ISO 45001's risk-based thinking model.
  - Worker Participation in Safety Audits: Workers actively participate in safety evaluations and improvement programs, supporting the Plan-Do-Check-Act (PDCA) cycle.
  - Preventive Action via Predictive Maintenance: Machine data is analyzed to prevent system failures that could lead to human injury, adhering to the standard's emphasis on prevention over response.
- **Implementation of robotic safety measures outlined by the International Federation of Robotics (IFR):** The International Federation of Robotics (IFR) provides best practices and safety recommendations for industrial robot integration in collaborative workspaces. Relevant IFR safety provisions include:
  - Dynamic Safety Zones: Robots detect human presence using LIDAR and machine vision and adjust their operational speed or halt motion within defined geofenced areas.
  - Fail-Safe Programming and Redundancy Protocols: Collaborative robots (cobots) are programmed to default to a safe state in case of a malfunction, minimizing the likelihood of injury.
  - Human-Robot Interaction (HRI) Metrics: Smart factories are increasingly employing HRI analytics to evaluate proximity, frequency, and quality of interaction, driving continuous ergonomic optimization.

Psychological and Emotional Well-being

As smart factories become increasingly automated and data-driven, the human workforce faces new psychological and emotional challenges. The rapid pace of technological change, constant interaction with AI and robotics, and evolving role expectations can lead to cognitive overload, technostress, and feelings of alienation or inadequacy. Addressing these issues is not just a matter of corporate responsibility but also of strategic importance. A resilient, mentally healthy workforce is better positioned to adapt, innovate, and collaborate in complex, cyber-physical environments. Thus, fostering psychological well-being and inclusion must be integral to the design and management of human-machine ecosystems.

- **Managing Technostress and Anxiety:** In highly automated environments like smart factories, employees often face technostress, characterized by anxiety, fatigue, and cognitive overload arising from constant interaction with advanced technologies (Tarafdar, Cooper, & Stich, 2019). These effects may lead to reduced job satisfaction, lowered productivity, and increased turnover.

In smart factories, addressing technostress-stress caused by the rapid adoption of digital tools and automation-requires a structured, multi-stage approach that moves from assessment to prevention.

Assessment

The assessment stage is the starting point, where organizations evaluate the levels and sources of technostress among their employees. This process may involve a combination of anonymous surveys, wearable device data, one-on-one interviews, or the use of stress monitoring tools to capture real-time physiological responses. Key indicators often include signs of burnout, mental fatigue, frustration with digital interfaces, or work-life imbalance due to hyperconnectivity and constant notifications. For example, a smart factory might deploy a digital well-being survey and analyze data from wearable fitness trackers to identify spikes in stress during periods of major software upgrades or when new IoT systems are introduced.

Coping

Once stressors are identified, the coping phase begins. Here, both individuals and organizations implement coping strategies to help employees manage digital strain. These strategies might include mindfulness sessions on the factory floor, time management workshops, or the introduction of simplified user interfaces for complex machinery. In some smart factories, employees are encouraged to schedule regular digital detox breaks or use "focus mode" features on their devices to minimize distractions. Coping mechanisms empower workers to regain control over their work environment and reduce the negative impact of technostress.

Support

Once stressors are identified, the coping phase begins. Here, both individuals and organizations implement coping strategies to help employees manage digital strain. These strategies might include mindfulness sessions on the factory floor, time management workshops, or the introduction of simplified user interfaces for complex machinery. In some smart factories, employees are encouraged to schedule regular digital detox breaks or use "focus mode" features on their devices to minimize distractions. Coping mechanisms empower workers to regain control over their work environment and reduce the negative impact of technostress.

Prevention

Finally, the prevention stage closes the loop by introducing proactive mechanisms based on earlier findings. Preventive measures may include human-centered technology design that prioritizes usability, the implementation of mental health policies tailored for digital workplaces, and continuous monitoring of technostress indicators. For example, a smart factory might redesign machine interfaces to be more intuitive, offer ongoing digital literacy training, and regularly review stress data to anticipate and address emerging issues before they escalate. This comprehensive, cyclical approach ensures that technostress is not only managed but systematically minimized, supporting both employee well-being and organizational productivity in the age of Industry 4.0.

The below image Technostress Management Cycle illustrates a cyclical model for managing technostress in smart factory. Tarafdar et al. (2019) do not label these as a cycle, but their work supports this flow conceptually.

Figure 2. Technostress management cycle

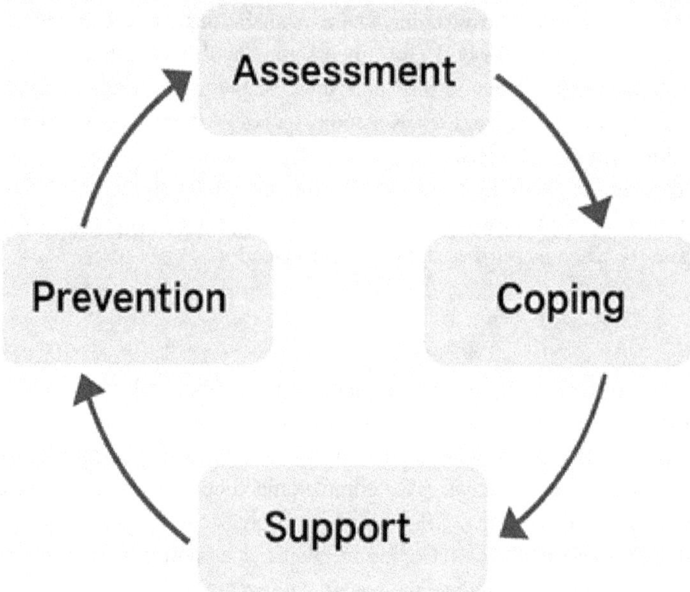

In smart factories, where AI, robotics, and digital systems are integral, this model helps in proactively managing mental well-being and ensuring that employees stay engaged, mentally resilient, and productive. To address this:

- **Employee Assistance Programs (EAPs):** These should be tailored to the context of smart factories. EAPs can provide access to mental health professionals, facilitate workshops on burnout prevention, and offer digital stress diagnostics using AI-enabled chatbots. By integrating real-time emotional analytics from wearables or workplace IoT systems, companies can proactively identify employees at risk.
- **Mindfulness and Resilience Training:** Programs such as digital mindfulness sessions and emotional regulation modules improve employees' capacity to cope with tech-driven change. Apps like Headspace, Calm, or enterprise-level platforms like Unmind can be embedded into daily workflows. These have been shown to enhance focus, reduce anxiety, and build "techno-resilience" (Tarafdar et al., 2019), especially among shift workers or those engaged in repetitive cobot interactions.

In smart manufacturing contexts where digital twins and real-time dashboards dominate, cognitive demands can be reduced by integrating user-centric interfaces, adaptive task load management systems, and ambient computing that auto-adjusts to workers' stress levels.

- **Fostering a Culture of Inclusion:** Technological transitions can marginalize employees unfamiliar with digital tools, exacerbating feelings of exclusion or incompetence.

Ensuring psychological safety — where workers feel confident to ask questions, raise concerns, and admit mistakes — is crucial in such environments. Key initiatives include:

- **Inclusive Onboarding:** Tailored onboarding modules should include basic robotics literacy, gamified tutorials for interacting with AI interfaces, and hands-on experiences with collaborative robots (cobots). This demystifies technology for non-tech-savvy staff and reduces tech apprehension.
- **Trust-building Initiatives:** Implementing human-robot collaboration simulations, where cross-functional teams engage in problem-solving tasks with cobots, can foster shared understanding and confidence. These interactions enhance interpersonal trust and improve acceptance of machines as cooperative partners (Siau & Wang, 2018).
- **Peer Mentorship Programs:** Assigning digital mentors to new or returning employees can bridge the confidence gap in adapting to evolving technologies.

As smart factories often utilize a hybrid workforce model, fostering psychological safety ensures smoother transitions during digital upskilling, reduces resistance to automation, and improves the ROI of tech adoption initiatives.

ADDRESSING BARRIERS TO SEAMLESS COLLABORATION: TRUST, COMMUNICATION, AND ADAPTABILITY

Effective human-machine collaboration in smart factories relies not only on technological compatibility but also on human-centric considerations. To unlock the full potential of Industry 4.0, organizations must address the emotional, cognitive, and cultural barriers that hinder smooth integration. Central to this endeavor is the cultivation of trust, transparent communication, and inclusive design processes.

Building Trust in Human-Machine Partnerships

Trust forms the foundation for effective interaction between humans and intelligent systems. Without trust, workers may resist AI deployment, overlook system recommendations, or feel disempowered (Siau & Wang, 2018).

1. **Transparent AI and Explainable Automation:** One of the most significant enablers of trust in automation is the ability to understand it. Providing end-users with interpretable reasoning behind AI decisions boosts confidence and reduces fear of loss of control. Explainable AI involves the design of systems that provide transparent, interpretable, and user-friendly justifications for their decisions. For instance, when an AI flags a defect in a product, the system should clearly indicate the data points and logic it used to reach that conclusion. This demystification process reduces the fear of "black-box" decisions and empowers workers to feel part of the quality assurance loop.
2. **Demonstrating Reliability:** Another key to fostering trust is involving employees early in the design, testing, and deployment of AI systems. Engaging employees in the design and training phases of AI systems to ensure relevance and acceptance. This participatory approach ensures that tools are adapted to the actual workflows, comfort levels, and insights of those

who will use them daily. Workers are more likely to embrace technologies they have helped shape, leading to better adoption and reduced resistance.

Enhancing Communication in Hybrid Work Environments

Human-machine collaboration thrives on real-time communication that is multimodal, intuitive, and inclusive of diverse user capabilities.

1. **Training Employees in Digital Communication Tools:** To facilitate meaningful interaction with AI interfaces, robots, and dashboards, employees must possess basic digital competencies. Tailored training programs that introduce AI concepts, system navigation, and troubleshooting techniques can bridge the skill gap and demystify complex technologies. This reduces anxiety and builds confidence in navigating automated environments.
2. **Feedback Mechanisms:** Communication should not rely on a single sensory channel. Combining visual alerts (e.g., dashboard indicators), audio prompts (e.g., alarms or spoken messages), and haptic feedback (e.g., vibration-based signals on wearable devices) ensures that information is accessible and promptly received, especially in noisy or visually busy factory settings. Such layered communication reduces errors and enables faster, more intuitive responses.

By addressing these collaboration barriers, organizations can ensure smoother integration of automation while maintaining a motivated workforce.

Human Resistance to Automation and the Fear of Job Displacement

The integration of automation and artificial intelligence (AI) into manufacturing—often termed the advent of "smart factories"—has raised significant concerns among employees about job security and the potential displacement of human labor. Addressing these apprehensions requires a comprehensive understanding of the underlying factors contributing to resistance and the implementation of effective mitigation strategies. One of the most formidable barriers in the transition to smart factories is human resistance to automation, primarily driven by fears of job loss and skill redundancy (Frey & Osborne, 2017). While technological advancement promises increased efficiency, productivity, and innovation, it simultaneously generates psychological unease and economic anxiety among employees (Frey & Osborne, 2017).

Understanding Resistance to Automation

Psychological Factors

Fear of Redundancy and Skill Obsolescence. Many workers perceive automation as a direct threat to their job security. Studies by the *World Economic Forum* (World Economic Forum, 2023) show that 40% of employees worry their job will be obsolete within five years. Employees often worry that automation will render their current skills obsolete, leading to job loss. This anxiety can result in decreased morale and productivity. A study published by Kim, Kakdol and

Solomon (2023) highlights that such fears are prevalent among workers facing technological changes, emphasizing the need for proactive measures to address these concerns.

Anxiety Over Losing Control to AI-Driven Decision-Making. As AI systems take over decision-making in production, logistics, and quality control, workers fear the erosion of autonomy and influence over outcomes. The delegation of decision-making processes to AI systems can lead to feelings of loss of control and uncertainty among employees. This anxiety is compounded by a lack of understanding of how AI systems operate and make decisions.

Economic and Social Considerations

Wage Disparities Between Traditional and Automated Job Roles. Automation can lead to significant wage disparities. An article, Dizikes (2022), published in MIT News indicates that automation has been a substantial driver of income inequality, with less-educated workers experiencing wage declines due to displacement by automated systems.

Perception of Automation as a Threat to Job Stability. The introduction of AI and automation is often perceived as a direct threat to job stability. [16] Reports suggest that approximately 27% of jobs are at high risk of automation, fuelling fears of widespread job displacement (MCColl,2023).

Mitigation Strategies

1. **Job Redesign and Role Evolution:** To address resistance, companies must reimagine roles rather than eliminate them.
    - Hybrid roles that integrate human creativity with AI efficiency: Developing roles that combine human ingenuity with AI capabilities can enhance job satisfaction and productivity. For instance, in the financial sector, AI can handle data analysis while human employees focus on strategic decision-making and client relations. For example, in Bosch's smart factories, line workers are trained to monitor AI-enabled machines, while still applying their domain knowledge to optimize production. These "cyber-physical" roles blend human intuition with machine precision
    - New career pathways in AI ethics, machine supervision, and human-robot coordination: As automation becomes more prevalent, new roles emerge that focus on overseeing AI systems, ensuring ethical AI use, and facilitating human-robot collaboration. Educational institutions and companies are increasingly offering training programs to prepare workers for these new opportunities.
2. **Transparent Communication from Leadership:** Leadership must acknowledge concerns and frame automation as an augmentation, not a replacement.
    - Clearly defining how automation enhances rather than replaces human capabilities. Leadership should communicate that automation is intended to augment human work, not replace it. It should also be noted that emphasis should be given by using AI to shift human roles toward decision-making and supervisory tasks, illustrating a commitment to integrating AI in a supportive manner (Dines, 2018).
    - Providing job transition support for workers in vulnerable roles. Offering reskilling and upskilling programs demonstrates an organization's investment in its employees' futures. For instance, companies can establish partnerships with educational institutions to provide training in emerging fields related to automation and AI.

Siemens runs a program in Germany where workers are offered vocational reskilling to transition into tech-support roles within the same factory. Amazon's Career Choice Program offers funding for training in robotics, healthcare, and cloud computing for employees affected by automation.

By proactively addressing these fears, organizations can foster a future-focused workforce that sees automation as an opportunity rather than a threat.

FUTURE RESEARCH DIRECTIONS

As smart factories evolve, the convergence of advanced technologies and human-centric values must become a strategic priority. The following future directions and recommendations aim to ensure that automation enhances, rather than replaces, human capabilities—leading to sustainable, inclusive, and psychologically resilient workplaces.

Adopt Predictive Learning and Analytics

Organizations should move beyond reactive training and embrace AI-powered learning analytics to identify emerging skill gaps and personalize learning pathways. Platforms that track employee performance and predict future training needs will be critical to maintaining workforce relevance in high-tech environments.

Expand Emotional Intelligence in Human-Machine Interfaces

Smart factories must invest in emotion-aware AI systems that can detect employee stress, fatigue, or disengagement through biometric feedback. This enables real-time support and adaptive workflows that reduce burnout and improve overall job satisfaction.

Design for Neurodiversity and Inclusion

Technological interfaces and training programs must be tailored to include neurodiverse employees, enabling access to automation-based roles for all cognitive profiles. Inclusive design enhances innovation and reduces exclusion-related turnover.

CONCLUSION

The rise of smart factories signals more than a technological revolution—it marks a turning point in how we perceive work, roles, and relationships between humans and machines. As we have explored throughout this chapter, resistance to automation is deeply rooted in fear—fear of redundancy, loss of purpose, and the erosion of familiar structures. But this fear, while powerful, is not immutable.

By recognizing the psychological anxieties and socio-economic implications that fuel this resistance, organizations can move from reactive disruption to proactive transformation. The key lies not in battling these fears, but in understanding them empathetically, responding to them strategically, and redesigning systems with people at the center.

Human-machine collaboration in smart factories offers unprecedented opportunities for efficiency, innovation, and growth. However, these benefits can only be realized through:

- Strategic workforce development that equips employees for hybrid roles.
- A deep commitment to employee well-being and psychological safety.
- An open culture of trust, inclusion, and transparent communication.

This is where the Design Thinking framework becomes indispensable. Its focus on empathy, ideation, co-creation, and iteration enables organizations to not just introduce automation—but to do so in a way that is human-centered and adaptive. From creating tailored training programs to co-designing workflows, Design Thinking provides the roadmap for sustainable and inclusive transformation.

By investing in upskilling, fostering a culture of belonging and safety, and addressing legitimate concerns around job displacement, organizations can empower their workforce to not just adapt, but thrive in the era of smart manufacturing.

"The future is not about man versus machine, but man with machine. *– Klaus Schwab, Founder, World Economic Forum*

This quote captures the essence of what lies ahead. It is not a battle between human and machine—it is a collaboration. The success of smart factories will not be measured solely by the speed of robots or the scale of automation, but by how well we integrate these advancements with the creativity, empathy, and adaptability of human beings.

The factories of the future are not just built with steel and circuits—they are shaped by trust, driven by learning, and sustained by purpose. Let us move forward with that vision—where technology does not diminish our humanity but elevates it.

Table 5. From fear to future-readiness: Transforming resistance to smart factory automation

Dimension	Fear-Based Approach	Future-Ready Approach
Mindset	"Machines will replace us."	"Machines will support and enhance our roles."
Communication	Opaque, top-down announcements	Transparent, consistent, and inclusive dialogue
Job Roles	Job loss, redundancy concerns	Hybrid roles combining human creativity with AI efficiency
Employee Involvement	Excluded from automation decisions	Co-creators of change through workshops and feedback
Training and Reskilling	Minimal or reactive training	Proactive learning pathways aligned with evolving roles
Mental Health and Well-being	Anxiety, fear of obsolescence	Psychological safety, growth mindset, and support systems
Organizational Culture	Distrust, resistance	Trust, innovation, and inclusivity
Framework for Change	Ad hoc adjustments	Design Thinking: empathize, define, ideate, prototype, test

Finally, to bring this journey to life, the timeline-style graphic below illustrates how organizations can move from fear and resistance to a place of confidence and collaboration. It's a simple reminder that change does not happen overnight—but with awareness, empathy, and the right strategies, we can guide our people through uncertainty towards a future where humans and machines thrive together.

Figure 3. From fear to future

REFERENCES

Bersin, J. (2021). *The big reset: Learning and skills in the flow of work.* Josh Bersin Company.

Boller, S., & Fletcher, L. (2020). *Design thinking for training and development: Creating learning journeys that get results.* ATD Press.

Brown, T. (2009). *Change by design: How design thinking creates new alternatives for business and society.* Harper Business.

Dines, D. (2018). *CEO commentary on AI augmentation.* UiPath.

Dizikes, P. (2022). *How automation exacerbates inequality.* MIT News.

Ejsmont, K., Gladysz, B., & Kluczek, A. (2020). Impact of Industry 4.0 on sustainability—Bibliometric analysis. *Sustainability (Basel), 12*(14), 5650. DOI: 10.3390/su12145650

Frey, C. B., & Osborne, M. A. (2017). The future of employment: How susceptible are jobs to computerization? *Technological Forecasting and Social Change, 114*, 254–280. DOI: 10.1016/j.techfore.2016.08.019

Gorecky, D., Schmitt, M., Loskyll, M., & Zühlke, D. (2014). Human-machine-interaction in the Industry 4.0 era. In *Proceedings of the 12th IEEE International Conference on Industrial Informatics (INDIN),* 289–294. DOI: 10.1109/INDIN.2014.6945523

Kim, J., Kakdol, K., & Solomon, P. (2023). Employee anxiety and automation: Understanding workforce resistance in the age of AI. *Journal of Organizational Psychology, 23*(1), 55–68.

Lasi, H., Fettke, P., Kemper, H.-G., Feld, T., & Hoffmann, M. (2014). Industry 4.0. *Business & Information Systems Engineering, 6*(4), 239–242. DOI: 10.1007/s12599-014-0334-4

McColl, K. (2023). The new reality of automation and labor disruption. *The Economist Intelligence Unit.*

McKinsey & Company. (2020, February 12). *How companies are reskilling to address skill gaps | McKinsey.* Www.mckinsey.com. https://www.mckinsey.com/capabilities/people-and-organizational-performance/our-insights/beyond-hiring-how-companies-are-reskilling-to-address-talent-gaps

Norman, D. A. (2013). *The design of everyday things* (Revised and expanded edition). Basic Books.

OECD. (2021). *OECD skills outlook 2021: Learning for life.* OECD Publishing.

PwC. (2022). *What does virtual reality and the metaverse mean for training?* PricewaterhouseCoopers.

Siau, K., & Wang, W. (2018). Building trust in artificial intelligence, machine learning, and robotics. *Cutter Business Technology Journal, 31*(2), 47–53.

Tarafdar, M., Cooper, C. L., & Stich, J. F. (2019). The technostress trifecta—Techno eustress, distress, and the dark side of technology use. *Information Systems Journal, 29*(1), 6–42. DOI: 10.1111/isj.12169

UNESCO. (2022). *Global education monitoring report: Building bridges for learning.*

World Economic Forum. (2023). *The future of jobs report 2023.*

KEY TERMS AND DEFINITIONS

Cobots (Collaborative Robots): Safety-enhanced robots designed to work alongside humans, assisting in tasks that require both human oversight and robotic precision.

Design Thinking: A human-centered problem-solving methodology that emphasizes empathy, ideation, prototyping, and testing to co-create solutions that align technology with human needs.

Digital Literacy: The competence to use digital tools and platforms effectively, including basic coding, understanding AI dashboards, and navigating IoT systems—essential in Industry 4.0 environments.

Digital Twin: A virtual simulation of a real-world system, enabling predictive analysis, training, and system optimization in smart factory operations.

Emotional Intelligence (EI): The ability to recognize, understand, and manage emotions—both in oneself and in others—which is critical for leadership, collaboration, and adapting to change in technology-driven environments.

Human-Machine Collaboration: An interactive partnership between humans and intelligent machines (like AI or robots) that combines decision-making, precision, and creativity for optimized operations.

Microlearning: A bite-sized learning approach that delivers information in small, focused segments—ideal for on-the-job training in smart factory settings where employees need fast, efficient upskilling.

Reskilling: The process of training employees in entirely new skill sets to transition them into different roles—often necessary due to automation-related job shifts.

Technostress: A condition marked by mental fatigue and anxiety due to frequent or complex interactions with digital technology, especially in high-automation settings.

Upskilling: Improving or expanding current employee skills to meet evolving job requirements, such as learning AI supervision or data analytics.

Chapter 18
Transformational Leadership and Change Management in Human–Machine Collaboration for Smart Factories

Bhagawan Chandra Sinha
https://orcid.org/0000-0003-2466-1316
Sharda University, Greater Noida, India

Nitu Singhi
Kirloskar Institute of Management, Pune, India

Bhardwaj Shukla
https://orcid.org/0009-0007-6270-5654
SOL, University of Delhi, India

Kishore Kumar Talukdar
Sharda University, Greater Noida, India

B. Sumbul
https://orcid.org/0009-0009-7142-8809
Accurate Institute of Management and Technology, India

Vivek Sharma
Gopal Narayan Singh University, India

Pankaj Sharma
https://orcid.org/0009-0000-0968-1816
School of Open Learning, University of Delhi, India

Hppater acob Hine
https://orcid.org/0009-0002-7234-5553
Ministry of Agriculture, Liberia

B. Eswaran
School of Commerce and Management, Sanjivani University, Kopargaon, India

ABSTRACT

The development of smart factories heralds a paradigm shift in industrial systems, where automation, human labour, and artificial intelligence (AI) come together to produce intelligent, flexible manufacturing settings. This chapter examines how the smooth integration of human and machine collaboration in smart factories is made possible by transformational leadership and organized change management frameworks. It offers a leadership road map for advancing

DOI: 10.4018/979-8-3373-2985-7.ch018

digital transformation, bringing cultural values into line with changes in technology, and building trustworthy, productive human-machine teams. The chapter provides an integrated strategy to managing complexity, guaranteeing ethical alignment, and facilitating sustainable innovation in smart manufacturing ecosystems by drawing on industry cases, theoretical models, and international best practices.

BACKGROUND

The accelerating wave of Industry 4.0—fueled by artificial intelligence (AI), the Internet of Things (IoT), cyber-physical systems (CPS), big data analytics, cloud computing, and advanced robotics—is fundamentally redefining industrial production. The emergence of smart factories represents a new frontier in manufacturing where machines, sensors, and human operators form a tightly interlinked network capable of real-time monitoring, predictive decision-making, and self-optimization.

Unlike traditional manufacturing environments, smart factories are hyper-connected ecosystems. They can integrate operational data from machinery, supply chains, and customer demand to autonomously adjust production schedules, predict equipment failures, and optimize workflows with minimal human intervention. These capabilities promise unprecedented productivity, efficiency, and responsiveness—allowing organizations to meet rapidly changing market conditions with agility and precision.

However, technological capability alone does not guarantee success. As these systems become more sophisticated, they also blur the boundaries between human and machine roles. Machines are no longer merely passive tools directed by human operators—they are emerging as intelligent collaborators, capable of learning from data, making complex decisions, and initiating actions without direct human command. This shift brings with it profound organizational, cultural, and leadership challenges. While much of the discourse around smart factories emphasizes technological infrastructure—connectivity standards, automation architecture, and AI algorithms—the human dimension of transformation often receives less focus. Yet, it is precisely this dimension that determines whether technology adoption thrives or falters. Leaders in the smart factory era must navigate a dual challenge:

1. Harnessing the potential of intelligent systems without undermining human contribution.
2. Guiding people through disruptive change while maintaining morale, trust, and shared purpose.

This requires a move away from traditional command-and-control leadership models, which often emphasize compliance and hierarchy, towards transformational and inclusive leadership styles that inspire vision, cultivate adaptability, and promote ethical decision-making. Transformational leaders can articulate a compelling narrative about how technology empowers rather than threatens, ensuring employees view machines as partners rather than competitors.

At the same time, the dynamic and ongoing nature of smart factory transformation demands structured, resilient change management frameworks. Unlike one-off technological rollouts, human–machine collaboration evolves continuously—new systems are integrated, workflows change, and employee skill requirements shift. Without a clear change management strategy, resistance can build, adoption rates may lag, and cultural alignment can be lost. A robust change management approach ensures that:

- Resistance is managed through transparency and involvement.
- Workforces are reskilled to match the needs of digitally integrated processes.
- Cultural values are aligned with innovation, collaboration, and ethical AI usage.
- Continuous improvement is embedded into the organization's DNA.

This chapter emerges at the intersection of these two critical elements: transformational leadership and change management. Their synergy is the key to unlocking the full potential of smart factories—not only by enabling technological adoption but by ensuring it is sustainable, ethical, and human-centered. Drawing on global case studies, theoretical frameworks, and industry research, this chapter presents a comprehensive roadmap for leaders tasked with navigating the transition from traditional manufacturing to intelligent, adaptive, human–machine collaborative environments. It examines how leaders can inspire, motivate, and mentor teams through technological disruption while simultaneously using structured change management practices to anchor transformation into daily operations. In doing so, it aims to answer a central question for Industry 4.0 and beyond: How can leaders build organizations where machines amplify human capabilities and humans guide machines with wisdom, ethics, and vision?

FOCUS OF THE CHAPTER

IoT sensors, AI, advanced analytics, and cyber-physical systems power smart factories, which are the industrial industry's future. Unprecedented scalability, accuracy, and efficiency are promised by these technologies. Global manufacturing leaders' experience, however, demonstrates that human leadership—rather than technology or algorithms—is the real differentiator in fulfilling this promise. Automation can make processes more efficient, but only leaders can motivate, bring people together, and help them navigate the operational and cultural upheaval that comes with technological innovation. This chapter emphasizes that the key success factors in the smart factory revolution are leaders who can effectively handle opposition, spur change, and unite teams around a common goal. The emphasis is on:

- Recognizing how transformational leadership may inspire and mentor individuals through technology disruption: Employees' sense of identity, stability, and professional relevance are frequently called into question by technological upheaval. In this ambiguity, transformational leaders serve as guiding lights, managing tasks but also influencing attitudes. Their captivating vision of a future in which technology enhances rather than replaces human skill inspires. Through individualized coaching, they assist people transition into new roles, acquire new abilities, and gain confidence while engaging with intelligent systems. They serve as role models for flexibility, demonstrating that accepting change is a collaborative process rather than a directive from above. To put it simply, transformational leadership transforms disturbance into a chance for development, creativity, and a revitalized mission.

- Utilizing change management techniques to guarantee that smart factory practices are adopted sustainably: Introducing smart factory practices—AI-assisted decision-making, predictive maintenance, or autonomous material handling—is not a one-time project; it is a continuous evolution. Without a structured approach, even the most promising technologies risk rejection or underuse. Change management provides the frameworks, tools, and communication strategies to make adoption stick. It helps identify potential sources of resistance, address them early, and maintain momentum beyond initial implementation. By creating clear roadmaps, building coalitions of support, and celebrating early wins, leaders ensure that adoption is not only fast but sustainable. A well-managed change process transforms the fear of "machines taking over" into a shared excitement for "machines helping us excel."
- Establishing cooperative ecosystems in which machines and people work together as learning partners: In a truly smart factory, humans and machines do not operate in isolation; they learn from and with each other. Machines provide data-driven insights, automate repetitive tasks, and enhance accuracy, while humans contribute judgment, creativity, and ethical reasoning. Leaders are responsible for designing workflows where this cooperation is natural and productive. They encourage "reciprocal learning," where AI systems are refined by human input, and human teams gain insights and decision-support from AI. They foster a culture where collaboration with machines is viewed as a skill, not a threat, positioning employees as *co-creators* of intelligent systems. When done right, the smart factory becomes not just a site of production, but a mutual learning environment—constantly evolving through the combined intelligence of people and machines.

INTRODUCTION

Manufacturing has undergone a profound transformation with the advent of Industry 4.0, ushering in an era where smart factories are redefining production processes. By integrating cutting-edge technologies such as advanced robotics, artificial intelligence (AI), the Internet of Things (IoT), augmented reality (AR), and big data analytics, these factories are evolving into flexible, hyper-connected, and adaptive ecosystems. In these environments, machines are no longer isolated units of automation—they are intelligent collaborators capable of analyzing data, making autonomous decisions, and optimizing operations in real time.

At the heart of this transformation lies human–machine cooperation. Rather than replacing human effort, intelligent systems are designed to augment human strengths—enhancing productivity through precision, improving creativity by freeing humans from repetitive tasks, and increasing responsiveness through predictive insights. This synergy ensures that technology amplifies uniquely human capabilities such as critical thinking, emotional intelligence, problem-solving, and ethical judgment.

However, achieving this integration is not simply a matter of installing new machinery or deploying advanced software. The real challenge lies in aligning human potential with digital innovation—a process that demands visionary leadership and strategic change management. Technology may set the pace of transformation, but it is leadership that determines whether the workforce embraces, adapts to, and thrives within this new paradigm. Leaders must not only champion the benefits of technological adoption but also guide cultural change, address employee concerns, and cultivate an environment where learning and innovation are continuous. This leadership imperative becomes even more pronounced in the context of the emerging Industry 5.0 (I5.0) paradigm. Unlike its predecessor, which largely emphasized automation and efficiency, Industry 5.0 shifts the focus to human-centric manufacturing.

As noted by Adel (2022), it seeks to place humans back at the center of the manufacturing process—not as passive operators but as active, creative partners in collaboration with intelligent systems. This philosophy contrasts sharply with earlier industrial revolutions, where technological advancement often sidelined the human role. Instead, I5.0 champions personalization, sustainability, resilience, and societal well-being as core drivers of manufacturing innovation. In this light, the success of modern manufacturing depends not only on technological prowess but also on the ability of leaders to foster trust, inclusivity, and ethical decision-making in human–machine collaborations. The factories of the future will not be defined merely by how smart their machines are, but by how effectively humans and machines learn, adapt, and innovate together.

In this new paradigm, machines are no longer mere tools. They are evolving into collaborative partners, capable of learning, adapting, and interacting with human workers. This integration of human and machine intelligence marks a pivotal point in industrial evolution. However, this transformation cannot be driven by technology alone. The real differentiator lies in leadership—particularly transformational leadership—and robust change management strategies that align people, processes, and technology.

Inspiring, empowering, and innovative, transformational leadership stands out as a critical enabler for organizations seeking to navigate the complex terrain of smart manufacturing. In an era defined by rapid technological disruption, this leadership style transcends traditional managerial roles by articulating a compelling vision, fostering shared purpose, and empowering individuals to take ownership of change. Transformational leaders do not merely manage processes; they ignite intrinsic motivation, inspire creativity, and cultivate resilience, ensuring that employees perceive change as an opportunity rather than a threat. When coupled with robust change management frameworks, transformational leadership ensures not only the smooth integration of advanced technologies but also the active participation and adaptability of the workforce. Change management provides the structured methodology—encompassing clear communication, stakeholder engagement, reskilling programs, and continuous feedback loops—that transforms vision into tangible outcomes. The synergy between these two domains allows organizations to bridge the gap between technological potential and human readiness, which is often the most significant determinant of success in digital transformation initiatives.

The transition to smart manufacturing is not merely a technical upgrade; it is a deep organizational shift. As Fernandez-Vidal et al. (2022) emphasize, digital transformation demands leaders who can guide their organizations from their current operational state toward a desired future state—often in contexts where organizational structures are ambiguous, digital competencies are uneven, and the pace of change is unrelenting. This journey requires leaders who can navigate uncertainty, foster a culture of continuous learning, and encourage experimentation without fear of failure.

Moreover, transformational leaders in smart manufacturing must balance the dual imperatives of operational efficiency and human-centric development. While technology promises unprecedented gains in productivity, it is the human capacity to learn, adapt, and innovate that ultimately determines sustainable success. By prioritizing opportunities for skill development, cross-functional collaboration, and knowledge sharing, leaders can ensure that the workforce remains agile, engaged, and future-ready—transforming potential resistance into proactive contribution. In essence, transformational leadership, when reinforced by disciplined change management, becomes the driving force that converts technological disruption into strategic advantage. It enables organizations to not only adopt smart manufacturing systems effectively but also to elevate human potential in ways that enhance creativity, adaptability, and long-term competitiveness.

Drawing on Deloitte's (2025) extensive research on human–machine collaboration, which underscores the shift from automation as a tool to automation as a collaborative partner, and insights from Research Features' exploration of leadership in the age of machine intelligence, which emphasize the need for visionary, adaptive, and ethical leadership, this chapter delves into the strategic nexus of transformational leadership and change management in the context of smart factories. By synthesizing findings from peer-reviewed academic literature, industry white papers, and global case studies, we aim to present a comprehensive and actionable guide for executives tasked with leading through the disruptive yet opportunity-rich environment of Industry 4.0. The analysis presented here is not confined to theory; it integrates theoretical frameworks—including transformational leadership models, Kotter's 8-Step Change Model, and Prosci's ADKAR methodology—with practical, field-tested strategies used by leading manufacturing organizations. We explore real-world tactics such as reskilling programs, human–machine task design, trust-building initiatives, and culture-shaping interventions that have proven effective in ensuring sustainable adoption of advanced manufacturing systems.

To deepen practical understanding, the chapter also features case studies from global manufacturing leaders who have successfully balanced technological integration with human capital development. These cases illustrate how transformational leaders inspire shared vision, create psychological safety, and foster an environment where humans and intelligent machines work as learning partners—maximizing efficiency while retaining human creativity and judgment. Furthermore, the discussion incorporates ethical considerations, including data privacy, algorithmic transparency, and the preservation of human dignity in machine-mediated workflows. By embedding these considerations into leadership and change management practices, we argue that executives can future-proof their organizations, ensuring not just immediate productivity gains but long-term resilience and trust within the workforce.

Ultimately, this chapter offers leaders a thorough manual for navigating human–machine collaboration—equipping them with the mindset, methodologies, and moral compass needed to lead effectively in smart factory ecosystems. In doing so, it bridges the gap between technological innovation and human empowerment, ensuring that the promise of Industry 4.0 is realized not only in operational performance but also in organizational well-being and societal benefit.

HUMAN-MACHINE COLLABORATION: THREE MODES

Based on Deloitte's insights (2020), human-machine collaboration generally occurs in three modes:

Machines as Tools or Assistants

In this foundational mode, machines act as extensions of human capability, executing predefined instructions and performing repetitive, precision-oriented, or physically demanding tasks. Humans retain full decision-making control, using machines primarily to amplify efficiency, accuracy, and speed. Examples include AI-enabled maintenance scheduling, robotic arms that handle hazardous materials, or IoT dashboards that display real-time production metrics. While this mode offers quick productivity gains and relatively low change resistance, it underutilizes the full potential of smart technologies. Leaders in this stage need to focus on building trust in AI-driven outputs, training employees to leverage these tools effectively, and fostering a mindset that sees machines as partners rather than threats.

Machines as Supervisors

Here, machines move beyond execution to guiding or controlling human actions. Advanced analytics, AI-driven quality control systems, and predictive maintenance platforms can detect anomalies and issue instructions—sometimes overriding human decisions for greater safety, quality, or efficiency. For instance, AI-based vision systems can identify microscopic defects invisible to the human eye and halt production lines automatically. While this mode improves consistency, compliance, and error reduction, it also challenges traditional authority structures. Human operators may feel disempowered or skeptical if machine recommendations contradict their experience. Leaders must actively manage the emotional and cultural implications by explaining the rationale behind machine-guided decisions, encouraging human oversight as a safeguard, and positioning technology as a collaborative coach rather than a replacement.

Machines as Teammates

The most advanced and aspirational mode envisions machines as intelligent collaborators that can co-create solutions, learn from human interaction, and adapt dynamically to new situations. This is the teammate model, where humans and machines engage in shared decision-making, problem-solving, and even creative design. For example, AI-powered generative design tools can propose multiple engineering solutions, which human experts refine based on contextual knowledge, safety considerations, and customer needs. In this mode, machines contribute analytical depth and speed, while humans bring ethical judgment, intuition, and empathy—creating a synergistic loop of continuous learning.

While the tool/assistant and supervisor modes have proven operational benefits, they often stop short of unlocking the transformative potential of Industry 4.0 and the emerging Industry 5.0 paradigm, which emphasizes human-centric manufacturing. The teammate model—when implemented effectively—enables:

- Shared responsibility and innovation, where solutions emerge from human creativity enhanced by machine intelligence.
- Adaptive resilience, as humans and machines can quickly reconfigure workflows in response to disruptions.

- Deep personalization of products and processes, driven by human insight plus AI-enabled customization.

However, the success of this model depends on three critical enablers:

- Trust – Workers must believe that machine decisions are accurate, transparent, and aligned with shared goals.
- Adaptability – Both humans and machines must be capable of learning from each other and adjusting to evolving tasks.
- Cultural readiness – Organizations must create an environment where collaboration across human–machine boundaries is normalized and celebrated.

These enablers are directly influenced by leadership style and change management maturity. Transformational leaders, with their focus on inspiration, empowerment, and shared vision, play a decisive role in shaping attitudes, mitigating fear, and embedding collaborative practices. Meanwhile, structured change management ensures that the transition between collaboration modes—from tools to supervisors to teammates—is executed in a way that is sustainable, ethical, and embraced by the workforce.

TRANSFORMATIONAL LEADERSHIP: CATALYZING HUMAN-MACHINE SYNERGY

In the era of smart manufacturing, where technological disruption is constant and organizational boundaries are fluid, leadership must transcend the limitations of traditional command-and-control systems. Research by Tagscherer & Carbon (2023) emphasizes that visionary thinking, coupled with flat organizational hierarchies, employee empowerment, digital skill cultivation, and the promotion of teamwork, forms the foundation for leadership success in digital transformation initiatives. Visionary thinking enables leaders to anticipate industry shifts, foresee the long-term implications of human–machine integration, and articulate a clear, compelling roadmap that inspires both confidence and commitment across the organization. This vision serves as the navigational compass in a volatile, uncertain, complex, and ambiguous (VUCA) manufacturing environment, ensuring that both human and technological resources are aligned toward a common goal.

Flat hierarchies reduce bureaucratic bottlenecks and encourage faster decision-making, which is crucial in environments where data-driven insights from machines must be acted upon in real time. By dismantling rigid chains of command, leaders create agile structures that allow for cross-functional collaboration between engineers, operators, IT specialists, and AI systems designers—accelerating innovation cycles and responsiveness. Employee empowerment is not merely a motivational tool; it is a strategic necessity in Industry 4.0 and the emerging Industry 5.0 context. As intelligent systems assume more operational tasks, human roles shift toward complex problem-solving, creative thinking, and ethical decision-making. Empowering leadership styles delegate meaningful responsibility, encourage autonomy, and trust employees to make technology-enabled decisions. This not only boosts job satisfaction but also enhances the sense of ownership in transformation initiatives, making resistance to change less likely.

In this regard, Zhang et al. (2023) provide empirical evidence that empowering leadership significantly enhances employee creativity. By granting autonomy, acknowledging contributions, and providing access to learning opportunities, leaders enable individuals to generate novel ideas—a critical advantage in manufacturing contexts where innovation determines competitive survival. Employee creativity, when supported by a technologically enriched environment, can lead to process improvements, product customization, and entirely new manufacturing paradigms. Teamwork and collaboration remain at the heart of this leadership model. In a human–machine collaborative ecosystem, effective teams are not only composed of human members but also integrate machines as active contributors to decision-making and execution. Leaders must cultivate a culture where co-creation between humans and intelligent systems is viewed as normal and desirable, breaking down silos between digital and human competencies.

Furthermore, digital skills form the bridge between empowerment and actual capability. Empowerment without digital literacy risks frustration and inefficiency; thus, leaders must ensure continuous reskilling and upskilling programs so employees can confidently engage with AI tools, IoT platforms, and advanced robotics. In essence, leadership that embraces vision, flattened hierarchies, empowerment, teamwork, and digital fluency creates an organizational climate where creativity thrives, technology is trusted, and transformation momentum is sustained. This aligns directly with the human-centric values of Industry 5.0, ensuring that technological advancement enhances—not replaces—the human contribution.

According to Burns (1978) and Bass (1985), transformational leadership is rooted in the idea that leaders inspire followers to place the needs of the group above individual self-interest. This leadership style is not transactional—focused solely on exchanges of rewards for performance—but instead redefines goals, raises aspirations, and fosters deeper commitment to shared organizational objectives. Transformational leaders achieve this through four interconnected pillars:

Idealized Influence

Transformational leaders act as ethical role models, consistently demonstrating integrity, transparency, and alignment with the organization's values. In the context of smart manufacturing, where rapid automation and AI integration can trigger uncertainty and skepticism, leaders who embody moral responsibility reassure employees that technological change serves the greater good. For example, a factory leader introducing predictive maintenance AI might openly address its purpose, limitations, and safeguards—reinforcing trust in both leadership and technology.

Inspirational Motivation

Leaders provide a compelling, future-focused vision that frames digital transformation as an exciting journey rather than a disruptive threat. In smart factories, this vision might highlight the potential for human–machine synergy to improve product quality, reduce repetitive labor, and open up more meaningful, creative roles for employees. By articulating clear goals and purpose, leaders motivate teams to approach transformation with enthusiasm and shared ownership.

Intellectual Stimulation

Transformational leaders encourage innovation, experimentation, and critical thinking, urging employees to question outdated methods and embrace novel solutions. In a smart manufacturing environment, this might mean inviting cross-functional teams to co-design AI-assisted workflows or explore new applications for IoT data. By creating a safe space for exploration and failure, leaders nurture a workforce that actively participates in shaping the digital future, rather than passively adapting to it.

Individualized Attention

Every employee experiences digital transformation differently, influenced by their skills, comfort with technology, and career aspirations. Transformational leaders recognize these differences, offering personalized coaching, mentorship, and learning opportunities. In smart factories, this may involve designing customized reskilling paths for machine operators transitioning into roles such as robotics technicians, data analysts, or quality assurance specialists. By valuing each person's unique contribution, leaders sustain morale and engagement during periods of rapid change.

In smart manufacturing, transformational leadership bridges the gap between human interaction and technological innovation. While machines may handle operational precision and data processing, it is human vision, empathy, and adaptability that ensure technology adoption is inclusive, ethical, and sustainable. Leaders must actively cultivate a culture of continuous learning, where employees see technological evolution as an opportunity to grow their capabilities rather than as a threat to their roles.

Equally, transformational leaders instil trust in new systems by ensuring that AI, robotics, and automation are implemented transparently and with clear benefits for both the organization and its workforce. They are also adept at-risk management, anticipating the social, cultural, and operational challenges of digital transformation and addressing them proactively—be it resistance to change, skill gaps, or ethical concerns around automation. Ultimately, transformational leadership in smart manufacturing does more than manage change—it humanizes the digital shift, ensuring that the technological leap forward is matched by an equally significant evolution in organizational culture and human potential.

THE PRACTISE OF TRANSFORMATIONAL LEADERSHIP

Transformational leadership in the context of Industry 4.0 and Industry 5.0 is not an abstract concept—it is a hands-on, strategic practice that bridges advanced technology adoption with the human factors essential for long-term success. By aligning cutting-edge innovations with organizational goals and employee well-being, transformational leaders ensure that the digital evolution of manufacturing remains both competitive and human-centered.

As Research Features (2023) emphasizes, "Leadership in the age of machine intelligence requires a new kind of fluency—one that blends emotional depth with algorithmic understanding." In other words, leaders must be equally comfortable discussing predictive maintenance algorithms as they are understanding the anxieties of a worker learning to operate alongside collaborative robots.The practice of transformational leadership in smart manufacturing typically manifests through three critical applications:

Visionary Alignment

At the heart of transformational leadership lies visionary communication—the ability to articulate how technological changes serve a bigger, inspiring picture. In smart factories, this means explaining to employees not just what is changing, but why it matters. For instance, when implementing IoT-enabled sensors and AI-driven analytics, leaders must connect these technologies to tangible benefits: faster production cycles, predictive quality control, and quicker response to market shifts. Example: A transformational leader might show teams how real-time supply chain data enables proactive adjustments in production scheduling, reducing customer wait times and minimizing inventory waste. This future-focused narrative transforms technology adoption from a top-down mandate into a shared organizational mission.

Such visionary alignment builds strategic clarity, helping employees understand how their roles evolve in parallel with technological upgrades, and how the entire organization benefits from embracing these tools.

Workforce Empowerment

No matter how advanced a technology is, it will underperform without skilled, confident human operators. This is where workforce empowerment—through reskilling and upskilling—becomes a central pillar of transformational leadership. According to Deloitte's 2025 Human Capital Trends report, 70% of manufacturing workers will require new skills to work effectively with smart systems. These skills range from robotics programming and data analytics to human–machine interface management. Transformational leaders do more than approve training budgets—they:

- Advocate for continuous learning as a cultural norm.
- Ensure that skill development pathways are inclusive and cater to different learning paces.
- Pair technical training with soft skills development, such as adaptability and problem-solving, which are crucial in a rapidly changing environment.

By empowering employees, leaders reduce resistance to change, build confidence, and foster a workforce that actively seeks to optimize new systems rather than merely adapting to them.

Synergistic Collaboration

Perhaps the most visible—and measurable—impact of transformational leadership in smart manufacturing is in fostering synergy between humans and machines. This goes beyond integration; it's about creating partnerships where each side complements the other's strengths.

Siemens has implemented collaborative robots (cobots) in several factories. These cobots handle precision-intensive tasks, such as fine assembly, while human operators focus on problem-solving and customization. Results from certain Siemens plants show up to a 15% increase in output, as cobots not only reduce human fatigue but also improve overall accuracy and consistency. In these environments, transformational leaders ensure that human dignity and creativity remain central—machines handle the repetitive and hazardous work, while humans drive innovation, judgment, and relationship management.

One of the most compelling illustrations of transformational leadership in action is General Electric's (GE) Brilliant Factory program. This initiative sought to integrate digital twins—virtual replicas of physical assets—into manufacturing processes. GE's leaders didn't simply deploy the technology; they:

- Explained its purpose: Digital twins would allow real-time simulation and optimization of factory operations.
- Involved employees early: Teams were trained to use and interpret digital twin data, enabling them to troubleshoot issues before they impacted production.
- Linked adoption to measurable benefits: Within a few years, GE factories reported a 25% reduction in production downtime and significant improvements in product quality.

The critical factor here wasn't just technological investment, but leadership investment—motivating employees to actively engage with the tools, experiment with their capabilities, and take ownership of optimization processes.

BUILDING PSYCHOLOGICAL SAFETY

In the high-stakes environment of Industry 4.0 and Industry 5.0, where technologies evolve faster than job descriptions, psychological safety becomes the foundation for sustainable innovation. Coined by Harvard scholar Amy Edmondson, psychological safety refers to an environment in which individuals feel safe to take interpersonal risks—to voice ideas, admit mistakes, and ask questions—without fear of ridicule or reprisal.

In smart factories, where human–machine collaboration is the norm, this safety net is essential. Employees need the confidence to experiment with AI systems, IoT-enabled devices, and robotics platforms, knowing that mistakes are seen not as failures, but as stepping stones toward better solutions. According to Deloitte's 2024 Human Performance Insights, trust is the key enabler here: "When employees believe they can explore and innovate without career penalties, they are more likely to uncover breakthrough efficiencies." In practical terms, trust transforms hesitation into action and compliance into creativity. Transformational leaders play a pivotal role in embedding psychological safety by leveraging three interconnected practices:

Promoting Experimentation

Transformational leaders actively encourage trial-and-error learning through structured experimentation. Instead of rolling out a technology with rigid rules, they create pilot projects or "sandbox" environments where employees can try, fail, adjust, and try again. For example, in a smart factory adopting AI-driven predictive maintenance, a leader might invite technicians to customize the system's alert parameters during a three-month pilot. This hands-on freedom builds both technical skill and personal ownership of the system. By linking experimentation to learning goals rather than performance penalties, leaders reduce the fear factor often associated with disruptive technologies. Such initiatives reflect the intellectual stimulation aspect of transformational leadership—pushing teams to question existing processes and think beyond traditional solutions.

Open Communication

A psychologically safe environment thrives on transparent and two-way communication. In smart manufacturing, this means establishing dedicated channels—such as digital suggestion platforms, cross-functional town halls, or machine–human workflow review sessions—where employees can share experiences and concerns without bureaucratic delays. For example,when introducing autonomous mobile robots in material handling, leaders may schedule weekly "robot review" meetings where warehouse workers can raise operational or safety concerns directly with implementation teams. This approach not only surfaces valuable ground-level insights but also reduces resistance by showing employees that their voices directly influence technology integration.

This practice embodies individualized consideration, as it acknowledges and addresses unique employee concerns rather than imposing one-size-fits-all solutions.

Recognizing Contributions

Acknowledgment is a powerful tool for reinforcing psychological safety. When leaders publicly recognize individual and team efforts in experimenting with or adopting new technologies, it signals that initiative and engagement are valued as much as technical success. For example, in one of the Bosch smart factory, leaders launched "Innovation Labs"—dedicated spaces where frontline workers collaborate with engineers to test new automation tools. Contributions from these labs are celebrated during quarterly meetings, with participants receiving both symbolic recognition and opportunities for career advancement. The result: Bosch reported a 20% increase in employee engagement with smart technologies, as workers began to see innovation not as a top-down imposition but as their own achievement.

Recognition strengthens the idealized influence pillar of transformational leadership, where leaders act as role models who value inclusivity, creativity, and mutual respect. For example, at Bosch's smart factories, leaders implemented "innovation labs" where workers collaborate with engineers to test new automation tools. This approach, rooted in intellectual stimulation and individualized consideration, has led to a 20% increase in employee engagement with smart technologies.

CHANGE MANAGEMENT: NAVIGATING THE TRANSITION TO SMART FACTORIES

The journey toward becoming a smart factory is as much about people as it is about technology. While the allure of AI-driven analytics, IoT-enabled monitoring, and collaborative robotics is undeniable, the real challenge lies in guiding human capital through the transition. Change management, in this context, serves as the bridge between current operational realities and the envisioned state of digitally enabled, human–machine symbiosis.

According to Silva et al. (2020), successful transformation hinges on more than just installing advanced systems; it requires building organizational capabilities for continuous learning, reskilling, knowledge sharing, and innovation. Without this, even the most sophisticated automation may fail to deliver its promised value. Nahavandi (2019) further warns that employee reluctance—rooted in fear, uncertainty, or lack of trust—can undermine the largest potential gains of smart manufacturing initiatives.

In other words, the real bottleneck is not the *technology*, but the *mindset*. As Carlsson (2023) observes, neglecting human-centric perspectives in change and innovation management can result in expensive, underutilized systems and diminished ROI.

Drawing from Kotter's 8-Step Change Model and ADKAR (Awareness, Desire, Knowledge, Ability, Reinforcement), smart factory transitions can be mapped as follows:

Table 1. Mapping Smart factory transitions using Kotter's 8-step change model and ADKAR

Phase	Action	Leadership Role
Create Urgency	Highlight competitive pressures and benefits	Share success stories and market trends
Build Coalition	Form cross-functional AI task forces	Include both engineers and shop floor workers
Form Vision	Define human-machine synergy goals	Align with employee values and business KPIs
Communicate Vision	Use transparent messaging	Engage through storytelling and town halls
Empower Action	Train teams, remove silos	Invest in skilling and peer mentoring
Create Wins	Celebrate small automation successes	Recognize hybrid team efforts
Consolidate Gains	Scale across departments	Use metrics for feedback loops
Anchor in Culture	Embed digital thinking	Model behaviors from top leadership

While Kotter's model provides a roadmap, leaders must anticipate and address the unique barriers that arise when implementing advanced technologies in manufacturing:

Opposition to Automation

Workers may interpret automation as a prelude to layoffs. Transformational leaders must emphasize augmentation over replacement—highlighting how machines will take over hazardous, repetitive, or precision-heavy tasks, enabling employees to focus on higher-value work. At a Siemens plant, the introduction of cobots was framed as an ergonomics initiative, leading to a 15% productivity increase without job losses.

Skill Deficits

Industry 4.0 requires capabilities like data interpretation, robotics programming, and digital troubleshooting—skills not traditionally taught in manufacturing environments. Reskilling must be continuous and accessible, using blended learning (classroom, VR simulations, on-the-job practice) to keep pace with evolving tools.

Cultural Transformation

Smart factories demand a shift from rigid hierarchies to agile, cross-functional collaboration. This can be unsettling in organizations with decades of top-down decision-making traditions. Leaders must model collaborative behaviors, break down silos, and create structures where technicians, data scientists, and line supervisors co-create solutions.

Organizations can also use other the following tactics, which are based on Deloitte and other sources, to address these issues:

- **Establishing Urgency:** Leaders need to emphasize how smart factories are essential for competitive advantages like cost savings or quicker time to market. According to Deloitte (2025), companies that use smart technologies increase their operational efficiency by 30%.
- **Creating Coalitions:** Involving management, staff, and technology partners guarantees widespread support. Coalitions are key to accelerating change, according to Kotter's approach.
- **Upskilling Programs:** It's crucial to spend money on robotics, AI, and IoT training. According to Deloitte's 2024 research on entry-level positions, reskilling initiatives boost employee productivity in tech-driven workplaces by 18%.
- **Vision Communication:** Using narrative to highlight empowerment, leaders must explain how human-machine collaboration improves roles. For example, describing how cobots free up workers' time to concentrate on innovative problem-solving by eliminating monotonous activities.
- **Cultural Integration:** New behaviours must be in line with organizational values in order to be embedded. According to Deloitte's six signature traits (2020), inclusive leadership promotes cooperation and trust.

An excellent example of change management is Toyota's lean automation approach. Toyota's executives increased productivity across smart factories by 22% by combining robotics with lean concepts, ensuring that workers saw automation as an improvement through pilot projects and open communication.

SYNERGIZING TRANSFORMATIONAL LEADERSHIP AND CHANGE MANAGEMENT

In smart factories, transformational leadership and change management work hand in hand. While change management frameworks offer the structure to carry out change, leaders supply the inspiration and vision to propel it. While Deloitte's insights on human performance (2024) underscore the importance of trust in change adoption, McKinsey's 2025 report on superagency highlights how visionary leadership integrates AI activities with business goals. This collaboration shows up as:

Vision-Driven Change

Transformational leaders describe a future in which creativity is fueled by human-machine collaboration, directing change management initiatives.

Empowered Implementation

With the help of change management's emphasis on breaking down barriers, leaders enable staff members to carry out the vision.

Sustained Transformation

Long-term success is ensured by combining Kotter's methods for anchoring change with inspirational motivation.

Few examples of this synergy include:

- **Siemens' Digital Transformation**: Siemens' leaders articulated a vision for Industry 4.0, using digital twins and IoT to optimize production. Change management included pilot programs and reskilling initiatives, resulting in a 20% efficiency gain across plants.
- **Bosch's Industry 4.0 Strategy**: Bosch's transformational leaders fostered innovation through cross-functional teams, while change management ensured scalability via pilot projects and training. This led to a 15% reduction in production costs.
- **Tesla's Gigafactory**: Elon Musk's visionary leadership drove rapid automation adoption, supported by change management strategies like transparent communication and reskilling. Tesla's Gigafactory achieved a 30% increase in production capacity within two years.

ETHICAL LEADERSHIP IMPERATIVES

The rapid adoption of AI, IoT, and robotics in smart factories brings enormous potential for efficiency, precision, and innovation—but it also introduces a new frontier of ethical considerations. As manufacturing becomes increasingly data-driven, leaders must navigate complex issues involving fairness, transparency, privacy, and trust. Leaders who inspire, challenge, and empower their teams—are uniquely positioned to champion ethical technology integration. By embedding ethical principles into decision-making, they ensure that smart factories are not only technologically advanced, but also socially responsible and human-centered. In smart manufacturing, transformational leaders need to face ethical issues such as:

Transparency

One of the most pressing challenges in smart manufacturing is the opacity of AI-driven decision-making. Predictive analytics may determine machine maintenance schedules, automate quality inspections, or even recommend workforce allocation—but without clear explanations, employees and stakeholders may perceive these decisions as arbitrary or biased. The AI Institute at Deloitte (2024) emphasizes open governance frameworks where algorithmic processes are documented, interpretable, and regularly reviewed. Leaders must practice intellectual stimulation by inviting employees to question and understand AI outputs, creating a culture where curiosity and critical thinking are encouraged rather than suppressed.

Inclusivity

Without deliberate action, the benefits of Industry 4.0 risk being unevenly distributed—particularly excluding marginalized groups such as older workers, women in traditionally male-dominated trades, or employees in rural manufacturing plants with less access to training. Deloitte's framework for inclusive leadership calls for leaders to proactively provide equitable access to upskilling opportunities, ensure recruitment is free from bias, and redesign roles to be accessible for diverse abilities. Through individualized consideration, leaders can tailor development plans that meet the needs of varied employee profiles, ensuring no one is left behind in the digital shift.

Accountability

AI and automation systems can amplify biases if left unchecked—whether in production scheduling, performance evaluations, or safety monitoring. Ethical leaders take accountability for technological outcomes, even if the algorithm made the decision. McKinsey's AI Ethical Guidelines (2024) recommend continuous bias audits, integrating human oversight in critical decisions, and establishing escalation protocols when AI recommendations are contested. IBM's smart factory initiatives implemented bias-detection algorithms to flag anomalies in AI-based quality control, ensuring that automated defect rejection did not disproportionately affect certain suppliers or product lines. This contributed to a 25% increase in employee trust, as workers saw their input and oversight valued.

Ethics in smart manufacturing is not just a compliance requirement—it is a competitive advantage. Factories that protect privacy, foster fairness, and maintain transparency are more likely to retain talent, win customer trust, and avoid costly reputational risks. In a highly connected supply chain ecosystem, one ethical lapse can cascade into lost contracts, regulatory penalties, and public backlash. Transformational leaders, by embedding ethical considerations into every stage of smart factory transformation, future-proof both their workforce and their brand. They do so not by treating ethics as an afterthought, but by integrating it into vision-setting, communication, and innovation strategies from day one.

PRACTICAL FRAMEWORKS FOR IMPLEMENTATION

Effective leadership in smart factories requires not only a strategic vision but also the practical use of proven frameworks that bridge human and technological capabilities. The following tools and methodologies—drawn from both academic research and industry best practices—offer leaders concrete methods to design, monitor, and refine human–machine collaboration in digital transformation initiatives.

Transformational Leadership Framework

Effective leadership in the smart factory era cannot rely solely on intuition or legacy management models. Instead, it demands structured tools and agile frameworks that can navigate complexity, guide human-machine integration, and reinforce learning cultures. Below are key leadership tools and frameworks essential for driving transformation:

Digital Capability Maturity Model (DCMM)

The concept of capability maturity models was introduced by Watts Humphrey and later refined into the Capability Maturity Model (CMM) at Carnegie Mellon University's Software Engineering Institute in the late 1980s. The Digital Capability Maturity Model adapts these principles for digital transformation. Sector-specific versions, such as the Construction DCMM by Aghimien et al. (2021), assess readiness across technology, people, and process dimensions. DCMM is a framework used to assess an organization's readiness and maturity across digital transformation dimensions—people, technology, processes, and leadership. It identifies current digital capabilities and provides a roadmap for future growth.

- Benchmark current digital leadership capabilities (e.g., AI fluency, agile decision-making).
- Prioritize investments in workforce development, smart tools, and collaborative platforms.
- Track the maturity of human-AI collaboration practices over time.
- Assess alignment between leadership competencies and technological capabilities.
- Integrate findings into strategic leadership development programs.

Human-AI Interaction Playbook

This concept is widely referenced in human–computer interaction (HCI) research, including principles from Microsoft Research's "Guidelines for Human–AI Interaction" (Amershi et al., 2019). It emphasizes transparency, control, and mutual trust between humans and AI systems. A structured guide to understanding how humans and AI systems can collaborate effectively. It includes principles such as transparency, control and reciprocity.

- Design co-working environments where humans and machines interact in predictable, supportive ways.
- Evaluate interaction quality through metrics like trust scores, error recovery rates, and decision confidence.
- Promote ethical AI usage and ensure that machine decisions are explainable to human collaborators.
- Clear role boundaries (who decides, who assists, who monitors).
- Continuous user feedback loops to improve AI systems based on real-time human input.

Reciprocal Human–Machine Learning (RHML)

Proposed by Dov Te'eni et al. (2023), RHML extends the idea of mutual learning beyond humans adapting to machines—it includes machines adapting to human behaviors and preferences. AA concept that emphasizes mutual adaptation between humans and intelligent systems. Not only do humans learn to use machines better, but machines also adapt based on human behavior and feedback.

- Foster environments where machine systems evolve through user feedback, not just static programming.
- Encourage co-learning models—e.g., a robot adapting its assistance style based on an operator's working speed and comfort.
- Use RHML insights to drive iterative training, rather than one-off implementation.
- Promote a growth mindset and experimentation culture.
- Redefine KPIs from pure output to learning efficiency and adaptability.

Organizational Network Analysis (ONA)

Developed in organizational sociology and refined in the 1990s by scholars like Rob Cross and Andrew Parker, ONA maps information flow and collaboration patterns inside organizations. ONA is a method of studying communication and information flow within an organization. It uses data from digital tools, surveys, and workflows to map collaborative patterns.

- Identify informal influencers (change agents) who can champion human-machine collaboration.
- Understand bottlenecks, silos, and areas where AI adoption might lag.
- Use network data to promote cross-functional teaming, especially between engineers, operators, and data scientists.
- Implement peer-to-peer mentoring in AI-enabled roles.
- Design collaboration spaces based on actual behavioral networks, not reporting hierarchies.

Agile Leadership Framework

Rooted in the Agile Manifesto (Beck et al., 2001) and later expanded in leadership literature (e.g., Simon Hayward, 2018), this framework emphasizes adaptability, servant leadership, and iterative change. A leadership approach rooted in the principles of Agile development, emphasizing iteration, flexibility, and responsiveness to change.

- Lead by adaptive decision-making in fast-changing technological environments.
- Use short feedback cycles (sprints) to experiment with new tech workflows.
- Apply servant leadership principles—prioritizing team needs, removing roadblocks, and facilitating autonomy.
- Run AI/automation pilots in short iterations and co-refine based on operator feedback.
- React to machine learning performance and retrain both systems and humans dynamically.

People-Centric AI Governance Model

Draws from AI ethics research by bodies like the European Commission's High-Level Expert Group on AI (2019), OECD AI Principles, and IEEE's Ethically Aligned Design. AA governance model that ensures AI deployment aligns with human values, safety, and fairness.

- Transparency: AI decisions are explainable and auditable.
- Fairness: Avoid algorithmic bias in hiring, scheduling, or resource allocation.
- Accountability: Clear lines of responsibility when machines make decisions.
- Establish ethical review boards for AI use.
- Train staff in AI literacy and ethics.
- Monitor AI deployments to ensure compliance with societal and workplace norms.

Leadership in smart factories isn't about knowing everything—it's about knowing how to orchestrate systems of people, machines, and insights. These tools and frameworks offer repeatable, scalable, and ethical structures for decision-making, behavior shaping, and cultural transformation. By using these instruments:

- Leaders shift from managing technology adoption to enabling human-machine flourishing.
- Teams move from compliance to engagement.
- Organizations evolve from disconnected efforts to a coherent digital-first leadership strategy.

CHANGE MANAGEMENT FRAMEWORK

The transition to smart factories requires more than technology installation—it necessitates a profound organizational shift in mindset, capabilities, culture, and workflows. A structured change management framework ensures this transformation is sustainable, inclusive, and aligned with business strategy.

Assessment of Preparedness

Armenakis and Harris (2009) emphasized that organizational readiness is a critical precursor to successful change implementation, especially in tech-heavy transformations. Before embarking on change, leaders must diagnose current capabilities. A robust tool like SWOT Analysis (Strengths, Weaknesses, Opportunities, Threats) allows organizations to assess:

- **Technical preparedness**: Is the existing IT infrastructure capable of supporting AI/IoT deployments?
- **Cultural hurdles**: Are employees open to digital change or apprehensive about AI?
- **Talent gaps**: Do teams possess skills like data literacy, systems thinking, and AI collaboration?

Pilot Projects

Pilot projects are low-risk experimental zones where smart technologies can be tested, and human-machine interactions fine-tuned. Siemens' Amberg factory runs small-scale AI deployments on digital twins before scaling across sites, ensuring local context and employee adaptation are considered (Siemens AG, 2022).

- Start with one process (e.g., predictive maintenance or automated quality inspection).
- Collect both quantitative performance data and qualitative employee feedback.
- Use agile cycles (short sprints, feedback loops) to iterate and refine.

Communication Strategy

Change fails when communication fails. Leaders must use multi-modal, inclusive communication strategies, such as:

- Town halls for emotional alignment and vision-sharing.
- Internal digital platforms (e.g., Yammer, Teams) for continuous dialogue.
- Visual dashboards showing transformation metrics (uptime, training hours, error reduction).
- Video testimonials from frontline adopters of new technologies.

Sustained Change

Sustainability comes from institutionalizing the change:

- Embed new behaviors and workflows into SOPs and policies.
- Use recognition systems to reward tech adoption and upskilling.
- Align the transformation with company values—for instance, reframing automation not as cost-cutting but as safety-enhancing and growth-enabling.
- Regular evaluations (via surveys, KPIs, digital maturity assessments) should guide continuous refinement.

THE INTEGRATED FRAMEWORK

An integrated change management framework unites visionary leadership, human–machine integration, and cultural transformation into one cohesive strategy. In the context of smart factories, it's not enough to have cutting-edge technology—success depends on aligning why we change, how we change, and who we change with into a single roadmap.

Strategic Alignment

At the heart of transformation is leadership clarity. The aims of change management must be anchored in the organization's leadership vision so every stakeholder understands why the change matters.

When leaders articulate a compelling purpose—such as "We are not just automating; we are augmenting human potential"—they foster shared purpose and collective commitment. This is not just theory: a Harvard Business Review (2019) study found that companies with aligned digital visions experience 70% faster ROI on transformation initiatives compared to those without such clarity (Westerman, Bonnet, & McAfee, 2014). For smart factories, the vision must move beyond cost-cutting to focus on:

- Enhancing human creativity through automation that removes repetitive, low-value tasks.
- Improving decision quality by leveraging AI insights for precision and foresight.
- Building a future-ready, adaptive organization that thrives amid uncertainty.

By aligning every change initiative with a broader human–machine collaboration vision, leaders ensure that transformation is both purpose-driven and sustainable.

Collaborative Teams

Transformation in smart factories is not a solo act—it's a cross-functional performance. Engineers, HR professionals, IT experts, production staff, and change agents must work as one unit to drive successful integration.

- Tech teams bring system expertise, ensuring the right tools and automation are implemented.
- Operators offer practical user insights, ensuring the technology fits the workflow reality. HR facilitates reskilling and talent redeployment.
- Leaders model adaptability and openness, setting the cultural tone.

When diverse perspectives are included, transformation avoids the pitfall of top-down imposition and builds collective ownership. At Bosch's Industry 4.0 labs, AI developers work directly with machinists to co-design automation systems—ensuring the end product is both technologically advanced and intuitively usable (Bosch Connected Industry, 2021).

Continuous Improvement with Real-Time Analytics

A smart factory is a constantly evolving ecosystem, generating immense data streams from sensors, machines, and human interactions. Leaders can leverage this data to fuel continuous improvement:

- Process analytics measure cycle time, defect rates, and throughput efficiency.
- Behavioral analytics track how operators engage with AI tools and automation systems.
- Sentiment analysis gauges employee trust, concerns, and adoption readiness.

This turns the factory into a living laboratory, where every action, interaction, and decision become an opportunity for learning and refinement. At Siemens' smart factories, real-time analytics guide ongoing adjustments to processes, layouts, and training programs—ensuring the roadmap is not static but responsive to emerging needs.

Leaders should treat the factory like a living lab, adjusting the transformation roadmap based on real-world signals. As seen in Siemens' smart factories, use real-time analytics to track results and modify plans. Smart factories of the future will be defined not by the quantity of automation, but by the quality of collaboration. Success metrics are shifting from:

- Output Outcome – Value creation, innovation, and customer impact matter more than sheer production volume.
- Efficiency Adaptability – The ability to reconfigure processes quickly is more valuable than marginal gains in speed.
- Control Trust – Empowered teams and transparent systems outperform rigid command-and-control models.

The new currency of smart manufacturing is shared intelligence between humans and machines—facilitated by leaders who combine vision, empathy, and bold execution.

FROM AUTOMATION TO COLLABORATION: A PARADIGM SHIFT

Smart factories are moving beyond the early vision of fully "automated" production lines—where machines simply replaced human tasks—toward collaborative ecosystems where humans and intelligent machines work in synergy. In this new paradigm, the relationship between people and technology is no longer transactional but mutually empowering. Machines bring precision, speed, and data-driven insights; humans bring creativity, judgment, and adaptability. This requires changing success metrics:

Table 2. Shifting success metrics in smart factory transitions

Traditional Metric	Future-Oriented Metric
Output (volume)	Outcome (value created)
Efficiency (speed)	Adaptability (resilience)
Control (top-down)	Trust (shared agency)

"The new currency of manufacturing is shared intelligence—not just smart machines, but smart partnerships between humans and machines." *(Research Features, 2023).* To lead in this new environment, leaders must go beyond managing processes—they must design relationships between humans and machines that are productive, ethical, and sustainable. This requires a shift in mindset and behavior:

Empowers rather than enforces Leaders delegate decision making authority to teams and equip them with AI-driven tools, creating ownership and innovation at every level.

Listens before launching – Change initiatives are preceded by active engagement with operators, engineers, and data specialists to understand real needs and concerns.

Shares success across roles – Achievements in production, problem-solving, and innovation are recognized collectively, reinforcing the idea that both humans and machines contribute to the win.

In short, collaboration replaces command, and shared intelligence becomes the competitive advantage. Factories that adopt this model are not just faster—they are more adaptable, resilient, and prepared for the future's uncertainties.

RECOMMENDATION

Embed Ethics into Leadership Training

The transformation of smart factories is a psychological and ethical shift in addition to a technical one. To ensure that managers are prepared to lead ethically in a digital-first world, leadership development programs should incorporate courses on AI ethics, inclusion, and transparent communication.

Prioritize Workforce Reskilling and Adaptability

To prevent digital exclusion, organizations must make regular investments in reskilling programs for robots, data analytics, and human–machine interface management, paying special attention to underrepresented populations.

Foster Psychological Safety as a Strategic Imperative

It is not appropriate to consider psychological safety a "soft" add-on. Leaders should establish open dialogue platforms, safe-to-fail pilots, and innovation laboratories to enable staff members to try out new technologies without worrying about failing. The rate of adoption and the results of innovation are directly impacted by this cultural transformation.

Adopt a Phased Change Management Approach

Smart factory transformation should be approached by firms as a steady culture evolution using models like Kotter's Eight-Step Framework. In order to maintain momentum and lessen employee opposition, leaders must make sure that short-term successes are widely shared.

Institutionalize Human–Machine Collaboration Models

With a long-term focus on the teammate model, where human creativity and machine intelligence merge for dynamic decision-making, leaders should purposefully build organizational procedures to harness the three modes of collaboration (machines as tools, supervisors, and colleagues).

SCOPE FOR FUTURE STUDY

While the transformative potential of smart factories is evident, the field remains under-explored in several areas:

Leadership Behaviors in AI-Driven Contexts

Future research should examine how transformational leadership and other leadership philosophies, such as servant leadership and adaptive leadership, compare in terms of promoting creativity and trust in human-machine partnerships.

Cross-Cultural Perspectives on Digital Transformation

The majority of study focuses on manufacturing contexts in East Asia and the West. Studies that compare emerging economies could provide insight into how cultural values affect workers' acceptance of AI and automation.

Metrics for Measuring Human–Machine Synergy

The main focus of current research is on reducing downtime and increasing productivity. To assess the true effects of smart factories, future research might look at holistic metrics like job happiness, employee well-being, ethical compliance, and innovation quality.

Ethics and Governance in Industry 4.0

There is little empirical study on the practical application of ethical AI frameworks in factories, despite the existence of standards (e.g., Deloitte, McKinsey). Whether these frameworks actually increase trust and lessen bias might be investigated through longitudinal research.

CONCLUSION

The ability of leaders to strike a balance between technological innovation and human-centric values will determine the future of smart factories. With its emphasis on empowerment and inspiration, transformational leadership prepares businesses to handle the challenges of human-machine collaboration. The framework for implementing these changes is provided by change management, which guarantees organizational resilience and worker alignment.

Leaders must continue to be flexible as emerging technologies like generative AI and sophisticated robotics increasingly revolutionize smart factories. By 2035, 85% of manufacturing processes will be entirely automated, according to the World Economic Forum (2023), highlighting the necessity of ongoing reskilling and cultural adaption. In order to create environments where humans and technology collaborate to promote innovation, transformational leaders must foresee these changes.

As the fourth industrial revolution enters its next stage, smart factories—which are made up of robotics, AI, real-time data, linked technology, and, most importantly, people—are becoming living ecosystems rather than theoretical aspirations. Machines bring endurance, speed, and accuracy, but human leadership offers vision, purpose, and flexibility. As this chapter has shown, effective human-machine collaboration requires transformational leadership and strategic change management that promote acceptance, alignment, and innovation rather than just technological preparedness.

The cultural, strategic, and emotional environment in which smart technologies function is greatly influenced by transformational leaders. The foundation for overcoming uncertainty is their capacity to empower teams, model new behaviours, build trust, and convey a compelling vision. They do more than just supervise automation projects; they give them a human face, making sure that the use of robotics and AI enhances rather than replaces human ability. By doing this, they transform opposition into participation, fear into opportunity, and fragmentation into cooperation.

Smart factories require change management that goes beyond conventional training and communication strategies. It needs to be highly compassionate, flexible, and iterative. Change is no more a one-time occurrence but rather a continual capability, as demonstrated by the frameworks described, such as the integrated strategy of integrating leadership vision with cross-functional collaboration and continuous feedback loops. The operational levers of effective transformation are pilot initiatives, stakeholder discussions, skills transformation, and open communication tactics; they are not checkboxes.

Furthermore, the new standards for success in this era are changing. We are transitioning from command-and-control to distributed trust, from efficiency to adaptability, and from output to results. These changes necessitate a new paradigm for leadership, one that emphasizes ethical decision-making in AI systems, reciprocal learning between humans and machines, and resilient cultures that welcome complexity.

In the long run, smart factories that automate the most tasks won't be the most successful. They will be the ones who use collective intelligence, in which human directs technology with knowledge, morality, and empathy, and machines enhance human abilities. Companies that foster these ecosystems will not only attain operational excellence but also establish themselves as accountable leaders in a world that is becoming more and more influenced by intelligent machines.

The development of smart manufacturing involves both technology and human efforts. Organizations may guarantee that technology adoption results in inclusive development, sustainable innovation, and meaningful work by putting transformational leadership and dynamic change management at the centre of the digital transformation agenda. In addition to technological advancements, the development of smart factories necessitates transformational, moral, and visionary leadership that balances human potential with machine intelligence. Organizations may realize Industry 4.0's full potential by addressing ethical imperatives, promoting psychological safety, and effectively managing change. In order to guarantee that the smart factories of the future are not only effective but also inclusive, sustainable, and incredibly human-centric, future research must continue to examine the human, cultural, and ethical aspects of this transition.

In conclusion, transformational leadership and change management are the cornerstones of success in smart factories. By inspiring vision, empowering employees, and strategically managing change, leaders can unlock the full potential of human-machine collaboration, creating a future of unparalleled efficiency, innovation, and sustainability.

REFERENCES

Adel, A. (2022). Future of Industry 5.0 in society: Human-centric solutions, challenges and prospective research areas. *Journal of Cloud Computing (Heidelberg, Germany), 11*(1), 1–15. DOI: 10.1186/s13677-022-00314-5 PMID: 36101900

Armenakis, A. A., & Harris, S. G. (2009). Reflections: Our journey in organizational change research and practice. *Journal of Change Management, 9*(2), 127–142. DOI: 10.1080/14697010902879079

Bass, B. M. (1985). *Leadership and performance beyond expectations*. Free Press.

Burns, J. M. (1978). *Leadership*. Harper & Row.

Carlsson, D. (2023). Human-centricity in digital manufacturing: Integrating innovation capabilities and organizational change. *Technology Innovation Management Review, 13*(3), 14–24.

Deloitte. (2025). *Future of work in manufacturing.*

Fernandez-Vidal, J., Perotti, F. A., Gonzalez, R., & Gascó, J. (2022). Managing digital transformation: The view from the top. *Journal of Business Research, 152*, 29–41. DOI: 10.1016/j.jbusres.2022.07.020

Harvard Business Review. (2019). *Digital transformation is not about technology*. Harvard Business Publishing.

Kotter, J. P. (1996). *Leading change*. Harvard Business School Press.

McKinsey & Company. (2024). The state of AI in early 2024: Gen AI adoption spikes and and starts to generate value.

Nahavandi, A. (2019). *The art and science of leadership* (7th ed.). Pearson Education.

Research Features. (2023). Unlocking the future of leadership: Navigating human-machine collaboration in the age of machine. https://researchfeatures.com/unlocking-future leadership-human-machine-collaboration-machine-learning/

Silva, F. J. G., Campilho, R. D. S. G., & Sá, J. C. (2020). Digital transformation in manufacturing industries: Challenges, enablers and roadmap. *Procedia Manufacturing, 51*, 1183–1190.

Tagscherer, U., & Carbon, C. C. (2023). Digital leadership and innovation culture in smart manufacturing firms. *Technological Forecasting and Social Change, 189*, 122343.

Te'eni, D., Yahav, I., Zagalsky, A., Schwartz, D., Silverman, G., Cohen, D., Mann, Y., & Lewinsky, D. (2023, November 14). Reciprocal Human-Machine Learning: A Theory and an Instantiation for the Case of Message Classification. *Management Science*, mnsc.2022.03518. Advance online publication. DOI: 10.1287/mnsc.2022.03518

Westerman, G., Bonnet, D., & McAfee, A. (2014). *Leading digital: Turning technology into business transformation*. Harvard Business Review Press.

World Economic Forum. (2023). *The future of jobs report 2023*. Geneva, Switzerland: World Economic Forum.

Zhang, W., Liu, Y., & Yang, Q. (2023). Empowering leadership and employee creativity in digital contexts: The role of psychological safety and learning orientation. *Journal of Organizational Behavior*, *44*(1), 72–88.

KEY TERMS AND DEFINITIONS

Cyber-Physical Systems (CPS): Integrated systems in which physical processes are monitored and controlled by computer-based algorithms, tightly integrated with the internet and its users.

Internet of Things (IoT): A network of physical devices embedded with sensors, software, and connectivity to exchange data with other devices and systems over the internet.

Digital Twin: A virtual representation of a physical object, process, or system that enables simulation, monitoring, and optimization in real time.

Human-in-the-Loop (HITL): A model in which humans are actively involved in training, monitoring, and validating AI decision-making to ensure accountability and ethical outcomes.

Workforce Transformation: The process of equipping employees with the skills, tools, and mindsets required to succeed in digitally advanced and AI-integrated environments.

Employee Experience (EX): The holistic perception employees have of their journey within the organization, influenced by workplace culture, technology, leadership, and purpose.

Digital Fluency: The ability to effectively and confidently use digital tools, data, and platforms to enhance work outcomes and innovation.

AI Ethics: A set of values, principles, and techniques designed to guide the development and implementation of AI technologies in a way that respects human rights, fairness, transparency, and accountability.

Organizational Agility: The capability of a business to rapidly respond to internal and external changes by adapting quickly and effectively.

Psychological Safety: A workplace environment where employees feel safe to take risks, express ideas, and engage in experimentation without fear of punishment or humiliation.

Sociotechnical Systems Theory: An approach to organizational design that recognizes the interaction between people and technology in workplaces, emphasizing joint optimization.

Digital Transformation Strategy: A comprehensive plan that leverages digital technologies to fundamentally change business operations, culture, and customer value delivery.

Change Agents: Individuals within the organization who act as advocates and drivers of change, influencing others and bridging communication between leadership and staff.

Innovation Culture: An organizational atmosphere that encourages creativity, experimentation, and the willingness to take calculated risks in pursuit of improvement.

Augmented Workforce: A human workforce enhanced by technologies such as AI, AR, and automation, enabling them to perform tasks more efficiently or safely.

Compilation of References

Abdelali, H. A., Derrouz, H., Zennayi, Y., Thami, R. O. H., & Bourzeix, F. (2021). Multiple hypothesis detection and tracking using deep learning for video traffic surveillance. *IEEE Access : Practical Innovations, Open Solutions, 9*, 164282–164291. DOI: 10.1109/ACCESS.2021.3133529

Abid, N., Ceci, F., & Ikram, M. (2022). Green growth and sustainable development: Dynamic linkage between technological innovation, ISO 14001, and environmental challenges. *Environmental Science and Pollution Research International, 29*(17), 25428–25447. DOI: 10.1007/s11356-021-17518-y PMID: 34843051

Accenture. (2020). AI: Built to scale - Rewire your business for AI. Accenture. https://www.accenture. com/us - en/insights/artificial - intelligence/ai - investments - built - to - scale

Acemoglu, D., & Restrepo, P. (2019). Artificial intelligence, automation, and work. NBER Working Paper No. 24196.

Acemoglu, D., & Restrepo, P. (2020). Robots and jobs: Evidence from US labor markets. *Journal of Political Economy, 128*(6), 2188–2244. DOI: 10.1086/705716

Acharya, S., Lin, V., & Dhingra, N. (2018). The role of health in achieving the sustainable development goals. *Bulletin of the World Health Organization, 96*(9), 591–591A. DOI: 10.2471/BLT.18.221432 PMID: 30262936

Acikgoz, Y. (2019). Employee recruitment and job search: Towards a multi-level Integration. *Human Resource Management Review, 29*(1), 1–13. DOI: 10.1016/j.hrmr.2018.02.009

Adams, B. (2019). *Green development: Environment and sustainability in a developing world.* Routledge. DOI: 10.4324/9780203386033

Adel, A. (2022). Future of Industry 5.0 in society: Human-centric solutions, challenges and prospective research areas. *Journal of Cloud Computing (Heidelberg, Germany), 11*(1), 1–15. DOI: 10.1186/s13677-022-00314-5 PMID: 36101900

Adewuyi, A. O., & Awodumi, O. B. (2021). Environmental pollution, energy import, and economic growth: Evidence of sustainable growth in South Africa and Nigeria. *Environmental Science and Pollution Research International, 28*(12), 14434–14468. DOI: 10.1007/s11356-020-11446-z PMID: 33215283

Agrawal, R., Priyadarshinee, P., Kumar, A., Luthra, S., Garza-Reyes, J. A., & Kadyan, S. (2025). Are emerging technologies unlocking the potential of sustainable practices in the context of a net-zero economy? An analysis of driving forces. *Environmental Science and Pollution Research International, 32*(12), 7130–7148. DOI: 10.1007/s11356-023-26434-2 PMID: 36934193

Agustina, R., Dartanto, T., Sitompul, R., Susiloretni, K. A., Suparmi, , Achadi, E. L., Taher, A., Wirawan, F., Sungkar, S., Sudarmono, P., Shankar, A. H., Thabrany, H., Agustina, R., Dartanto, T., Sitompul, R., Susiloretni, K. A., Suparmi, , Achadi, E. L., Taher, A., & Khusun, H. (2019). Universal health coverage in Indonesia: Concept, progress, and challenges. *Lancet, 393*(10166), 75–102. DOI: 10.1016/S0140-6736(18)31647-7 PMID: 30579611

Ahmad, M., & Jabeen, G. (2023). Do economic development and tourism heterogeneously influence ecological sustainability? Implications for sustainable development. *Environmental Science and Pollution Research International, 30*(37), 87158–87184. DOI: 10.1007/s11356-023-28543-4 PMID: 37420157

Ahmed, H. G. M.-D., Sajjad, M., Li, M., Azmat, M. A., Rizwan, M., Maqsood, R. H., & Khan, S. H. (2019). Selection Criteria for Drought-Tolerant Bread Wheat Genotypes at Seedling Stage. *Sustainability (Basel), 11*(9), 2584. DOI: 10.3390/su11092584

Ahuja, V., Yang, J., & Shankar, R. (2009). Study of ICT adoption for building project management in the Indian construction industry. *Automation in Construction, 18*(4), 415–423. DOI: 10.1016/j.autcon.2008.10.009

Akçakanat, Ö. (2024). Yapay zekâ kaygısının teknoloji kaynaklı işsizlik endişesi üzerine etkisi: Muhasebe meslek mensupları üzerine bir araştırma. *Kocatepe İİBF Dergisi, 26*(special issue), 53–67. DOI: 10.33707/akuiibfd.1458358

Akkalkot, A. I., Mahawar, S., Modak, K. C., Srivastava, P., Shavkatov, N., & Dhanraj, J. A. (2025). Brain-Computer Interface-Based Real-Time Movement Control for Virtual Learning Environments. In Darwish, D., & Pandey, D. (Eds.), *Concepts and Applications of Brain-Computer Interfaces* (pp. 417–432). IGI Global Scientific Publishing., DOI: 10.4018/979-8-3693-7427-6.ch025

Akkoyun, B. (2021). Bilgi ve teknoloji becerileri ile sistem kullanılabilirliklerinin, işletmelerin yeni teknolojiyi kabul düzeyi üzerinden verimliliklerine etkisi, *İşletme Araştırmaları Dergisi, 13* (1), 739 - 759.

Akkoyun, B. (2016). Çalışanların teknoloji yönetimi algılarının demografik değişkenlere göre farklılaşması üzerine bir araştırma. *Elektronik Sosyal Bilimler Dergisi, 15*(56), 70–82. DOI: 10.17755/esosder.13489

Akman, A. Z. (2023). Türkiye'de "teknostres" konusunda gerçekleştirilen çalışmalara yönelik bir inceleme. *International Journal of Entrepreneurship and Management Inquiries, 2*(special issue), 71–86. DOI: 10.55775/ijemi.1326272

Alam, M. A. (2016). Techno - stress and productivity: Survey evidence from the aviation industry. *Journal of Air Transport Management, 50*, 62–70. DOI: 10.1016/j.jairtraman.2015.10.003

Alemu, S., Kosmowski, F., Stevenson, J. R., Mallia, P., Taye, L., & Macours, K. (2024). Investigationing inclusive innovation with the right data: An empirical illustration from Ethiopia. *Agricultural Systems, 219*, 103988. Advance online publication. DOI: 10.1016/j.agsy.2024.103988

Ali, M., Kirikkaleli, D., Sharma, R., & Altuntaş, M. (2022). The nexus between remittances, natural resources, technological innovation, economic growth, and environmental sustainability in Pakistan. *Environmental Science and Pollution Research International*, *29*(50), 75822–75840. DOI: 10.1007/s11356-022-21228-4 PMID: 35661303

Almatrodi, I., Li, F., & Alojail, M. (2023). Organizational resistance to automation success: How status quo bias influences organizational resistance to an automated workflow system in a public organization. *Systems*, *11*(4), 191. DOI: 10.3390/systems11040191

Alvesson, M., & Sveningsson, S. (2015). *Changing organizational culture: Cultural change work in progress*. Routledge. DOI: 10.4324/9781315688404

Alzubaidy, H. A. K., & Ibrahim, A. A. (2023). Implementing deep learning and CNNs for vehicle detection to alleviate highway traffic congestion. *2023 7th International Symposium on Innovative Approaches in Smart Technologies (ISAS)*, 1–6. https://doi.org/DOI: 10.1109/ISAS60782.2023.10391509

Amalia, K. (2023). *Advancements in Robotic Systems and Human Robot Interaction for Industry 4.0*. .DOI: 10.53759/9852/JRS202301010

Ambler, S. W., & Constantine, L. (2001). *The Unified Process Transition and Production Phases*. CRC Press. DOI: 10.1201/9781482280746

Ambrose, M., & Kulik, C. (1993). Old friends, new faces: Motivation research in the 1990's. *Journal of Management*, *25*(3), 231–292. DOI: 10.1177/014920639902500302

Appelbaum, S. H. (1997). Socio-technical systems theory: An intervention strategy for organizational development. *Management Decision*, *35*(6), 452–463. DOI: 10.1108/00251749710173823

Armenakis, A. A., & Harris, S. G. (2009). Reflections: Our journey in organizational change research and practice. *Journal of Change Management*, *9*(2), 127–142. DOI: 10.1080/14697010902879079

Arnetz, B. B., & Clairy, W. (1997). Technological stress: Psychophysiological symptoms in modern offices. *Journal of Psychosomatic Research*, *43*(1), 35–42. DOI: 10.1016/S0022-3999(97)00083-4

Arputharaj, J. V., & Pal, S. K. (2024). Transforming Industry 5.0: Real time monitoring and decision making with IIOT. In *Sustainability in Industry 5.0* (pp. 76-106). CRC Press.

Aryal, J. P., Sapkota, T. B., Rahut, D. B., & Jat, M. L. (2020). Agricultural sustainability under emerging climatic variability: The role of climate-smart agriculture and relevant policies in India. *International Journal of Innovation and Sustainable Development*, *14*(2), 219–245. DOI: 10.1504/IJISD.2020.106243

Ashford, S. J., Lee, C., & Bobko, P. (1989). Content, causes, and consequences of job insecurity: A theory - based measure and sunstantive test. *Academy of Management Journal*, *32*(4), 803–829. DOI: 10.2307/256569

Aswini, J., Sudha, K., Ganesh, G., Subramanian, S., & Ghinea, G. (2024). AI-powered parallel computing architecture and its applications. In *The convergence of self-sustaining systems with AI and IoT* (pp. 23 39). IGI Global Scientific Publishing. DOI: 10.4018/979-8-3693-1702-0.ch002

Autor, D. H. (2015). Why are there still so many jobs? The history and future of workplace automation. *The Journal of Economic Perspectives, 29*(3), 3–30. DOI: 10.1257/jep.29.3.3

Avolio, B. J., Walumbwa, F. O., & Weber, T. J. (2009). Leadership: Current theories, investigation, and future directions. In *Annual Review of Psychology* (Vol. 60, pp. 421–449). DOI: 10.1146/annurev.psych.60.110707.163621

Avunduk, Z. B., & Kazan, H. (2019). *New trends and ındustrial changes in production technology*. in *Sosyal Bilimlerde Yeni Araştırmalar,* (Ed.s, Doç. Dr. Emel İslamoğlu & Doç. Dr. Sinem Yıldırımalp). Ankara: Berikan Yayınevi, 453 - 465.

Avunduk, Z.B. (2023). Endüstri 4.0 ve inovasyon etkileşiminin uluslararası literatürdeki yansımaları: Bibliyometrik bir analiz. *Girişimcilik ve İnovasyon Yönetimi Dergisi/Journal of Entrepreneurship and Innovation Management, 12* (2), 59 - 80.

Awamleh, R., Stephens, M., & Salem, F. (2022). Agile government — Emerging perspectives in public management. *Agile Government*, 3-19. DOI: 10.1142/9789811239700_0001

Azar, A. T., Serrano, F. E., & Ahmad Kamal, N. (2021). Synchronization and Anti-Synchronization of Unidirectional and Bidirectional Coupled Chaotic Systems by Terminal Sliding Mode Control. In Azar, A., & Kamal, N. (Eds.), *Handbook of Research on Modeling, Analysis, and Control of Complex Systems* (pp. 399–433). IGI Global Scientific Publishing., DOI: 10.4018/978-1-7998-5788-4.ch016

Azim, H., Johnston, A. L., Nixon, M., Woodliffe, J. L., Theunissen, R., Suresh, R., Sivapalan, S., Bobo, J., & Licence, P. (2025). Collaborating for Impact: Navigating Partnerships and Overcoming Challenges across the Sustainable Development Goals. *ACS Sustainable Chemistry & Engineering, 13*(3), 1164–1173. DOI: 10.1021/acssuschemeng.4c10171 PMID: 39886474

Aziz, G., Strielkowski, W., Sarwar, S., & Tiwari, A. K. (2024). Implications of circular economy, digitalization and technological innovation to achieve sustainable environmental goal: Pre and post-vision 2030. *Heliyon, 10*(10), e30978. DOI: 10.1016/j.heliyon.2024.e30978 PMID: 38770279

Aziz, G., & Waheed, R. (2023, August 18). (2023). Relevance of oil price, military expenditure and exports to achieve the Saudi Arabian sustainable economic and environmental objectives: A non-linear ARDL approach. *Environmental Science and Pollution Research International.* Advance online publication. DOI: 10.1007/s11356-023-28979-8 PMID: 37592070

Azmani, A., & Taoufik, A. O. (2025). A consolidated conceptual framework of a smart e-government ecosystem: A scoping review. *Electronic Government, an International Journal, 21*(1). DOI: 10.1504/EG.2025.10063682

Badía, G., Pina, V., & Torres, L. (2019). Financial Performance of Government Bond Portfolios Based on Environmental, Social and Governance Criteria. *Sustainability (Basel), 11*(9), 2514. DOI: 10.3390/su11092514

Baena, P. A., Brunel, A., & Fernández-de-Larrinoa, Y.. (2023). In Brief: The White/Wiphala Paper on Indigenous Peoples' Food Systems. 2023 Jan 2. In von Braun, J., Afsana, K., & Fresco, L. O. (Eds.), *Science and Innovations for Food Systems Transformation* [Internet]. Springer., Available from https://www.ncbi.nlm.nih.gov/books/NBK599602/, DOI: 10.1007/978-3-031-15703-5_13

Bai, L., Wang, X., Zhang, L., Feng, J., Liao, J., Chen, B., Wang, P., & Zhang, X. (2024). A systematic study of interactions between sustainable development goals (SDGs) in Hainan Island. *Scientific Reports, 14*(1), 26613. DOI: 10.1038/s41598-024-77984-5 PMID: 39496804

Bai, X., Chen, J., & Shi, P. (2012). Landscape urbanization and economic growth in China: Positive feedbacks and sustainability dilemmas. *Environmental Science & Technology, 46*(1), 132–139. DOI: 10.1021/es202329f PMID: 22103244

Bajic, B., Cosic, I., Katalinic, B., Moraca, S., Lazarevic, M., & Rikalovic, A. (2019). EDGE COMPUTING VS. CLOUD COMPUTING: CHALLENGES AND OPPORTUNITIES IN INDUSTRY 4.0. *Annals of DAAAM & Proceedings, 30.*

Bajic, B., Suzic, N., Simeunovic, N., Moraca, S., & Rikalovic, A. (2020). Toward Real-time Data Analytics Application for Industry 4.0. In *Industrial Innovation in Digital Age* (pp. 71–78). Springer International Publishing.

Balai, P. S., Sheikh, A., Rabha, G., Das, S., Kuli, B. K., & Raj, M. (2025). Revolutionizing Agricultural Machinery: The Role of AI, IoT, and Renewable Energy in Enhancing Efficiency and Sustainability. *International Journal of Scientific Research in Science and Technology, 12*(2), 813–830. DOI: 10.32628/IJSRST251222626

Baniya, A. K., Pant, S. S., Paudel, D., Gupta, A., Singh, S., & Mohapatra, H. (2025). Load balancing in cloud computing ensuring fault tolerance, high availability, and security. *Risk-Based Approach to Secure Cloud Migration*, 253-284. DOI: 10.4018/979-8-3693-7137-4.ch010

Baroudi, S., Abdulaal, D., & Rampur, M. (2018). Training effectiveness during technological changes in Dubai based companies: Perspectives from managers and employees. *Development and Learning in Organizations, 32*(3), 1–3. DOI: 10.1108/DLO-10-2017-0085

Bashir, M. F., Sadiq, M., Talbi, B., Shahzad, L., & Adnan Bashir, M. (2022). An outlook on the development of renewable energy, policy measures to reshape the current energy mix, and how to achieve sustainable economic growth in the post COVID-19 era. *Environmental Science and Pollution Research International, 29*(29), 43636–43647. DOI: 10.1007/s11356-022-20010-w PMID: 35416580

Basili, V., Shull, F., & Lanubile, F. (1999). Building knowledge through families of experiments. *IEEE Transactions on Software Engineering, 25*(4), 456–473. DOI: 10.1109/32.799939

Bass, B. M. (1985). *Leadership and performance beyond expectations.* Free Press.

Bassi, L., Carpenter, R., & McMurrer, D. (2012). *HR analytics handbook.* McBassi & Company.

Becker, P. R. (2009). Technology management degree programs: Meeting the needs of employers. (In) *PICMET'09 - 2009 Portland International Conference on Management of Engineering & Technology*, 2171 - 2183.

Belgasmia, M. (2021). *Numerical Methods for Solving Dynamic Equilibrium Equations.* IGI Global Scientific Publishing., DOI: 10.4018/978-1-7998-4399-3.ch007

Benbya, H., Davenport, T. H., & Pachidi, S. (2020). Artificial intelligence in organizations: Current state and future opportunities. *MIS Quarterly Executive, 19*(4).

Benson, S. G., & Dundis, S. P. (2003). Understanding and motivating health care employees: Integrating Maslow's hierarchy of needs, training and technology. *Journal of Nursing Management, 11*(5), 315–320. DOI: 10.1046/j.1365-2834.2003.00409.x PMID: 12930537

Berdal, Q., Pacaux-Lemoine, M.-P., Trentesaux, D., & Chauvin, C. (2018). *Human-Machine Cooperation in Self-organized Production Systems: A Point of View*. Springer., DOI: 10.1007/978-3-030-03003-2_9

Bergman, Z., Bergman, M. M., Fernandes, K., Grossrieder, D., & Schneider, L. (2018). The Contribution of UNESCO Chairs toward Achieving the UN Sustainable Development Goals. *Sustainability (Basel), 10*(12), 4471. DOI: 10.3390/su10124471

Bersin, J. (2018). HR technology disruptions for 2018: Productivity, design, and intelligence reign. Deloitte Insights. https://www2. deloitte. com/insights/us/en/focus/human - capital - trends/2018/hr - technology - disruptions. html

Bersin, J. (2020). Building digital talent: The new workforce strategy. Josh Bersin Academy. https://joshbersin. com/2020/01/building - digital - talent - the - new - workforce - strategy/

Bersin, J. (2021). *The big reset: Learning and skills in the flow of work*. Josh Bersin Company.

Bhandari, G. P., & Gupta, R. (2021). An overview of cloud and edge computing architecture and its current issues and challenges. *Research Anthology on Architectures, Frameworks, and Integration Strategies for Distributed and Cloud Computing*, 2340-2370.

Bharadwaj, A., El Sawy, O. A., Pavlou, P. A., & Venkatraman, N. (2013). Digital business strategy: Toward a next generation of insights. *Management Information Systems Quarterly, 37*(2), 471–482. DOI: 10.25300/MISQ/2013/37:2.3

Bibri, S. E., Alexandre, A., Sharifi, A., & Krogstie, J. (2023). Environmentally sustainable smart cities and their converging AI, IoT, and big data technologies and solutions: An integrated approach to an extensive literature review. *Energy Informatics, 6*(1), 9. DOI: 10.1186/s42162-023-00259-2 PMID: 37032812

Bi, Z. M., Chen, B., Xu, L., Wu, C., Malott, C., Chamberlin, M., & Enterline, T. (2022). Security and safety assurance of collaborative manufacturing in industry 4.0. *Enterprise Information Systems, 16*(12), 2008512. DOI: 10.1080/17517575.2021.2008512

Bocken, N. M. P., Short, S. W., Rana, P., & Evans, S. (2014). A literature and practice review to develop sustainable business model archetypes. Journal of Cleaner Production, 65, 42 - 56. . jclepro. 2013. 11. 039DOI: <ALIGNMENT.qj></ALIGNMENT>10. 1016/j

Boller, S., & Fletcher, L. (2020). *Design thinking for training and development: Creating learning journeys that get results*. ATD Press.

Bonekamp, L., & Sure, M. (2015). Consequences of Industry 4.0 on Human Labour and Work Organisation. *Journal of Business and Media Psychology, 6*(1), 33–40. https://journal-bmp.de/wp-content/uploads/04_Bonekamp-Sure_final.pdf

Bonnet, D. (2022). 3 stages of a successful digital transformation. *Harvard Business Review.* https://hbr.org/2022/0 9/3 - *stages - of - a - successful - digital - transformation.*

Boon, C. T., Low, D. F. L., Lim, S. Y., Ng, X. Q., & Wong, P. P. (2011). *The influence of core self - evaluation (CSE) on employees' motivation in continuing education.* UTAR.

Brancu, C., Turcu, O., Stefan, M., & Popp, M. (2024). Agile approaches, for E-businEss solutions strategy, at the government level. *Journal of Research and Innovation for Sustainable Society, 6*(2), 263–272. DOI: 10.33727/JRISS.2024.2.29:263-272

Braun, V., & Clarke, V. (2006). Using thematic analysis in psychology. *Qualitative Research in Psychology, 3*(2), 77–101. DOI: 10.1191/1478088706qp063oa

Brettel, M., Friederichsen, N., Keller, M., & Rosenberg, M. (2014). How virtualization, decentralization and network building change the manufacturing landscape: An industry 4.0 perspective. *International Journal of Mechanical, Aerospace, Industrial, Mechatronic and Manufacturing Engineering, 8*(1), 37-44. (Focuses on the broader Industry 4.0 context.)

Brief, A. P., & Weiss, H. M. (2002). Organizational behavior: Affect in the workplace. *Annual Review of Psychology, 53*(1), 279–307. DOI: 10.1146/annurev.psych.53.100901.135156

Brod, K. (1984). On the uniqueness of solution for all wavenumbers in acoustic radiation. *The Journal of the Acoustical Society of America, 76*(4), 1238–1243. DOI: 10.1121/1.391418

Brodny, J., & Tutak, M. (2021). Assessing the level of digitalization and robotization in the enterprises of the European Union Member States. *PLoS One, 16*(7), e0254993. DOI: 10.1371/journal.pone.0254993 PMID: 34293023

Brougham, D., & Haar, J. (2018). Smart technology, artificial intelligence, robotics, and algorithms (STARA): Employees' perceptions of our future workplace. *Journal of Management & Organization, 24*(2), 239–257. DOI: 10.1017/jmo.2016.55

Brown, T. (2009). *Change by design: How design thinking creates new alternatives for business and society.* Harper Business.

Bryman, A., & Bell, E. (2011). *Business research methods* (3rd ed.). Oxford University Press.

Brynjolfsson, E., & McAfee, A. (2014). *The second machine age: Work, progress, and prosperity in a time of brilliant technologies.* W.W. Norton & Co.

Buchanan, D. A., & Huczynski, A. A. (2019). *Organizational behaviour* (10th ed.). Pearson Education.

BULGURCU, B. & MASHATAN, A. A. (2024). Environmental Factors that Hinder an O0rganization's Ability to Learn from Cyber Incidents: A Case Study on SolarWinds.

Buonocore, F., Annosi, M. C., de Gennaro, D., & Riemma, F. (2024). Digital transformation and social paradigm shift: Leadership strategies for responsible innovation. *Journal of Engineering and Technology Management - JET-M, 74.* DOI: 10.1016/j.jengtecman.2024.101843

Burns, J. M. (1978). *Leadership.* Harper & Row.

Çakır, Ö. (2007). *İşini kaybetme kaygısı: İş güvencesizliği.* Çalışma ve Toplum.

Calderon-Monge, E., & Ribeiro-Soriano, D. (2023). The role of digitalization in business and management: A systematic literature review. *Review of Managerial Science,* •••, 1–43. Advance online publication. DOI: 10.1007/s11846-023-00647-8

Candemir, A., Duvaleix, S., & Latruffe, L. (2021). Agricultural cooperatives and farm sustainability–A literature review. *Journal of Economic Surveys*, *35*(4), 1118–1144. DOI: 10.1111/joes.12417

Cappelli, P., & Tavis, A. (2018). HR goes agile. *Harvard Business Review*, *96*(2), 46–52.

Carlsson, D. (2023). Human-centricity in digital manufacturing: Integrating innovation capabilities and organizational change. *Technology Innovation Management Review*, *13*(3), 14–24.

Carson, D., Gilmore, A., Perry, C., & Gronhaug, K. (2001). *Qualitative marketing research.* Sage. DOI: 10.4135/9781849209625

Centre for Science and Environment (CSE). (2021). Tirupur's water - wise success: A zero liquid discharge model for the textile industry. https://www. cseindia. org/tirupur - zld - textile - model

Challenges and opportunities for Indigenous Peoples' sustainability. (2021). *UN Department of Economic and Social Affairs (DESA) Policy Briefs*. DOI: 10.18356/df7eed76-en

Chaparro-Banegas, N., Ibañez Escribano, A. M., Mas-Tur, A., & Roig-Tierno, N. (2023). Innovation facilitators and sustainable development: A country comparative approach. *Environment, Development and Sustainability*, *26*(4), 1–29. Advance online publication. DOI: 10.1007/s10668-023-03055-w PMID: 37363019

Charalambous, G., Fletcher, S. R., & Webb, P. (2013). *Human-automation collaboration in manufacturing: identifying key implementation factors*. https://dspace.lib.cranfield.ac.uk/handle/1826/9480

Charran, R. S., & Dubey, R. K. (2022). Two-wheeler vehicle traffic violations detection and automated ticketing for Indian road scenario. *IEEE Transactions on Intelligent Transportation Systems*, *23*(11), 22002–22007. DOI: 10.1109/TITS.2022.3186679

Chaudhary, N. K., Rath, A., Babbar, G., Verma, A., Sinha, S. D., & Mohapatra, H. (2025). A critical analysis on edge computing in smart city applications. *Risk-Based Approach to Secure Cloud Migration*, 1-28. DOI: 10.4018/979-8-3693-7137-4.ch001

Chaudhry, N. I., Jariko, M. A., Mushtaque, T., Mahesar, H. A., & Ghani, Z. (2017). Impact of Working Environment and Training & Development on Organization. *European Journal of Training and Development Studies*, *4*(2), 33–48.

Chelladurai, S. P., Krishna, S. H., Sarkar, P., Lami, Z. F., Lourens, M., & Al-Shaikhli, T. R. (2025). Smartphones With Brain-Computer Interface Control for HR Secure Data Management. In B. Pandey, A. George, S. Tiwari, S. Albermany, & H. Hung (Eds.), *Brain-Computer Interfaces and Applications in Business* (pp. 215-232). IGI Global Scientific Publishing. https://doi.org/DOI: 10.4018/979-8-3373-5122-3.ch013

Chen, A., Li, L., & Shahid, W. (2024). Digital transformation as the driving force for sustainable business performance: A moderated mediation model of market-driven business model innovation and digital leadership capabilities. *Heliyon*, *10*(8), e29509. Advance online publication. DOI: 10.1016/j.heliyon.2024.e29509 PMID: 38655293

Chen, F., & Li, R. (2024). Improvement and replacement: The dual ımpact of automation on employees' job satisfaction. *Systems*, *12*(2), 46. DOI: 10.3390/systems12020046

Chen, K., Huang, T., Liu, E., Lu, Y., Huo, Z., Mi, L., Zhang, W., & Frazer, I. (2020). The Innovation: A Journal to See the Unseen and Change the Unchanged. *Innovation (Cambridge (Mass.))*, *1*(1), 100014. DOI: 10.1016/j.xinn.2020.04.014 PMID: 34557702

Chen, M., Chen, L., Li, Y., & Xian, Y. (2022). Developing computable sustainable urbanization science: Interdisciplinary perspective. *Computational Urban Science*, *2*(1), 17. DOI: 10.1007/s43762-022-00048-9 PMID: 35755716

Chikazunga, D., & Paradza, G. (2012). Smallholder farming: A panacea for employment creation and rural development in South Africa? PLAAS, University of the Western Cape.

Chinthamu, N., Priya, W. D., Srinivasan, K., Vidyasagar, G., Sivarajan, S., & Srihari, T. (2025). Identification of Human-Centric Designs with Advanced Robotics in Industry 4.0 Through Deep Learning. *Social Science Research Network*. DOI: 10.2139/ssrn.5086682

Chiumento. (2007). Happiness at Work. Research Report 2007. Retrieved from https://www.scribd. com/document/72220871/Happiness - at - Work - Index - 2007

Chiwawa, J., Naidoo, V., & Darwish, S. (2025). The Influence of Agricultural Cooperatives on the Environmental Sustainability of Rural Livelihoods in KwaZulu-Natal, South Africa. In *Sustainable Data Management: Navigating Big Data, Communication Technology, and Business Digital Leadership* (Vol. 2, pp. 31–41). Springer Nature Switzerland. DOI: 10.1007/978-3-031-83915-3_3

Chiwawa, J., Naidoo, V., & Rena, R. (2024). Role of farming cooperatives on the social sustainability of rural livelihoods in KwaZulu-Natal, South Africa. *Research World*, *15*(2), 133–141.

Choi, D., Chung, C. Y., & Young, J. (2019). An Economic Analysis of Corporate Social Responsibility in Korea. *Sustainability (Basel)*, *11*(9), 2676. DOI: 10.3390/su11092676

Christian, M., Hans, T., & Gidi, L. (2018). Factors affecting performance of youth and womens agricultural cooperatives: A case of some selected cooperative societies in Eastern Cape, South Africa. *Journal of Development and Agricultural Economics*, *10*(12), 394–403. DOI: 10.5897/JDAE2018.0932

Chui, M., Manyika, J., & Miremadi, M. (2016). Where machines could replace humans—And where they can't (yet). *The McKinsey Quarterly*.

Chuong, H. N., Uyen, V. T. P., Ngan, N. D. P., Tram, N. T. B., Han, N. D. M., & Duyen, P. H. K. (2025). The impact of globalization, renewable energy, and labor on sustainable development: A cross-country analysis. *PLoS One*, *20*(2), e0315273. DOI: 10.1371/journal.pone.0315273 PMID: 39982918

Ciancarini, P., Giancarlo, R., & Grimaudo, G. (2024). Scrum@PA: Tailoring an Agile methodology to the digital transformation in the public sector. *Information (Basel)*, *15*(2), 110. DOI: 10.3390/info15020110

Cieslak, V., & Valor, C. (2025). Moving beyond conventional resistance and resistors: An integrative review of employee resistance to digital transformation. *Cogent Business & Management*, *12*(1), 2442550. Advance online publication. DOI: 10.1080/23311975.2024.2442550

Çini, M. A., Erdirençelebi, M., & Akman, A. Z. (2023). The effect of organisation employees' perspective on digital transformation on their technostress levels and performance: The public institutions example. *Central European Business Review, 12*(4), 1–25. DOI: 10.18267/j.cebr.331

Cirera, X., Cruz, M., Lee, K. M., Nogueira, C., Lakhtakia, S., & Weiss, S. F. (2023). *Understanding Firm-level Adoption of Technology in Ethiopia* *.

Cirsostomo, B. C., Fidelino, J. I. R., Landicho, D. K. D., Pareja, J. E. E., del Rosario, J. R. B., & Bandala, A. A. (2023). Development of a CNN-based intelligent traffic control system for emergency vehicle detection and intersectional traffic flow. *2023 IEEE 15th International Conference on Humanoid, Nanotechnology, Information Technology, Communication and Control, Environment, and Management (HNICEM)*, 1–5. https://doi.org/DOI: 10.1109/HNICEM60674.2023.10589128

Civelek, M. E., & Pehlivanoğlu, M. Ç. (2020). Technological unemployment anxiety scale development. *Eurasian Business & Economics Journal, 22*, 64–76. DOI: 10.17740/eas.econ.2020.V22 - 05

Cleopatra Enterprise. (2023). Top HR challenges in the fourth industrial revolution. https://www.cleopatraenterprise. com/hr - challenges - in - the - 4ir/

Colombo, F., García-Goñi, M., & Schwierz, C. (2016). Addressing multimorbidity to improve healthcare and economic sustainability. *Journal of Comorbidity, 6*(1), 21–27. DOI: 10.15256/joc.2016.6.74 PMID: 29090168

Curry, G. N., Nake, S., Koczberski, G., Oswald, M., Rafflegeau, S., Lummani, J., Peter, E., & Nailina, R. (2021). Disruptive innovation in agriculture: Socio-cultural factors in technology adoption in the developing world. *Journal of Rural Studies, 88*, 422–431. DOI: 10.1016/j.jrurstud.2021.07.022

Dafflon, B., Moalla, N., & Ouzrout, Y. (2021). The challenges, approaches, and used techniques of CPS for manufacturing in Industry 4.0: A literature review. *International Journal of Advanced Manufacturing Technology, 113*(7-8), 2395–2412. DOI: 10.1007/s00170-020-06572-4

Dahabiyeh, L., Najjar, M. S., & Wang, G. (2022). Online teaching during COVID - 19 crisis: The role of technostress and emotional dissonance on online teaching exhaustion and teaching staff productivity. *The International Journal of Information and Learning Technology, 39*(2), 97–121. DOI: 10.1108/IJILT-09-2021-0147

Dahlin, E. (2024). Who says artificial intelligence is stealing our jobs? *Socius: Sociological Research for a Dynamic World, 10*, 1–13. srd.sagepub.com. DOI: 10.1177/23780231241259672

Dai, X., Chen, Y., Zhang, C., He, Y., & Li, J. (2023). Technological revolution in the field: Green development of Chinese agriculture driven by digital information technology (DIT). *Agriculture, 13*(1), 199. DOI: 10.3390/agriculture13010199

Dana, L. P., Salamzadeh, A., Mortazavi, S., & Hadizadeh, M. (2022). Investigating the impact of international markets and new digital technologies on business innovation in emerging markets. *Sustainability (Basel), 14*(2), 983. DOI: 10.3390/su14020983

Darwish, D. (2024). Human-AI Collaboration in Industry 5. In Kumar, N., Pal, S., Agarwal, P., Rosak-Szyrocka, J., & Jain, V. (Eds.), *Human-Machine Collaboration and Emotional Intelligence in Industry 5.0* (pp. 44–70). IGI Global Scientific Publishing., DOI: 10.4018/979-8-3693-6806-0.ch003

Davenport, T. H., & Bean, R. (2025). The GenAI focus shifts to innovation at Colgate-Palmolive. *MIT Sloan Management Review.*

Davenport, T. H., & Ronanki, R. (2018). Artificial intelligence for the real world. *Harvard Business Review*, *96*(1), 108–116.

Dawane, M. K., & Malwatkar, G. M. (2025). Comparative Study on Implementation of PID Control Architecture and Sliding Mode Control With IoT Integration on Inverted Pendulum Systems. In D. Pandey, B. Muniandi, B. Pandey, & A. George (Eds.), *Interdisciplinary Approaches to AI, Internet of Everything, and Machine Learning* (pp. 49-72). IGI Global Scientific Publishing. https://doi.org/DOI: 10.4018/979-8-3373-1032-9.ch003

de Andrade, L. C., Borges-Pedro, J. P., Gomes, M. C. R. L., Tregidgo, D. J., do Nascimento, A. C. S., Paim, F. P., Marmontel, M., Benitz, T., Hercos, A. P., & do Amaral, J. V. (2021). The sustainable development goals in two sustainable development reserves in central amazon: Achievements and challenges. *Discover Sustainability*, *2*(1), 54. DOI: 10.1007/s43621-021-00065-4 PMID: 35425916

Deloitte. (2019). 3D opportunity for production: Additive manufacturing makes its (business) case. Deloitte Insights. https://www2. deloitte. com/insights/us/en/focus/3d - opportunity/overview. html

Deloitte. (2020). *Global Human Capital Trends: The social enterprise at work—Paradox as a path forward.* Deloitte Insights.

Deloitte. (2021). 2021 Global Human Capital Trends: The social enterprise in a world disrupted. https://www2. deloitte. com/global/en/pages/human - capital/articles/gx - human - capital - trends. html

Deloitte. (2025). *Future of work in manufacturing.*

Deng, C., Gulseren, D., Isola, C., Grocutt, K., & Turner, N. (2023). Transformational leadership Influence: An evidence-based primer. *Human Resource Development International*, *26*(5), 627–641. DOI: 10.1080/13678868.2022.2135938

Deng, D., Zhang, J., Wang, J., & Zong, X. (2025). Sustainable performance evaluation of pharmaceutical companies: Sustainable balanced scorecard and hybrid MCDM approach. *Frontiers in Public Health*, *12*, 1495156. DOI: 10.3389/fpubh.2024.1495156 PMID: 39845657

Department of Agriculture, Land Reform and Rural Development (DALRRD). (2023). *Economic Review of South African Agriculture for the Year Ended 30 June 2023.* Government of South Africa.

Department of Agriculture. Forestry and Fisheries. (2015). *Annual report 2014/2015.* Pretoria, South Africa: Department of Agriculture, Forestry and Fisheries.

Department of Small Business Development (2020). Annual Report; *Republic of South Africa.*

Department of Trade and Industry. (2004). A Cooperative Development Policy. *Pretoria: Republic of South Africa*. Available at: https://www.pmg.org.za/docs/2004/appendices

Dessler, G. (2019). *Human Resource Management*. Pearson.

Dhaoui, I. (2022). E-Government for Sustainable Development: Evidence from MENA Countries. *Journal of the Knowledge Economy, 13*(3), 2070–2099. DOI: 10.1007/s13132-021-00791-0 PMID: 40477231

Dhinakaran, D., Kumar, N. J., & Ponnuviji, N. P. (2025). Safeguarding confidentiality and privacy in cloud-enabled healthcare systems with spectrasafe encryption and dynamic k-anonymity algorithm. *Expert Systems with Applications, 279*, 127584. DOI: 10.1016/j.eswa.2025.127584

DIGITAL ETHIOPIA2025. (2020). *DIGITAL ETHIOPIA 2025-A STRATEGY FOR ETHIOPIA INCLUSIVE PROSPERITY*.

Dinçer, Ö. (2008). *Örgüt geliştirme teori, uygulama ve teknikleri* (2 b.). İstanbul: Alfa Yayınları.

Dines, D. (2018). *CEO commentary on AI augmentation*. UiPath.

Dizikes, P. (2022). *How automation exacerbates inequality*. MIT News.

Doğrular, M. (2019). *Teknostresin verimlilik üzerine etkisi* (Published PhD dissertation). Marmara University, İstanbul.

Dou, F., Ye, J., Yuan, G., Lu, Q., Niu, W., Sun, H., . . . Song, W. (2023). Towards artificial general intelligence (agi) in the internet of things (iot): Opportunities and challenges. *arXiv preprint arXiv:2309.07438*.

Dragano, N., & Lunau, T. (2020). Technostress at work and mental health: Concepts and research results. *Current Opinion in Psychiatry, 33*(4), 407–413. DOI: 10.1097/YCO.0000000000000613

Drucker, P. F. (2007). *Management: Tasks, responsibilities, practices*. Transaction Publishers.

Dubé, L., Jha, S., Faber, A., Struben, J., London, T., Mohapatra, A., Drager, N., Lannon, C., Joshi, P. K., & McDermott, J. (2014). Convergent innovation for sustainable economic growth and affordable universal health care: Innovating the way we innovate. *Annals of the New York Academy of Sciences, 1331*(1), 119–141. DOI: 10.1111/nyas.12548 PMID: 25294668

Duc, Q.-A. N., ... (2024). Optimizing traffic light control using YOLOv8 for real-time vehicle detection and traffic density. *2024 9th International Conference on Integrated Circuits, Design, and Verification (ICDV)*, 119–124. https://doi.org/DOI: 10.1109/ICDV61346.2024.10616901

Duháček Šebestová, J., & Popescu, C. (Eds.). (2025). *Building a Green Future Through Essential Decision-Making Competencies*. IGI Global., DOI: 10.4018/979-8-3373-0189-1

Duraimutharasan, N., Deepan, A., Swadhi, R., Velmurugan, P. R., & Varshney, K. R. (2025). Enhancing Control Engineering Through Human-Machine Collaboration: AI for Improved Efficiency and Decision-Making. In Mellal, M. (Ed.), *Harnessing AI for Control Engineering* (pp. 155–176). IGI Global Scientific Publishing., DOI: 10.4018/979-8-3693-7812-0.ch008

Duvenaud, D., Maclaurin, D., Aguilera-Iparraguirre, J., Gómez-Bombarelli, R., Hirzel, T. D., Aspuru-Guzik, A., & Adams, R. P. (2015). Convolutional networks on graphs for learning molecular fingerprints. *Neural Information Processing Systems, 28*, 2224–2232. https://proceedings.neurips.cc/paper/2015/file/f9be311e65d81a9ad8150a60844bb94c-Paper.pdf

Dworschak, B., & Zaiser, H. (2014). Competences for cyber - physical systems in manufacturing - first findings and scenarios. *Procedia CIRP, 25*, 345–350. DOI: 10.1016/j.procir.2014.10.048

Ebadinezhad, S., Dereboylu, Z., & Ever, E. (2019). Clustering-Based Modified Ant Colony Optimizer for Internet of Vehicles (CACOIOV). *Sustainability (Basel), 11*(9), 2624. DOI: 10.3390/su11092624

Efthymiou, K., Pagoropoulos, A., Papakostas, N., Mourtzis, D., & Chryssolouris, G. (2012). Manufacturing systems complexity review: Challenges and outlook. *Procedia CIRP, 3*, 644–649. DOI: 10.1016/j.procir.2012.07.110

Ejsmont, K., Gladysz, B., & Kluczek, A. (2020). Impact of Industry 4.0 on sustainability—Bibliometric analysis. *Sustainability (Basel), 12*(14), 5650. DOI: 10.3390/su12145650

El Zaatari, S., Marei, M., Li, W., & Usman, Z. (2019). The impact of Industry 4.0 technologies on supply chains. [Connects collaborative manufacturing to supply chain implications.]. *Journal of Manufacturing Technology Management, 30*(5), 851–879.

Elastic, P., Bakhtiar, T., & Jaharuddin. (2020). An Optimal Control Problem of Knowledge Dissemination. In E. Idemudia (Ed.), *Handbook of Research on Social and Organizational Dynamics in the Digital Era* (pp. 461-482). IGI Global Scientific Publishing. https://doi.org/.DOI: 10.4018/978-1-5225-8933-4.ch022

Elavarasan, R. M., Pugazhendhi, R., Shafiullah, G. M., Kumar, N. M., Arif, M. T., Jamal, T., Chopra, S. S., & Dyduch, J. (2022). Impacts of COVID-19 on Sustainable Development Goals and effective approaches to maneuver them in the post-pandemic environment. *Environmental Science and Pollution Research International, 29*(23), 33957–33987. DOI: 10.1007/s11356-021-17793-9 PMID: 35032263

Elhosseini, M. A. (2020). Using Fuzzy Logic to Control Combined Cycle Gas Turbine During Ambient Computing Environment. [IJACI]. *International Journal of Ambient Computing and Intelligence, 11*(3), 106–130. DOI: 10.4018/IJACI.2020070106

Elzein, B. (2024). Nano Revolution: "Tiny tech, big impact: How nanotechnology is driving SDGs progress". *Heliyon, 10*(10), e31393. DOI: 10.1016/j.heliyon.2024.e31393 PMID: 38818162

Erwin, D. G., & Garman, A. N. (2010). Resistance to organizational change: Linking research and practice. *Leadership and Organization Development Journal, 31*(1), 39–56. DOI: 10.1108/01437731011010371

Eurostat (2021). ICT specialists in employment, (https://ec.europa.eu/eurostat/statisticsexplained/index.php?title=ICT_specialists_in_employment#Relative_share_of_ICT_specialists_in_the_total_workforce, (Retrieved: 3.04.2025).

Faizal, S., Jaffar, N., & Mohd Nor, A. S. (2022). Integrate the adoption and readiness of digital technologies amongst accounting professionals towards the fourth industrial revolution. *Cogent Business & Management*, *9*(1), 1–14. DOI: 10.1080/23311975.2022.2122160

Fang, X., & Cao, Y. (2023). Spatial Association Network Evolution and Variance Decomposition of Economic Sustainability Development Efficiency in China. *International Journal of Environmental Research and Public Health*, *20*(4), 2966. DOI: 10.3390/ijerph20042966 PMID: 36833661

Fanni, S. C., Febi, M., Aghakhanyan, G., & Neri, E. (2023). Natural language processing. In *Introduction to artificial intelligence* (pp. 87–99). Springer International Publishing. DOI: 10.1007/978-3-031-25928-9_5

FAOSTAT. (2018). *Food and Agriculture Organization of the United Nations Statistical Database*. Retrieved from https://www.fao.org/faostat

Felten, E., Manav, R., & Robert, S. (2023). How will language modelers like ChatGPT affect occupations and industries? *arXiv*. (Retrieved: 4.04.2025). https://arxiv.org/ abs/2303.01157

Fernandez-Vidal, J., Perotti, F. A., Gonzalez, R., & Gascó, J. (2022). Managing digital transformation: The view from the top. *Journal of Business Research*, *152*, 29–41. DOI: 10.1016/j.jbusres.2022.07.020

Fertő, I., & Harangozó, G. (2025). Trade-offs and synergies when balancing economic growth and globalization for sustainable development goals achievement. *Scientific Reports*, *15*(1), 8634. DOI: 10.1038/s41598-025-93360-3 PMID: 40082545

Filho, W. L., Sierra, J., Price, E., Eustachio, J. H. P. P., Novikau, A., Kirrane, M., Dinis, M. A. P., & Salvia, A. L. (2024). The role of universities in accelerating the sustainable development goals in Europe. *Scientific Reports*, *14*(1), 15464. DOI: 10.1038/s41598-024-65820-9 PMID: 38965303

Finniston, M. (1982). Technology: The future determinant of society. *Technology in Society*, *4*(1), 1–13. DOI: 10.1016/0160-791X(82)90010-0

Firoznia, G., Salman, M., & Gharani Arani, B. (2021). The Impact of Rural Production Cooperatives on Sustainable Rural Development: A Case of Kashan's Rural Production Cooperatives. *Journal of Sustainable Rural Development*, *5*(1), 121–134.

Fojtik, R. (2011). Extreme programming in development of specific software. *Procedia Computer Science*, *3*, 1464–1468. DOI: 10.1016/j.procs.2011.01.032

Forbes Insights. (2022). 2022 state of manufacturing report. Forbes & Oracle NetSuite. https:// www. forbes. com/sites/forbestechcouncil/2022/08/16/the - state - of - manufacturing - 2022/? Sh=6b9823666b3b

Ford Motor Company. (2022). Sustainability report 2022: Driving the future of sustainability. https://corporate. ford. com/microsites/sustainability - report - 2022

Ford, M. (2016). *Rise of the robots: Technology and the threat of a jobless future*. Basic Books.

Frey, C. B., & Osborne, M. A. (2017). The future of employment: How susceptible are jobs to computerisation? *Technological Forecasting and Social Change*, *114*, 254–280. DOI: 10.1016/j.techfore.2016.08.019

Frichi, Y., & Jawab, F. (Eds.). (2025). *Multi-Stakeholder Collaboration for Sustainable Supply Chain*. IGI Global Scientific Publishing., DOI: 10.4018/979-8-3693-8925-6

Frick, N. R. J., Mirbabaie, M., Stieglitz, S. & Salomon, J. (2021). Maneuvering through the stormy seas of digital transformation: The impact of empowering leadership on the AI readiness of enterprises. *Journal of Decision Systems, 30* (2 - 3), 235 - 258. .DOI: 10.1080/12460125.2020.1870065

Gadre, M., & Deoskar, A. (2020). Industry 4.0–digital transformation, challenges and benefits. *International Journal of Future Generation Communication and Networking, 13*(2), 139–149.

Gaie, C., & De Lacerda, R.GAIE. (2022). Ensuring project resilience and public services in a pandemic context: Two French case studies. *International Journal of Computational Systems Engineering, 7*(1), 1. DOI: 10.1504/IJCSYSE.2022.10054831

Galaske, N., Arndt, A., Friedrich, H., Bettenhausen, K. D., & Anderl, R. (2017). *Workforce Management 4.0 - Assessment of Human Factors Readiness Towards Digital Manufacturing*. Springer., DOI: 10.1007/978-3-319-60474-9_10

Galin, R., & Meshcheryakov, R. (2019). *Automation and robotics in the context of Industry 4.0: the shift to collaborative robots. 537*(3), 032073. .DOI: 10.1088/1757-899X/537/3/032073

Gallicchio, C., & Micheli, A. (2020). Ring Reservoir Neural Networks for Graphs. *International Joint Conference on Neural Network*, 1–7. https://doi.org/DOI: 10.1109/IJCNN48605.2020.9206723

Gallup. (2021). State of the Global Workplace: 2021 Report. https://www. gallup. com/workplace/349484/state - of - the - global - workplace - 2021 - report. aspx

Ganesan, I., Ponnuviji, N. P., Kumar, A. S., Nithya, M., Jambulingam, U., & Lalitha, S. D. (2024). Real-Time Event Detection and Predictive Analytics Using IoT and Deep Learning. In Industry Applications of Thrust Manufacturing: Convergence with Real-Time Data and AI (pp. 1-41). IGI Global Scientific Publishing. DOI: 10.4018/979-8-3693-4276-3.ch001

Ganesh, M., & Indradevi, R. (2015). Importance and effectiveness of training and development. *Mediterranean Journal of Social Sciences, 6*(1), 334–338. DOI: 10.5901/mjss.2015.v6n1p334

Gartner. (2021). Hype cycle for edge computing. Gartner Research. https://www. gartner. com/en/documents/4009496

Gartner. (2021). Top 5 priorities for HR leaders in 2021. https://www. gartner. com/en/human - resources/trends/top - priorities - for - hr - leaders

Gartner. (2022). redesigning work for well - being: The future of employee experience. https://www. gartner. com/en/articles/redesigning - work - for - well - being

Gebresenbet, G., Bosona, T., Patterson, D., Persson, H., Fischer, B., Mandaluniz, N., Chirici, G., Zacepins, A., Komasilovs, V., Pitulac, T., & Nasirahmadi, A. (2023). A concept for application of integrated digital technologies to enhance future smart agricultural systems. *Smart Agricultural Technology, 5*, 100255. DOI: 10.1016/j.atech.2023.100255

Geetha, B. T., Tripathi, S. P., Saxena, C. R., Chouhan, K., Shavkatov, N., & Dadheech, P. (2025). Brain-Computer Interface-Controlled Smartphones for Streamlined Organizational Data Management. In B. Pandey, A. George, S. Tiwari, S. Albermany, & H. Hung (Eds.), *Brain-Computer Interfaces and Applications in Business* (pp. 183-198). IGI Global Scientific Publishing. https://doi.org/DOI: 10.4018/979-8-3373-5122-3.ch011

Gely, C., Trentesaux, D., Pacaux-Lemoine, M.-P., & Sénéchal, O. (2020). *Human-Machine Co-operation with Autonomous CPS in the Context of Industry 4.0: A Literature Review*. Springer., DOI: 10.1007/978-3-030-69373-2_23

Gemeda, H. K., & Lee, J. (2020a). Leadership styles, work engagement and outcomes among information and communications technology professionals: A cross-national investigation. *Heliyon*, *6*(4), e03699. Advance online publication. DOI: 10.1016/j.heliyon.2020.e03699 PMID: 32280799

Genç, N. (2005). *Yönetim ve organizasyon*. Seçkin Yayınları.

Gh. Popescu, C. R., & Ban a, V. C. (2019). Performance Evaluation of the Implementation of the 2013/34/EU Directive in Romania on the Basis of Corporate Social Responsibility Reports. *Sustainability*, *11*(9), 2531. DOI: 10.3390/su11092531

Ghanghorkar, Y., & Pillai, R. (2024). *Human–Robot Coordination and Collaboration in Industry 4.0*. Springer Nature., DOI: 10.1007/978-981-99-8118-2_9

Ghasabeh, M. S. (2021). Transformational Leadership: Implementing a Cultural Approach in Corporates. In *The Journal of Values-Based Leadership* (Vol. 14, Burning matter 1).

Ghobakhloo, M. (2018). The future of manufacturing industry: A strategic roadmap toward Industry 4.0. *Journal of Manufacturing Technology Management*, *29*(6), 910–936. DOI: 10.1108/JMTM-02-2018-0057

Ghobakhloo, M. (2020). Industry 4.0, digitization, and opportunities for sustainability. *Journal of Cleaner Production*, *252*, 119869. DOI: 10.1016/j.jclepro.2019.119869

Ghosh, R. (2020). Human capital management in the age of automation. *Journal of Human Resource and Sustainability Studies*, *8*(2), 191–204.

Gill, A., & Mathur, A. (2024). Emotional Intelligence in the Age of AI: Enhancing Workforce Development for Human-Machine Collaboration. In Sharma, A., Moses, O., Sharma, R., & Gupta, S. (Eds.), *Sustainable Innovation for Industry 6.0* (pp. 271–294). IGI Global Scientific Publishing., DOI: 10.4018/979-8-3693-3140-8.ch014

Gill, R. (2011). *Theory and practice of leadership*. SAGE Publications.

Gini, K. B., & Agala, H. O. (2023). The Impact of Macro - Environmental Factors on Business Performance. *International Journal of Research and Innovation in Social Science*, *7*(1), 1837–1843. DOI: 10.47772/IJRISS.2023.7011145

Goalla, A., Shefali, P., & Sushmitha, K. (2024). Human-AI Collaboration in Environmental Decision Making: Ethical Considerations. In Kannan, H., Rodriguez, R., Paprika, Z., & Ade-Ibijola, A. (Eds.), *Exploring Ethical Dimensions of Environmental Sustainability and Use of AI* (pp. 330–345). IGI Global Scientific Publishing., DOI: 10.4018/979-8-3693-0892-9.ch016

Gökbulut, B., & Dindaş, S. (2022). Öğretmenlerin mesleki tükenmişlik ve teknostres düzeylerinin incelenmesi. [IJOESS]. *International Journal of Eurasia Social Sciences, 13*(47), 42–59. DOI: 10.35826/ijoess.3096

Gokhale, P., Bhat, O., & Bhat, S. (2018). Introduction to IOT. *International Advanced Research Journal in Science. Engineering and Technology, 5*(1), 41–44.

Goleman, D. (2006). *Emotional intelligence: Why it can matter more than IQ*. Bantam Books.

Gorecky, D., Schmitt, M., Loskyll, M., & Zühlke, D. (2014). Human-machine-interaction in the industry 4.0 era. *International Conference on Industrial Informatics*, 289–294. DOI: 10.1109/INDIN.2014.6945523

Gourisaria, M. K., Agrawal, R., Harshvardhan, G. M., Pandey, M., & Rautaray, S. S. (2021). Application of machine learning in industry 4.0. *Machine learning: Theoretical foundations and practical applications*, 57-87.

Graetz, G., & Michaels, G. (2018). Robots at work. *The Review of Economics and Statistics, 100*(5), 753–768. DOI: 10.1162/rest_a_00754

Greenhalgh, T., Rosen, R., Shaw, S. E., Byng, R., Faulkner, S., Finlay, T., Grundy, E., Husain, L., Hughes, G., Leone, C., Moore, L., Papoutsi, C., Pope, C., Rybczynska-Bunt, S., Rushforth, A., Wherton, J., Wieringa, S., & Wood, G. W. (2021). [eXtyles1.1]. Planning and Evaluating Remote Consultation Services: A New Conceptual Framework Incorporating Complexity and Practical Ethics. *Frontiers in Digital Health, 3*(3). Advance online publication. DOI: 10.3389/fdgth.2021.726095

Gubbi, J., Buyya, R., Marusic, S., & Palaniswami, M. (2013). Internet of Things (IoT): A vision, architectural elements, and future directions. *Future Generation Computer Systems, 29*(7), 1645–1660. DOI: 10.1016/j.future.2013.01.010

Guenzi, P., & Nijssen, E. J. (2021). The impact of digital transformation on salespeople: An empirical investigation using the JD - R model. *Journal of Personal Selling & Sales Management, 41*(2), 130–149. DOI: 10.1080/08853134.2021.1918005

Guo, H., Chen, F., Tang, Y., Ding, Y., Chen, M., Zhou, W., Zhu, M., Gao, S., Yang, R., Zheng, W., Fang, C., Lin, H., Roders, A. P., Cigna, F., Tapete, D., & Xu, B. (2023). Progress toward the sustainable development of world cultural heritage sites facing land-cover changes. *Innovation (Cambridge (Mass.)), 4*(5), 100496. DOI: 10.1016/j.xinn.2023.100496 PMID: 37663934

Guo, H., Huang, L., & Liang, D. (2022). Further promotion of sustainable development goals using science, technology, and innovation. *Innovation (Cambridge (Mass.)), 3*(6), 100325. DOI: 10.1016/j.xinn.2022.100325 PMID: 36193207

Guo, S., Diao, Y., & Du, J. (2022). Coupling Coordination Measurement and Evaluation of Urban Digitalization and Green Development in China. *International Journal of Environmental Research and Public Health, 19*(22), 15379. DOI: 10.3390/ijerph192215379 PMID: 36430095

Gupta, A., Bharadwaj, N., & Rastogi, V. (2021). Computational Framework of Various Semi-Active Control Strategies for Road Vehicles Thorough Bondgraphs. [IJSDA]. *International Journal of System Dynamics Applications, 10*(4), 1–29. DOI: 10.4018/IJSDA.20211001.oa9

Hamad, A. A., Jha, S., & Al-Badri, K. (Eds.). (2025). *Integrating Intelligent Control Systems With Sensor Technologies*. IGI Global Scientific Publishing., DOI: 10.4018/979-8-3373-0330-7

Hammer, M., & Stanton, S. (1999). How process enterprises really work. *Harvard Business Review*, *77*(6), 108–118. PMID: 10662000

Hanelt, A., Bohnsack, R., Marz, D., & Antunes Marante, C. (2021). A systematic review of the literature on digital transformation: Insights and implications for strategy and organizational change. *Journal of Management Studies*, *58*(5), 1159–1197. DOI: 10.1111/joms.12639

Hao, W., Rasul, F., Bhatti, Z., Hassan, M. S., Ahmed, I., & Asghar, N. (2021). A technological innovation and economic progress enhancement: An assessment of sustainable economic and environmental management. *Environmental Science and Pollution Research International*, *28*(22), 28585–28597. DOI: 10.1007/s11356-021-12559-9 PMID: 33544345

Harari, Y. N. (2018). *21 Lessons for the 21st Century*. Jonathan Cape.

Hariharan, K., & Anand, V. (2023). Transformational leadership and learning flows. *The Learning Organization*, *30*(3), 309–325. DOI: 10.1108/TLO-09-2021-0115

Hartmann, S., Brock, J., Kühn, A., & Dumitrescu, R. (2024). Applying Artificial Intelligence in the Smart Factory: Lessons Learned from real-world use cases. *Procedia CIRP*, *58*(27), 100–105. DOI: 10.1016/j.procir.2024.10.062

Harvard Business Review. (2019). *Digital transformation is not about technology*. Harvard Business Publishing.

Hatamlou, A. (2018). Numerical Optimization Using the Heart Algorithm. [IJAEC]. *International Journal of Applied Evolutionary Computation*, *9*(2), 33–37. DOI: 10.4018/IJAEC.2018040103

Hatch, N. W., & Dyer, J. H. (2004). Human capital and learning as a source of sustainable competitive advantage. *Strategic Management Journal*, *25*(12), 1155–1178. DOI: 10.1002/smj.421

Helldén, D., Weitz, N., Nilsson, M., & Alfvén, T. (2022). Situating Health Within the 2030 Agenda-A Practical Application of the Sustainable Development Goals Synergies Approach. *Public Health Reviews*, *43*, 1604350. DOI: 10.3389/phrs.2022.1604350 PMID: 35465141

Heller, F. (1997). Sociotechnology and the environment. *Human Relations*, *50*(5), 605–624. DOI: 10.1177/001872679705000507

He, W. W., He, S. L., & Hou, H. L. (2024). Digital economy, technological innovation, and sustainable development. *PLoS One*, *19*(7), e0305520. DOI: 10.1371/journal.pone.0305520 PMID: 39042638

Hewitt, A. (2015). Trends in Global Employee Engagement - Making engagement happen. Aon Hewitt cooperation. https://www. aon. com/attachments/human - capital - consulting/2015 - trends - in - global - employee - engagement - report. pdf, Hill, C. W., Schilling, M. A. & Jones, G. R. (2020). Strategic management: an integrated approach: theory and cases, Cengage Learning: Boston.

Hidayat-ur-Rehman, I., & Alsolamy, M. (2023). A SEM-ANN analysis to investigate sustainable performance in SMEs: The moderating role of transformational leadership. *Journal of Open Innovation*, *9*(4), 100166. Advance online publication. DOI: 10.1016/j.joitmc.2023.100166

Hoda, R., Salleh, N., Grundy, J., & Tee, H. M. (2017). Systematic literature reviews in Agile software development: A tertiary study. *Information and Software Technology*, *85*, 60–70. DOI: 10.1016/j.infsof.2017.01.007

Hollnagel, E., & Woods, D. D. (2005). *Joint cognitive systems: Foundations of cognitive systems engineering*. CRC press. DOI: 10.1201/9781420038194

Holm, S., & Hovland, J. (1999). Waiting for the other shoe to drop: Help for the job - insecure employee. *Journal of Employment Counseling*, *36*(December), 36, 156–164. DOI: 10.1002/j.2161-1920.1999.tb01018.x

Hopcan, S., Türkmen, G., & Polat, E. (2023). Exploring the artificial intelligence anxiety and machine learning attitudes of teacher candidates. *Education and Information Technologies*. Advance online publication. DOI: 10.1007/s10639-023-12086-9

Hordofa, D. F. (2023). Impacts of external factors on Ethiopia's economic growth: Insights on foreign direct investment, remittances, exchange rates, and imports. *Heliyon*, *9*(12), e22847. DOI: 10.1016/j.heliyon.2023.e22847 PMID: 38058435

Houti, M., Abbadi, L. E., & Abouabdellah, A. (2019). CSFs for the Implementation of the Hybrid Lean ERP System. Stakeholders Interactions. *Advances in Science, Technology and Engineering Systems Journal*, *4*(6), 443–447. DOI: 10.25046/aj040655

Huang, L., Swain, R. B., Jeppesen, E., Cheng, H., Zhai, P., Gu, B., Barceló, D., Lu, J., Wei, K., Luo, L., Wang, F., Wang, H., Zeng, J., & Guo, H. (2024). Harnessing science, technology, and innovation to drive synergy between climate goals and the SDGs. *Innovation (Cambridge (Mass.))*, *5*(6), 100693. DOI: 10.1016/j.xinn.2024.100693 PMID: 39399226

Huang, Y., Lu, Y., Wang, F., Fan, X., Liu, J., & Leung, V. C. (2018, October). An edge computing framework for real-time monitoring in smart grid. In *2018 IEEE International conference on industrial internet (ICII)* (pp. 99-108). IEEE. DOI: 10.1109/ICII.2018.00019

Huang, Z., Yu, J., He, W., Yu, J., Deng, S., Yang, C., Zhu, W., & Shao, X. (2023). AI-enhanced chemical paradigm: From molecular graphs to accurate prediction and mechanism. *Journal of Hazardous Materials*, *465*, 133355. DOI: 10.1016/j.jhazmat.2023.133355 PMID: 38198864

Hui, Y. (2024). *Using Artificial Intelligence Solutions Towards Asset Performance Management System in Smart Manufacturing: A Case Study with Capgemini on Service Integration for Operation and Maintenance Purposes* (Master's thesis, UIS).

Hundie, Z., & Habtewold, E. (2024). The Influence of Transformational, Transactional, and Laissez-Faire Leadership Styles on Employees' Level of Performance: The Case of Hospital in Oromia Region, Ethiopia. *Journal of Healthcare Leadership*, *16*, 67–82. DOI: 10.2147/JHL.S450077 PMID: 38380132

Hu, Y., Jia, Q., Yao, Y., Lee, Y., Lee, M., Wang, C., Zhou, X., Xie, R., & Yu, F. R. (2024). Industrial internet of things intelligence empowering smart manufacturing: A literature review. *IEEE Internet of Things Journal*, *11*(11), 19143–19167. DOI: 10.1109/JIOT.2024.3367692

Hye, Q. M. A., Ul-Haq, J., Visas, H., & Rehan, R. (2023). The role of eco-innovation, renewable energy consumption, economic risks, globalization, and economic growth in achieving sustainable environment in emerging market economies. *Environmental Science and Pollution Research International*, *30*(40), 92469–92481. DOI: 10.1007/s11356-023-28945-4 PMID: 37491494

Inkulu, A. K., Bahubalendruni, M. V. A. R., Dara, A., & Sankaranarayanasamy, K. (2021). Challenges and opportunities in human robot collaboration context of Industry 4.0 - a state-of-the-art review. *Information Retrieval*. Advance online publication. DOI: 10.1108/IR-04-2021-0077

Investopedia. (2023). How generative AI is revolutionizing manufacturing. https://www. investopedia. com/how - generative - ai - is - revolutionizing - manufacturing - 7480630

Irajifar, L., Chen, H., Lak, A., Sharifi, A., & Cheshmehzangi, A. (2023). The nexus between digitalization and sustainability: A scientometrics analysis. *Heliyon*, *9*(5), e15172. DOI: 10.1016/j.heliyon.2023.e15172 PMID: 37153424

Islam, M. Z., Rahaman, S. H., & Chen, F. (2024). How do R&D and remittances affect economic growth? Evidence from middle-income countries. *Heliyon*, *10*(9), e30160. DOI: 10.1016/j.heliyon.2024.e30160 PMID: 38694034

Iwasokun, G. B., & Alimi, S. A. (2022). Genetic Algorithm Model for Stock Management and Control. [IJSDS]. *International Journal of Strategic Decision Sciences*, *13*(1), 1–20. DOI: 10.4018/IJSDS.309119

Jacobson, D., & Hartley, J. (1991). *Mapping the context*, in *Job insecurity coping with jobs at risk*, (Ed.s: J. Hartley, D. Jacobson, B. Klandermans & T. V. Vuuren), London: Sage Publications, 1 - 22.

Jagatheesaperumal, S. K., Rahouti, M., Ahmad, K., Al-Fuqaha, A., & Guizani, M. (2021). The duo of artificial intelligence and big data for industry 4.0: Applications, techniques, challenges, and future research directions. *IEEE Internet of Things Journal*, *9*(15), 12861–12885. DOI: 10.1109/JIOT.2021.3139827

Jalil, A., Rauf, A., Sikander, W., Yonghong, Z., & Tiebang, W. (2021). Energy consumption, economic growth, and environmental sustainability challenges for Belt and Road countries: A fresh insight from "Chinese Going Global Strategy". *Environmental Science and Pollution Research International*, *28*(46), 65987–65999. DOI: 10.1007/s11356-021-15549-z PMID: 34324149

Jamtsho, Y.. (2020). Real-time license plate detection for non-helmeted motorcyclist using YOLO. *ICT Express*. Advance online publication. DOI: 10.1016/j.icte.2019.11.001

Janssen, M., & Estevez, E. (2013). Lean government and platform-based governance—Doing more with less. *Government Information Quarterly*, *30*, S1–S8. DOI: 10.1016/j.giq.2012.11.003

Javaid, M., Haleem, A., Singh, R. P., & Suman, R. (2022). Artificial intelligence applications for industry 4.0: A literature-based study. *Journal of Industrial Integration and Management*, *7*(01), 83–111. DOI: 10.1142/S2424862221300040

Javaid, M., Haleem, A., Singh, R. P., & Suman, R. (2022). Enhancing smart farming through the applications of Agriculture 4.0 technologies. *International Journal of Intelligent Networks*, *3*, 150–164. DOI: 10.1016/j.ijin.2022.09.004

Jeni, F., & AlAmin, M. (2021). The impact of training and development on employee performance and productivity:an empirical study on private bank of Noakhali region in Bangladesh. *South Asian Journal of Social Studies and Economics*, *1*(1), 1–18. DOI: 10.9734/sajsse/2021/v9i230234

Jeong, S., Roh, S., & Sohn, K. (2021). Multi-regime analysis for computer vision-based traffic surveillance using a change-point detection algorithm. *IEEE Access : Practical Innovations, Open Solutions*, *9*, 40980–40995. DOI: 10.1109/ACCESS.2021.3064603

Jia, W., Xu, S., Liang, Z., Zhao, Y., Min, H., Li, S., & Yu, Y. (2021). Real-time automatic helmet detection of motorcyclists in urban traffic using improved YOLOv5 detector. *IET Image Processing*, *15*(14), 3623–3637. Advance online publication. DOI: 10.1049/ipr2.12295

Jinga, A. A., Omer Hussen, J., Gezahagn Negash, H., & Bezabih Estifanos, A. (2024). Leadership Behavior as well as organizational Paradigm shift Management in Selected Public Universities of Ethiopia: Exploring the Influence of Leadership Influences and Paradigm shift Processes. *Heliyon*, *37149*. Advance online publication. DOI: 10.1016/j.heliyon.2024.e37149 PMID: 39398013

John Joseph, F., Chinnusamy, K., Jeganathan, J., & Obaid, J. A., & Rajest, S. (Eds.). (2026). *Machine Learning, Predictive Analytics, and Optimization in Complex Systems*. IGI Global Scientific Publishing. https://doi.org/DOI: 10.4018/979-8-3373-5203-9

Joo, Y. J., Lim, K. Y., & Kim, N. H. (2015). The effects of secondary teachers' technostress on the intention to use technology in South Korea. *Computers & Education*, *95*, 114–122. DOI: 10.1016/j.compedu.2015.12.004

Joshi, P. R., & Islam, S. (2018). E-government maturity model for sustainable e-government services from the perspective of developing countries. *Sustainability (Basel)*, *10*(6), 1882. DOI: 10.3390/su10061882

Jovanovic, B., & Rousseau, P. L. (2005) *General purpose technologies*, in *Handbook of economic growth*, (Ed.s: Philippe A. & Steven N. D. 1181 - 1224, London: Elsevier DOI: 10.3386/w11093

Jumbri, I. A., Ikeda, S., Jimichi, M., Saka, C., & Managi, S. (2019). Inequality of health stock and the relation to national wealth. *International Journal for Equity in Health*, *18*(1), 188. DOI: 10.1186/s12939-019-1096-x PMID: 31791346

Junaid, M. D., & Gokce, A. F. (2024). Global agricultural losses and their causes. *Bulletin of Biological and Allied Sciences Research*, *2024*(1), 66–66. DOI: 10.54112/bbasr.v2024i1.66

Kabir, M. A., Nandi, S. K., Suma, A. Y., Abdul Kari, Z., Mohamad Sukri, S. A., Wei, L. S., Al Mamun, A., Seguin, P., Herault, M., Khoo, M. I., & Téllez-Isaías, G. (2024). The Potential of Fish Protein Hydrolysate Supplementation in Nile Tilapia Diets: Effects on Growth and Health Performance, Disease Resistance, and Farm Economic Analysis. *Applied Biochemistry and Biotechnology*, *196*(10), 7145–7167. DOI: 10.1007/s12010-024-04913-7 PMID: 38489116

Kalfas, D., Kalogiannidis, S., Papaevangelou, O., Melfou, K., & Chatzitheodoridis, F. (2024). Integration of technology in agricultural practices towards agricultural sustainability: A case study of Greece. *Sustainability (Basel)*, *16*(7), 2664. DOI: 10.3390/su16072664

Kaplan, A., & Haenlein, M. (2020). Rulers of the world, unite! The challenges and opportunities of artificial intelligence. *Business Horizons*, *63*(1), 37–50. DOI: 10.1016/j.bushor.2019.09.003

Kaplan, J. (2016). *Artificial Intelligence: What everyone needs to know*. Oxford University press. DOI: 10.1093/wentk/9780190602383.001.0001

Katiyar, S., & Farhana, A. (2021). Smart agriculture: The future of agriculture using ai and iot. *Journal of Computational Science*, *17*(10), 984–999. DOI: 10.3844/jcssp.2021.984.999

Kaur, J. (2024). The Nexus of Care: Human-Machine Collaboration Redefining Healthcare Delivery. In T. Murugan, J. W., & V. P. (Eds.), *Technologies for Sustainable Healthcare Development* (pp. 230-253). IGI Global Scientific Publishing. https://doi.org/DOI: 10.4018/979-8-3693-2901-6.ch013

Kaushal, I., Ma, J., & Dallasega, P. (2025). Sustainability Assessment of Smart Mobile Factory and Centralized Factory - related Logistics for Linear Infrastructure Projects: A Comparative Investigation. *Procedia Computer Science*, *253*, 1981–1990. DOI: 10.1016/j.procs.2025.01.260

Kavitha, C. R., Varalatchoumy, M., Mithuna, H. R., Bharathi, K., Geethalakshmi, N. M., & Boopathi, S. (2023). Energy Monitoring and Control in the Smart Grid: Integrated Intelligent IoT and ANFIS. In Arshad, M. (Ed.), *Applications of Synthetic Biology in Health, Energy, and Environment* (pp. 290–316). IGI Global Scientific Publishing., DOI: 10.4018/978-1-6684-6577-6.ch014

Kayongo, A., Guloba, A., Muvawala, J., & Ssali, A. (2022). A Framework for Strengthening and Sustaining Cooperatives for Socio-Economic Transformation in Uganda. *Applied Economics and Finance*, *9*(2), 79–92. DOI: 10.11114/aef.v9i2.5554

Kellogg, K. C., Valentine, M. A., & Christin, A. (2020). Algorithms at work: The new contested terrain of control. *The Academy of Management Annals*, *14*(1), 366–410. DOI: 10.5465/annals.2018.0174

Kenwright, B. (Ed.). (2025). *Cutting-Edge Techniques for Effective Data Mining and Machine Learning*. IGI Global Scientific Publishing., DOI: 10.4018/979-8-3693-1606-1

Khalid, S. (2025). Advanced Electric Propulsion Systems for Hybrid Electric Vehicles: Design, Control, and Performance Optimization. In Khalid, S., Ali, J., & Yadav, D. (Eds.), *Hybrid Electric Vehicles and Distributed Renewable Energy Conversion: Control and Vibration Analysis* (pp. 1–54). IGI Global Scientific Publishing., DOI: 10.4018/979-8-3693-5797-2.ch001

Khan, A., & Turowski, K. (2016, April). A perspective on industry 4.0: From challenges to opportunities in production systems. In *International Conference on Internet of Things and Big Data* (Vol. 2, pp. 441-448). SciTePress. DOI: 10.5220/0005929704410448

Khan, K. A., Ahmad, W., Zubair, A. O., Subhan, M., & Shah, M. I. (2025). From invention to progress: Energy technology innovation and sustainable development in OECD economies. *PLoS One*, *20*(2), e0310104. DOI: 10.1371/journal.pone.0310104 PMID: 39946336

Khan, M. K., Hussain, P., Hussain, M. J., & Michael, K. (2025). Innovation at the nexus of technology and sustainability: CEO leadership and digital transformation in China. *Sustainable Futures : An Applied Journal of Technology, Environment and Society*, *9*, 100494. Advance online publication. DOI: 10.1016/j.sftr.2025.100494

Khan, N., Ray, R. L., Sargani, G. R., Ihtisham, M., Khayyam, M., & Ismail, S. (2021). Current progress and future prospects of agriculture technology: Gateway to sustainable agriculture. *Sustainability (Basel), 13*(9), 4883. DOI: 10.3390/su13094883

Khan, R. A. G., Khan, F. A., & Khan, M. A. (2011). Impact of training and development on organisational performance. *Global Journal of Management and Business Research, 11*(7), 63–68.

Kieny, M. P., Bekedam, H., Dovlo, D., Fitzgerald, J., Habicht, J., Harrison, G., Kluge, H., Lin, V., Menabde, N., Mirza, Z., Siddiqi, S., & Travis, P. (2017). Strengthening health systems for universal health coverage and sustainable development. *Bulletin of the World Health Organization, 95*(7), 537–539. DOI: 10.2471/BLT.16.187476 PMID: 28670019

Kim, J., Kakdol, K., & Solomon, P. (2023). Employee anxiety and automation: Understanding workforce resistance in the age of AI. *Journal of Organizational Psychology, 23*(1), 55–68.

Kishore, S., & Fonceca, C. M. (2023). Impact of Training and Development on Employee Performance and Productivity. *Journal of Academia and Industrial Research, 11*(3), 43–47.

Korstanje, M. E. (Ed.). (2017). *Threat Mitigation and Detection of Cyber Warfare and Terrorism Activities*. IGI Global Scientific Publishing., DOI: 10.4018/978-1-5225-1938-6

Kotter, J. P. (1996). *Leading change*. Harvard Business Press.

Koval, S. (2015). Employee training and development [Undergraduate thesis, University College Prague]. IS UCP. https://is. ucp. cz/th/u9zm0/

Kranz, J. J., Hanelt, A., & Kolbe, L. M. (2016). Understanding the influence of absorptive capacity and ambidexterity on the process of business model change. *Technological Forecasting and Social Change, 103*, 104–113.

Krishnaveni, S., & Schwartz, E. B. (2024). *The Role of Human-Centric Solutions in Tackling Challenges and Unlocking Opportunities in Industry 4.0*. 365–375. .DOI: 10.1201/9781032694375-20

Küçükcivil, B., Gargalık, E., & Koçyiğit, M. (2024). Üniversitelerde dijital eğitim - öğretim faaliyetleri ve teknostres. *Erciyes İletişim Dergisi, 11*(1), 105–134. DOI: 10.17680/erciyesiletisim.1352212

Kumar, N., Pal, S. K., Agarwal, P., Rosak-Szyrocka, J., & Jain, V. (Eds.). (2024). *Harnessing Artificial Emotional Intelligence for Improved Human-Computer Interactions*. IGI Global Scientific Publishing., DOI: 10.4018/979-8-3693-2794-4

Kustepeli, Y., Gulcan, Y., Yercan, M., & Yıldırım, B. (2023). The role of agricultural development cooperatives in establishing social capital. *The Annals of Regional Science, 70*(3), 681–704. DOI: 10.1007/s00168-019-00965-4

Kvasnička, V., & Pospíchal, J. (1991). Application of neural networks in chemistry. Prediction of product distribution of nitration in a series of monosubstituted benzenes. *Journal of Molecular Structure THEOCHEM, 235*(3-4), 227–242. DOI: 10.1016/0166-1280(91)85103-E

La Torre, G., Esposito, A., Sciarra, I., & Chiappetta, M. (2019). Definition, symptoms and risk of techno - stress: A systematic review. *International Archives of Occupational and Environmental Health, 92*(1), 13–35. DOI: 10.1007/s00420-018-1352-1

Lachgar, M., Hrimech, H., & Kartit, A. (2023). Unmanned aerial vehicle-based applications in smart farming: A systematic review. *International Journal of Advanced Computer Science and Applications*, *14*(6).

Landers, R. N., & Marin, S. (2021). Theory and design of digital nudges for behavior change: Implications for AI in HRM. *Applied Psychology*, *70*(3), 889–921.

Lasi, H., Fettke, P., Kemper, H. G., Feld, T., & Hoffmann, M. (2014). Industry 4.0. *Business & Information Systems Engineering*, *6*(4), 239–242. DOI: 10.1007/s12599-014-0334-4

Leal Filho, W., Vidal, D. G., Chen, C., Petrova, M., Dinis, M. A. P., Yang, P., Rogers, S., Álvarez-Castañón, L., Djekic, I., Sharifi, A., & Neiva, S. (2022). An assessment of requirements in investments, new technologies, and infrastructures to achieve the SDGs. *Environmental Sciences Europe*, *34*(1), 58. DOI: 10.1186/s12302-022-00629-9 PMID: 37128271

Lee, J., Bagheri, B., & Kao, H. A. (2015). A cyber-physical systems architecture for industry 4.0-based manufacturing systems. *Manufacturing Letters*, *3*, 18–23. DOI: 10.1016/j.mfglet.2014.12.001

Lee, K. H., Noh, J., & Khim, J. S. (2020). The Blue Economy and the United Nations' sustainable development goals: Challenges and opportunities. *Environment International*, *137*, 105528. DOI: 10.1016/j.envint.2020.105528 PMID: 32014791

Lee, M. F. R. (2023). A review on intelligent control theory and applications in process optimization and smart manufacturing. *Processes (Basel, Switzerland)*, *11*(11), 3171. DOI: 10.3390/pr11113171

Le, H. T. T., Dao, Q. T. M., Pham, V. C., & Tran, D. T. (2019). Global trend of open innovation investigation: A bibliometric analysis. *Cogent Business & Management*, *6*(1), 1633808. Advance online publication. DOI: 10.1080/23311975.2019.1633808

Lengnick-Hall, C. A., & Beck, T. E. (2005). Adaptive fit versus robust transformation: How organizations respond to environmental change. *Journal of Management*, *31*(5), 738–757. DOI: 10.1177/0149206305279367

Leora, A. (2021). GEN MZ TOWARD SUSTAINABLE FASHION PRACTICES - ALTERNATIVE SERVICES AND BUSINESS MODELS. *Iadisportal.org*, 61–70. https://www.iadisportal.org/digital-library/gen-mz-toward-sustainable-fashion-practices-alternative-services-and-business-models

Levy, F., & Murnane, R. J. (1992). US earnings levels and earnings inequality: A review of recent trends and proposed explanations. *Journal of Economic Literature*, *30*(3), 1333–1381.

Levy, F., & Murnane, R. J. (2013). *Dancing with robots: Human skills for computerized work.* Third Way Foundation.

Leymann, G., Lundan, S., & Van Assche, A. (2022). Global challenges for Agile government: Global connectivity policy from the perspective of the international business literature. *Agile Government*, 307-320. DOI: 10.1142/9789811239700_0016

Li, Y., Ou, L., You, J., Li, N., & Zheng, X. (2023). Real-time traffic accident detection based on the derivative of traffic parameters. *2023 35th Chinese Control and Decision Conference (CCDC)*, 1470–1473. https://doi.org/DOI: 10.1109/CCDC58219.2023.10327413

Liang, D., Guo, H., Nativi, S., Kulmala, M., Shirazi, Z., Chen, F., Kalonji, G., Yan, D., Li, J., Duerler, R., Luo, L., Han, Q., Deng, S., Wang, Y., Kong, L., & Jelinek, T. (2023). A future for digital public goods for monitoring SDG indicators. *Scientific Data, 10*(1), 875. DOI: 10.1038/s41597-023-02803-x PMID: 38062062

Liao, S. H. (2005). Technology management methodologies and applications: A literature review from 1995 to 2003. *Technovation, 25*(4), 381–393. DOI: 10.1016/j.technovation.2003.08.002

Liao, Z. (2023). Supply chain optimization for environmental sustainability and economic growth. *Environmental Science and Pollution Research International, 30*(58), 121599–121613. DOI: 10.1007/s11356-023-30521-9 PMID: 37957491

Li, D., Bae, J. H., & Rishi, M. (2023). Sustainable Development and SDG-7 in Sub-Saharan Africa: Balancing Energy Access, Economic Growth, and Carbon Emissions. *European Journal of Development Research, 35*(1), 112–137. DOI: 10.1057/s41287-021-00502-0 PMID: 35106034

Lina, Z. (2024). Towards the sustainable financial management in green economic growth through big data technology innovation; Case study in China. *Heliyon, 10*(19), e35517. DOI: 10.1016/j.heliyon.2024.e35517 PMID: 39403525

Li, S., Wu, W., Wang, Y., Zhang, N., Sun, F., Jiang, F., & Wei, X. (2023). Production Data Management of Smart Farming Based on Shili Theory. *Agriculture, 13*(4), 751. DOI: 10.3390/agriculture13040751

Liu, B., Lam, C.-T., & Ng, B. K. (2021). Improved real-time traffic congestion detection with automatic image cropping using online camera images. *2021 IEEE 21st International Conference on Communication Technology (ICCT)*, 1117–1122. https://doi.org/DOI: 10.1109/ICCT52962.2021.9657853

Liu, L., Yu, B., & Wu, W. (2019). The Formation and Effects of Exploitative Dynamic Capabilities and Explorative Dynamic Capabilities: An Empirical Study. *Sustainability (Basel), 11*(9), 2581. DOI: 10.3390/su11092581

Liu, Y., Yang, Y., Zhang, X., & Yang, Y. (2024). The impact of technological innovation on the green digital economy and development strategies. *PLoS One, 19*(4), e0301051. DOI: 10.1371/journal.pone.0301051 PMID: 38662690

Lokshina, I. V., & Lanting, C. J. (2020). Analysis of Queueing Networks in Equilibrium: Numerical Steady-State Solutions of Markov Chains. [IJITN]. *International Journal of Interdisciplinary Telecommunications and Networking, 12*(4), 1–17. DOI: 10.4018/IJITN.2020100101

Lueck, T. (2014). Internet archive: Digital library of free books, movies, music, and Wayback machine/The internet archive companion. *American Journalism, 31*(2), 299–301. DOI: 10.1080/08821127.2014.905381

Luo, L., Wang, X., & Guo, H. (2022). Contribution of UNESCO designated sites to the achievement of Sustainable Development Goals. *Innovation (Cambridge (Mass.)), 3*(3), 100227. DOI: 10.1016/j.xinn.2022.100227 PMID: 35373170

Luo, S., & Liu, J. (2024). Enterprise service-oriented transformation and sustainable development driven by digital technology. *Scientific Reports*, *14*(1), 10047. DOI: 10.1038/s41598-024-60922-w PMID: 38698122

MacPherson, J., Voglhuber-Slavinsky, A., Olbrisch, M., Schöbel, P., Dönitz, E., Mouratiadou, I., & Helming, K. (2022). Future agricultural systems and the role of digitalization for achieving sustainability goals. A review. *Agronomy for Sustainable Development*, *42*(4), 70. DOI: 10.1007/s13593-022-00792-6 PMID: 35818482

Ma, D., Gausemeier, J., Fan, X., & Grafe, M. (Eds.). (2011). *Virtual reality & augmented reality in industry*. Springer Berlin Heidelberg. DOI: 10.1007/978-3-642-17376-9

Madhumitha, S., & Asrafi, S. (2024). The Role of Technology in Revolutionizing Training & Development. *International Journal of Business and Management Invention, 13*(4), 134–137.

Mahajan, S. L., Obiene, S., Ojwang, L., Olwero, N., Valdivia, A., Wosu, A., Adrid, E., Andradi-Brown, D. A., Andriamalala, G., Ban, N. C., Bennett, N. J., Blythe, J., Cheng, S. H., Darling, E., De Nardo, M., Drury O'Neill, E., Epstein, G., Fidler, R. Y., Fisher, K., Geldmann, J., … Ahmadia, G. (2024). Introducing Elinor for monitoring the governance and management of area-based conservation. *Conservation biology: the journal of the Society for Conservation Biology, 38*(2), e14213. https://doi.org/.DOI: 10.1111/cobi.14213

Maheshwaran, T., Muthumarilakshmi, S., Suganya, K., Maheswari, B., & Girija, P. (2024). Securing E-Commerce Strategies With Cloud, Blockchain, AI, and ML. In *Strategies for E-Commerce Data Security: Cloud, Blockchain, AI, and Machine Learning* (pp. 470-500). IGI Global.

Maier, C., Laumer, S., & Eckhardt, A. (2015). Information technology as daily stressor: Pinning down the causes of burnout. *Journal of Business Economics*, *85*(4), 349–387. DOI: 10.1007/s11573-014-0759-8

Mamo, Y. A., Haji, J., Legesse, B., & Sassi, M. (2021). Challenges Of Agricultural Cooperatives And Their Impact On Profitability Of Potato Growers In Eastern Ethiopia. *Review of Agricultural and Applied Economics*, 24(1340-2021-1620), 62-77.

Mantere, S., & Ketokivi, M. (2013). Reasoning in Organization Science. *Academy of Management Review*, *38*(1), 70–89. DOI: 10.5465/amr.2011.0188

Manyanga, M., Pedzisa, T., & Hanyani-Mlambo, B. (2023). Adoption of agroecological intensification practices in Southern Africa: A scientific review. *Cogent Food & Agriculture*, *9*(1), 2261838. DOI: 10.1080/23311932.2023.2261838

Manyika, J., Lund, S., Chui, M., Bughin, J., Woetzel, J., Batra, P., & Sanghvi, S. (2017). Jobs lost, jobs gained: Workforce transitions in a time of automation. McKinsey Global Institute. https://www. mckinsey. com/featured - insights/future - of - work/jobs - lost - jobs - gained - what - the - future - of - work - will - mean - for - jobs - skills - and - wages

Maota, T., & Naidoo, V. (2025). The Impact of Artificial Intelligence on Human Resource Management: Barriers and Benefits. In *Sustainable Management Practices for Employee Retention and Recruitment* (pp. 311–342). IGI Global Scientific Publishing. DOI: 10.4018/979-8-3693-8679-8.ch016

Ma, Q., Yang, R., & Lin, Y. (2024). Dynamic role of clean energy and sustainable economic growth in coastal region: Novel observations from China. *Environmental Research*, *251*(Pt 2), 118659. DOI: 10.1016/j.envres.2024.118659 PMID: 38462089

Marcoulides, G. A. (1989). Measuring computer anxiety: The computer anxiety scale. *Educational and Psychological Measurement*, *49*(3), 733–739. DOI: 10.1177/0013164489049003028

Marr, B. (2020). *Artificial intelligence in practice: How 50 successful companies used AI and machine learning to solve problems*. Wiley.

Martins, H. S., & Guerreiro, S. L. (2021). Access Control Challenges in Enterprise Ecosystems: Blockchain-Based Technologies as an Opportunity for Enhanced Access Control. In I. Management Association (Ed.), *Research Anthology on Blockchain Technology in Business, Healthcare, Education, and Government* (pp. 503-528). IGI Global Scientific Publishing. https://doi.org/DOI: 10.4018/978-1-7998-5351-0.ch029

Mataba, L. D., & Komba, C. K. (2021). *Social protection in co-operatives: evidence from arusha women in business savings and credit cooperative society.*

Matits, P. (2022). Industry 4.0 to your industrial campus. *Synchroinfo Journal*, *8*(5), 26–33. DOI: 10.36724/2664-066X-2022-8-5-26-33

Matsui, T., Suzuki, K., Ando, K., Kitai, Y., Haga, C., Masuhara, N., & Kawakubo, S. (2022). A natural language processing model for supporting sustainable development goals: Translating semantics, visualizing nexus, and connecting stakeholders. *Sustainability Science*, *17*(3), 969–985. DOI: 10.1007/s11625-022-01093-3 PMID: 35136451

McBride, K., Kupi, M., & Bryson, J. J. (2021). Untangling Agile government: On the dual necessities of structure and agility. DOI: 10.31235/osf.io/qwjcx

McColl, K. (2023). The new reality of automation and labor disruption. *The Economist Intelligence Unit.*

McKinsey & Company. (2020). Diversity wins: How inclusion matters. https://www. mckinsey. com/featured - insights/diversity - and - inclusion/diversity - wins - how - inclusion - matters

McKinsey & Company. (2020). the internet of things: Catching up to an accelerating opportunity. https://www. mckinsey. com/business - functions/mckinsey - digital/our - insights/the - internet - of - things - catching - up - to - an - accelerating - opportunity

McKinsey & Company. (2020, February 12). *How companies are reskilling to address skill gaps | McKinsey*. Www.mckinsey.com. https://www.mckinsey.com/capabilities/people-and-organizational-performance/our-insights/beyond-hiring-how-companies-are-reskilling-to-address-talent-gaps

McKinsey & Company. (2021). The future of work after [McKinsey Global Institute.]. *COVID*, •••, 19.

McKinsey & Company. (2024). The state of AI in early 2024: Gen AI adoption spikes and and starts to generate value.

Mehta, S., Gupta, S. K., Aljohani, A. A., & Khayyat, M. (Eds.). (2024). *Impact and Potential of Machine Learning in the Metaverse*. IGI Global Scientific Publishing., DOI: 10.4018/979-8-3693-5762-0

Memon, M. A., Salleh, R., Mirza, M. Z., Cheah, J. H., Ting, H., Ahmad, M. S., & Tariq, A. (2021). Satisfaction matters: The relationships between HRM practices, work engagement and turnover intention. *International Journal of Manpower*, *42*(1), 21–50. DOI: 10.1108/IJM-04-2018-0127

Mezgár, I., & Pedone, G. (2021). Cloud-Based Manufacturing (CBM) Interoperability in Industry 4.0. In I. Management Association (Ed.), *Research Anthology on Cross-Industry Challenges of Industry 4.0* (pp. 540-567). IGI Global Scientific Publishing. https://doi.org/DOI: 10.4018/978-1-7998-8548-1.ch029

Mi, C., Chen, Y., Cheng, C.-S., Uwanyirigira, J. L., & Lin, C.-T. (2019). Exploring the Determinants of Hot Spring Tourism Customer Satisfaction: Causal Relationships Analysis Using ISM. *Sustainability (Basel)*, *11*(9), 2613. DOI: 10.3390/su11092613

Michalos, G., Karagiannis, P., Dimitropoulos, N., Andronas, D., & Makris, S. (2022). *Human Robot Collaboration in Industrial Environments*. Springer., DOI: 10.1007/978-3-030-78513-0_2

Milić, J. V. (2024). Sustainable Development Goals in Chemistry in Switzerland. *Chimia*, *78*(6), 369–371. DOI: 10.2533/chimia.2024.369 PMID: 38946407

Milojevic, I., & Davidow, S. (2018). The futures of work. *TEXT Specia*, *52*, 1–12.

Milousi, M., Souliotis, M., Arampatzis, G., & Papaefthimiou, S. (2019). Evaluating the Environmental Performance of Solar Energy Systems Through a Combined Life Cycle Assessment and Cost Analysis. *Sustainability (Basel)*, *11*(9), 2539. DOI: 10.3390/su11092539

Minos, D., Butzlaff, I., Demmler, K. M., & Rischke, R. (2016). Economic Growth, Climate Change, and Obesity. *Current Obesity Reports*, *5*(4), 441–448. DOI: 10.1007/s13679-016-0234-7 PMID: 27714641

Moalosi, R., & Rapitsenyane, Y. (Eds.). (2023). *African industrial design practice: perspectives on ubuntu philosophy*. Taylor & Francis. DOI: 10.4324/9781003270249

Mohamed, M. (2018). Challenges and benefits of industry 4.0: An overview. *International Journal of Supply and Operations Management*, *5*(3), 256–265.

Mohammed, A. A., & AL-Abrrow, H. (2023). The influence of empowering as well as transformational leadership on organizational performance and innovation: The mediating role of shared leadership and moderating role of organizational culture in the Iraqi healthcare sector. *The International Journal of Organizational Analysis*, *31*(7), 3532–3552. DOI: 10.1108/IJOA-08-2022-3380

Mohammed, A. J., & Ghebreyesus, T. A. (2018). Healthy living, well-being and the sustainable development goals. *Bulletin of the World Health Organization*, *96*(9), 590–590A. DOI: 10.2471/BLT.18.222042 PMID: 30262935

Mohanan, P. V. (2025). Artificial Intelligence and Biological Sciences.

Mohapatra, H., Pramanik, S., & Ranjan Mishra, S. (2025). Revolutionizing software development: The transformative influence of machine learning integrated SDLC model. *Artificial Intelligence-Enhanced Software and Systems Engineering*, 41-69. DOI: 10.1007/978-3-031-88188-6_3

Moka, S. S. P., Pilla, S. M., & Radhika, S. (2020). Real-time density-based traffic surveillance system integrated with acoustic-based emergency vehicle detection. *2020 4th International Conference on Computer, Communication and Signal Processing (ICCCSP)*, 1–7. https://doi.org/DOI: 10.1109/ICCCSP49186.2020.9315209

Mokoena, O. P., Ntuli, T. S., Ramarumo, T., & Seeletse, S. M. (2023). Sustainability of Rural Small-Scale Farmers Using a Thematic Content-Fed Analytic Hierarchy Process. *Sustainability (Basel)*, *15*(15), 11983–11983. DOI: 10.3390/su151511983

Mokyr, J. & Strotz, RH (1998). The Second industrial revolution, 1870 - 1914. *Storia dell'ekonomia Mondiale,* 1 - 16.

Mokyr, J. (1990). *The Lever of Riches: Technological Creativity and Economic Progress*. Oxford University Press.

Mokyr, J. (1998). The Second Industrial Revolution, 1870 - 1914. In Floud, R., & Johnson, P. (Eds.), *the Cambridge Economic History of Modern Britain* (Vol. 2, pp. 1–29). Cambridge University Press.

Moon, S., & Lee, S. H. (2020). A strategy for sustainable development of cooperatives in developing countries: The success and failure case of agricultural cooperatives in musambira sector, Rwanda. *Sustainability (Basel)*, *12*(20), 86–132. DOI: 10.3390/su12208632

Moore, M. H. (1995). *Creating public value: Strategic management in government*. Harvard University Press.

Morrar, R., Arman, H., & Mousa, S. (2017). [eXtyles1.1]. The Fourth Industrial Revolution (Industry 4.0): A Social Innovation Perspective. *Technology Innovation Management Review*, *7*(11), 12–20. DOI: 10.22215/timreview/1117

Mosia, P., & Khalo, T. (2021). The Role of Agricultural Cooperatives in Promoting Sustainable Development in South Africa. *Sustainable Development*, *29*(2), 241–250.

Mubuli, A., Nigdeli, S. M., & Bekdaş, G. (2021). Control of Full-Size Frame Structures via Optimum Tuned Mass Dampers. [IJDIBE]. *International Journal of Digital Innovation in the Built Environment*, *10*(1), 47–61. DOI: 10.4018/IJDIBE.2021010104

Mueller, J. R., & Baum, B. (2011). The definitive guide to hiring right, Journal of applied. *Business Economics (Cleveland, Ohio)*, *12*(3), 140–153.

Müller, S. D., Konzag, H., Nielsen, J. A., & Sandholt, H. B. (2024). Digital transformation leadership competencies: A contingency approach. *International Journal of Information Management*, *75*, 102734. Advance online publication. DOI: 10.1016/j.ijinfomgt.2023.102734

Mundia, M., Nkonde, C., Simui, F., & Imasiku, J. (2022). Exploration of Agriculture Cooperative Enterprising Performance in Kabwe District, Zambia. *World Journal of Research and Review*, *1*(5), 1–5. DOI: 10.31871/WJRR.15.3.5

Murshed, M. (2024). Is digitalization essential for abating carbon emission growth in South Asia? *Heliyon, 10*(20), e39012. DOI: 10.1016/j.heliyon.2024.e39012 PMID: 39497997

Mushonga, M., Arun, T. G., & Marwa, N. W. (2019). The cooperative movement in South Africa: Can financial cooperatives become sustainable enterprises? *Strategic Change, 28*(4), 259–271. DOI: 10.1002/jsc.2268

Muzekenyi, M., Zuwarimwe, J., & Kilonzo, B. M. (2022). Utilizing small-scale commercial farming to enhance local economic development in South Africa. *Journal of Contemporary Management, 19*(2), 22–39. DOI: 10.35683/jcm20007.155

Nadler, D. A., & Tushman, M. L. (1997). *Competing by design: The power of organizational architecture.* Oxford University Press. DOI: 10.1093/acprof:oso/9780195099171.001.0001

Nagarsheth, S. H., & Sharma, S. N. (2021). The Impact of Fractional-Order Control on Blood Pressure Regulation. [IJEHMC]. *International Journal of E-Health and Medical Communications, 12*(3), 38–54. DOI: 10.4018/IJEHMC.20210501.oa3

Nagy, B., Doka, J., Racz, S., Szabo, G., Pelle, I., Czentye, J., Toka, L., & Sonkoly, B. (2019). Towards Human-Robot Collaboration: An Industry 4.0 VR Platform with Clouds Under the Hood. *International Conference on Network Protocols*, 1–2. DOI: 10.1109/ICNP.2019.8888107

Nahavandi, A. (2019). *The art and science of leadership* (7th ed.). Pearson Education.

Naidoo, V. (2025). Employee Engagement: A Key Catalyst Within an Organization During COVID - 19 and Beyond the Pandemic - The South African Perspective. In T. Kapoor, A. Singh, K. Sahay, H. Kaur, & R. Verma (Eds.), Sustainable Management Practices for Employee Retention and Recruitment (pp. 127 - 150). IGI Global Scientific Publishing. . ch007DOI: 10.4018/979-8-3693-8679-8

Naidoo, V. (2020). *Creativity and Innovation for Entrepreneurs in the Circular Economy, IGI Global, Handbook of Research on Entrepreneurship Development and Opportunities in Circular Economy* (Baporikar, N., Ed.).

Nakayiso, E., & Andrew, N. (2023). A historical review on the global evolution, benefits, challenges and performance of cooperatives. *Journal of the Humanities and Social Sciences, 8*(1), 51.

Nasir, A. W., Kasireddy, I., & Singh, A. K. (2020). Fractional-Order-Based Inferential Control: A Case Study for the Level Control of Three Interacting Tank Processes. In Matouk, A. (Ed.), *Advanced Applications of Fractional Differential Operators to Science and Technology* (pp. 322–341). IGI Global Scientific Publishing., DOI: 10.4018/978-1-7998-3122-8.ch014

National Academy of Medicine, & the National Academies of Sciences, Engineering, and Medicine. Health and Medicine Division; Board on Health Sciences Policy; Committee on Creating a Framework for Emerging Science, Technology, and Innovation in Health and Medicine (2023). Toward Equitable Innovation in Health and Medicine: A Framework. Washington (DC): National Academies Press (US); 2023 Aug 22. 3, The Innovation Life Cycle in Health and Medicine and the Challenge of Equity. Available from: https://www.ncbi.nlm.nih.gov/books/NBK599705/

National Academy of Medicine, & the National Academies of Sciences, Engineering, and Medicine. Health and Medicine Division; Board on Health Sciences Policy; Committee on Creating a Framework for Emerging Science, Technology, and Innovation in Health and Medicine (2023). Toward Equitable Innovation in Health and Medicine: A Framework. Washington (DC): National Academies Press (US); 2023 Aug 22. Available from: https://www.ncbi.nlm.nih.gov/books/NBK599708/ DOI: 10.17226/27184

Nazir, M., Zaman, K., Khan, S., Nassani, A. A., Khan, H. U. R., & Haffar, M. (2023). Economic growth and carbon emissions in Pakistan: The effects of China's Logistics Industry. *Environmental Science and Pollution Research International*, *30*(18), 53778–53795. DOI: 10.1007/s11356-023-26150-x PMID: 36867335

Ndauka, F., & Matotola, S. (2023). The role of technology adopted by Agricultural Marketing Cooperative Society (AMCOS) on performance of grapevine farmers in Tanzania. Cogent Business & Management, 10(2), 2226420.75.

Nelson, D. L., & Kletke, M. G. (1990). Individual adjustment during technological innovation: A research framework. *Behaviour & Information Technology*, *9*(4), 257–271. DOI: 10.1080/01449299008924242

Ngonghala, C. N., Enright, H., Prosper, O., & Zhao, R. (2024). Modeling the synergistic interplay between malaria dynamics and economic growth. *Mathematical Biosciences*, *372*, 109189. DOI: 10.1016/j.mbs.2024.109189 PMID: 38580079

Nguyen, L. L. H., Khuu, D. T., Halibas, A., & Nguyen, T. Q. (2024). Factors that influence the intention of smallholder rice farmers to adopt cleaner production practices: An empirical study of precision agriculture adoption. *Evaluation Review*, *48*(4), 692–735. DOI: 10.1177/0193841X231200775

Nguyen, V. Q., Turner, N., Barling, J., Axtell, C. M., & Davies, S. (2023). Reconciling general transformational leadership and safety-specific transformational leadership: A paradox perspective. *Journal of Safety Research*, *84*, 435–447. DOI: 10.1016/j.jsr.2022.12.006 PMID: 36868673

Niedenthal, P. M., & Brauer, M. (2012). Social functionality of human emotion. *Annual Review of Psychology*, *63*(1), 259–285. DOI: 10.1146/annurev.psych.121208.131605

Niekerk, V. (1988). *Co-operative theory and practice*. Promedia Publications.

Nilashi, M., Ali Abumalloh, R., Mohd, S., Nurlaili Farhana Syed Azhar, S., Samad, S., Hang Thi, H., Alghamdi, O. A., & Alghamdi, A. (2023). COVID-19 and sustainable development goals: A bibliometric analysis and SWOT analysis in Malaysian context. *Telematics and Informatics*, *76*, 101923. DOI: 10.1016/j.tele.2022.101923 PMID: 36510580

Nissoul, S., Pacaux-Lemoine, M.-P., & Chaabane, S. (n.d.). *Exploring Human-Machine Relations and Approaches for Task Management in Dynamic Environments: A Comprehensive Literature Review*. 263–273. .DOI: 10.1007/978-3-031-53445-4_22

Niu, W., & Fan, M. (2024). Control and Research of Computer Virus by Multimedia Technology. [IJISSCM]. *International Journal of Information Systems and Supply Chain Management*, *17*(1), 1–17. DOI. 10.4018/IJISSCM.333896

Noe, R. A. (2017). *Employee Training and Development*. McGraw - Hill Education.

Nof, S. Y., & Chen, F. F. (2007). Collaborative control theory for e-Work, e-Production, and e-Service. *Computers & Industrial Engineering*, *53*(2), 252–268.

Norman, D. A. (2013). *The design of everyday things* (Revised and expanded edition). Basic Books.

OECD. (2019). *Preparing for the future of work.* OECD Publishing.

OECD. (2021). *OECD skills outlook 2021: Learning for life.* OECD Publishing.

Öğüt, A. (2007). *Bilgi çağında yönetim.* Çizgi Kitapevi.

Oreg, S. (2006). Personality, context, and resistance to organizational change. *European Journal of Work and Organizational Psychology*, *15*(1), 73–101. DOI: 10.1080/13594320500451247

Orlikowski, W. J. (2000). Using technology and constituting structures: A practice lens for studying technology in organizations. *Organization Science*, *11*(4), 404–428. DOI: 10.1287/orsc.11.4.404.14600

Ortmann, G. F., & King, R. P. (2006). Small-scale farmers in South Africa: Can agricultural cooperatives facilitate access to input and product markets? 170(1), 2016-139140.

Othman, M. H., & Razali, R. (2013). Key contributing factors towards successful electronic government systems interoperability. *2013 International Conference on Research and Innovation in Information Systems (ICRIIS)*, 302-307. DOI: 10.1109/ICRIIS.2013.6716726

Özbalcı, A. A., & Çevik, O. (2020). *Sağlık çalışanlarının bilişim sistemlerindeki değişimlere karşı gösterdikleri direnç ile demografik özellikleri arasındaki ilişki,* in *Current Debates on Social Sciences 5, Human Studies.* (Ed.s: Salih Çeçen, Zeynel Karacagil, Ali Osman Tiro, Ömer Kahya & Şafak Bozgun). Bilgin Kültür Sanat Yayın Dağıtım Pazarlama Ltd. Şti.

Ozturk, I., Alqassimi, O., & Ullah, S. (2024). Digitalization and SMEs development in the context of sustainable development: A China perspective. *Heliyon*, *10*(6), e27936. DOI: 10.1016/j.heliyon.2024.e27936 PMID: 38510016

Palau-Pinyana, E., Llach, J., & Bagur-Femenías, L. (2023). Mapping enablers for SDG implementation in the private sector: a systematic literature review and research agenda. *Management Review Quarterly*, 1–30. Advance online publication. https://doi.org/.DOI: 10.1007/s11301-023-00341-9

Pandey, A., & Joseph, J. (2025). Human-Machine Interaction: Ethical and Legal Considerations. In Chakraborty, S. (Ed.), *Exploring AI Implications on Law, Governance, and Industry* (pp. 177–200). IGI Global Scientific Publishing., DOI: 10.4018/979-8-3373-3384-7.ch007

Pandey, D., Muniandi, B., Pandey, B. K., & George, A. (Eds.). (2025). *Interdisciplinary Approaches to AI, Internet of Everything, and Machine Learning.* IGI Global Scientific Publishing., DOI: 10.4018/979-8-3373-1032-9

Pansera, M., & Rizzi, F. (2020). Furbish or perish: Italian social cooperatives at a crossroads. *Organization*, *27*(1), 17–35. DOI: 10.1177/1350508418760980

Pan, X., Guo, Y., & Men, A. (2010). Traffic surveillance system for vehicle flow detection. *2010 Second International Conference on Computer Modeling and Simulation*, 314–318. https://doi.org/DOI: 10.1109/ICCMS.2010.75

Paramarthalingam, A., Janarthanan, A., Arivunambi, A., Ariyangavu, S. S., Senthamaraikannan, H., & Ganapathy, R. (2025). AI-Powered Virtual Mouse Control Through Hand Gestures With Computer Vision. In Mobo, F. (Ed.), *Harnessing AI in Geospatial Technology for Environmental Monitoring and Management* (pp. 61–76). IGI Global Scientific Publishing., DOI: 10.4018/979-8-3693-8104-5.ch003

Parasuraman, R., & Riley, V. (1997). Humans and automation: Use, misuse, disuse, abuse. *Human Factors, 39*(2), 230–253. DOI: 10.1518/001872097778543886

Parikh, P., Diep, L., Hofmann, P., Tomei, J., Campos, L. C., Teh, T. H., Mulugetta, Y., Milligan, B., & Lakhanpaul, M. (2021). Synergies and trade-offs between sanitation and the sustainable development goals. *UCL Open Environment, 3*, e016. DOI: 10.14324/111.444/ucloe.000016 PMID: 37228800

Park, H., Kim, J., Kim, M., Yoon, J., Hwang, K., Park, J., & So, J. (2025). Traffic crash countermeasure recommendations using deep neural network: A decision support tool for traffic safety engineer. *IEEE Access : Practical Innovations, Open Solutions, 13*, 53718–53730. DOI: 10.1109/ACCESS.2025.3543641

Pei, J. (2022). Approaches Toward Building the Digital Enterprise and Sustainable Economic Development: The Moderating Role of Sustainability. *Frontiers in Psychology, 13*, 835602. DOI: 10.3389/fpsyg.2022.835602 PMID: 35360590

Peiperl, M. (2005). *Resistance to change*, in *The Blackwell encyclopedia of management. Organizational behaviour*, (Ed.s: N. Nicholson, P. G. Audia, and M. M. Pillutula), 348 - 349. Blackwell.

Peng, W. (2025). Shaping sustainable perceptions: The role of metaphors in Olympic news discourse. *PLoS One, 20*(1), e0317380. DOI: 10.1371/journal.pone.0317380 PMID: 39804915

Pereira, V., Hadjielias, E., Christofi, M., & Vrontis, D. (2023). A systematic literature review on the impact of artificial intelligence on workplace outcomes: A multi - process perspective. *Human Resource Management Review, 33*(1), 100857. DOI: 10.1016/j.hrmr.2021.100857

Pfeffer, J. (1998). *The human equation: Building profits by putting people first.* Harvard Business School Press.

Phaal, R., Farrukh, C., & Probert, D. (2006). Technology management tools: Concept, development and application. *Technovation, 26*(3), 336–344. DOI: 10.1016/j.technovation.2005.02.001

Pillai, R. (2020). Digital learning trends in the workplace: Personalization, gamification, and immersive technologies. *Journal of Workplace Learning, 32*(3), 181–198. DOI: 10. 1108/JWL - 01 - 2020 - 0010

Ponmalar, A., Vijayakumar, K., Lakshmipriya, C., Karthikeyan, M., Gracelin Sheena, B., Beslin Pajila, P. J., & Siva Subramanian, R. (2024). Meta-Heuristics and Machine Learning Applications in Complex Systems.

Ponnada, V. T. (Ed.). (2026). *Innovative Machine Learning Applications in the Aerospace Industry*. IGI Global Scientific Publishing., DOI: 10.4018/979-8-3693-7525-9

Ponnusamy, P. N., Monickaraj, V. P., & Periyathambi, E. (2022). Efficient intrusion detection and prevention model in cloud environment using Sgd-LSTM and C2HA. *Studies in Informatics and Control, 31*(2), 95–104. DOI: 10.24846/v31i2y202209

Ponnuviji, N. P., Nirmala, G., Kokila, M. S., & Priyadharshini, S. I. (2025). Optimizing Image Retrieval in Cloud Servers with TN-AGW: A Secure and Efficient Approach. *Journal of The Institution of Engineers (India): Series B, 106*(2), 459-473.

Ponnuviji, N. P., Sumaiya Begum, A., Gnanasekaran, T., Rajendiran, M., & Shenbaga Ezhil, S. (2023, April). Leveraging Random Forest for Intelligent IoT Systems in Industrial Environments. In *International Conference on Internet of Things* (pp. 46-57). Cham: Springer Nature Switzerland.

Popescu, C. (Ed.). (2025). *Building Inclusive Global Knowledge Societies for Sustainable Development.* IGI Global., DOI: 10.4018/979-8-3693-6298-3

Popescu, C. R. G. (2019). Corporate Social Responsibility, Corporate Governance and Business Performance: Limits and Challenges Imposed by the Implementation of Directive 2013/34/EU in Romania. *Sustainability (Basel), 11*(19), 5146. DOI: 10.3390/su11195146

Popescu, C. R. G. (2020). Sustainability Assessment: Does the OECD/G20 Inclusive Framework for BEPS (Base Erosion and Profit Shifting Project) Put an End to Disputes Over The Recognition and Measurement of Intellectual Capital? *Sustainability (Basel), 12*(23), 10004. DOI: 10.3390/su122310004

Popescu, C. R. G. (2025a). Editorial: Durable, Inclusive, Sustainable Economic Growth and Challenge. *Journal of Risk and Financial Management, 18*(5), 257. DOI: 10.3390/jrfm18050257

Popescu, C. R. G. (2025b). Preface to the Special Issue on "Recent Advances in Business and Industry: Mathematical Analysis, Sustainability Assessment Instruments and Methods". *Mathematics, 13*(7), 1207. DOI: 10.3390/math13071207

Popescu, C. R. G., & Popescu, G. N. (2019). An Exploratory Study Based on a Questionnaire Concerning Green and Sustainable Finance, Corporate Social Responsibility, and Performance: Evidence from the Romanian Business Environment. *Journal of Risk and Financial Management, 12*(4), 162. DOI: 10.3390/jrfm12040162

Porter, M. E., & Heppelmann, J. E. (2014). How smart, connected products are transforming competition. *Harvard Business Review, 92*(11), 64–88.

Prisecaru, P. (2017). The challenges of the industry 4.0. *Global Economic Observer, 5*(1), 66–72.

Pullagura, L., Brahma, B., Kumari, N. V., Ravikumar, L., Katta, S. K. G., & Chiwariro, R. (2024). *Industry 4.0 Design Principles, Technologies, and Applications.* 357–388. .DOI: 10.1201/9781003581963-11

PwC. (2020). Seeing is believing: How virtual reality and augmented reality are transforming business and the economy. PricewaterhouseCoopers. https://www. pwc. com/gx/en/technology/seeing - is - believing. html

PwC. (2022). *What does virtual reality and the metaverse mean for training?* PricewaterhouseCoopers.

Qaim, M. (2020). Role of new plant breeding technologies for food security and sustainable agricultural development. *Applied Economic Perspectives and Policy*, *42*(2), 129–150. DOI: 10.1002/aepp.13044

Qian, L., Luo, Z., Du, Y., & Guo, L. (2009, December). Cloud computing: An overview. In *IEEE international conference on cloud computing* (pp. 626–631). Springer Berlin Heidelberg.

Qi, S., Huang, D., Ma, Q., & Zhou, M. (2023). Factors Influencing Sustainable Development Literacy among Engineering Undergraduates in China: Based on the College Impact Model. *International Journal of Environmental Research and Public Health*, *20*(2), 1249. DOI: 10.3390/ijerph20021249 PMID: 36673998

Qu, X. (2016). Journal of Materials Chemistry B top picks web collection: "seeing the unseen: advances in bioimaging and biosensors". *Journal of Materials Chemistry. B*, *4*(26), 4500–4501. DOI: 10.1039/C6TB90078J PMID: 32263391

Qu, X. (2018). Journal of Materials Chemistry B Editor's choice web collection: "Seeing the unseen updated: advances in biosensing". *Journal of Materials Chemistry. B*, *6*(19), 2922–2923. DOI: 10.1039/C8TB90055H PMID: 32254329

Raghav, Y. Y., & Kait, R. (2024). Edge computing empowering distributed computing at the edge. In *Emerging Trends in Cloud Computing Analytics, Scalability, and Service Models* (pp. 67-83). IGI Global Scientific Publishing. DOI: 10.4018/979-8-3693-0900-1.ch003

Rahman, M. M., Hasan, M. J., Chandra Deb, B., Rahman, M. S., & Kabir, A. S. (2023). The effect of social media entrepreneurship on sustainable development: Evidence from online clothing shops in Bangladesh. *Heliyon*, *9*(9), e19397. DOI: 10.1016/j.heliyon.2023.e19397 PMID: 37662716

Rainnie, A. & Dean, M. (2020). Industry 4.0 and the future of quality work in the global digital economy. *Labour & Industry: A journal of the social and economic relations of work, 30* (1), 16 - 33.

Raj, P., Mohapatra, H., Ranjan, R., & Maithili, P. (2025). Innovative strategies for sustainable waste disposal. *Leveraging Urban Computing for Sustainable Urban Development*, 117-144. DOI: 10.4018/979-8-3373-0761-9.ch006

Rajaram, C., & Kumar, R. P. (2021). Numerical Modeling of Near Fault Seismic Ground Motion for Denali Fault. [IJGEE]. *International Journal of Geotechnical Earthquake Engineering*, *12*(2), 1–22. DOI: 10.4018/IJGEE.2021070104

Ramakuri, S. K., Chithaluru, P., & Kumar, S. (2019). Eyeblink Robot Control Using Brain-Computer Interface for Healthcare Applications. [IJMDWTFE]. *International Journal of Mobile Devices, Wearable Technology, and Flexible Electronics*, *10*(2), 38–50. DOI: 10.4018/IJMDWTFE.2019070103

Raman, R., Sreenivasan, A., Kulkarni, N. V., Suresh, M., & Nedungadi, P. (2025). Analyzing the contributions of biofuels, biomass, and bioenergy to sustainable development goals. *iScience*, *28*(4), 112157. DOI: 10.1016/j.isci.2025.112157 PMID: 40230525

Ranawaka, A., Alahakoon, D., Sun, Y., & Hewapathirana, K. (2024). Leveraging the Synergy of Digital Twins and Artificial Intelligence for Sustainable Power Grids: A Scoping Review. *Energies*, *17*(21), 5342. DOI: 10.3390/en17215342

Raniga, T. (2021). Role of economic development cooperatives in improving the livelihoods of women in Gauteng, South Africa. *Development Southern Africa*, *38*(4), 591–606. DOI: 10.1080/0376835X.2021.1906628

Ranjbari, M., Shams Esfandabadi, Z., Zanetti, M. C., Scagnelli, S. D., Siebers, P. O., Aghbashlo, M., Peng, W., Quatraro, F., & Tabatabaei, M. (2021). Three pillars of sustainability in the wake of COVID-19: A systematic review and future research agenda for sustainable development. *Journal of Cleaner Production*, *297*, 126660. DOI: 10.1016/j.jclepro.2021.126660 PMID: 34785869

Rasheed, M., & Jianhua, L. (2023). Unleashing the power of innovation promoters for sustainable economic growth: A global perspective. *Environmental Science and Pollution Research International*, *30*(45), 100979–100993. DOI: 10.1007/s11356-023-29313-y PMID: 37642909

Rawashdeh, A. (2023). The consequences of artificial intelligence: An investigation into the impact of AI on job displacement in accounting. *Journal of Science and Technology Policy Management*. .DOI: 10.1108/JSTPM-02-2023-0030

Raygoza-L., M. E., Orduño-Osuna, J. H., Mercado-Herrera, A., Jimenez-Sanchez, R., & Murrieta-Rico, F. N. (Eds.). (2025). *Exploring Psychology, Social Innovation and Advanced Applications of Machine Learning*. IGI Global Scientific Publishing., DOI: 10.4018/979-8-3693-6910-4

Reddy, A. S., Ashraf, M., Manideepak, T., & Mohapatra, H. (2025). A comprehensive analysis of governance and regulatory frameworks for cloud-based initiatives for smart cities. *Organizational Sociology in the Digital Age*, 127-146. DOI: 10.4018/979-8-3693-7398-9.ch008

Reddy, C. K. K., Reddy, A. V., Doss, S., & Priyanka, K. (2024). Human-Machine Collaboration and Emotional Intelligence in Industry 6.0: Concepts, Challenges, and Future Directions. In Nijalingappa, P., Gururaj, T., Goyal, S., Shukla, V., & Bruno, A. (Eds.), *Examining the Metaverse in Healthcare: Opportunities, Challenges, and Future Directions* (pp. 221–246). IGI Global Scientific Publishing., DOI: 10.4018/979-8-3693-1515-6.ch009

Rehman Khan, S. A., Ahmad, Z., Sheikh, A. A., & Yu, Z. (2022). Digital transformation, smart technologies, and eco-innovation are paving the way toward sustainable supply chain performance. *Science Progress*, *105*(4), 368504221145648. DOI: 10.1177/00368504221145648 PMID: 36573795

Rehmani, M., Farheen, N., Khokhar, M. N., Khalid, A., Dalain, A. F., & Irshad, H. (2023). How Does Transformational Leadership Stimulate Employee Innovative Behavior? A Moderated Mediation Analysis. *SAGE Open*, *13*(3), 21582440231198401. Advance online publication. DOI: 10.1177/21582440231198401

Rena, R. (2021). Managing Cooperatives in South Africa: An Economic Perspective. *Cooperatives in the Global Economy*, 175-191

Research Features. (2023). Unlocking the future of leadership: Navigating human-machine collaboration in the age of machine. https://researchfeatures.com/unlocking-future leadership-human-machine-collaboration-machine-learning/

Rezazadegan, R., & Sharifzadeh, M. (2022). Applications of artificial intelligence and big data in industry 4.0 technologies. *Industry 4.0 Vision for Energy and Materials: Enabling Technologies and Case Studies*, 121-158.

Ridder, H. (2012). Rezensionen: Case study research. Design and methods. *German Journal of Human Resource Management: Zeitschrift für Personalforschung*, *26*(1), 93–95. DOI: 10.1177/239700221202600101

Rifkin, J. (2011). *The Third Industrial Revolution: How Lateral Power Is Transforming Energy, the Economy, and the World.* Palgrave Macmillan.

Risal, N. (2020). Role of Cooperatives for Sustainable Livelihood in Bagmati Province, Nepal. *The International Research Journal of Management Science*, *5*(1), 75–92.

Rizzo, G., Migliore, G., Schifani, G., & Vecchio, R. (2024). Key factors influencing farmers' adoption of sustainable innovations: A systematic literature review and research agenda. *Organic Agriculture*, *14*(1), 57–84. DOI: 10.1007/s13165-023-00440-7

Robert, L. P., Pierce, C., Marquis, L., Kim, S. & Alahmad, R. (2020). Designing fair AI for managing employees in organizations: A review, critique, and design agenda. *Human - Computer Interaction, 35* (5 - 6), 545 - 575. .DOI: 10.1080/07370024.2020.1735391

Rogan, M., & Alfers, L. (2020). Employment and Poverty Alleviation in South Africa: The Role of Cooperatives. *Development Southern Africa*, *37*(4), 549–564.

Rogers, E. M. (1983). *Diffusion of Innovations* (3rd ed.). The Free Press.

Romero, D., Bernus, P., Noran, O., Stahre, J., & Fast-Berglund, Å. (2016). The Operator 4.0: Human Cyber-Physical Systems & Adaptive Automation Towards Human-Automation Symbiosis Work Systems. *IFIP Advances in Information and Communication Technology*, *488*, 677–686. DOI: 10.1007/978-3-319-51133-7_80

Rossignoli, C., Mauro, G., & Rocco, A. (2016). Introducing and discussing information and technology management for organizational innovation and change. *Springer International Publishing.*, *13*, 1–7. DOI: 10.1007/978-3-319-22921-8_1

Roundtable on Environmental Health Sciences, Research, and Medicine; Board on Population Health and Public Health Practice; Institute of Medicine (2013). Public Health Linkages with Sustainability: Workshop Summary. Washington (DC): National Academies Press (US); 2013 Jul 19. 2, Overview Of The Links Between Sustainability And Human Health. Available from: https://www.ncbi.nlm.nih.gov/books/NBK202304/

Rožman, M., Tominc, P., & Milfelner, B. (2023). Maximizing employee engagement through artificial intelligent organizational culture in the context of leadership and training of employees: Testing linear and non - linear relationships. *Cogent Business & Management*, *10*(2), 1–30. DOI: 10.1080/23311975.2023.2248732

Rübmann, M., Lorenz, M., Gerbert, P., Waldner, M., Justus, J., Engel, P., & Harnisch, M. (2015). Industry 4.0: The future of productivity and growth in manufacturing industries. *Boston Consulting Group, 9* (1), 54 - 89.

Ruth, J. A., Mahesh, V. G., Visalakshi, P., Uma, R., & Meenakshi, A. (Eds.). (2024). *Machine Learning and Cryptographic Solutions for Data Protection and Network Security*. IGI Global Scientific Publishing., DOI: 10.4018/979-8-3693-4159-9

Sacchetti, S., & Tortia, E. C. (2021). Governing cooperatives in the context of individual motives. *International Journal of Social Economics*, *48*(2), 181–203. DOI: 10.1108/IJSE-09-2019-0579

Saeedikiya, M., Salunke, S., & Kowalkiewicz, M. (2025). The nexus of digital transformation and innovation: A multilevel framework and investigation agenda. *Journal of Innovation and Knowledge*, *10*(1), 100640. Advance online publication. DOI: 10.1016/j.jik.2024.100640

Şahin, Y. L., & Çoklar, A. N. (2009). Social networking users' views on technology and the determination of technostress levels. *Procedia: Social and Behavioral Sciences*, *1*(1), 1437–1442. DOI: 10.1016/j.sbspro.2009.01.253

Sahoo, A. K., Meher, S. K., Panda, T. C., Sahu, S., Begum, R., & Barik, N. C. (2020). Critical review on cooperative societies in agricultural development in India. *Current Journal of Applied science and technology, 39*(22), 114-121.

Said, Y., Alassaf, Y., Ghodhbani, R., Alsariera, Y. A., Saidani, T., Rhaiem, O. B., Makhdoum, M. K., & Hleili, M. (2024). AI-based helmet violation detection for traffic management system. *Computer Modeling in Engineering & Sciences*, *141*(1), 733–749. DOI: 10.32604/cmes.2024.052369

Sajil, C. K., & Nair, A. S. (2020). Role of Acoustic Properties in Biomedical Active Noise Control. [IJBCE]. *International Journal of Biomedical and Clinical Engineering*, *9*(1), 48–60. DOI: 10.4018/IJBCE.2020010104

Salanova, M., Llorens, S., & Cifre, E. (2013). The dark side of technologies: Technostress among users of information and communication technologies. *International Journal of Psychology*, *48*(3), 422–436. DOI: 10.1080/00207594.2012.680460

Salin, E. D., & Winston, P. H. (1992). Machine Learning and Artificial Intelligence. *Analytical Chemistry*, *64*(1).

Salisbury, D. F. (1996). *Five technologies for educational change*. Educational Technology Publications.

Saluja, A., & Mongia, A. (2025). Human-Machine Collaboration: Augmenting Human Abilities With Robotic Assistance in the Workplace. In Kaur, J. (Ed.), *Technological Enhancements for Improving Employee Performance, Safety, and Well-Being* (pp. 145–170). IGI Global Scientific Publishing., DOI: 10.4018/979-8-3693-9631-5.ch007

Sandini, G., Sciutti, A., & Morasso, P. (2024). Collaborative Robots with Cognitive Capabilities for Industry 4.0 and beyond. *AI*, *5*(4), 1858–1869. DOI: 10.3390/ai5040092

Sanjiv, S. T. E., & Baskar, M. (2024). YOLOv4-powered traffic intelligence: From detection to optimization. *2024 IEEE International Conference on Information Technology, Electronics and Intelligent Communication Systems (ICITEICS)*, 1–4. https://doi.org/DOI: 10.1109/ICITEICS61368.2024.10624983

SANS Institute. (2021). Security awareness report: Managing human cyber security risk. https://www. sans. org/white - papers/security - awareness - report/

Santarius, T., Bieser, J. C. T., Frick, V., Höjer, M., Gossen, M., Hilty, L. M., Kern, E., Pohl, J., Rohde, F., & Lange, S. (2023). Digital sufficiency: Conceptual considerations for ICTs on a finite planet. *Annales des Télécommunications, 78*(5-6), 277–295. DOI: 10.1007/s12243-022-00914-x PMID: 37593439

Santiteerakul, S., Sopadang, A., Yaibuathet Tippayawong, K., & Tamvimol, K. (2020). The role of smart technology in sustainable agriculture: A case study of wangree plant factory. *Sustainability (Basel), 12*(11), 4640. DOI: 10.3390/su12114640

Saraç, H. (2022). Otomasyonun yükselişi. Kitlesel işsizlik mi? Yeni istihdam mı? İşgücü piyasasına etkisine dair bir değerlendirme. *Çalışma İlişkileri Dergisi, 13* (2). 55 - 76.

Saraçoğlu, M. & Duran, C. (2009). Yaratıcı girişimcinin gelişiminde çevrenin rolü. *Elektronik Sosyal Bilimler Dergisi,* 131 - 151.

Sari, F. T., Nigdeli, S. M., Bekdaş, G., & Geem, Z. W. (2024). A Review of Optimization of Structural Control Systems: Usage of Metaheuristics in Structural Control. In Ahmad, A., & Camp, C. (Eds.), *Advanced Optimization Applications in Engineering* (pp. 152–171). IGI Global Scientific Publishing., DOI: 10.4018/979-8-3693-2161-4.ch007

Sasidharan, S. (2022). Technostress in the workplace: A social network perspective. *Information Technology & People, 35*(4), 1219–1238. DOI: 10.1108/ITP-09-2020-0649

Saunders, B., Kisanga, P., Carvalho, G. H. S., & Woungang, I. (2024). *A Graph Convolutional Networks-Based DDoS Detection Model.* https://doi.org/.DOI: 10.1109/SySCon61195.2024.10553611

Schmitz, M. L., Antonietti, C., Consoli, T., Cattaneo, A., Gonon, P., & Petko, D. (2023). Transformational leadership for technology integration in schools: Empowering teachers to use technology in a more demanding way. *Computers & Education, 204,* 104880. Advance online publication. DOI: 10.1016/j.compedu.2023.104880

Schoenherr, J. R. (2022). *Ethical artificial intelligence from popular to cognitive science: trust in the age of entanglement.* Routledge. DOI: 10.4324/9781003143284

Schwab, K. (2017). *The Fourth Industrial Revolution.* Crown Business.

Schwettmann, J. (2022). *Cooperatives in the social and solidarity economy: Sustainable development and decent work in Africa's informal economy* (Doctoral dissertation, Manchester Metropolitan University).

Searle, R. H. (2006). New technology: The potential impact of surveillance techniques in Recruitment practices. *Personnel Review, 35*(3), 336–351. DOI: 10.1108/00483480610656720

Sebastian, I. M., Ross, J. W., Beath, C., Mocker, M., Moloney, K. G., & Fonstad, N. O. (2020). How big old companies navigate digital transformation. In *Strategic information management* (pp. 133–150). Routledge. DOI: 10.4324/9780429286797-6

Šebestová, J. D., & Popescu, C. R. G. (2022). Factors Influencing Investments into Human Resources to Support Company Performance. *Journal of Risk and Financial Management, 15*(1), 19. DOI: 10.3390/jrfm15010019

Seeniappan, K., Akram, C. M., Soundararajan, S., & Natrayan, L. (2023). Modelling and Development of Energy Systems Through Cyber Physical Systems With Optimising Interconnected With Control and Sensing Parameters. In T. R., S. Kaliappan, & C. Jegadheesan (Eds.), *Cyber-Physical Systems and Supporting Technologies for Industrial Automation* (pp. 280-295). IGI Global Scientific Publishing. https://doi.org/DOI: 10.4018/978-1-6684-9267-3.ch016

Sellare, J., Meemken, E. M., Kouamé, C., & Qaim, M. (2020). Do sustainability standards benefit smallholder farmers also when accounting for cooperative effects? Evidence from Côte d'Ivoire. *American Journal of Agricultural Economics*, *102*(2), 681–695. DOI: 10.1002/ajae.12015

Şen, A. T. (2024). Kamu çalışanlarının teknoloji yönetimi algılarının demografik değişkenler esasında analiz edilmesi: Kastamonu örneği. [Manisa Celal Bayar University Journal of Social Sciences]. *Manisa Celal Bayar Üniversitesi Sosyal Bilimler Dergisi*, *22*(4), 182–200. DOI: 10.18026/cbayarsos.1514391

Senior, B. (2002). *Organisational change*. Prentice Hall.

Serra, R., & Davidson, K. A. (2020). Selling Together: The Benefits of Cooperatives to Women Honey Producers in Ethiopia. *Journal of Agricultural Economics*, *72*(1), 202–223. DOI: 10.1111/1477-9552.12399

Shah, W. (2024). Human-AI Collaboration: Designing Intelligent Applications for Enhanced Productivity. *Journal of AI Range*, *1*(3), 37–48.

Sharma, V. K., Joshi, S. K., & Bora, A. (2025). Quantum Control of Networked Robotic Systems. In Gajbhiye, P., Kannan, H., Rodriguez, R., & Rojas-Méndez, J. (Eds.), *The Quantum AI Era of Neuromarketing* (pp. 357–368). IGI Global Scientific Publishing., DOI: 10.4018/979-8-3693-7673-7.ch015

Shawkat, S. A., & Thivagar, M. L. (2024). Fundamentals of Numerical. In Hamad, A., & Jha, S. (Eds.), *Coding Dimensions and the Power of Finite Element, Volume, and Difference Methods* (pp. 67–91). IGI Global Scientific Publishing., DOI: 10.4018/979-8-3693-3964-0.ch003

Shen, C. C., Yeh, C. C., & Lin, C. N. (2022). Using the perspective of business information technology technicians to explore how information technology affects business competitive advantage. *Technological Forecasting and Social Change*, *184*(1), 121973. DOI: 10.1016/j.techfore.2022.121973

Sherwani, F., Asad, M. M., & Ibrahim, B. S. K. K. (2020, March). Collaborative robots and industrial revolution 4.0 (ir 4.0). In *2020 international conference on emerging trends in smart technologies (ICETST)* (pp. 1-5). IEEE.

Sherwani, F., Asad, M. M., & Kader Ibrahim, B. S. K. (2020). *Collaborative Robots and Industrial Revolution 4.0 (IR 4.0)*. 1–5. .DOI: 10.1109/ICETST49965.2020.9080724

Shiomi, Y.. (2021). International analysis on social and personal determinants of traffic violations and accidents employing logistic regression with elastic net regularization. *IATSS Research*. Advance online publication. DOI: 10.1016/j.iatssr.2021.12.004

Shoeman, N. (2006). The Co-operative as an Appropriate Form of Enterprise for Black Economic Empowerment, Published thesis at the *University of Free State*.

Siau, K., & Wang, W. (2018). Building trust in artificial intelligence, machine learning, and robotics. *Cutter Business Technology Journal, 31*(2), 47–53.

Sibanda, M., & Ndlela, B. (2020). Agricultural exports and economic growth: Empirical evidence from South Africa. *Journal of African Trade, 7*(1–2), 20–34.

Siddiqui, D. A., & Sahar, N. (2019). The Impact of Training & Development and Communication on Employee Engagement. A Study of Banking Sector. *Business Management and Strategy, 10*(1), 23. DOI: 10.5296/bms.v10i1.14592

Sileshi, G. W., Stewart, Z. P., Odhong, J., Mhlanga, B., Amede, T., Aynekulu, E., ... & Snapp, S. (2025). A review of organic inputs to inform soil health advice for African smallholder farmers: localization matters. *npj Sustainable Agriculture, 3*(1), 20.

Silva, F. J. G., Campilho, R. D. S. G., & Sá, J. C. (2020). Digital transformation in manufacturing industries: Challenges, enablers and roadmap. *Procedia Manufacturing, 51*, 1183–1190.

Silvestri, S., Richard, M., Edward, B., Dharmesh, G., & Dannie, R. (2021). Going digital in agriculture: How radio and SMS can scale-up smallholder participation in legume-based sustainable agricultural intensification practices and technologies in Tanzania. *International Journal of Agricultural Sustainability, 19*(5-6), 583–594. DOI: 10.1080/14735903.2020.1750796

Simbi, C. H., Yao, F., Zhang, J., Tenaw, D., Sugira Murekezi, J., Magati, M. T., Hirwa, H., Al-Sakkaf, A. S., Kayiranga, A., & Peng, J. (2024). Balancing growth and preservation: Unravelling Africa's carbon-economic nexus through the environmental kuznets curve. *Heliyon, 10*(20), e39269. DOI: 10.1016/j.heliyon.2024.e39269 PMID: 39640728

Şimşek, T. (2025). Human-Machine Collaboration: How Do Employees Interact With Artificial Intelligence? In Demircioğlu, A. (Ed.), *AI Use in Social Sciences* (pp. 157–180). IGI Global Scientific Publishing., DOI: 10.4018/979-8-3373-2612-2.ch006

Singh, R. K., Berkvens, R., & Weyn, M. (2021). AgriFusion: An architecture for IoT and emerging technologies based on a precision agriculture survey. *IEEE Access : Practical Innovations, Open Solutions, 9*, 136253–136283. DOI: 10.1109/ACCESS.2021.3116814

Singh, S., & Tretten, P. (2020). *Operator 4.0 Within the Framework of Industry 4.0.* IGI Global., DOI: 10.4018/978-1-7998-3904-0.ch007

Smaniotto, C., Battistella, C., Brunelli, L., Ruscio, E., Agodi, A., Auxilia, F., Baccolini, V., Gelatti, U., Odone, A., Prato, R., Tardivo, S., Voglino, G., Valent, F., Brusaferro, S., Balzarini, F., Barchitta, M., Carli, A., Castelli, F., Coppola, C., & Sisi, S. (2020). Sustainable Development Goals and 2030 Agenda: Awareness, Knowledge and Attitudes in Nine Italian Universities, 2019. *International Journal of Environmental Research and Public Health, 17*(23), 8968. DOI: 10.3390/ijerph17238968 PMID: 33276530

Soori, M., Arezoo, B., & Dastres, R. (2023). Internet of things for smart factories in industry 4.0, a review. *Internet of Things and Cyber-Physical Systems, 3*, 192–204. DOI: 10.1016/j.iotcps.2023.04.006

Soumyalatha, S. G. H. (2016, May). Study of IoT: Understanding IoT architecture, applications, issues and challenges. In 1st International Conference on Innovations in Computing & Net-working (ICICN16), CSE, RRCE. *International Journal of Advanced Networking & Applications*, *478*, 14.

Spanaki, K., Sivarajah, U., Fakhimi, M., Despoudi, S., & Irani, Z. (2022). Disruptive technologies in agricultural operations: A systematic review of AI-driven AgriTech research. *Annals of Operations Research*, *308*(1), 491–524. DOI: 10.1007/s10479-020-03922-z

Sreenivasan, A., Suresh, M., Nedungadi, P., & R, R. R. (2023). Mapping analytical hierarchy process research to sustainable development goals: Bibliometric and social network analysis. *Heliyon*, *9*(8), e19077. DOI: 10.1016/j.heliyon.2023.e19077 PMID: 37636443

Statistics South Africa. (2024). *Agricultural Statistics*. Government of South Africa.

Stearns, P. N. (2013). *The Industrial Revolution in World History* (4th ed.). Westview Press.

tefan, M. (2022). Agile approaches to developing E-BusinEss solutions in a secure cyber environment. *Proceedings of the International Conference on Business Excellence*, *16*(1), 239-250. DOI: 10.2478/picbe-2022-0023

Stolterman, E., & Fors, A. (2004). *Information technology and the good life*, in *Information systems research. IFIP ınternational federation for ınformation processing*, (Ed.s: B. Kaplan, D. P. Truex, D. Wastell, A. T. Wood - Harper, & J. I. DeGross), 687 - 692. Springer. DOI: 10.1007/1-4020-8095-6_45

Sundaram, A. & L. V., N. (2024). Collaborate and Energize: Human-Machine Synergy in Advanced Energy Industries (AEI) Within the Framework of Industry 5.0. In N. Kumar, S. Pal, P. Agarwal, J. Rosak-Szyrocka, & V. Jain (Eds.), *Human-Machine Collaboration and Emotional Intelligence in Industry 5.0* (pp. 71-89). IGI Global Scientific Publishing. https://doi.org/.DOI: 10.4018/979-8-3693-6806-0.ch004

Susskind, R., & Susskind, D. (2015). *The future of the professions: How technology will transform the work of human experts*. Oxford University Press. DOI: 10.1093/oso/9780198713395.001.0001

Swaned, M., Javid, S., Humaney, S., Sachan, A., Chauhan, N. S., & Kumar, N. (2024). Enhancing traffic management through advanced vehicle detection for congestion prevention. *2024 IEEE 21st International Conference on Mobile Ad-Hoc and Smart Systems (MASS)*, 623–628. https://doi.org/DOI: 10.1109/MASS62177.2024.00099

Tagscherer, U., & Carbon, C. C. (2023). Digital leadership and innovation culture in smart manufacturing firms. *Technological Forecasting and Social Change*, *189*, 122343.

Tahat, M., & Alananbeh, M., K., A. Othman, Y., & I. Leskovar, D. (. (2020). Soil health and sustainable agriculture. *Sustainability*, *12*(12), 48–59. DOI: 10.3390/su12124859

Tahmid, T., & Hossain, E. (2017). Density-based smart traffic control system using canny edge detection algorithm for congregating traffic information. *2017 3rd International Conference on Electrical Information and Communication Technology (EICT)*, 1–5. https://doi.org/DOI: 10.1109/EICT.2017.8275131

Tan, E. (2014). Human capital theory: A holistic criticism. *Review of Educational Research*, *84*(3), 411–445. DOI: 10.3102/0034654314532696

Tan, L., Pan, Y., Wu, J., Zhou, J., Jiang, H., & Deng, Y. (2020). A New Framework for DDoS Attack Detection and Defense in SDN Environment. *IEEE Access : Practical Innovations, Open Solutions*, *8*, 161908–161919. DOI: 10.1109/access.2020.3021435

Tao, F., Qi, Q., Zhang, D., & Zhang, H. (2018). Digital twin and Industry 4.0: What are they, and how to apply them? [Essential for understanding digital twin technology.]. *Applied Sciences (Basel, Switzerland)*, *8*(10), 1819.

Tao, F., Xiao, B., Qi, Q., Cheng, J., & Ji, P. (2022). Digital twin modeling. *Journal of Manufacturing Systems*, *64*, 372–389. DOI: 10.1016/j.jmsy.2022.06.015

Tao, F., Zhang, H., Liu, A., & Nee, A. Y. (2018). Digital twin in industry: State-of-the-art. *IEEE Transactions on Industrial Informatics*, *15*(4), 2405–2415. DOI: 10.1109/TII.2018.2873186

Tapscott, D., & Tapscott, A. (2016). *Blockchain revolution: How the technology behind bitcoin is changing money, business, and the world.* Portfolio.

Tarafdar, M., Cooper, C. L., & Stich, J. F. (2019). The technostress trifecta—Techno eustress, distress, and the dark side of technology use. *Information Systems Journal*, *29*(1), 6–42. DOI: 10.1111/isj.12169

Tarafdar, M., Tu, Q., Ragu-Nathan, B. S., & Ragu-Nathan, T. S. (2007). The impact of technostress on role stress and productivity. *Journal of Management Information Systems*, *24*(1), 301–328. DOI: 10.2753/MIS0742-1222240109

Tatarinov, K. A., Anikienko, N. N., Savchenko, I. A., & Muzyka, S. M. (n.d.). *Collaborating with robots in industry 4.0. Èkonomika i Upravlenie: Problemy.* Rešeniâ., DOI: 10.36871/ek.up.p.r.2023.10.01.001

Te'eni, D., Yahav, I., Zagalsky, A., Schwartz, D., Silverman, G., Cohen, D., Mann, Y., & Lewinsky, D. (2023, November 14). Reciprocal Human-Machine Learning: A Theory and an Instantiation for the Case of Message Classification. *Management Science*, mnsc.2022.03518. Advance online publication. DOI: 10.1287/mnsc.2022.03518

Teklu, A., Simane, B., & Bezabih, M. (2023). Effect of climate smart agriculture innovations on climate resilience among smallholder farmers: Empirical evidence from the Choke Mountain Watershed of the blue nile highlands of Ethiopia. *Sustainability (Basel)*, *15*(5), 4331. DOI: 10.3390/su15054331

Teng, Y., Cox, A., & Chatziantoniou, I. (2021). Environmental degradation, economic growth and tourism development in Chinese regions. *Environmental Science and Pollution Research International*, *28*(26), 33781–33793. DOI: 10.1007/s11356-021-12567-9 PMID: 33625700

Thaba, S. C., & Mbohwa, C. (2015). The nature, role and status of cooperatives in South African context. *In Proceedings of the world congress on engineering and computer sciences*, 2, 21-23.

Thakur, T., Srivastava, L., Panda, A., & Mohapatra, H. (2025). Smart parking in foggy condition real-time efficiency with PSO, ACO, and IoT sensors. *Revolutionizing Urban Development and Governance With Emerging Technologies*, 233-260. DOI: 10.4018/979-8-3373-1375-7.ch008

Thakur, V. (2020). Framework for PESTEL dimensions of sustainable healthcare waste management: Learnings from COVID - 19 outbreak. *Journal of Cleaner Production, 287*(1), 125562. PMID: 33349739

Thanigaivelan, R., Rajan, N., & Argul, T. (Eds.). (2022). *Advanced Manufacturing Techniques for Engineering and Engineered Materials*. IGI Global Scientific Publishing., DOI: 10.4018/978-1-7998-9574-9

Thomas, D. R. (2006). A General Inductive Approach for Analyzing Qualitative Evaluation Data. *The American Journal of Evaluation, 27*(2), 237–246. DOI: 10.1177/1098214005283748

Tikadar, S., Rodrigues, J. J., Shaw, R., Hassan, A., & Patra, B. K. (Eds.). (2026). *Driving Smarter Human-Computer Interaction With Multidisciplinary Personalized Systems*. IGI Global Scientific Publishing., DOI: 10.4018/979-8-3373-1444-0

Tiwari, S., Bahuguna, P. C., & Srivastava, R. (2023). Smart manufacturing and sustainability: A bibliometric analysis. *Benchmarking, 30*(9), 3281–3301. DOI: 10.1108/BIJ-04-2022-0238

Tote, A. S., Pardeshi, S. S., & Patange, A. D. (2022). Automatic number plate detection using TensorFlow in Indian scenario: An optical character recognition approach. *Materials Today: Proceedings*. Advance online publication. DOI: 10.1016/j.matpr.2022.09.165

Toyota Motor Corporation. (2022). Sustainability Data Book 2022. https://global. toyota/en/sustainability/report/sdb

Türen, U., Erdem H. & Kalkın, G. (2015). İşyerinde tekno - stres ölçeği: Havacılık ve bankacılık sektöründe bir araştırma techno - stress at work scale: A research in aviation and banking sectors. *Çalışma İlişkileri Dergisi,* 6 (1), 1 - 19.

Türen, U., & Erdem, H. (2017). Örgütsel adalet algısının işgörenlerin bilgi teknolojileri verimliliği üzerindeki etkisi: Havacılık sektöründe bir araştırma. *Uluslararası İktisadi ve İdari İnceleme Dergisi, 1*(1), 89–110.

Tybout, A. M., & Artz, N. (1994). Consumer psychology. *Annual Review of Psychology, 45*(1), 131–169. DOI: 10.1146/annurev.ps.45.020194.001023

U. S. Environmental Protection Agency. (2020). Sustainable manufacturing: Principles and practices. https://www. epa. gov/sustainability/sustainable - manufacturing

Uddin, M. S., Das, A. K., & Taleb, M. A. (2015). Real-time area-based traffic density estimation by image processing for traffic signal control system: Bangladesh perspective. *2015 International Conference on Electrical Engineering and Information Communication Technology (ICEEICT)*, 1–5. https://doi.org/DOI: 10.1109/ICEEICT.2015.7307377

Ugwu, C. N., Ugwu, O. P., Alum, E. U., Eze, V. H. U., Basajja, M., Ugwu, J. N., Ogenyi, F. C., Ejemot-Nwadiaro, R. I., Okon, M. B., Egba, S. I., & Uti, D. E. (2025). Sustainable development goals (SDGs) and resilient healthcare systems: Addressing medicine and public health challenges in conflict zones. *Medicine, 104*(7), e41535. DOI: 10.1097/MD.0000000000041535 PMID: 39960902

Ul Rehman, S. (Ed.). (2025). *Challenges and Solutions for Cybersecurity and Adversarial Machine Learning*. IGI Global Scientific Publishing., DOI: 10.4018/979-8-3373-2200-1

Ullah, A., Pinglu, C., Ullah, S., & Hashmi, S. H. (2021). Nexus of regional integration, socio-economic determinants and sustainable development in belt and road initiative countries. *PLoS One*, *16*(7), e0254298. DOI: 10.1371/journal.pone.0254298 PMID: 34242342

Ulrich, D., Younger, J., Brockbank, W., & Ulrich, M. (2019). *HR Transformation: Building Human Resources from the Outside In*. McGraw - Hill Education.

UNESCO. (2022). *Global education monitoring report: Building bridges for learning*.

UNESCO. (2025). UNESCO and Sustainable Development Goals. Retrieved https://www.unesco.org/en/sdgs[REMOVED HYPERLINK FIELD]. Accessed on the 15th of June 2025.

United Nations Department of Economic and Social Affairs, Masaquiza Jerez, M. (2021). *Challenges and Opportunities for Indigenous Peoples' Sustainability.* UN Department of Economic and Social Affairs (DESA) Policy Briefs.

United Nations Environment Program (UNEP). (2025a). UNEP and the Sustainable Development Goals. Retrieved https://www.unep.org/explore-topics/sustainable-development-goals. Accessed on the 15th of June 2025.

United Nations Environment Program (UNEP). (2025b). Sustainable Development Goals. Understand how the environmental SDGs are measured, the progress being made and the current gaps that exist. Retrieved https://sdgs.unep.org/[REMOVED HYPERLINK FIELD]. Accessed on the 15th of June 2025.

United Nations Environment Program (UNEP). Department of Economic and Social Affairs Sustainable Development (2025). United Nations Environment Program (UNEP). Retrieved https://sdgs.un.org/un-system-sdg-implementation/united-nations-environment-programme-unep-34570[REMOVED HYPERLINK FIELD]. Accessed on the 15th of June 2025.

United Nations Environment Program (UNEP). Finance Initiative (2025). SDGs and Impact. Holistic impact analysis and management to close the SDG financing gap. Retrieved https://www.unepfi.org/impact/positive-impact/[REMOVED HYPERLINK FIELD]. Accessed on the 15th of June 2025.

Ünsal, E. (2009). Dinamik bir yetenek olarak teknoloji yönetimi: Teknoloji yönetimi yeteneği. *Savunma Bilimleri Dergisi*, *8*(2), 167–189.

Valentine, N. B., Koller, T. S., & Hosseinpoor, A. R. (2016). Monitoring health determinants with an equity focus: A key role in addressing social determinants, universal health coverage, and advancing the 2030 sustainable development agenda. *Global Health Action*, *9*(1), 34247. DOI: 10.3402/gha.v9.34247 PMID: 27989275

van Driel, M., Biermann, F., Kim, R. E., & Vijge, M. J. (2024). The impact of the Sustainable Development Goals on global policies on sustainable consumption and production. *Globalizations*, *21*(8), 1366–1382. DOI: 10.1080/14747731.2024.2351301 PMID: 39634121

Van Rijmenam, M. (2019). *The Organisation of Tomorrow: How AI, blockchain and analytics turn your business into a data organisation*. Routledge. DOI: 10.4324/9780429279973

Van Tulder, R., Rodrigues, S. B., Mirza, H., & Sexsmith, K. (2021). The UN's Sustainable Development Goals: Can multinational enterprises lead the Decade of Action? *Journal of International Business Policy*, *4*(1), 1–21. DOI: 10.1057/s42214-020-00095-1

Vázquez, J. E., Martin-Ortiz, M., Olmos-Pineda, I., & Olvera-Lopez, A. (2019). Wheelchair Control Based on Facial Gesture Recognition. [IJITSA]. *International Journal of Information Technologies and Systems Approach*, *12*(2), 104–122. DOI: 10.4018/IJITSA.2019070106

Vega, J., & Frenz, P. (2013). Integrating social determinants of health in the universal health coverage monitoring framework. *Revista panamericana de salud publica = Pan. American Journal of Public Health*, *34*(6), 468–472. PMID: 24569977

Venson, E., Da Costa Figueiredo, R. M., & Canedo, E. D. (2024). Leveraging a startup-based approach for digital transformation in the public sector: A case study of Brazil's startup gov.br program. *Government Information Quarterly*, *41*(3), 101943. DOI: 10.1016/j.giq.2024.101943

Verdejo Espinosa, Á., Lopez Ruiz, J., Mata Mata, F., & Estevez, M. E. (2021). Application of IoT in Healthcare: Keys to Implementation of the Sustainable Development Goals. *Sensors (Basel)*, *21*(7), 2330. DOI: 10.3390/s21072330 PMID: 33810606

Vial, G. (2019). Understanding digital transformation: A review and a research agenda. *The Journal of Strategic Information Systems*, *28*(2), 118–144. DOI: 10.1016/j.jsis.2019.01.003

Vigneash, L., Samraj, S., Bhuvaneswari, S., Saravanan, T., & Saritha, G. (2023). Vehicle density and emergency vehicle detection for smart traffic control. *2023 3rd International Conference on Technological Advancements in Computational Sciences (ICTACS)*, 542–549. https://doi.org/ DOI: 10.1109/ICTACS59847.2023.10389918

Wade, M., & Shan, J. (2020). Covid - 19 has accelerated digital transformation, but may have made it harder not easier. *MIS Quarterly Executive*, *19*(3), 213–220. DOI: 10.17705/2msqe.00034

Waheed, R. (2023). Energy Challenges, Green Growth, Blue Indicators, and Sustainable Economic Growth: A Study of Saudi Arabia. *Evaluation Review*, *47*(6), 983–1024. DOI: 10.1177/0193841X221134653 PMID: 36281482

Wang, L., Gao, R. X., & Sivardière, F. (2018). Collaborative manufacturing: State-of-the-art and future trends. *Journal of Manufacturing Systems, 48*(Part B), 283-294. (Provides a comprehensive overview and future directions.)

Wang, C.-N., Tsai, T.-T., Hsu, H.-P., & Nguyen, L.-H. (2019). Performance Evaluation of Major Asian Airline Companies Using DEA Window Model and Grey Theory. *Sustainability (Basel)*, *11*(9), 2701. DOI: 10.3390/su11092701

Wang, F., Harindintwali, J. D., Yuan, Z., Wang, M., Wang, F., Li, S., Yin, Z., Huang, L., Fu, Y., Li, L., Chang, S. X., Zhang, L., Rinklebe, J., Yuan, Z., Zhu, Q., Xiang, L., Tsang, D. C. W., Xu, L., Jiang, X., & Chen, J. M. (2021). Technologies and perspectives for achieving carbon neutrality. *Innovation (Cambridge (Mass.))*, *2*(4), 100180. DOI: 10.1016/j.xinn.2021.100180 PMID: 34877561

Wang, M., Zhu, X., & Yin, S. (2023). Spatial-temporal coupling coordination and interaction between digitalization and traditional industrial upgrading: A case study of the Yellow River Basin. *Scientific Reports, 13*(1), 17997. DOI: 10.1038/s41598-023-44995-7 PMID: 37865702

Wang, W., Zhang, P., & He, T. (2025). Climate investment and financing, economic growth target constraints and urban green development efficiency. *Journal of Environmental Management, 382*, 125369. DOI: 10.1016/j.jenvman.2025.125369 PMID: 40254006

Wang, Y. Y., & Wang, Y. S. (2022). Development and validation of an artificial intelligence anxiety scale: An initial application in predicting motivated learning behavior. *Interactive Learning Environments, 30*(4), 619–634. DOI: 10.1080/10494820.2019.1674887

Wang, Z., & Zhao, Y. (2021). Energy Strategy for Sustainable Development of Rural Areas Based on the Analysis of Sustainable Digital Economy. *Frontiers in Psychology, 12*, 788026. DOI: 10.3389/fpsyg.2021.788026 PMID: 34987452

Wei, H. H., Zhang, Y., Sun, X., Chen, J., & Li, S. (2023). Intelligent robots and human-robot collaboration in the construction industry: A review. *Journal of Intelligent Construction, 1*(1), 1–12. DOI: 10.26599/JIC.2023.9180002

Weil, M., & Rosen, L. (1997). *Technostress: Coping with technology at work at home at play.* John Wiley.

Weir, S., Arstein-Kerslake, A., Eadie, T., & McVilly, K. (2023). Realizing economic and social rights for children with communication and swallowing disability: Sustainable Development Goals 1, 8 and 10. *International Journal of Speech-Language Pathology, 25*(1), 37–41. DOI: 10.1080/17549507.2022.2153166 PMID: 36562595

Westerman, G., Bonnet, D., & McAfee, A. (2014). *Leading digital: Turning technology into business transformation.* Harvard Business Review Press.

Westerman, G., Bonnet, D., & McAfee, A. (2014). *Leading Digital: Turning Technology into Business Transformation.* Harvard Business Review Press.

Wichert, I. (2001). *Job insecurity and work intensification, The effects of an health and well - being*, in *Job Insecurity & Work Intensification.* (Ed.s: Brendan Burchell, David Ladipo & Frank Wilkinson). London: Routledge.

World Economic Forum. (2020). the Future of Jobs Report 2020. https://www. weforum. org/reports/the - future - of - jobs - report - 2020

World Economic Forum. (2021). the future of jobs report 2020. World Economic Forum. https://www. weforum. org/reports/the - future - of - jobs - report - 2020

World Economic Forum. (2023). *The future of jobs report 2023.*

World Economic Forum. (2023). *The future of jobs report 2023.* Geneva, Switzerland: World Economic Forum.

Wuest, T., Romero, D., & Stahre, J. (2017). Introducing 'Operator 4.0,' a tech-augmented human worker. *The Conversation.* https://theconversation.com/introducing-operator-4-0-a-tech-augmented-human-worker-74117

Wu, Z., Wang, J., Du, H., Jiang, D., Kang, Y., Li, D., Pan, P., Deng, Y., Cao, D.-S., Hsieh, C.-Y., & Hou, T. (2023). Chemistry-intuitive explanation of graph neural networks for molecular property prediction with substructure masking. *Nature Communications*, *14*(1), 2585. Advance online publication. DOI: 10.1038/s41467-023-38192-3 PMID: 37142585

Wyszomirski, A. (2023). THE ECONOMIC MACRO - ENVIRONMENT AS A FACTOR SHAPING THE USE OF NEW TECHNOLOGIES. SELECTED ASPECTS. Management & Quality/Zarządzanie i Jakość, 5 (4).

Xin, L., Lang, S., & Mishra, A. R. (2022). Evaluate the challenges of sustainable supply chain 4. 0 implementation under the circular economy concept using new decision making approach. *Operations Management Research : Advancing Practice Through Research*, *15*(1), 773–792. DOI: 10.1007/s12063-021-00243-7

Xu, M., David, J. M., & Kim, S. H. (2018). The fourth industrial revolution: Opportunities and challenges. *International journal of financial research, 9*(2), 90-95.

Xu, J., She, S., & Liu, W. (2022). Role of digitalization in environment, social and governance, and sustainability: Review-based study for implications. *Frontiers in Psychology*, *13*, 961057. DOI: 10.3389/fpsyg.2022.961057 PMID: 36533022

Xu, X. (2009). *Integrating Advanced Computer-Aided Design, Manufacturing, and Numerical Control: Principles and Implementations*. IGI Global Scientific Publishing., DOI: 10.4018/978-1-59904-714-0

Xu, X., Wang, S., Li, J., & Qiao, T. (2024). Environmental regulatory intensity, green finance and corporate green sustainable development performance. *Heliyon*, *10*(9), e30114. DOI: 10.1016/j.heliyon.2024.e30114 PMID: 38694050

Yamin, A. E., & Mason, E. (2019). Why accountability matters for universal health coverage and meeting the SDGs. *Lancet*, *393*(10176), 1082–1084. DOI: 10.1016/S0140-6736(19)30434-9 PMID: 30894258

Yam, K. C., Tang, P. M., Jackson, J. C., Su, R., & Gray, K. (2022). The rise of robots increases job ınsecurity and maladaptive workplace behaviors: Multimethod evidence. *The Journal of Applied Psychology*. Advance online publication. DOI: 10.1037/apl0001045

Yaqoob, N., Ali, S. A., Kannaiah, D., Khan, N., Shabbir, M. S., Bilal, K., & Tabash, M. I. (2023). The effects of Agriculture Productivity, Land Intensification, on Sustainable Economic Growth: A panel analysis from Bangladesh, India, and Pakistan Economies. *Environmental Science and Pollution Research International*, *30*(55), 116440–116448. DOI: 10.1007/s11356-021-18471-6 PMID: 35165846

Yeniçeri, Ö. (2002). Örgütleri etkinleştirme aracı olarak bilgi ve bilgi yönetimi. *Standard Ekonomik ve Teknik Dergi,* 71 - 79.

Yip, W. S., Zhou, H., & To, S. (2023). A critical analysis on the triple bottom line of sustainable manufacturing: Key findings and implications. *Environmental Science and Pollution Research International*, *30*(14), 41388–41404. DOI: 10.1007/s11356-022-25122-x PMID: 36631618

You, X., Liu, H., Wang, T., Feng, S., & Lang, C. (2022). *Object detection by crossing relational reasoning based on graph neural network. 33*(1), 1–14. .DOI: 10.1007/s00138-021-01257-8

Young, K. (2015). *The evolution of internet addiction disorder,* in *Internet addiction: Neuro-scientific approaches and therapeutical interventions,* (Ed.s: C. Montag & M. Reuter), 3 - 17. Springer Science + Business Media. DOI: 10.1007/978-3-319-07242-5_1

Yuan, Q., Zhang, G., Wang, R., Ma, X., & Niu, J. (2024). Does technological innovation in National Sustainable Development Agenda Innovation Demonstration Zones promote green development?-the case from Chengde City, China. *PLoS One, 19*(5), e0300315. DOI: 10.1371/journal.pone.0300315 PMID: 38805430

Zafar & Kantola. (2020). The effect of macro environmental and firm - level factors on open innovations in product development and their impact on firms' performance. In Advances in Human Factors, Business Management and Leadership: Proceedings of the AHFE 2019 International Conference on Human Factors, Business Management and Society, and the AHFE International Conference on Human Factors in Management and Leadership, July 24 - 28, 2019, Washington DC, USA 10 (pp. 61 - 71). Springer International Publishing.

Zahoor, N., & Adomako, S. (2023). Be open to failure: Open innovation failure in dynamic environments. *Technological Forecasting and Social Paradigm shift, 193.* DOI: 10.1016/j.techfore.2023.122632

Zang, S., & Ye, M. (2015). Human Resource Management in the Era of Big Data. *Journal of Human Resource and Sustainability Studies, 3*(01), 41–45. DOI: 10.4236/jhrss.2015.31006

Zantsi, S. (2021). Why do agricultural co-operatives fail to attract youth and create rural employment? Evidence from a case study of Zanokhanyo in Butterworth, Eastern Cape. *South African Journal of Agricultural Extension, 49*(1), 183–197. DOI: 10.17159/2413-3221/2021/v49n1a10785

Zarte, M., Pechmann, A., & Nunes, I. L. (2020). *Principles for Human-Centred System Design in Industry 4.0 – A Systematic Literature Review*. Springer., DOI: 10.1007/978-3-030-51369-6_19

Zero, N., & Summers, J. D. (2024). Aligning a collaborative resistance model with a change management process: A case study in automation in industry. *Journal of Engineering Design, 35*(2), 1–25. DOI: 10.1080/09544828.2024.2387218

Zhang, C. H., Benjamin, W. A., & Miao, W. A. N. G. (2021). The contribution of cooperative irrigation scheme to poverty reduction in Tanzania. *Journal of Integrative Agriculture, 20*(4), 953–963. DOI: 10.1016/S2095-3119(21)63634-1

Zhang, J. (2024). Research on the impact of digitalization on green development: An empirical analysis from the low-carbon strategy perspective. *PLoS One, 19*(3), e0300288. DOI: 10.1371/journal.pone.0300288 PMID: 38457442

Zhang, Q., Dai, W., Chen, J., Gu, Y., & Zhao, Y. (2025). The 'side effects' of digitalization: A study on role overload and job burnout of employees. *PLoS One, 20*(4), e0322112. DOI: 10.1371/journal.pone.0322112 PMID: 40305484

Zhang, Q., Qu, Y., & Zhan, L. (2023). Great transition and new pattern: Agriculture and rural area green development and its coordinated relationship with economic growth in China. *Journal of Environmental Management, 344*, 118563. DOI: 10.1016/j.jenvman.2023.118563 PMID: 37418914

Zhang, Q., & Wang, Q. (2023). Digitalization, Electricity Consumption and Carbon Emissions-Evidence from Manufacturing Industries in China. *International Journal of Environmental Research and Public Health, 20*(5), 3938. DOI: 10.3390/ijerph20053938 PMID: 36900949

Zhang, W., Liu, Y., & Yang, Q. (2023). Empowering leadership and employee creativity in digital contexts: The role of psychological safety and learning orientation. *Journal of Organizational Behavior, 44*(1), 72–88.

Zhang, Y., Lin, Y., Zheng, G., Liu, Y., Sukiennik, N., Xu, F., Xu, Y., Lu, F., Wang, Q., Lai, Y., Tian, L., Li, N., Fang, D., Wang, F., Zhou, T., Li, Y., Zheng, Y., Wu, Z., & Guo, H. (2025). MetaCity: Data-driven sustainable development of complex cities. *Innovation (Cambridge (Mass.)), 6*(2), 100775. DOI: 10.1016/j.xinn.2024.100775 PMID: 39991486

Zhan, J. X., & Santos-Paulino, A. U. (2021). Investing in the Sustainable Development Goals: Mobilization, channeling, and impact. *Journal of International Business Policy, 4*(1), 166–183. DOI: 10.1057/s42214-020-00093-3

Zhao, J., & Kumar, V. (Eds.). (2022). *Handbook of Research on Technologies and Systems for E-Collaboration During Global Crises*. IGI Global Scientific Publishing., DOI: 10.4018/978-1-7998-9640-1

Zhao, W., Yin, C., Hua, T., Meadows, M. E., Li, Y., Liu, Y., Cherubini, F., Pereira, P., & Fu, B. (2022). Achieving the Sustainable Development Goals in the post-pandemic era. *Humanities & Social Sciences Communications, 9*(1), 258. DOI: 10.1057/s41599-022-01283-5 PMID: 35967482

Zhao, X., Xia, Q., & Huang, W. (2020). Impact of technostress on productivity from the theoretical perspective of appraisal and coping processes. *Information & Management, 57*(8), 1–11. DOI: 10.1016/j.im.2020.103265

Zhou, R. (2022). Sustainable Economic Development, Digital Payment, and Consumer Demand: Evidence from China. *International Journal of Environmental Research and Public Health, 19*(14), 8819. DOI: 10.3390/ijerph19148819 PMID: 35886669

Zhuge, K., Lin, W., Yuan, Y., He, H., & Zhang, Y. (2023). Does Digital Capability Promote Sustainable Development of New Ventures? The Dual Impact of Green Knowledge Creation and Green Pressure. *International Journal of Environmental Research and Public Health, 20*(3), 2274. DOI: 10.3390/ijerph20032274 PMID: 36767640

Zhu, P., Ahmed, Z., Pata, U. K., Khan, S., & Abbas, S. (2023). Analyzing economic growth, eco-innovation, and ecological quality nexus in E-7 countries: Accounting for non-linear impacts of urbanization by using a new measure of ecological quality. *Environmental Science and Pollution Research International, 30*(41), 94242–94254. DOI: 10.1007/s11356-023-29017-3 PMID: 37531055

Zhu, W., & Yang, G. (2025). Research on the balanced, coordinated and sustainable development of China manufacturing industry. *PLoS One, 20*(4), e0322400. DOI: 10.1371/journal.pone.0322400 PMID: 40294056

Zuboff, S. (2019). *The age of surveillance capitalism: The fight for a human future at the new frontier of power*. PublicAffairs.

Zyoud, S. H. (2023). Analyzing and visualizing global research trends on COVID-19 linked to sustainable development goals. *Environment, Development and Sustainability*, *25*(6), 5459–5493. DOI: 10.1007/s10668-022-02275-w PMID: 35340818

About the Contributors

Vannie Naidoo is a full time staff member at the University of KwaZulu-Natal, South Africa, in the Faculty of Management and Law. Her field of teaching is in management, corporate strategy, project management, marketing, project management research methodology and entrepreneurship. The research areas pursued by Prof. Naidoo are in the areas of management, management, marketing, service marketing, entrepreneurship and social media and e-learning or learning with new technologies. She has presented papers at international conferences throughout the world and has written on various issues in management and marketing in journal articles and books.

Veena Prasad Vemuri is an accomplished and highly respected academic professional, particularly in the fields of education and research. With her extensive experience as the Principal of NSM Degree College in Mumbai, India and her contribution to Mumbai University, India, she stands out as an expert and leader in her field. Her active involvement in academic publishing, including authoring and co-authoring textbooks, as well as presenting research at national and international conferences, demonstrates her dedication to advancing knowledge. Additionally, her role on the editorial board of many journals shows her influence in shaping academic discourse. Her recognition through state and national awards speaks volumes about the impact she's had in education and research. The successful completion of many PhD students under her guidance reflects her mentorship capabilities, marking her as a significant contributor to the academic community.

Rahul R. Verma is an accomplished academician and researcher in the field of Commerce, with extensive experience spanning over 15 years at the University of Delhi, India. Holding a Ph.D. in Commerce, Dr. Verma has cultivated expertise in areas including Marketing, International Business, and Finance. A prolific contributor to academia, he has participated in over 80 research engagements - encompassing papers, chapters, and presentations at both national and international levels. He has also edited more than 18 books with esteemed publishers such as IGI Global and Apple Academic Press. In addition, Dr. Verma actively contributes to the academic community as a member of editorial boards for several peer - reviewed journals.

***

B. Mythili is an Assistant Professor in the Department of Artificial Intelligence and Data Science at RMK Engineering College, Kavaraipettai. She is dedicated to teaching and guiding students in emerging technologies related to AI and data-driven research. Her academic interests include artificial intelligence, machine learning, and data science applications.

B. Yamini is an Associate Professor in the Department of Networking and Communications, School of Computing, at SRM Institute of Science and Technology, Tamil Nadu, India. Her research interests include computer networks, wireless communication, and emerging technologies in computing. She has contributed to several research publications and actively engages in academic and professional development activities.

Yuvasri B is an faculty member at R.M.K. College of Engineering and Technology, Puduvoyal, Tamil Nadu. She is committed to academic teaching and student development in her department. Her professional interests include research, innovation, and applications in computer science and engineering.

Ranjan Banerjee is an Assistant Professor in the Department of Computer Science and Engineering at Brainware University. With a strong academic foundation, he holds a B. Tech in Computer Science and Engineering, an M.Tech in Computer Technology, and an MBA specializing in Digital Marketing. Currently, he is pursuing a Ph.D. in Computer Science and Engineering. His multidisciplinary expertise spans technical research, digital innovation, and pedagogy, positioning him as a dynamic educator and scholar in the evolving landscape of computer science and digital technologies.

Julius Chiwawa a recently graduated PHD student at University of KwaZulu-Natal.He is pursuing a career in academia.

Jacob Hinneh is a seasoned professional with over a decade of experience in administration, logistics, management, and as a business development specialist. He currently serves as the Facility Officer at the Program Management Unit (PMU) within Liberia's Ministry of Agriculture, where he is responsible for managing logistical operations, fleet coordination, and facilities maintenance across multiple program sites. His work ensures operational efficiency and supports the Ministry's broader development goals. Mr. Hinneh holds a bachelor's degree in business administration from the African Methodist Episcopal Zion University, Liberia. He is a 2025 graduate of the Sharda School of Business Studies, Sharda University, India, where he earned an MBA in Logistics and Supply Chain Management. He also holds a Certificate in Advanced Project Management Program from the prestigious Indian Institute of Management – Rohtak. Over the course of his career, he has held several key positions, including Office Supervisor, Administrative Officer, and Legal Secretary, giving him broad-based experience in program planning, office operations, facility oversight, and public sector coordination. He has also undergone specialized training in sustainable agribusiness, field epidemiology, and rural energy development through leading global institutions such as the World Bank, the U.S. Centers for Disease Control (CDC), and the Ministry of Commerce, P.R. China. From 2024-2025, Mr. Hinneh served in a leadership role within the Association of African Students in India (AASI). During his tenure, he played a pivotal role in advancing academic excellence, professional development, and personal growth among African students across India. He was instrumental in forming strategic partnerships and launching initiatives that promoted employability, entrepreneurship, and industrial exposure. In addition to his administrative and leadership roles, Mr. Hinneh is a dedicated researcher with peer-reviewed publications. His academic work focuses on logistics systems, supply chain resilience, human resource development, and rural development, contributing to practical, evidence-based policy solutions in both African and global contexts. Widely respected for his problem-solving abilities, collaborative leadership, and innovative thinking, Mr. Hinneh is recognized as a valuable contributor in both public and private sectors. He also played an active role in community health efforts, including health promotion and disease surveillance during Liberia's Ebola outbreak, reflecting his ongoing commitment to public service and national development.

Shilpa Hiremath received her B. E (Electronics and Communication Engineering) degree from Visvesvaraya Technological University (VTU), Belgaum in the year 2008 and M. Tech (Communication Systems) degree from Visvesvaraya Technological University (VTU), Belgaum in the year 2013. She is currently pursuing her Phd. from Visvesvaraya Technological University (VTU), Belgaum and also working as Assistant Professor in the Department of ECE at BMS Institute of Technology and Management, Bengaluru, India.

Nivaskumar V is a faculty member at Jawahar Engineering College, Chennai. He is actively involved in teaching and guiding students in the field of engineering and technology. His academic interests include modern advancements in computer science and fostering innovation among learners.

L. Dharani is an Assistant Professor in the Department of Computer Science and Engineering at S.A. Engineering College. Her academic interests include teaching and research in emerging areas of computer science. She is committed to guiding students in both technical knowledge and practical applications.

Sushil Laddhu holds a Ph.D. in Management, is UGC-NET qualified, and has an MBA in Marketing. With 13 years of industry experience and 9 years in academia at various institutions, he is currently an Assistant Professor in the Department of Management at Medi-Caps University, Indore. He has published numerous research papers in Scopus, ABDC, and UGC Care-listed journals, authored several book chapters, and holds multiple patents in marketing, advertising, and corporate management. Dr. Laddhu has also attended various Faculty Development Programs (FDPs), including one at IIM Ahmedabad.

|Pravin Landge - Contributing Author|He is a assistant professor at SVKM'S NMIMS MPSTME.

Jenifer Lawrence an Associate Professor of Management Department at Woldia University, Ethiopia. He has a rich experience in teaching and research activities. He received his Ph.D in Management from Bharathiar University in 2016. His research interest includes the advancements and the new concepts pertaining to Marketing and Human Resources areas. Dr.Jenifer Lawrence has published many research articles and book chapters in several peer-reviewed, and Scopus indexed journals. He has also published many book chapters in Springer and IGI Scopus indexed journals. He has secured a patent from the Patent Office, Government Of India for "Learning Tablet for Kids". Dr.Jenifer Lawrence teaches postgraduate courses on Marketing and Human Resources courses. He has supervised more than 80 MBA theses and served on 14 PhD adjudication as an external examiner for various universities in India. He is also an editorial board member for i-manager's "Journal on Management", and IJMSRT, and acts as a reviewer for different journals.

Dinesh M. G. is a faculty member at EASA College of Engineering and Technology, Coimbatore, Tamil Nadu. He is dedicated to advancing education and research in engineering disciplines. With a passion for academic excellence, he actively contributes to student development and innovative learning.

Hitesh Mohapatra received the B.E. degree in Information Technology from Biju Patnaik University of Technology (BPUT), Odisha in 2006, and the MTech. Degree in CSE from Odisha University of Technology and Research (OUTR) University, Odisha in 2009. He received his Ph.D. in Computer Science & Engineering in 2021 from Veer Surendra Sai University of Technology (VSSUT), Burla, India. He has contributed 50+ SCI and Scopus indexed research articles. He has authored two books on Software Engineering and C Programming respectively. He has 15 years of teaching experience both in industry and academia. He has served the research community with various capacities like session chair, technical chair, keynote speaker etc. His research interests include wireless sensor network, smart city, smart grid and smart water. Currently he is working as Associate Professor at School of Computer Engineering, KIIT (Deemed to be) University, Bhubaneswar, Odisha, India.

Cristina Raluca Gh. Popescu is Full-Professor Habil. at the University of Bucharest, Bucharest, Romania, PhD. Supervisor and Doctoral School Member at The Bucharest University of Economic Studies, Bucharest, Romania, and an award-winning Scientific Researcher. Having more than 24 years of experience, focuses on helping students learn and apply the subjects more effectively, and centers on improving scientific research excellence. Values a special environment where her in-depth knowledge of creative thinking and academic writing can help deliver accurate research results. She is focused on becoming an outstanding researcher, centered on interdisciplinary topics, in a multicultural environment, constantly driven by passion in promoting the Sustainable Development Goals (SDGs). Showing a positive attitude in exploring the endless possibilities of the transformative power of individuals deeply concerned with health and well-being, education, innovation, and responsible production and consumption may be found among the top priorities on her agenda. Her most important skills are: analytical ability and foresight, attention to detail, collaborative spirit, communication skills, creativity, critical thinking, flexibility, intellectual courage, objectivity, persistence, time management, and thirst for knowledge. She is deeply centered on adopting a healthy and balanced lifestyle, thus improving efficiency and productivity in order to be a good and valuable researcher. She is skilled in Business Administration, Economics, Management, Accounting, and Audit. Collaborations: The National Institute for Research and Development in Environmental Protection (I.N.C.D.P.M.), Romania; National Research and Development Institute for Gas Turbines COMOTI, Romania; University of Craiova, Romania; Ministry of Research and Innovation, Romania. She is Member of Chamber of Financial Auditors of Romania (CAFR), Romania as Financial Auditor, and Member of Authority for the Public Oversight of the Statutory Audit Activity (ASPAAS), Ministry of Public Finance, Romania as Financial Auditor. Single author and coauthor of more than 400 scientific works, with national and international publishers, having national and international coauthors: books, book chapters, articles, congress, conference, and symposium papers. She has more than 3000 citations in national and international publications. Single author and coauthor of numerous scientific works indexed in WorldCat – the world's largest library catalog.

Duraipandian R, Dean and Research supervisor, Department of Management, Saveetha School of Management, Saveetha University, Chennai, India, has many years of excellence in teaching and with an outstanding research guidance and had worked in many top most institutions.

Pushpalatha. S Research Scholar, in the Department of Management, Saveetha School of Management, Saveetha University Chennai, India.

Milan Sahoo is a passionate technologist and researcher currently pursuing a Diploma in Computer Science & Engineering at Brainware University, Barasat. With a strong interest in emerging technologies, he has contributed to academia through published works, including a research paper titled "Virtual Reality and Human Experience" in the Journal of Modern Research and a book chapter in "Artificial Intelligence Everywhere: Innovations Across Disciplines" (SSS Publications). He completed an internship at Agnirath Aerospace and Defence Research Pvt. Ltd., where his performance led to a promotion to the Research & Development team. Currently, he works in R&D at Agnirath, exploring cutting-edge advancements in aerospace and defense technologies. His work reflects a commitment to bridging theoretical knowledge with practical innovation, driving progress at the intersection of artificial intelligence, virtual reality, and engineering.

Salankara Sarkar, Student of Computer Science & Engineering in Brainware University completed schooling from the Newtown Girls' High School, Coochbehar

Pankaj Sharma Assistant Professor, Department of Commerce, School of Open Learning, University of Delhi (Oct 2023 - Present) Assistant Professor, Department of Commerce and Management, Sunrise University, Alwar, Rajasthan (March 2023 - October 2023) Assistant Professor, Department of Commerce, Presidency College, Alwar, Rajasthan (September 2018 - February 2023) Passionate about empowering students through innovative teaching, rigorous research, and mentorship, my expertise spans finance, business research, and management. I'm eager to collaborate with fellow academicians, researchers, and industry professionals to advance knowledge and drive positive impact. Let's connect to explore opportunities for academic collaborations, research projects, and knowledge sharing.

Upasana Sharma, BBA (Jammu University), MBA (Jagannath University), pursuing PhD from Sharda University, Greater Noida. She has one year of experience as an HR in Cazzano Apparels Private Limited

Bhardwaj Shukla is currently an Assistant Professor of Commerce at the School of Open Learning, University of Delhi, where he also serves as the Teacher-in-Charge of the Department of Commerce. Prior to this, he held the same position at Sarala Birla University, Ranchi. He earned his Ph.D. from Ranchi University in 2020 with a focus on self-reliant India, exploring the role of Swadeshi organizations like Patanjali Ayurved and Vikas Bharti in the socio-economic development of Jharkhand. With a strong academic foundation that includes an M.Com, B.Ed., and a Post Graduate Diploma in Rural Development, Dr. Shukla actively engages in organizing and participating in seminars, workshops, and training programs. His research is regularly published, with several articles featured in Scopus-indexed and CARE-listed journals, alongside multiple book chapters. His expertise lies in Human Resource Management and Business Management. He is the author of Management Dynamics, Principles of Marketing Management, Mind Management and holds a design patent for an AI-based Human Resource Management device. His commitment to academics and innovation has earned him recognition and awards across various platforms.

T. Thilagam is an academician at Vel Tech Rangarajan Dr. Sagunthala R&D Institute of Science and Technology, Chennai, Tamil Nadu. She is actively engaged in teaching and research in her department, contributing to advancements in science and technology. Her professional interests focus on academic research, innovation, and higher education development.

Anish T. P is an Assistant Professor in the Department of Computer Science and Engineering at R.M.K. College of Engineering and Technology, Puduvoyal. He specializes in computer science domains with a focus on emerging technologies and research applications. His teaching approach integrates theoretical foundations with practical insights to enhance student learning. He actively contributes to research, publications, and academic development within the institution. Residing in Tamil Nadu, he continues to inspire students in pursuing excellence in computer science and engineering.

V. Rekha is an Assistant Professor in the Department of Artificial Intelligence and Data Science at Panimalar Engineering College. She focuses on imparting knowledge in AI and data-driven technologies while guiding students toward academic and professional excellence. Her teaching and research interests emphasize innovation and practical applications in computing.

K. Venkatesh Guru is an Assistant Professor in the School of Computing at SRM Institute of Science and Technology, Trichy. He specializes in teaching and research in emerging areas of computer science and technology. With a strong academic and research background, he is committed to fostering innovation and excellence in the field.

Getnet Worku, PhD Scholar, College of Business and Economics, Department of Management, Bule Hora University, Ethiopia.

Index

Symbols

A

B

C

D

E

F

G

H

I

J

L

M

O

P

R

S

T

W

Y

www.ingramcontent.com/pod-product-compliance
Lightning Source LLC
LaVergne TN
LVHW080613300926
849162LV00005B/1145